AF781472

Aquaculture: Farming Aquatic Animals

Aquaculture: Farming Aquatic Animals

Edited by **Olando Martin**

SYRAWOOD
PUBLISHING HOUSE
New York

Published by Syrawood Publishing House,
750 Third Avenue, 9th Floor,
New York, NY 10017, USA
www.syrawoodpublishinghouse.com

Aquaculture: Farming Aquatic Animals
Edited by Olando Martin

© 2016 Syrawood Publishing House

International Standard Book Number: 978-1-68286-340-4 (Hardback)

This book contains information obtained from authentic and highly regarded sources. Copyright for all individual chapters remain with the respective authors as indicated. All chapters are published with permission under the Creative Commons Attribution License or equivalent. A wide variety of references are listed. Permission and sources are indicated; for detailed attributions, please refer to the permissions page and list of contributors. Reasonable efforts have been made to publish reliable data and information, but the authors, editors and publisher cannot assume any responsibility for the validity of all materials or the consequences of their use.

The publisher's policy is to use permanent paper from mills that operate a sustainable forestry policy. Furthermore, the publisher ensures that the text paper and cover boards used have met acceptable environmental accreditation standards.

Trademark Notice: Registered trademark of products or corporate names are used only for explanation and identification without intent to infringe.

Printed in China.

Contents

Preface

Aquaculture focuses on harvesting of aquatic organisms and is a widely practiced economic activity. The various advancements in the techniques of aquaculture are glanced at and their applications as well as ramifications on aquafarming are looked at in detail within this book. Topics such as algaculture, mariculture, methods of aquaculture, etc. are covered in detail. The chapters included in the text will serve as a reference to a broad spectrum of readers. Scientists and students actively engaged in this field will find this book full of crucial and unexplored concepts.

This book unites the global concepts and researches in an organized manner for a comprehensive understanding of the subject. It is a ripe text for all researchers, students, scientists or anyone else who is interested in acquiring a better knowledge of this dynamic field.

I extend my sincere thanks to the contributors for such eloquent research chapters. Finally, I thank my family for being a source of support and help.

Editor

The sediment characteristic and benthic macro-invertebrate fauna of some fish ponds in Ife north local government area (LGA), Nigeria

A. A. Adedeji*, I. F. Adeniyi and O. R. Adetokunbo

Department of Zoology, Obafemi Awolowo University, Ile-Ife, Nigeria.

This study investigated the sediment characteristics as well as the taxonomic composition of macrobenthic invertebrate faunae of some ponds in Ife north local government area,(LGA) Nigeria. The sediments were generally sandy, moderately acidic (pH range of 3.8 to 5.8) and well drained. The organic carbon of the sediment were within the usual range of 0.5 to 5% organic carbon while the nutrients (NO_3^-, PO_4^{3-}, SO_4^{2-}) varied from very low to moderate concentration in all the ponds. The failure for a buildup of organic material and nutrients to a very high level indicated that the wastes from the fish and exogenous feed were rapidly and completely broken down and thus does not develop to poor sediment characteristic. Macro-invertebrate fauna recorded were dominated by the molluscs typical of freshwater macrobenthic communities. Orders Architaeniaglossa and opisthobranchia were encountered with *Melanoides (Melanoides) tuberculata* of the order Architaeniaglossa being the most dominant taxon in the ponds. The abundance of pulmonates in these ponds showed that they were mesosaprobic – eutrophic water bodies that are slightly polluted and nutrient rich.

Key words: Poor sediment, organic carbon, acidic pH, nutrients, macrobenthic invertebrate.

INTRODUCTION

Nigeria is greatly endowed with natural and environmental conditions for rapid development of aquaculture in terms availability of suitable aquacultural sites in all the geographical zones of the country (Adikwu, 1999), optimum temperature, optimum annual rainfall and water availability in more than 95% of the land area (FAO, 1994). Nigeria has about 12.5 million hectares of inland water mass plus over 20 million hectares of swamps; lagoons and estuaries which can be used for productive fish farming (Ita, 1980). Though fish farming in Nigeria has a very good investment prospect, it is faced with a lot of problems which make its success unrealizable. Previous studies in Nigeria have shown ineffective management practices coupled with poor supervision skills as the main cause of the poor culture fish production in the country (Egborge, 1999; Anadu et al., 1990; Ajayi et al., 2001). Bad management practices such as excessive exogenous feed input, utilization of wrong feedstuff as well as the application of organic and inorganic fertilizers could lead to increase in the load of nutrients, low dissolved oxygen content, high ammonia concentration and poor bottom sediment conditions. The investigation of this problem of the poor bottom sediment and their biological resources in ponds at Ife North Local Government Area of Osun State is the main concern of this research.

Area of study

Ponds investigated were located on the farm owned by Niger Feeds and Agricultural Operations Limited (NIFAGOL). The company was established in 1984 for

*Corresponding author. E-mail: bbkadedeji@yahoo.com.

Abbreviations: NIFAGOL, Niger Feeds and Agricultural Operations Limited; **LGA,** Local Government Area.

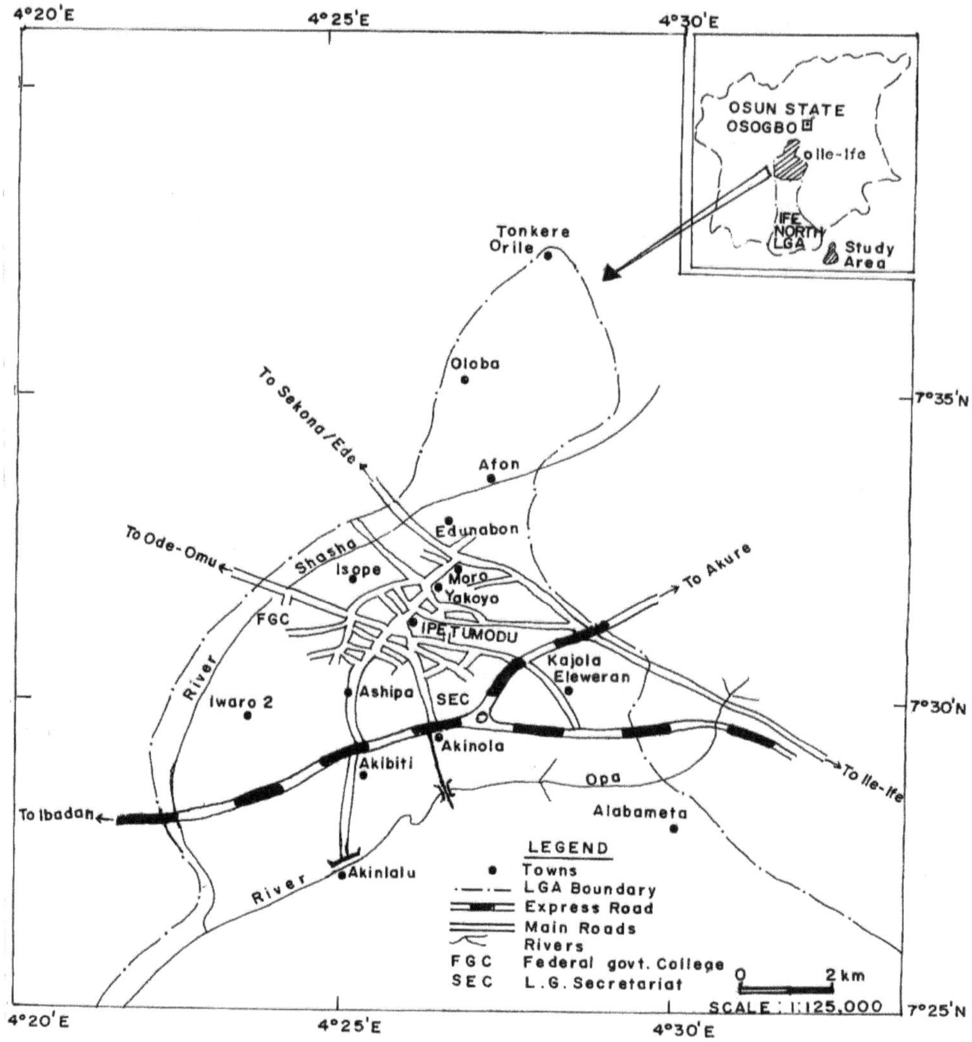

Figure 1. Map of the northern part of Ife north LGA showing the area of study.

the production of fingerlings and table size fish for sale to the public. The ponds are located between latitudes 07° 32.285' N to 07° 32.487' N and longitude 004° 26.699' E and 004° 26.856'E at an altitude between 254to 259 m (Figure 1). The climatic condition of the studied area is governed mainly by two dominant winds, the south-west winds and the North-East winds which blow over most parts of West Africa. The dry season extends from November to March while the wet or rainy season occurs from April to October. There is abundant sunshine almost throughout the year and temperature is more or less equable with very low range of variation. The soil is well drained clayey soil which is being considered as the best soil for pond and fish rearing. The vegetation of the area consists mainly of low-land rainforest and some areas of grassland.

MATERIALS AND METHODS

Sediment samples were taken once in two months between January through July, 2006 using an improvised mad grabber (metallic plates of cross sectional area of 15.2 by 15.2 cm) from four fish ponds located within NIFAGOL farm, Ipetumodu. Half of the grabbed sediment samples were sieved through 0.5 mm mesh sieve using the pond water. The macro-invertebrate benthic organisms found were removed (using a pair of long forceps) and preserved inside an adequately labeled collecting tubes containing 5% buffered formalin solution (APHA, 1995). Identification of the benthic macro invertebrate fauna collected was done using identification keys prepared by IOWATER (2005) and Kohl (2005). Although the identification keys are not specific for tropical freshwater, they satisfy standard international scope for identification of benthic macrobenthic fauna. The other half of the sediment samples were transferred to the laboratory, air dried crushed with a pestle in a porcelain mortar. The sediment was then sieved through a 2 mm mesh sieve before analysis of selected physico-chemical parameters were carried out (Royse, 1970). Colour, particle size analysis, textural composition, pH, organic matter, total nitrogen, pH, sulphate and phosphate levels in the sediment were determined according to Boyd (1995) because natural physicochemical sediment differences were found to be the main determinants in species assemblage patterns among sites (MacFarlane and Booth, 2001).

Table 1. The physiochemical parameters of the pond sediments for the months investigated.

Parameter	February				May				July			
	A	B	C	D	A	B	C	D	A	B	C	D
Value/chroma	3/1	5/6	4/2	2/2	3/3	4/4	2/0	4/2	3/3	4/2	4/2	4/2
Hue	10 year	10 year	7.5 year	5 year	5 year	10 year	2.5 year	5 year	5 year	10 year	2.5 year	2.5 year
Colour	Very dark grey	Yellowish brown	Dark brown	Dark reddish brown	Dark red brown	Dark yellowish brown	Black	Dark reddish grey	Dark reddish brown	Dark greyish brown	Dark greyish brown	Dark greyish brown
Porosity	50	50	48	50	50	48	43	40	44	51	44	41
% Sand	31	83	63	69	77	61	89	87	89	89	89	55
% Silt	38	6	14	24	11	16	4	7	6	4	4	20
% Clay	31	11	23	7	12	23	7	6	5	7	7	25
pH H_2O 1:1	5.4	5.7	6.2	6.4	6.0	5.7	3.8	6.2	6.2	5.8	5.8	6.2
pH $CaCl_2$ 1:2	5.2	5.0	5.4	6.2	5.4	5.2	3.6	5.8	5.4	5.0	5.6	5.2
Redox potential	+105	+90	+50	+60	+80	+90	+200	+60	+75	+80	+40	+70
Org. carbon %	3.74	0.78	1.25	2.38	1.48	1.87	1.21	1.13	0.94	1.01	0.70	0.43
Org. matter %	6.44	1.34	2.15	4.09	2.55	3.22	2.08	2.01	1.61	1.74	1.21	0.74
Phosphate (ppm)	4.98	3.87	18.72	12.48	7.74	7.42	15.08	5.13	3.87	7.19	12.64	11.53
Total nitrogen (ppm)	0.20	0.03	0.06	0.14	0.10	0.09	0.06	0.05	0.03	0.04	0.04	0.05
SO_4^{2-} (ppm)	543.61	171.07	87.43	190.07	167.26	266.10	1334.31	254.7	144.46	247.09	790.70	912.35
Sediment type	Silty mud	Clayey sand	Sandy mud	Silty sand	Silty sand	Sandy mud	Sand	Sand	Sand	Sand	Sand	Silty mud

RESULTS

The sediments were either silty mud or sandy. The colour of silty mud sediment was very dark grey while the sandy ones varied from yellowish brown through dark brown to dark reddish brown to black (Table 1). Sediments of all the four ponds were relatively porous. Their pH was acidic ranging from 3.60 to 6.40. Sulphate concentrations were generally above 50 ppm (1334.31 to 144.46 ppm) while phosphate concentration ranged from 3.87 to 18.7 ppm (< 20 ppm). Total nitrogen concentration was very low ranging from 0.03 to 0.14 ppm. The sediments of all the four ponds were rich in organic carbon (0.43 to 3.74%) which is a factor of the organic matter concentration (0.74 to 6.44%) in the various ponds. The organic carbon and matter decreased from February through July in the ponds except in pond B where there was an increase between February and May (Table 1).

The standard deviation of organ carbon (%), organic matter (%), porosity, sand (%), silt (%), clay (%), redox potential, phosphate, sulphate and total nitrogen were characterized by high scattering around their respective mean values while that of pH is generally low and is characterized by low scattering around its mean value (Table 2). Three gastropod species, one Annelid species and one nymphal form of an insect were recorded. The checklist and outline classification of the fauna is given in Table 3. *Melanoides* (*Melanoides*) *tuberculata* dominated the four ponds and were found in abundance in all the ponds. A few numbers of apple snails (Pomacea) and sinistral snails (Physidae) were found in pond B.

DISCUSSION

The particle size variation recorded in the ponds during the study period revealed that the sediments were generally sandy during the period resulting in porosity of the sediments. The sediment colour is an indicator of the pattern of drainage in the ponds, the sediment with red, yellow and brown coloured sediments showed good drainage except pond A whose sediment was very dark grey in february indicating poor drainage during that period (Aldoufer, 1974). The pH of the sediment was moderately acidic and it was higher in May (early rainy season) and July (mid rainy season) than in February (dry season). This supported the observation of Richard and

Table 2. The mean and standard deviation of the various physiochemical parameters of the pond sediments.

Parameter	Pond A	Pond B	Pond C	Pond D
% Organic carbon	2.05±1.49	1.22±0.58	1.05±0.31	1.33±0.98
% Organic matter	3.53±2.56	2.10±0.99	1.81±0.52	2.28±1.69
Porosity	48.00±3.46	49.67±1.52	45.00±2.65	43.67±5.51
% Sand	65.67±30.61	77.67±14.74	80.33±15.01	70.33±16.04
% Silt	18.33±17.21	8.67±6.43	7.33±5.77	17.00±8.89
% Clay	16.00±13.45	13.67±8.33	12.33±9.23	12.67±10.69
pH $CaCl_2$ 1:2	5.33±0.12	5.07±0.09	4.87±1.10	5.73±0.50
pH H_2O 1:1	5.87±0.42	5.73±0.06	5.26±1.28	6.27±0.12
Redox potential	86.67±16.07	86.67±5.77	96.67±89.62	63.33±5.77
Phosphate (ppm)	5.53±1.99	5.16±1.99	15.48±3.05	9.71±4.00
Total nitrogen (ppm)	0.11±0.09	0.05±0.03	0.05±0.01	0.08±0.05
SO_4^{2-} (ppm)	285.11±224.15	228.09±50.28	889.54±404.51	452.37±399.66

Table 3. Checklist and outline classification of the macrobenthic invertebrates encountered.

	Category	Taxon
	Phylum	Mollusca
	Class	Gastropoda
1	Sub class	Eogastrophoda
	Sub order	*Neritaemorphi*
	Order	*Architaeniaglossa*
	Sub order	*Discopoda*
	Super family	Cerithiodea
a	Family	Thiaridae
	Genus	*Melanoides*
	Species	*Melanoides* (*Melanoides*)
	Sub species	*Melanoides* (*Melanoides*) *tuberculata* (Muller 1774)
	Order	Architaenioglossa
	Super family	Ampullarioidae
b	Family	Ampullarioidae
	Genus	*Pomacea*
	Species	*Pomacea* (*Pomacea*) *columellaris* (Gould 1848)
	Sub class	Orthogastropoda
	Super order	*Heterobranchia*
	Order	*Opisthobranchia*
	Sub order	*Basommtophora*
c	Super family	Physacea
	Family	Physidae
	Sub family	Physinae
	Genus	*Physa*
	Species	*Physa frotinalis*
	Phylum	*Annelida*
	Class	Hirudinea
	Order	*Rynchobdella*
	Family	Glossiphonidae
2	Species	*Glossiphonidae*
	Super phylum	Arthropoda
	Phylum	Insecta
	Order	Plecoptera

Bacon (1994) and Jackson (1997) that in most tropical inland lakes and reservoir, the sediments were acidic. The acidic pH was also observed in Lake Brolus (an Egyptian lake Delta) by Saad, (1980). Platts et al. (1983) suggested that the increase in the pH of the sediments during the wet season was due to the high inflow of flood and this is followed by the re-cycling and re-suspension of benthos material that accompanied it. On the other hand, Davies and Tawari (2010) suggested that the low pH recorded in the dry season was due to high temperature which could lead to high rate of decomposition of matter. This process reduces the amount of oxygen in the sediments thus low pH. The organic carbon of the sediment were within the usual range of 0.5 to 5% organic carbon (Boyd et al., 2002) but changed over time. These changes according to Munsiri et al. (1996) are greatly influenced by pond management inputs. Such as application of organic manure to maintain the ponds and even feeding the fish with dead chicken parts as observed during the period of study.

The very low concentration of total nitrogen in the pond sediments could be attributed to the fact that nitrogen concentration in sediment is controlled by the presence of organic matter as 90% of nitrogen in sediment exists in organic form (Martinova, 1993). Sulphate concentration the in pond sediments were generally low below 0.5% as its concentration in sediment is also known to be dependent on organic matter enrichment. The values of phosphate concentration (< 20 ppm) recorded were generally lower than the optimum range of 30 to 60 ppm (Boyd et al., 1994). This could probably be as a result of the acidic nature of the sediment. According to Boyd et al. (1994), acidic sediment contains primarily iron and aluminum phosphate which are less soluble in dilute acid. Macro-invertebrate fauna recorded were dominated by the molluscs which are very important in the breakdown of plant matter and other detritus that form the basis of many food chains. The rich occurrence of *M. tuberculata* could be attributed to their ability to strive well in the tropics (Derraik, 2008) with habitat preference temperature falling within 21 to 25°C (De Kock and Wolmarans, 2009). They were also known to become invasive on introduction to a new location due to their ability to reproduce parthenogenetically and even bring forth juvenile viviparously (Derraik, 2008).

Their high frequency in the ponds could be attributed to the muddy nature of the sediment, availability of decomposing material and more or less permanence of the water bodies as they are known to have poor ability for aestivation (De Kock and Wolmaran, 2009). Occurrence of *Physa frontinalis* and *Pomacea pomacea columellaris* in Pond 4 shows the level of organic pollution of this water body (Ersan et al., 2009). Moreover the immense abundance of the pulmonates in the water bodies reveals the high trophic state of the water bodies as well as deposits of allochthonous in the water bodies (Yildirim, 2004). In summary, it appears that the sediment at best can be classed as a rich soil. The failure for a buildup of organic material and nutrients to a very high level indicates that the wastes from the exogenous feed and fish are rapidly and completely broken down and thus does not result in poor bottom sediment condition.

REFERENCES

Adikwu IA (1999). Aquaculture in Nigeria: Prospects. Constraints J. Fish Technol., 1(1): 16–27.

Ajayi B, Olunuga T, Badejo N (2001). Soil and water quality management in Aquaculture Ponds. Fish Netw., 2(5/6): 11.

Aldorfer RB (1974). McGraw Hill Encyclopedia of Environmental Science: McGraw Hill Company, New York, pp. 543-545.

Anadu DI, Obioha A, Ejike C (1990). Water quality and plankton periodicity in two constrasting mine lakes in Jos, Nigeria. Hydrobiologia, 208: 17–25.

APHA, AWWA, WEF (1995). Standard methods for the examination of water and wastewaters 18th Edition. American Public Health Association, Washington D.C.; 1134 pp.

Boyd CE (1995). Bottom soil, sediment and pond aquaculture. Chapman and Hall U.S.A.

Boyd CE, Munsiri P, Hajek BF (1994). Composition of sediment from intensive shrimp ponds in Thailand. World Aquacult., 1: 53–55.

Boyd CE, Woods CW, Thunjai T (2002). Aquaculture Pond Bottom Soil Quality Management. Pond Dynamics/Aquaculture Collaborative Research Support Program. Oregon State University, Corvallis, 41 pp.

Davies OA, Tawari CC (2010). Season and tide effects on sediment characteristics of trans-okpoka creek, upper bonny Estuary, Nigeria. Agric. Biol. J. N. Am., 1(2): 89-96.

De Kock KN, Wolmarans CT (2009). Distribution and habitats of *Melanoides tuberculata* (Müller, 1774) and *M. victoriae* Dohrn (1865) (Mollusca: Prosobranchia: Thiaridae) in South Africa. Water South Afr., 35(5): 713-720.

Derraik JGB (2008). The potential significance to human health associated with the establishment of the snail *Melanoides tuberculata* in New Zealand. J. New Zealand Med. Assoc., 121(1280).

Egborge ABM (1999). Prospect and problems of limnology and their implication on fisheries development. J. Fish. Technol., 1 (1): 11–15.

Ersan E, Altinda A, Ahiska S, Ala A (2009). Zoobenthic fauna and seasonal changes of Mamasin dam lake (Central part of Turkey). Afr. J. Biotechnol., 8(18): 4702-4707.

FAO (1994). A strategic assessment of warm water fish farming potential in Africa. CIFA Technical Paper 27. FAO, Rome, Italy.

Ita EO (1980). A review of recent advances in warm water aquaculture research and a proposed experimental design for maximizing fish production in Nigerian fish ponds. Kainji Lake Research Institute Technical Report Series No 5. 48 pp.

IOWATER (2005). Benthic Macroinvertebrate Key. Volunteer water quality monitoring IOWA Department and Natural Resources, US.

Jackson DA (1997). Compositional data in community ecology; the paradigm or peril propositions. Ecology, 78: 929-940.

Kohl M (2005). Freshwater Molluscan shells http://members.aol.com/mokhl1/fwshells.html.

Macfarlane GR, Booth DJ (2001). Estuarine macrobenthic community structure in the Hawkesbury River, Australia: Relationships with sediment physicochemical and anthropogenic parameters. Environ. Monit. Assess., 72: 51-78.

Martinova MV (1993). Nitrogen and phosphorus compounds in bottom sediments: Mechanisms of accumulation, transformation and release. Hydrobiologia, 252: 1-22

Munsiri P, Boyd CE, Hajek BJ (1995). Physical and chemical characteristics of bottom soil profiles in ponds at Auburn, Alabama USA, and a proposed method for describing pond soil horizons. J. World Aquacult. Soc., 26: 346–377.

Platts WS, Megaham WF, Minshall GA (1983). Factors affecting micro-distribution of stream benthic insects. Limnol. Oceanogr., 23: 1030-1033.

Richard C, Bacon KL (1994). Influence of fine sediment on macrobenthic invertebrate colonisation of surface and hyperheic stream substrates. Hydrobiologia, 87: 208-219.
Royse CF (1970). An Introduction to Sediment Analysis. Arizona State University Press.
Saad AH (1980). Studies on the bottom deposit of Lake Brolus, a delta Egyptian Lake. Biol. Pollut., 3: 181-186.
Yildirim MZ (2004). The Gastropods of Lake Egirdir. Turk. J. Zool., 28: 97-102.

2

Multiple infections of Helminths in the alimentary system of *Clarias gariepinus* (Burchell, 1822) in a tropical reservoir

Ajala, O. Olumuyiwa and Fawole, O. Olatunde

Department of Pure and Applied Biology, Ladoke Akintola University of Technology, Ogbomoso, Oyo State, Nigeria.

The study was carried out in Erinle Reservoir accessed through Oore (Latitude 7° 58' 12" North; Longitude 4° 33' 36" East), in Osun State, Nigeria, between August 2011 and November 2013. A total of 103 live fishes purchased from local fishermen were examined. In the laboratory, morphometric parameters were measured, the sexes determined and the fish species *Clarias gariepinus* were dissected to separate the alimentary system. Parasites recovered were identified and counted and the data were subjected to statistical analyses. The parasites recovered were *Procamallanus laevionchus, Paracamallanus cyathopharynx* (Nematodes.), *Anomotaenia* sp., *Monobothrium* sp., *Polyonchobothrium clariae.* (Cestodes) and *Neoechinorynchus rutili* (Acanthocephalan). Using a non parametric (NPar) (Kolmogorov-Smirnov; K-S) test, infection was significant at p = 0.05 within seasons, and female fishes were more infected than male. *Monobothrium* sp. had the highest range of infection (0 to 44) and intensity (18.5±2.65) while *Anomotaenia* sp. had the least (0 to 1) and (1.00±0.02) respectively. Six parasites were found in the intestine, while four were in the stomach. There was high prevalence in medium and large sized fish and a direct linear relationship exist between length and intensity. The body weight was significant in relation to infection at p = 0.05 (K-S test) and was also significant in sex. Multiple infections were common, which showed a positive correlation between most of the parasites except *Anomotaenia* sp. which showed negative correlation with *P. clariae.*

Key words: *Clarias gariepinus,* morphometric, parasite, prevalence.

INTRODUCTION

In many parts of the world, studies of fish parasites and diseases are very advanced, resulting in the accumulation of an enormous amount of literature and information. Studies in Africa vary considerably from area to area and the parasites are mostly mentioned, as part of the fulfillment of the biology of the host fish species. Paperna (1996) published a concise update of the parasitic diseases of fish in Africa, which described the concurrence and geographical distribution, life cycles, pathology, epizootiology and control of the parasites. Khalil and Polling (1997) also produced a check list of the Helminthes parasites of African freshwater fishes. Numerous parasites were associated with *Tilapia zilli, Clarias anguillaris* and *Clarias gariepinus* during the study of the biology of the fish species in their natural environments (Eyo and Olatunde, 2001; Ovie and Ovie, 2002; Olofintoye, 2006).

From an aquaculture perspective, it is noteworthy that disease and parasite infestation have been poorly studied in Africa due to the low level intensity of aquaculture in the region. At present, the paucity of research in fish diseases in Africa is not seen as a factor that will negatively impact on aquaculture development and as such is not a target research area. The emanating need to culture fishes for protein consumption for the teeming rapidly growing populations in the developing countries like Nigeria have made it necessary to intensify studies on the parasite fauna of the African freshwater fishes (Akinsanya et al., 2007).

Fish parasites are important because they affect fish production particularly under culture systems, by decreasing their yield, aesthetic value, marketability, palatability and reproductive potential (Ibrahim et al., 2001; Oniye et al., 2004), and if left un-curtailed, may lead to mass mortality of fish, or in some cases, emergence of zoonotic species. They are therefore studied with a view to understanding their population biology and elucidate their life cycles in order to develop an efficient approach of controlling them (Olufemi, 2008).

The occurrences of Helminth parasites in fishes have been studied in various water bodies in Nigeria, with most of the work done primarily from the morphologic and morphometric descriptions, but factors that may limit the ability of parasites to co-exist in multiple infections in a host fish species had in most studies been neglected. This study is focused on the level of co-existence/concurrent infection between identified parasites in *C. gariepinus*, the factors that might limit this relationship and the effects it has on *C. gariepinus*.

MATERIALS AND METHODS

The study area was Erinle Reservoir accessed through Oore (Latitude 7° 58' 12" North; Longitude 4° 33' 36" East), in Odo-otin local government area of Osun State, Nigeria. Erinle Reservoir was impounded in 1958 and expanded in 1980. The tributaries are Oyan, Otin, and Erinle streams, in Osun State Nigeria. Two seasons are recognized in the study area - dry season from October to March and rainy season from April to September each year.

A total of 103 fishes were examined between August 2011 and November 2013. The fishes were purchased from fishermen, transferred into a plastic container with water and transported to the research laboratory. Transportation was done in the morning to avoid undue stress due to temperature rise. Dead fishes were removed from the collections and examined immediately while the live ones were preserved in a plastic aquaria containing water from the reservoir, and examined subsequently as the investigation progresses. The fish was killed by cervical dislocation. The standard / body length (from the tip of the snout to the end of the caudal peduncle) were measured in centimeters with the aid of a measuring board and recorded for each fish specimen. The weight for each fish was obtained by using a Metler balance. Sexes were determined, the male possess a distinct sexual papilla that is conspicuously located behind the anus, the sexual papillae are absent in females (Akinsanya and Otubanjo, 2006).

Dissection

The body of each fish was examined for abnormalities (if any), and placed on a dissecting board. The body cavity was opened with the aid of scissors and the mesentery and connective tissues, connecting loops of the gut and the liver were cut and the organs separated. The gut was then placed in a large Petri dish, stretched out and cut into two regions, that is, the stomach and the intestine. Each section was then placed in a separate labeled dish. The separated gut sections were opened by longitudinal incision to expose the inner surface which was washed with very little quantity of distilled water into labeled test tubes. A drop of the residue was placed on the slide, and observed under ×10 and ×40 objectives of the light microscope for the various parasites, a process that was repeated until the entire residue has been examined. Most of the parasites were easily identified by their wriggling movements, parasites found were counted, labeled with the serial number of the fish and placed in physiological saline overnight to allow them stretch and relax.

Cestodes, Nematode and Acanthocephalan parasites recovered were stained using the procedure of Khalil (1991). Identification of specimen to species level were undertaken and confirmed with the assistance of Oniye et al. (2004) and Akinsanya and Otubanjo (2006) who had earlier confirmed the identity of the parasites through the assistance of the British Museum (Natural history), United Kingdom. The terms prevalence and mean intensity were applied as defined by Margolis et al. (1982), while range refers to zero to maximum number of endoparasites retrieved from a fish host.

Statistical analysis

Infection of host by parasites was not normally distributed; significance of parasitic infection was tested using a non parametric (Npar.) statistical method, (Two independent sample Kolmogorov-Smirnov K-S test at $p=0.05$) and presented pictographically using SPSS version 10.

Pearson correlation analyses was carried out using the computer-based SPSS version 10 to determine if a pair of parasites in multiple and concurrent infection within hosts were linearly correlated and significant ($p = 0.01$).

RESULTS

Parasites recovered

The gastro-intestinal helminth parasites recovered comprised two nematodes *(Procamallanus laevionchus, Paracamallanus cyathopharynx.*), three cestodes

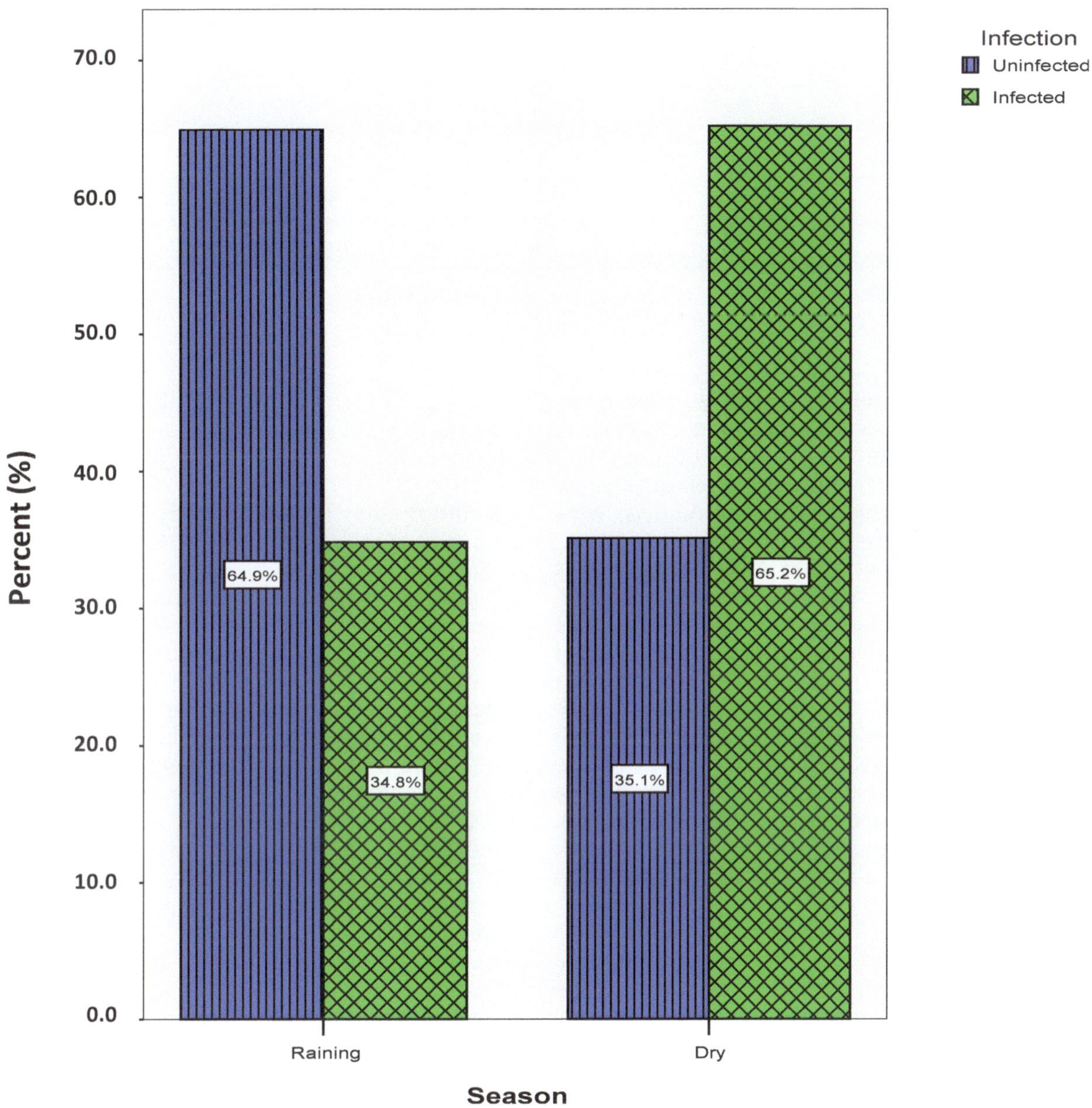

Figure 1. Differences in infection between rainy and dry seasons.

(*Anomotaenia* sp., *Monobothrium* sp., *Polyonchobothrium clariae.*) and one acanthocephalan (*Neoechinorynchus rutili*).

Prevalence and intensity of parasites in *C. gariepinus*

A total of 103 fishes were examined out of which 55 were infected giving 53.4% prevalence. There was no significant difference in the gastro-intestinal Helminth infection while Helminth intensity in the intestine was higher than in the stomach.

Seasonal distribution of parasites

There was a significant difference ($p = 0.05$; K-S test), in infected *C. gariepinus* over seasons as more fishes were infected in the dry season than the rainy season (Figure 1).

Distribution of parasites

Monobothrium sp. had the highest range of infection, with a maximum abundance of 44 in a single host specimen

Table 1. Prevalence, range and intensity of parasites in the stomach and intestine of *Clarias gariepinus*.

Parasite	Prevalence %		Range of parasites		Nos. of parasites		Intensity	
	STO	INT	STO	INT	STO	INT	STO	INT
Anomotaenia sp.	1.94	4.85	0 - 1	0 - 1	2	5	1.00±0.02	1.00±0.02
Neoechinorynchus rutili	0.00	12.62	0	0 - 4	0	25	0.00	1.92±0.07
Procamallanus laevionchus	0.00	8.74	0	0 - 3	0	14	0.00	1.56±0.05
Polyonchobothrium clariae	2.91	39.81	0 - 12	0 - 12	20	215	6.67±1.06	5.24±0.40
Paracamallanus cyathopharynx	16.50	38.83	0 - 12	0 - 20	91	434	5.35±1.04	10.85±1.45
Monobothrium sp.	36.89	41.75	0 - 20	0 - 44	406	796	10.68±1.35	18.5±2.65

STO = Stomach; INT = Intestine.

and an intensity of 18.5±2.65 in the sampled specimens. *Anomotaenia* sp. had the least range (0 to 1) and an intensity of 1.00±0.02 (Table 1).

Six parasites were found in the intestine, while only four were in the stomach. *P. laevionchus* and *N. rutili*, were absent in the stomach (Table 1).

Although *N. rutili* was not found in the stomach, its presence in the intestine varied with season and sex. Intensity of infection in dry season (0.263± 0.055) was higher than rainy season (mean 0.038±0.053), and female fishes (0.381±0.071) were more infected than male (0.220±0.079). Infection of the intestine by *P. laevionchus* was higher in the rainy season (0.334±0.043) than dry season (0.048±0.045). *P. clariae* infected both the stomach and intestine generally, but the infection was not affected by season. *Monobothrium* sp. has the highest occurrence, both in stomach and intestine; its distribution was normal in both seasons and sex.

Generally, the prevalence and intensity of parasitic infection in the stomach and intestine showed significant difference ($p = 0.05$; K-S test.) with a reduced level in the stomach. There was a significant difference between infected males and females ($p = 0.05$; K-S test). In all infected specimens examined, 43.8% were males while 56.2% were females.

Relationship between body length and degree of infection

There was high prevalence in medium (20 to 29.9 cm) and large (30 to 39.9 cm) sized *C. gariepinus* and a direct linear relationship between length and intensity (Figure 2). Most of the infected fish specimens had longer body lengths than uninfected specimens (the mean body lengths for uninfected and infected fishes were 20.14 cm ±0.40 and 24.38 cm ±0.40 respectively within sex; male uninfected 19.82 cm ±0.57, infected 24.01 cm ±0.63; female uninfected 20.45 cm ±0.67, infected 24.76 cm ±0.57). Also larger size fishes were mostly infected in the rainy (22.98 cm ±0.40) than the dry season (21.54 cm ±0.40).

Relationship between body weight (g), sex and degree of infection

Figure 3 showed that majority of sampled specimens infected with helminthes were within the same weight groups, (30 to 70 g). Infection of *C. gariepinus* by parasites results in reduction of weight of the fish and generally females have more weight than male in the sampled specimens. Within the same total length range (20 to 39.9 cm), the mean weight of uninfected (66.48 g ±1.97) samples was considerably higher than infected (50.70 g ±1.89). The female (uninfected, 70.04 g ±3.01; infected, 53.31 g ±2.53) samples were found to have more weight than their male (uninfected, 62.92 g ±2.55; infected, 48.08 g ±2.82) counterpart; the general mean weight was 64.40g ±2.07.

Concurrent/multiple parasitic infection

Multiple infection of host by the identified parasites was a common occurrence. Table 2 showed that ability to successfully co-inhabit with one another in the same host was positively correlated between most of the parasites. Although, *N. rutili* and *P. laeviochus* could not co-inhabit with one another, *Anomotaenia* sp. will only appear concurrently with only *P. laevionchus*.

DISCUSSION

Prevalence and intensities of parasites in the gut of *C. gariepinus*

Prevalence of Helminthes was higher in the intestine than the stomach, similar to the findings by Mohammed et al. (2009) and Bichi and Yelwa (2010). They argued that

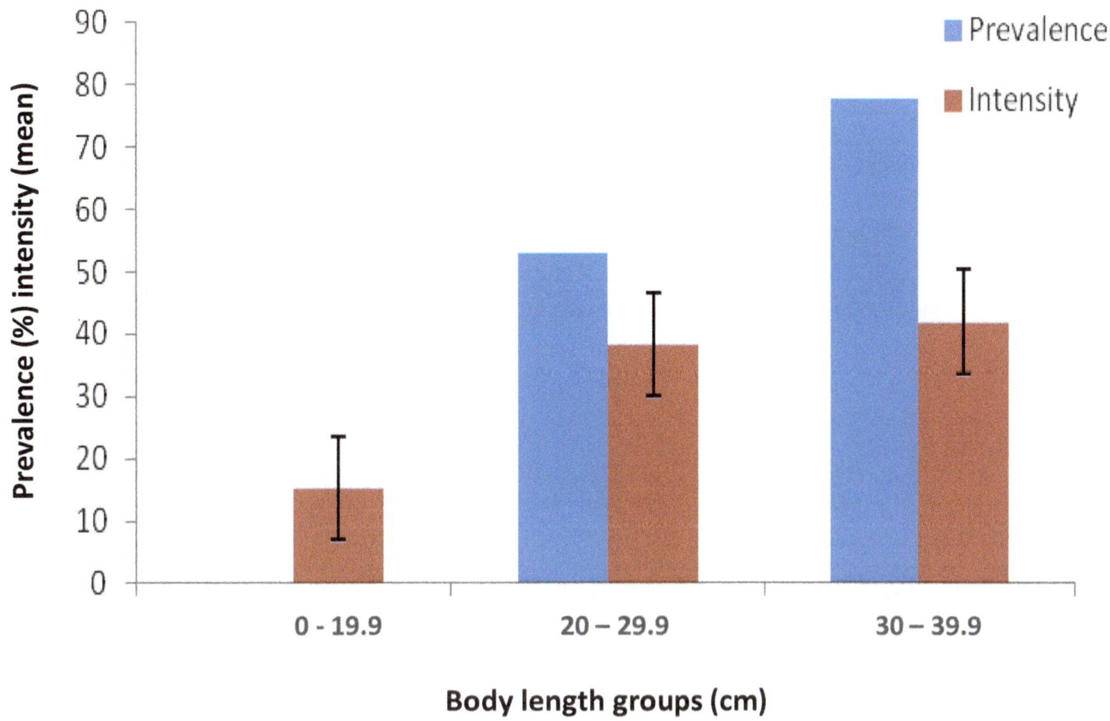

Figure 2. Relationship between body length, prevalence and intensity.

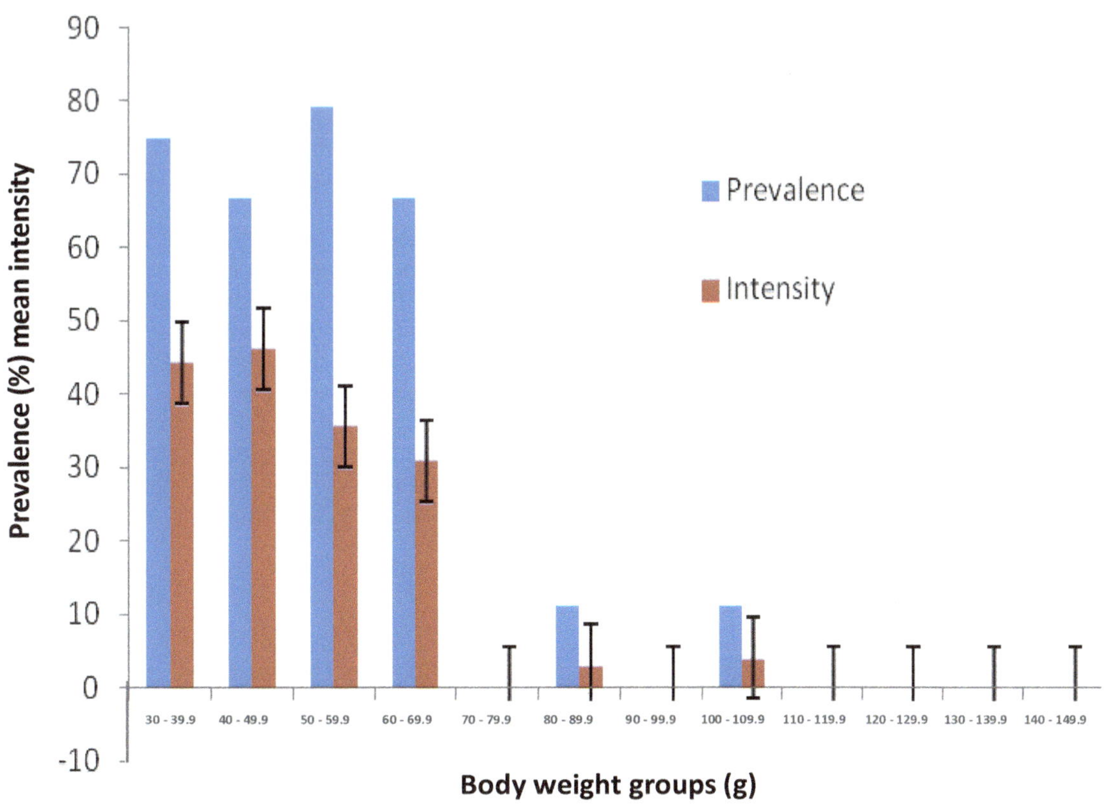

Figure 3. Relationship between body weight, prevalence and mean intensity.

Table 2. Simulation matrix for correlation between multiple infections by parasites in the stomach and intestine of *Clarias gariepinus* in Erinle Reservoir.

Parasite	*Anomotaenia* sp.	*N. rutili*	*P. laevionchus*	*P. clariae*	*P. cyathopharynx*
Neoechorynchus rutili	0.142				
Procamallanus laevionchus	0.256**	0.051			
Polyonchobothrium clariae	-0.006	0.361**	0.234**		
Paracamallanus cyathopharynx	0.005	0.354**	0.304**	0.726**	
Monobothrium sp.	0.025	0.429**	0.266**	0.726**	0.819**

Pearson correlation;** Correlation is significant at the 0.01 level (2-tailed).

regional localization in the gut can be attributed to several factors, such as hydrogen ion concentration, chemotactic response as well as food reserve. Onwuliri and Mgbemena (1989) reported that Helminthes differ in their nutritional and respiratory requirements which may influence their choice of habitat. However, the major factor that may have contributed to high parasitic prevalence in the intestine than in the stomach was that most parasites found in the reservoir were Cestodes that lacked digestive system. Obligatorily, they had to depend on the digested food in the intestine of their host, which they absorbed through their thin body tegument (that is syncytial and highly absorptive). These parasites may find an acidic medium as presented by the stomach not conducive.

The high prevalence is further supported by the findings from other studies that reported a high prevalence in the wild population of *C. gariepinus* (Anosike et al., 1992; Oniye et al., 2004; Olofintoye, 2006; Dankishiya and Zakari, 2007; Olufemi, 2008; Ajala and Fawole, 2012). The varying percentages reported from the West, through the Middle Belt to the far Northern part of Nigeria showed a pattern of decreasing prevalence from Western to Northern Nigeria. This may be attributed to a reduction in the parasite's intermediate host species diversity which may be dependent on types of forest/vegetation available in each geographical ecosystem. Western Nigeria is populated by tropical rain forest, the Middle Belt by savanna woodland/ Sudan savanna and Northern Nigeria by Sahel/dry savanna. Parasite intensity was also higher in the intestine than the stomach for majority of the parasites except Anomotaenia sp, which further confirmed the likely preference of the intestine by the parasites to the stomach.

Parasitic infection in relation to sex

There was a significant difference (p = 0.05) in the infection of male and female in the specimens examined. Emere (2000) reported differences in the incidence of infection between male and female fish, and argued that this may be due to differential feeding, either by quantity or quality of food eaten or as a result of different degrees of resistance and infection. The female samples had more weight than their male counterpart, which supported the opinion of differential feeding, either by quantity or quality of food eaten which predisposes the females to infection. Emere and Egbe (2006) reported that due to physiological state of the female, most gravid females could have reduced resistance to infection by parasites.

Effects of seasonal variation on parasitic infection

The prevalence and intensity of infection was higher in the dry than the rainy season. The factor responsible for this is eutrophication, which often raises parasitism because the associated increase in productivity will increase the abundance of the invertebrate intermediate hosts, mostly fresh water crustaceans (Lafferty and Kuris, 1999). Eutrophication leads to algal bloom at the peak of rainy season which results in increase in species variety and population of the parasite's intermediate host, towards the end of the rainy season. This may result in the infection of fishes that fed on them, and thus probably bring about the maturity of the parasites in the fish towards the dry season, depending on the life cycle of individual parasite. Another factor may be a drop in water level in the dry season exposing the invertebrates to their fish predators. Fawole and Akinsanya (2000) reported a similar prevalence of infection in *Sarotherodon galilaeus* by plerocercoid larvae of pseudophyllidean cestode in Opa Reservoir in Ile-Ife, Nigeria, and higher prevalence of infection was recorded in the dry than the rainy season.

Degree of infection of the gut of *C. gariepinus* by parasitic nematodes

Moravec (1974, 1975) studied the life cycles of *P. laevionchus* and *P. cyathopharynx* and obtained a development of the first three larval stages in *Mesocyclops leuckarti*. It was observed that the *callanidae* give birth to the first stage larvae, which are ingested by the copepods. The larvae of *P. laevionchus* reached the third stage after two moults in eight to nine days at temperatures of between 23 and 24°C. *P. cyathopharynx* is an ovoviviparous camallanid nematode whose larvae are liberated into the gut of the host and passed out with the faeces. The first moult of the parasite takes place in the Copepod intermediate host and the last two moults in the fish. It is speculated that the utilization of the copepods as food by the catfishes is responsible for their infection by the nematodes. Although Oniye et al. (2004) did not report the presence of *P. cyathopharynx*, the occurrence of *P. laevionchus* at a very low prevalence (0.83%) in the intestine was reported.

The intensity of *P. cyathopharynx* found in this study was comparable to that obtained by Ayanda (2009) who reported a similar occurrence in *C. gariepinus*. Although both parasites are nematodes, the absence of *P. laevionchus* in the stomach may be adduced to differences in physiological tolerance to the acidic medium prevalent in the stomach. Even though P. *cyathopharynx* was found both in the stomach and intestine, they preferred the intestine as portrayed by higher intensity and range displayed. Williams and Jones (1994) suggested that parasitism varied from one aquatic ecosystem to the other and this was influenced by the interplay of mixed biotic and abiotic factors. Akinsanya and Otubanjo (2006) also reported the presence of *P. cyathopharynx* in the intestine of *C. gariepinus* in the Lagos Lagoon; this suggested that *P. cyathopharynx* may also possess physiological adaptability/tolerance to salinity fluctuations.

P. laevionchus intensity was higher in the rainy than dry season, which may be deduced from the life cycle of the parasite as argued by Moravec (1975) - that the larvae reached the third stage after two moults in eight to nine days at temperatures of between 23 and 24°C in the Copepod intermediate host. This temperature range fell within the rainy season so that within nine days of reaching infective stage, fish host feeding on the copepods become infected. Dry season of higher temperature, may not support the development of the parasite larval stages in the invertebrate host, hence its low prevalence in the dry season.

Degree of infection by parasitic cestodes

Cestodes formed the bulk of the parasites recovered; Singh et al. (2013) also reported cestodes as having maximum prevalence in *C. batrachus* and *C. gariepinus*. Anatomically, cestodes lack digestive system and obligatorily have to depend on the digested food of their host, which they absorb through their body surfaces, hence they are better found in the part of the host intestine where their nutritional requirements can be met. Host body secretions and immune system may not be a threat as Buchmann and Lindenstrom (2002) suggested that parasites have an in-built molecular disguise to avoid the host hostile secretions present in its microhabitat.

P. clariae infection showed a higher prevalence in the intestine than the stomach but the intensity was a reversal. Although prevalence was low in the stomach, the intensity was high, which suggested *P. clariae*'s preference for both stomach and intestine. This observation was also reported by Oniye et al. (2004) and Barson and Avenant-Oldewage (2006). The high intensity in the stomach may be due to the anatomical structure of the stomach of *C. gariepinus,* which had a depression (at its exit to the intestine-pyloric region) in which already digested food from the stomach remain for some time before proceeding to the intestine (Oniye, Personal communication). The acidic level at this portion of the stomach may have become diluted to a value midway between stomach and intestinal pH. Here, food is not absorbed by host thus reducing the competition between parasite species residing in the intestine and host, thus giving the foraging cestodes from the intestine an ample opportunity to feed.

Bichi and Yelwa (2010) also reported *Anomotaenia* sp. and an unidentified species of *Procamallanus* in *C. gariepinus* as dominating the stomach only. *Anomotaenia* sp. a cestode which was expected to be prominent in the intestine than the stomach, as a result of their nutritional requirement (already digested food) seemed to be an exception.

Monobothrium sp's presence in stomach and intestine suggested an ability for a wide physiological adaptation and probably resistance to host defense system. This gave the parasite an opportunity for high prevalence and intensity. The difference in infection between the stomach and intestine could be attributed to high acidity of the stomach since the most resistant parasites occur comfortably there (Alfred-Ockiya, 1985).

Degree of infection by parasitic acanthocephalan

N. rutili was found in the intestine of the fish and its prevalence was low. It was not recovered in the stomach which signified the possibility of its not being able to survive in an acidic medium. This supported Oniye et al. (2004), Olofintoye (2006) and Ayanda (2009) who reported very low prevalence of the worm in the intestine of its host; while Nnadi (2012) reported the presence of its eggs in the intestine.

Body length in relation to parasitic infection and season

Infection was directly proportional to body length, bigger fish specimens were recovered in the rainy season which predisposes them to infection. Oniye et al. (2004), Olofintoye (2006) and Ayanda (2009) reported that prevalence of intestinal Helminth infection in *C. gariepinus* increases with their standard length. A plausible explanation for this may be the change in diet from weeds, seeds, phytoplankton and zooplankton as juveniles to insect larvae, crustaceans, worms and fish at sub-adult and adult stages. These invertebrates may serve as intermediate hosts to some of these parasites, while eutrophication which is seasonal will boost productivity in the rainy season, thereby increasing the population of those invertebrate organisms that served as food for the sub-adult and adult stages.

Body weight in relation to parasitic infection, season and sex

The mean weight of uninfected samples was considerably higher than infected which suggested a weight loss as a result of infection. This was evidenced in Bichi and Yelwa (2010) who reported the damage inflicted by infection of two different species of *Polyonchobothrium* (*Polyonchobothrium polypetri* and *Polyonchobothrium* sp.) in the gut of *C. gariepinus.* They were found blocking and attached firmly to the intestinal lining thus, inducing lesion at site of attachment. This also, likely suggested the absence of infective organisms in the type of diet of the fish while young. Omeji et al. (2013) observed heavier fishes as being infected more than low weighted fishes.

Generally female samples were found to have more weight than their male counterpart, which may have predisposed the females to a higher prevalence than their male counterpart due to their feeding habits. Furthermore, bigger fishes were recorded more in the rainy season which may be due to the high water level that made it difficult for fishermen to go fishing thereby giving the fish opportunity to grow coupled with eutrophication that increased invertebrate population and a change in diet on the part of the fish to more nutritive invertebrates.

Multiple and concurrent infection by parasites

Relationship between *Anomotaenia* sp. and *P. clariae,* suggested that *Anomotaenia* sp. will probably not survive in an acidic medium like the stomach and when found in the intestine with *P. clariae* an intense competition probably ensued, as a result of same physiological demand. *P. clariae* may likely have an advantage thereby diminishing the population of *Anomotaenia* sp. Relationship between *Anomotaenia* sp. and *P. cyathopharynx* was not significant though positive but very low; this also applied to *Monobothrium* sp. and *N. rutili. P. laevionchus* will easily co-exist with the remaining parasites but its relationship with *N. rutili,* though positive was low and not significant. Both parasites were absent in the stomach. This result likely pointed to similar physiological needs and sensitivity between the two parasites especially to certain existing factors in their immediate micro environment which may instigate fierce competition. Apart from the mentioned exceptions, positive linear association among the identified parasites occurred freely.

Conclusion

The gastro-intestinal helminth parasites in the gut of *C. gariepinus* include two nematodes *(P. laevionchus, P. cyathopharynx.)* three cestodes (*Anomotaenia* sp., *Monobothrium* sp., *P. clariae*) and one acanthocephalan (*N. rutili*). *Monobothrium* sp. had the greatest intensity of infection while *Anomotaenia* sp., had the least intensity. Sex and season were found to possibly influence the level of infection while diets; also, may likely contribute to the stage at which parasitic infection occurs in *C. gariepinus.* Fishes of small sizes and weight were not infected. *P. laevionchus* and *N. rutili* were absent in the stomach, maybe as a result of their sensitivity to an acidic environment as presented by the stomach. The cestodes formed the bulk of parasites found in the gut of *C. gariepinus.* The possibility of multiple and concurrent infection of different species of parasites in a fish was established and poses a health risk of zoonotic transmission to consumers.

REFERENCES

Ajala OO, Fawole OO (2012). A study of Helminth species assemblages at different host scales in *Clarias gariepinus* (Burchell, 1822) as a bio-indicator of aquatic water quality. Conf. Proc., World Acad. Sci. Eng. Technol. (WASET), Singapore, September 2012. 69:741-750.

Akinsanya B, Otubanjo OA (2006). Helminth Parasites of *Clarias gariepinus* (Clariidae) in Lekki Lagoon, Lagos, Nigeria. Revista de Biol. Trop. 54(1):93-99. PMid:18457179

Akinsanya B, Otubanjo OA, Hassan AA (2007). Helminth Parasites of *Malapterurus electricus* (Malapteruridae) from Lekki Lagoon, Lagos. Nig. J. Sci. 3(3):78-86.

Alfred-Ockiya JF (1985). Preliminary survey of parasites of Clarias in Niger Delta Area, Nigeria. Proceeding of the 4th Annual Conference of the Fisheries Society of Nigeria (FISON), Port Harcourt, 26th - 29th Nov 1985, pp. 210 - 215.

Anosike JC, Omoregie E, Ofojekwu PC, Nweke IE (1992). A survey of Helminth parasites of *Clarias gariepinus* in Plateau State, Nigeria. J. Aquat. Sci. 7:39-43.

Ayanda OI (2009). Comparison of parasitic helminthes infection between the sexes of *Clarias gariepinus* from Asa Dam Ilorin, north-central, Nigeria. Sci. Res. Essay 4(4):357-360.

Barson M, Avenant-Oldewage A (2006). Nematode parasites of *Clarias gariepinus* (Burchell, 1822) from Rietvlei dam, South Africa. Ondestepoort J. Vet. Res. 73:87-94. http://dx.doi.org/10.4102/ojvr.v73i2.154

Bichi AH, Yelwa SI (2010). Incidence of piscine parasites on the gills and gastro-intestinal tract of *Clarias gariepinus* at Bagauda fish farm

Kano. Bayero J. Pure Appl. Sci. 3(1):104-107.
Buchmann K, Lindenstrom T (2002). Interactions between monogenean parasites and their fish hosts. Int. J. Parasitol. 32:309-319. http://dx.doi.org/10.1016/S0020-7519(01)00332-0
Dankishiya AS, Zakari M (2007). Study on the gastrointestinal helminth parasites of *Clarias gariepinus* (Tuegels), in Gwagwalada, FCT, Nigeria. Best J. 4(2):79-81.
Emere MC (2000). Parasitic infection of the Nile perch Lates niloticus (L) in River Kaduna. J. Aquat. Sci. 15:51-54. http://dx.doi.org/10.4314/jas.v15i1.19988
Emere MC, Egbe NEL (2006). Protozoan parasites of Synodontis clarias (A fresh water fish) in River Kaduna. Best J. 3(3):58-64.
Eyo AA, Olatunde AA (2001). Protein and amino acid requirements of fish with particular reference to species cultured in Nigeria. In: Fish Nutrition and Fish Feed Technology (Ed.) Fisheries Society of Nigeria (FISON) Lagos, Nigeria, pp. 58-71.
Fawole OO, Akinsanya B (2000). Prevalence of parasitic infection on Sarothrodon galilaeus (Artedi) from Opa reservoir, Ile-Ife, Nigeria. World J. Biotechnol. 2:136-140.
Ibrahim CA, Needet SA, Aysel AOA, Ercument G (2001). Ecto-Endoparasite investigation on Mirror carp (*Cyprio carpio* L., 1758) captured from River Seyhan, Turkey. J. Fish Aquat. Sci. 18:87-90.
Khalil LF (1991). Techniques for identification and investigative helminthology. Helminthology Manual. International Institute of Parasitology, St. Albans, UK., pp. 156.
Khalil LF, Polling L (1997). Check list of the Helminthes parasites of African Freshwater Fishes. University of the North, Republic of South Africa.
Lafferty KD, Kuris AM (1999). How environmental stress affects the impacts of parasites. Limnol. Oceanogr. 44:925-931. http://dx.doi.org/10.4319/lo.1999.44.3_part_2.0925
Margolis L, Esch GW, Holmes JC, Kuris AM, Scad GA (1982). The use of ecological terms in parasitology. (Report of an Ad-hoc Committee of the American Society of Parasitology). J. Parasitol. 68(1):131-133. http://dx.doi.org/10.2307/3281335
Mohammed A, Kalu AU, Yem IY, Kolndacha O, Nyaku RE, Sanni AO (2009). Bacteriological quality of *Clarias gariepinus* (Burchell, 1822) in Lake Alau, Borno State, Nigeria. Best J. 6(1):15-18.
Moravec F (1974). The development of Paracamallanus cyathopharynx (Baylis, 1923) Nematoda: Camallanidae. Folia Parasitol. 21:333-343.
Moravec F (1975). The development of Procamallanus laevionchus (Wedl, 1862), (Nematoda Camallanidae), Vestnik Ceskoslovenske Spolecnosti Zoologicke 39:23-38.
Nnadi EI (2012). A study of parasitic infections of *Clarias gariepinus* in natural waters of Owerri, Imo state Nigeria. Int. J. Agric. Rural Dev. 15(2):976-981.
Olofintoye LK (2006). Parasitofauna in some freshwater fish species in Ekiti State, Nigeria. Pak. J. Nutr. 5(4):359-362. http://dx.doi.org/10.3923/pjn.2006.359.362
Olufemi DO (2008). Endoparasitic helminths of the upside-down Catfish, Synodontis Membranaceus in Jebba Lake, Nigeria. Int. J. Zool. Res. 4:181-188. http://dx.doi.org/10.3923/ijzr.2008.181.188
Omeji S, Solomon SG, Uloko C (2013). Comparative study on the Endo-parasitic infestation in *Clarias gariepinus* collected from earthen and concrete ponds in Makurdi, Benue State, Nigeria. J. Agric. Vet. Sci. 2(1):45-49.
Oniye SJ, Adebote DA, Ayanda OI (2004). Helminthes parasites of *Clarias gariepinus* (Tuegels) in Zaria. Nig. J. Aquat. 19(2):71-75.
Onwuliri COE, Mgbemena MO (1989). The parasite fauna of some fresh water fish from Jos, Plateau State, Nigeria. J. Appl. Fish. Hydrobiol. 2:33-37.
Ovie SI, Ovie SO (2002). Fish larval rearing: the effect of pure/mixed Zoopla nkton and artificial diet on the growth and survival of *Clarias anguillaris* (Linnaeus, 1758) larvae. J. Aquat. Sci. 17(1):67-73. http://dx.doi.org/10.4314/jas.v17i1.19915
Paperna I (1996). Parasites, infections and diseases of fishes in Africa: An update. CIFA Technical Paper. 31:157-170.
Singh H, Kaur P, Shrivastav R, Borana K, Manohar S, Qureshi TA (2013). Endo-parasitic Helminths of two dominating species of Family Clariidae. J. Chem. Biol. Phys. Sci. 3(2):1149-1154.
Williams H, Jones A (1994). Parasitic worms of fish. Taylor and Francis, Bristol, UK, P. 593.

Attitudinal analysis of women involvement in fish processing and marketing in Ondo State

S. D. Yomi Alfred*, O. O. Odefadehan and M. E. Ukut

Department of Agricultural Economics and Extension, Federal University of Technology, Akure, Ondo State, Nigeria.

Generally, it is believed that attitude, to an extent, determines participation in any vocation. This study sets out to investigate and analyze the attitude of women involved in fish processing and marketing in Ilaje Local Government Area (LGA) of Ondo state. The study also determined the profitability and constraints faced by the respondents in the business. The influence of selected respondent's socio-economic characteristics on fish processing and marketing were also determined. Data were collected with structured questionnaires. A total of 83 respondents were randomly selected across the wards in the LGA. The study revealed that the involvement of respondents in fish processing and marketing was significantly affected by age, marital status, household size and educational level. Furthermore, the study confirmed that women have negative attitude towards fish processing and marketing. The women's negative attitude was significantly determined by their perception that, fishing activities were stressful and repulsive and also by the use of crude methods of processing.

Key words: Attitudinal, women, involvement, fish, processing.

INTRODUCTION

Nigeria, with a population of about 150 million (2008 estimate), is the largest single consumer of fish and fish products in the African region (FOS, 1999) In Nigeria, like all coastal West African countries, fish is the cheapest form of animal protein in the diet of the people for many decades. Therefore, the importance of the fisheries products in satisfying the nutritional needs of the people as well as, providing employment cannot be overstated (Ajayi and Talabi, 1984). Entire family units; men, women and children in the fishing communities are engaged in the sector. Most of the coastal fishing is carried out by local fisher folks from canoes operating from surf-beaten beaches and riverine areas. In Nigeria, women play a crucial role in fish production, processing, distribution and marketing. In addition, some of the women are crucial in financing fish production units. Since the 1960s, the fisheries sector has contributed between 1.11 to 2.4% of Nigeria's Gross Domestic Product (GDP) and 3.0 to 6.0% of agricultural production. Fish as a major source of animal protein has continued to grow in importance over the years, particularly, as a substitute to beef. Recent studies confirmed the increasing shift from meat to fish consumption among households in Nigeria (CBN/NISER, 1992). The artisan fisheries sub-sector constitutes a major source of employment, especially, in the Southern coastal maritime areas. The estimated revenue from fish is over five hundred million naira. Fish in Nigeria is caught from the sea, inland and more recently from ponds and other artificially cultured systems (Lawal and Idega, 2004). A down turn in the economy of the country naturally affects all sectors. Consequently, the fishing sector is also being affected by the impact of the worsening economic situation. The artisan catches, which have therefore, started dwindling since 1983 got worse with the introduction of Structural Adjustment Programme and have continued unabated and thus, resulting in demand being far below supply (FOS, 1998).

Statement of the problem

Post-harvest fisheries' activities, especially, those

*Corresponding author. E-mail: yomialfred2003@yahoo.com.

involving preservation of products over several days, are an essential component of fisheries development in riverine areas in Nigeria. Indeed, what is the point of catching fish if it becomes unfit for consumption, resulting from lack of means of distribution, processing and preservation? Harvesting of fish is not usually on the basis fishermen-consumer (Lawal and Idega, 2004), therefore, price of fish changes as it passes through middlemen such that, it becomes expensive. Women are found to be more involved in the processing and marketing of fish while their men counterparts are found more in actual fishing (Samson, 1997). Though both men and women could be found in either of the fishing activity, relatively, each could be said to be more involved in one than the other. If fishing industry must be improved through the use of improved facilities, accessibility to credit and better information in the marketing structure for fish, the attitudes and perception of the women in the processing and marketing structure for fish, need to be more understood. It is against this background that this study investigated and determined the attitudes of women in fish processing and marketing in Ondo state, specifically the study is based on:

1. Examining the socio-economic characteristics of the respondents.
2. Identifying the processing and marketing methods used by the respondents.
3. Ascertaining the constraints faced by fish processors and marketers.
4. Determining the attitudes of respondents towards fish processing and marketing.

METHODOLOGY

The study was carried out in Ondo state of Nigeria. The state covers 20, 9955 s km. The state is made up of 18 local government areas. It is bounded in the South by the Atlantic Ocean, in the North by Kogi State, in the Western parts by Ogun and Osun States and in the Eastern side by Edo and Delta States. The study was focused purposively on Ilaje LGA along the coastal area of Ondo state. The predominant occupation of the people is fishing. Two communities who were predominantly fishermen/women were purposively selected. From each of the communities, five wards were selected, from each selected ward; ten women fish processors and marketers were randomly selected giving a total sample size of 100. Data were collected by means of structured questionnaire/ interview schedule, however, only 83 respondents' copies of questionnaire were found suitably completed to be included in the study and were analyzed using descriptive statistics and chi-square analysis (Ogunfiditimi, 1986). The attitudes of the respondents were measured using 4 and 5 point-Likert scale of "Very satisfied", "Just satisfied", "Don't know", and "Not satisfied, and "Strongly Agree", "Agree", "Undecided", "Disagree" and "Strongly Disagree" respectively. The former (4-point scale) was scored 4, 3, 2, and1 and the later (5-point scale) was scored 5, 4, 3, 2 and 1 for positive statements but were reversed for negative statements as 1, 2, 3, and 4, and 1, 2, 3, 4, and 5 respectively. The means obtained from this scoring was used to categorize the respondents' attitude into "unfavorable attitude," (<2.5) "indifferent attitude" (2.5 to 3.4) and "favorable attitude" (>3.4).

RESULTS AND DISCUSSION

Respondents' demographic characteristics

The demographic characteristics of the respondents identified included age, marital status, household size, educational level and religion, which are presented in Table 1. The age of respondents as presented in Table 1 shows that majority of the respondents (60.2%) were in the age range of 25 to 40 years, while 24.1% were below 25 years and only 15.7% of the respondents were above 40 years. This finding implies that most of the women involved in fish processing and marketing are in their youth or active age and as such have strength and vigor to carry out these activities.

The marital status of respondents also as presented in Table 1 indicates that 53% of the respondents were married, 25% were single and 6.0% were widowed. This implies that most of the women involved in marketing and processing of fish in Ondo state are married implying that marriage is a cherished value in the society. Those who were still singles may be due to their young age. The low percentage of divorcee is indicative of compliance with the African tradition which frowns at divorce and sees a divorced woman as being irresponsible. Furthermore, household size of the respondents as shown in Table 1 shows that 42.2% had between 6 to 10 people in their families. 27.7% of the respondents had between 1 to 5 people in their families, while only 2.4% had above 10 people in their household. This finding shows that most of these respondents had rather large household sizes. It could also be attributed to the fact that they got married or started child bearing at an early age. Large household size could imply that more hands would be available for labour in fish processing but may also imply that a large percentage of income would go for consumption.

A total of 16.9% of respondents had no formal education, 61.4% of the total respondents had primary school education as their highest level of education, and 18.1% had secondary education while only a paltry 3.6% had post-secondary education. This shows that majority of the women involved in fish processing and marketing have little or no formal education and can barely read or write. The implication of this finding is that many of them may not be able to keep records of their production or processing activities and income.

Table 1 showed that the majority of the respondents in the study area were Christians. This is a common trend in the Southern Nigeria where people are predominantly Christians. The religion or belief of the people does not likely to hold any restraint against fishing activities. This is because, acceptance of certain technologies and farming practices are often influenced by the individual's belief (Ekong, 2003).

Table 2 shows how respondents got into the fish business. Majority of them got involved in the business by inheritance from their mothers and or other female relations (79.5%), 16.9% of the respondents were trained

Table 1. Socio-economic characteristics of the respondents.

Variable		Frequency	Percent	Cumulative percent
Age				
	<25	20	24.1	24.1
Valid	25-40	50	60.2	84.3
	>40	13	15.7	100.0
	Total	83	100.0	
Marital status				
	Single	21	25.3	25.3
	Married	53	63.9	89.2
Valid	Separated	3	3.6	92.8
	Widowed	5	6.0	98.8
	Abandoned	1	1.2	100
	Total	83	100.0	
Household size				
	1-5	23	27.7	37.7
	6-10	35	42.2	95.1
	>10	2	2.4	98.4
Valid	Total	61	73.5	100.0
	Missing system	22	26.5	
	Total	83	100.0	
Educational level				
	No formal education	14	16.9	16.9
	Primary education	51	61.4	78.3
Valid	Secondary school education	15	18.1	96.4
	Tertiary education	3	3.6	100.0
	Total	83	100.0	
Religion				
Valid	Christianity	83	100.0	100.0

Field Survey (2010).

formally by NGOs and other bodies while 2.4% were apprentices. This shows that most of the women involved in the business did so through their families as it is being passed from one generation to the other. About 10.8% of the respondents were engaged in the processing of fish only, 16.9% were involved in the marketing of fish only, while the remaining 72.3% were involved in both marketing and processing of fish. This indicates that majority of the women were involved in both processing and marketing of fish in Ondo state.

Table 2 also shows that 28.9% of the respondents have been in the fishing business for between 1 to 5 years, 27.7% between 6 to 10 years while 43.3% were in it for more than 10 years. This result indicates that the women got into fish processing and marketing quite early in life. More importantly, it agrees with the earlier finding that they had little formal education (Table 1) because going by their ages (Table 1) and the number of years they have put into the business, they had little time for schooling.

Furthermore, Table 2 shows that 43.3% of the women were not involved in any other type of business but were engaged in either processing or marketing or both as their primary business, while 39.8 percent had other businesses apart from fish processing and marketing. In addition, 98.8% of the women involved in processing of fish identified smoking as the major means by which they processed their fish, while 1.2% used salting. Smoking is more popular with the women and could be because of its ease and longer shelf life of smoked fish compared with other methods. This could also be attributed to the more available fuel wood and charcoal in the study area when compared with the arid zone where fishing is also common but prefers other method of processing

Table 2. Fish processing and marketing business.

Variable		Frequency	Percent	Cumulative percent
How respondents got into business				
Valid	Inherited	66	79.5	80.5
	Trained	14	16.9	97.6
	Apprenticeship	2	2.4	100.0
	Total	82	98.8	
	Missing system	1	1.2	
	Total	83	100.0	
Area of specialization				
Valid	Processing	9	10.8	10.8
	Marketing	14	16.9	27.7
	Both processing and marketing	60	72.3	100.0
	Total	83	100.0	
Number of years in fish processing and marketing				
Valid	1-5	24	28.9	28.9
	6-10	23	27.7	56.6
	>10	36	43.4	100.0
	Total	83	100.0	
Processing and marketing of fish as the only business activity				
Valid	Yes	33	39.8	47.8
	No	36	43.4	100.0
	Missing system	14	16.9	
	Total	83	100.0	
Method of fish processing				
Valid	Smoking	82	98.8	98.8
	Salting	1	1.2	100.0
	Total	83	100.0	

Field Survey (2010).

as exemplified by Chad Basin, Majority of the women got their fish from fishermen and it was discovered during the survey that most of their husbands and sons were fishermen and as such, it was easy for them to get their fish, however, there were some who got their fish from fish dealers whose specialty is to get these fish from fishermen and then sell to those who would process them. It however, shows that getting supply of fish for processing is not a problem.

Table 3 shows that all the respondents agreed that fish processing and marketing is a profitable business. Although, all the women agreed that the business is profitable, but in Table 3, 27.7% of them said they would change the business if given an alternative, while the remaining 48.2% were not ready to change into another business. The majority's decision to stick to fish processing and marketing is a reflection of its being profitable, since that is the primary objective of any business. About 92% belongs to one market association or the other, as seen in Table 3. Being a member of an association gives them a voice for better coordination and planning of their activities.

About 36% of respondents identified fish dealers as their main source of patronage, 15.7% had individuals as customers while 26.5% had individuals and market groups as customers. These finding shows that respondents preferred to sell more to fish dealers in bulk who then retail in smaller quantities to others and also to industries who use them as raw materials for their products. The survey revealed that the respondents who realized between N25, 000 to N50, 000 were in the majority (56.6%). While 26.6% of the respondents realized less than N25, 000 weekly, 16.9% made over N50, 000 weekly.

Findings further showed that 90.4% of the respondents had no means of transport of their own and this made their movement especially, those involved in marketing rather cumbersome. Dearth of transportation facilities

Table 3. Perceived profitability analysis of fish processing and marketing.

Variable		**Frequency**	**Percentage**	**Cumulative percent**
Source of fish for processing				
Valid	Fishermen	26	31.3	31.3
	Fish dealers	28	33.7	65.1
	Both fishermen and fish dealers	29	34.9	100.0
	Total	83	100.0	
Interests in changing to alternative business				
Valid	Yes	23	27.7	36.5
	No	40	48.2	100.0
	Missing system	20	24.1	
	Total	83	100	
Profitability of fish processing and marketing				
Valid	Yes	83	100.0	100.0
Membership of marketing association				
Valid	Yes	76	91.6	97.4
	No	2	2.4	100.0
	Missing system	5	6.0	
	Total	83	100.0	
Buyers of fish from respondents				
Valid	Individuals	13	15.7	15.7
	Market group	8	9.6	25.3
	Fish dealers	30	36.1	61.4
	Individuals and fish dealers	22	26.5	88.0
	Market group and individuals	2	2.4	90.4
	Market group and fish dealers	8	9.6	100.0
	Total	83	100.0	
Weekly income (Naira)				
Valid	<25,000	22	26.5	26.5
	25,000-50,000	47	56.6	83.1
	>50,000	14	16.9	100.0
	Total	83	100.0	
Respondents owning means of transport				
Valid	Yes	7	8.4	8.5
	No	75	90.4	100.0
	Total	82	98.8	
	Missing system	1	1.2	
	Total	100	100.0	

Field Survey (2010).

might be as a result of high cost of vehicles to move on land and also as a result of ever increasing price of outboard engines for transportation on water. High cost of transportation could increase costs of operation which could be passed to the consumers.

Table 4 revealed that the respondents who were very satisfied with the business were 48.2%. While 44.65% were just satisfied 2.4% were not satisfied. This high percentage of respondents who were satisfied with the job could be said to be responsible for the equal percentage of those who were not ready to change to another type of business if the opportunity arises. It can

Table 4. Extent of satisfaction by respondents.

Variable		Frequency	Percentage	Cumulative percent
Valid	Very satisfied	40	48.2	48.8
	Just satisfied	37	44.6	93.9
	Don't know	3	3.6	97.6
	Not satisfied	2	2.4	100.0
	Missing system	1	1.2	
	Total	83	100.0	

Field Survey (2010).

Table 5. Women's attitude towards fish processing and marketing.

S/N	Item	Frequency					Mean (x)	Remark
		SA	A	U	D	SD		
1	People refuse to go into the business because it is a difficult job	7 (8.4)	3 (42.2)	4 (4.8)	3 (36.1)	7 (8.4)	2.59	U
2	It is a smelly and repulsive business	6 (7.2)	45 (54.2)	3 (3.6)	29 (34.9)	-	2.39	U
3	Processing and transportation requires large capital	30 (36.1)	5 (60.2)	-	3 (3.6)	-	1.67	D
4	Women prefer to market already processed fish than process the fish themselves	24 (28.9)	20 (24.1)	7 (8.4)	13 (15.7)	18 (21.7)	2.56	U
5	Rate of women leaving the business has increased due to hygiene and stress involved	10 (12.0)	28 (33.7)	12 (14.5)	28 (33.7)	5 (6.0)	2.77	U
6	The use of crude methods to process the fish makes other women run away from it	5 (6.0)	28 (33.7)	8 (3.6)	35 (42.2)	7 (8.4)	2.82	U
7	Fish processing and marketing is very profitable	59 (71.1)	21 (25.3)	2 (2.4)	-	-	4.65	SA
8	We are not being encouraged in the business by government	11 (13.3)	51 (61.4)	6 (7.2)	7 (8.4)	7 (8.4)	2.34	D
9	This business can be expanded if credit is made available	71 (85.5)	10 (12.0)	1 (1.2)	-	1 (1.2)	4.8	SA
10	I am just doing this business as a last resort	2 (2.4)	32 (38.6)	7 (8.4)	27 (32.5)	15 (18.1)	2.92	U
Total							29.51	

Total mean = 29.51; therefore $\frac{29.5}{10} = 2.95$

Computed data from field survey (2010).

be inferred that some women get involved in fishing business not out of interest but because of poverty. Okunmadewa (2001) concludes that the effect of poverty makes people to work on what does not even interest them.

Since 2.95<3.5 (Table 5), it shows that there was an indifferent attitude by women towards fish processing and marketing in Ondo state. Relationship between the respondent's socio-economic characteristics and their fishing activity (processing or marketing or both) was presented in Table 6, the χ^2 values for ages (27.93), marital status (114.89), household size (54.34) and educational level (63.07) all had significant relationship with respondents' fishing activity. Age was significant because it is an important factor in determining the type of activity one can engage in. The business of fish processing and marketing requires energy as could be obtained in women in their young and middle age. The older women may not be able to perform optimally be-cause of declining strength resulting from old age.

Marital status being significant could be attributed to the fact that fishing is a joint business

Table 6. Chi square analysis of relationship between respondents' demographic characteristics and their fishing activity (processing or marketing or both).

Demographic characteristic	Calculated χ^2	Tabulated χ^2	Degree of freedom	Decision
Age	27.93	9.21	2	S
Marital status	114.89	13.28	4	S
Household size	54.34	11.34	3	S
Educational level	63.07	11.34	3	S

Computed data from Field Survey (2010).

Table 7. Chi- square analysis of the relationship between factors affecting respondents' attitude and their participation in fishing activity.

Variable	Calculated χ^2	Tabulated χ^2	Degree of freedom	Decision
People refuse to go into the business because it is a difficult job	51.88	13.28	4	S
It is smelly and repulsive business	57.29	11.34	3	S
Processing and transportation require large capital	40.12	9.21	2	S
Women prefer to market already processed fish rather than process the fish themselves	10.56	13.28	4	N.S
Rate of women leaving the business has increased due to stress involved	27.66	13.28	4	S
The use of crude methods to process the fish makes other women run away from it	46.34	13.38	4	S
Fish processing and marketing is very profitable	61.63	9.21	2	S
We are not being encouraged in the business by government	92.15	13.28	4	S
This business can be expanded if credit is made available	164.85	11.34	3	S
I am just doing the business as a last resort	39.35	13.28	4	S

Computed data from field survey (2010); S – Significant, NS – Not significant.

between the members of the family, while the man is doing the fishing, the woman and other female members of the family engage in the processing and marketing. It is very likely that the larger the household size, the greater the production and consequently, the greater the income.

The reason why level of education was significant could be that, the higher the level of education the greater the quantity and quality of fish produced. Higher level of education affords the fish processor and marketers to take adequate record to maximize their profit and also higher access to improve processing methods and marketing information that could enhance production.

Chi- square analysis of the relationship between the factors affecting respondents' attitude and their participation in fishing activity was shown in Table 7. Findings revealed that there is a significant relationship between the respondents' perception of the factors that affect their attitudes and their participation in fishing activities, except for their preference for one or both of the activities (processing and marketing), where no significant relationship was found. The result agrees with Fazio (1986) who pointed out that one's behavior is dictated by one's attitude. The implication of this is that the women's perception of the factors affecting their attitude towards fish processing and marketing is influenced by the following; i) that it is a difficult job, ii) that it is a smelly and repulsive job, iii) that transportation/processing requires large capital, iv) the use of crude methods to process fish makes them run away from it, and v) they are not being encouraged in the business by the government.

The result of correlation in Table 8 shows that there was a high correlation between women's unfavorable attitude towards their fishing activity

Table 8. Correlation analysis showing the relationship between women's attitude and their involvement towards fishing activities.

Parameter	People refuse to go into the business because it is a difficult job	It is smelly and repulsive business	Processing and transportation requires large capital	Women prefer to market already processed fish than process the fish themselves	Rate of women leaving the business has increased due to hygienic and stress involved	The use of crude methods to process the fish makes other women run away from it	Fish processing and marketing is very profitable	We are not being encouraged in the business by government	This business can be expanded if credit is made available	I am just doing the business as a last resort
People refuse to go into the business because it is a difficult job	1									
It is smelly and repulsive business	0.269*	1								
Processing and transportation requires large capital	0.094	0.069	1							
Women prefer to market already processed fish than process the fish themselves	0.273*	0.160	0.400**	1						
Rate of women leaving the business has increased due to hygienic and stress involved	0.340**	0.175	0.486**	0.601**	1					
The use of crude methods to process the fish makes other women run away from it	0.460**	-0.019	0.025	0.135	0.318**	1				
Fish Processing and marketing is very profitable	0.168	-0.058	0.308**	0.239*	0.222*	0.195	1			
We are not being encouraged in the business by government	0.417**	0.066	0.069	0.173	0.141	0.262*	0.120	1		
This business can be expanded if credit is made available	-0.209	-0.315**	0.053	0.096	0.117	0.173	0.409**	-0.067	1	

Table 8. Contd.

I am just doing the business as a last resort	0.106	0.152	-0.073	0.362**	0.177	0.153	0.009	0.025	0.031	1

*Correlation is significant at 0.05 level (2 tailed); ** correlation is significant at 0.01 level (2 tailed).

and their perception that fishing business is a smelly and repulsive job. Unfavorable attitude of the respondents also correlates positively with their preference to market rather than processing of the fish, the stress involved, use of crude methods to process the fish and the non-assistance by the government. This result corroborates the view of Higgins (1996) who opines that one's attitude has effect on the actions and decisions one takes. Meanwhile, there is low correlation between the women's negative attitude and transportation difficulties, profitability of fish, fish business expansion and doing the business as a last resort. It therefore, implies that, women who would like to engage in fish processing and marketing would likely do so because of the profitability.

CONCLUSION AND RECOMMENDATION

The study has shown that although, majority of the women agree that fish processing and marketing is highly profitable, they still have an indifferent attitude towards fish processing and marketing. This is due to the constraints they face in the business and their negative perceptions towards the business in general. The government's nonchalant attitude towards providing necessary inputs to enhance the starting and sustenance in the business has not been encouraging and this is one of the problems faced by those in the business, however, there is urgent need to educate the women towards going into the business as it is a very profitable and viable business as this creates employment opportunities and increases revenue for the state and nation as a whole. In view of the findings from the study, the following are therefore recommended to improve the attitude of women's involvement in fish processing and marketing in Ondo state.

i. The women fish processor and marketers should be encouraged to form themselves into cooperatives. This will afford them the accessibility to government patronage and assistance such as government guaranteed credit facilities that would enable the women being able to overcome some teething problems like means of transportation and processing equipment.
ii. The women also need environmental sanitation education and orientation; this will enable them to reduce the unattractive environment that makes their operations smelly and repulsive.
iii. The fish marketers also need information on markets so that they can identify and locate markets where profit can be optimally maximized.

REFERENCES

Ajayi TO, Talabi SO (1984). The Potential and Strategies for Optimum Utilization of the Fisheries Resources of Nigeria. NIOMR Technical paper 18:5.

CBN/NISER (1992). The Impact of SAP on Nigerian Agriculture and Rural life, National Stud. 1:130.

Ekong EE (2003). An introduction into Rural Sociology, 2nd Edition. Dove Educational Publishers, Uyo. pp. 75-78.

Fazio RH (1986). How do attitude guide behavior? In: Sorrentino RM and Higgins ET (Eds.), The Handbook of Motivation and Cognition: Foundations of Social Behaviour. Guilford Press, New York. Retrieved from http://en.wikipedia.org/wiki/attitude_(psychology)

Federal Office of Statistics (1998). A Statistical Analysis of 1996/1997 National Consumer survey.

Federal Office of Statistics (1999). Poverty and the Agricultural Sector. Reports of FOS, Lagos.

Higgins ET (1996). Knowledge Activation: Accessibility, Applicability and Salience. Retrieved from http://en.wikipedia.org/wiki/attitude_(psychology).

Lawal WL, Idega EO (2004). Analysis of fish marketing in Benue State. A paper presented at 2001 Annual Conference of Association of Agricultural Economists (NAAE) held at ABU Zaria, November, 3-5.

Ogunfiditimi TO (1986). Community Survey Methods: Statistical Techniques and Computer Analysis. Nigeria's MaryGrant (Education). Ibadan. pp. 23-24.

Okunmadewa FY (2001). Poverty Reduction in Nigeria: A Four Point Demand. Living Book Series, Abuja, Nigeria.

Samson YA (1997). Introduction to Agriculture and Fisheries Management in Nigeria. Goal Educational Publishing, Abeokuta, Nigeria. pp. 10-24.

4

Population dynamics of the edible clam *Meretrix casta* (Chemnitz) (International Union for Conservation of Nature status: Vulnerable) from two estuaries of North Kerala, south west coast of India

P. Laxmilatha

Central Marine Fisheries Research Institute, P. B. NO. 1603., Ernakulam, Cochin 682 014, India.

The population parameters of the edible clam *Meretrix casta* (Chemnitz) from two estuaries along the coast of north Kerala, Southern India were estimated. *M. casta* has been cited as vulnerable as per al. (2006a). Growth of *M. casta* from the east coast has been reported by Abraham (1963), Durve (1970: 73) and Sreenivasan (1983). Jayawickrema and Wijeyratne (2009) have described the population dynamics of *M.* (IUCN) population status. The asymptotic length ($L\infty$) for *M. casta* from Chaliyar estuary was 24.24 mm. The growth coefficient (K) was 1.8^{-1}. The growth performance index (ф') was found to be 1.92 and t_0 was calculated at -0.04283547 years. The length weight relationship is given by Log Y = -1.259635861 + 3.090977418Log X, r = 0.903010085. The total mortality (Z) estimate for *M. casta* from Chaliyar estuary was 3.92 year^{-1}; fishing mortality (F) 2.12 year^{-1}. Exploitation level (E) was 0.54. The asymptotic length ($L\infty$) for *M. casta* from Kavvai estuary was 26.5 mm. The growth coefficient (K) was 2.0^{-1}. The growth performance index (ф') was found to be 2.03 and t_0 was calculated at -0.03441665 years. The length weight relationship is given by Log Y = -0.98701877 + 3.054878207 Log X, r = 0.613724895. The total mortality (Z) estimate for *M. casta* from Kavvai estuary was 2.98 year^{-1}; fishing mortality (F) 0.98 year^{-1}. Exploitation level (E) was 0.33. The *M. casta* is under exploited from both the estuaries. There is potential for increased exploitation of the edible clam from both estuaries since the present exploitation levels are far below the maximum sustainable yield (MSY).

Key words: *Meretrix casta*, population dynamics, Chaliyar, Kavvai estuaries, south west coast of India.

INTRODUCTION

Meretrix casta (Chemnitz) a venerid clam is edible and forms sustenance level fishery in the estuaries and backwaters of both east and west coast of India. Along west coast, it occurs in Goa, Aghnashini, Uppunda, Coondapur, Udyavara and Mulki estuaries in Karnataka and Ashtamudi, Chettuva, Beypore, Korapuzha, Moorad and Chaliyar estuaries in Kerala. This species contributes to livelihood of large number of fishers living along these estuaries (Laxmilatha et al., 2006a, b). Along the east coast, it forms a fishery in Vellar, Pulicat and Bhimunipatnam backwaters (Seshappa, 1971; Alagarswami and Meiyyapan, 1989b; Narasimham, 1991).

Biology and growth of the white clam has been studied by Abraham (1963), Durve (1964), Seshappa (1971), Krishnakumari et al. (1977), Sreenivasan (1983), Thangavelu and Sanjeevraj (1985), Thangavelu and Poovannan (1994), and Rao (1988). Age and growth of *M. casta* from west coast has been reported by Salih

(1973), Parulekar et al. (1973) and Harkantra (1975). Fishery of *M. casta* has been described by Laxmilatha et al. (2006a). Growth of *M. casta* from the east coast has been reported by Abraham (1963), Durve (1970: 73) and Sreenivasan (1983). Jayawickrema and Wijeyratne (2009) have described the population dynamics of *M.casta* in the Dutch canal of Sri Lanka. Influence of environmental variables on the abundance of *M. casta* in Southern Thailand has been detailed by Tanyaros and Tongnunui (2011).

M. casta has been cited as vulnerable (A 1c, 1d) by the International Union for Conservation of Nature (IUCN) based on the population decline in several estuaries due to decline in extent of occurrence area (bed), quality of habitat and number of mature individuals (Ananda-Rao et al., 1998). The knowledge of the population characteristics is important for the implementation of appropriate management and conservation measures for the sustainable exploitation of the stock. The objective of this study is to understand the population dynamics of *M. casta* from two geographically separate estuaries along the west coast of India, which will enable the development of conservation and management measures for sustaining the fishery.

MATERIALS AND METHODS

Monthly samples of the edible backwater clam *M. casta* (Chemnitz) were collected from two estuaries Chaliyar (11°31'29°N - 76°3'26°E), in Kozhikode district, north Kerala and Kavvai estuary (11°21'-15°58'N - 75°47'-70°27'E) located in Kasargod district of north Kerala, south west coast of India during 2008 (Figure 1). A total of 1560 numbers.

M. casta and 1559 numbers were collected from Chaliyar and Kavvai respectively, for the study. The samples of *M. casta* were cleaned of all encrusting organisms. The total length was measured using digital vernier calipers to the nearest 0.1 mm along the anterior-posterior axis. The total weight was recorded in an electronic balance to the nearest 0.01 g. The monthly length frequency distribution data of *M. casta* were grouped into 3 mm intervals and used to estimate growth and population parameters using the FiSAT software (Gayanilo et al., 1995). Asymptotic length (L_∞) and growth coefficient (K) of the von Bertalanffy growth function (VBGF) were estimated by means of ELEFAN –1 (Bertalanffy, 1934; Pauly, 1986). K-scan routine was conducted to assess a reliable estimate of the K value. The estimates of L_∞ and K were used to estimate the growth performance index (Φ') (Pauly and Munro, 1984) of *M. casta* using the equation $\Phi' = 2 \log_{10} L_\infty + \log_{10} K$.

The length – weight relationship ($W = a L^b$) was studied following Ricker (1975), where *W* is the weight (g), *L* the total length (mm), *a* the intercept (condition factor) and *b* is the slope (growth coefficient / relative growth rate). The parameters *a* and *b* were estimated by least squares linear regression on log-log transformed data: $\log_{10} W = \log_{10} a + b \log_{10} L$. The coefficient of determination (r^2) was used as an indicator of the quality of the linear regression. The inverse von Bertalanffy growth equation (Sparre and Venema, 1992) was used to find lengths of the *M. casta* at various ages. The VBGF was fitted to estimates of the length at curve using non-linear estimation procedure the VBGF is defined by the equation:

$$L_t = L_\infty [1-e^{-k(t-t_0)}]$$

Where L_t is the mean length at age L_∞ the asymptotic length, *K* the growth coefficient, *t* the age of the *M. casta* and t_0 is the hypothetical age at which the length is zero.

The total mortality (*Z*) was estimated by the length converted catch curve method (Pauly 1980, 1984). Natural mortality rate (*M*) for bivalves is considered approximately equal to the growth coefficient K (Gayanilo et al., 1997). The fishing mortality (*F*) was estimated using the relationship $F = Z - M$ where *Z* is the total mortality, *F* the fishing mortality and *M* is the natural mortality. The exploitation level (*E*) was obtained by the relationship: $E = F / Z$. Exploitation rate (*U*) was given by $U = F / Z * (1 - e^{-Z})$. The annual total stock and standing stock were estimated by Y / U and Y / F respectively, where "Y" is the annual average catch of the species (Beverton and Holt, 1957; Ricker, 1975). Maximum sustainable yield (MSY) was calculated by the equation for exploited stocks, $MSY = Z*0.5*B$ (Gulland, 1979). The recruitment pattern of the stock was determined by the backward projection on the length axis of the set of available length frequency data as in FiSAT.

RESULTS

Fishery

M. casta is collected by hand picking, from shallow areas during low tides by women and children, particularly during peak season and is used for local / domestic consumption. The other method of fishing is by using a scoop or bag net attached to a pole. Two to three men operate from a canoe. The net is pushed into the mud to rake up the clams and lifted up. The bag net is vigorously shaken in the water to clear the mud in the bag net and the clams are transferred to the canoe. The depth of the fishing area is about 1.5 to 2 m at high tide.

M. casta forms a sustenance level fishery in both the estuaries viz., Chaliyar and Kavvai. The annual production of *M. casta* in the Chaliyar estuary during 2008 was 1139 t, the total effort 6491 and the catch per unit effort of 175.5 Kg. The average monthly production was 94.9 t at an average monthly effort of 541. The maximum landings were during January to March (Table 1). The annual production of *M. casta* in the Kavvai estuary during 2008 was 2105 t, the total effort 12120 and the catch per unit effort of 173.7 Kg. The average monthly production was 175.4 t at an average monthly effort of 1010. The maximum landings were during March to May (Table 1).

Biology

Chaliyar estuary

The size range of *M. casta* contributing to the fishery in Chaliyar estuary was 15.8 to 33.4 mm. The modal classes were 20-22 (41.7%), 18-20 (24.2%), 22-24 (18.3%).The mean length was 21.5 ± 2.26 mm (n = 1560) and mean weight 3.17 ± 1.29 g. The sex ratio was 1:0.92. Over 50 % of the females were in mature condition. The mean condition index of *M. casta* during the year was 7.5%.

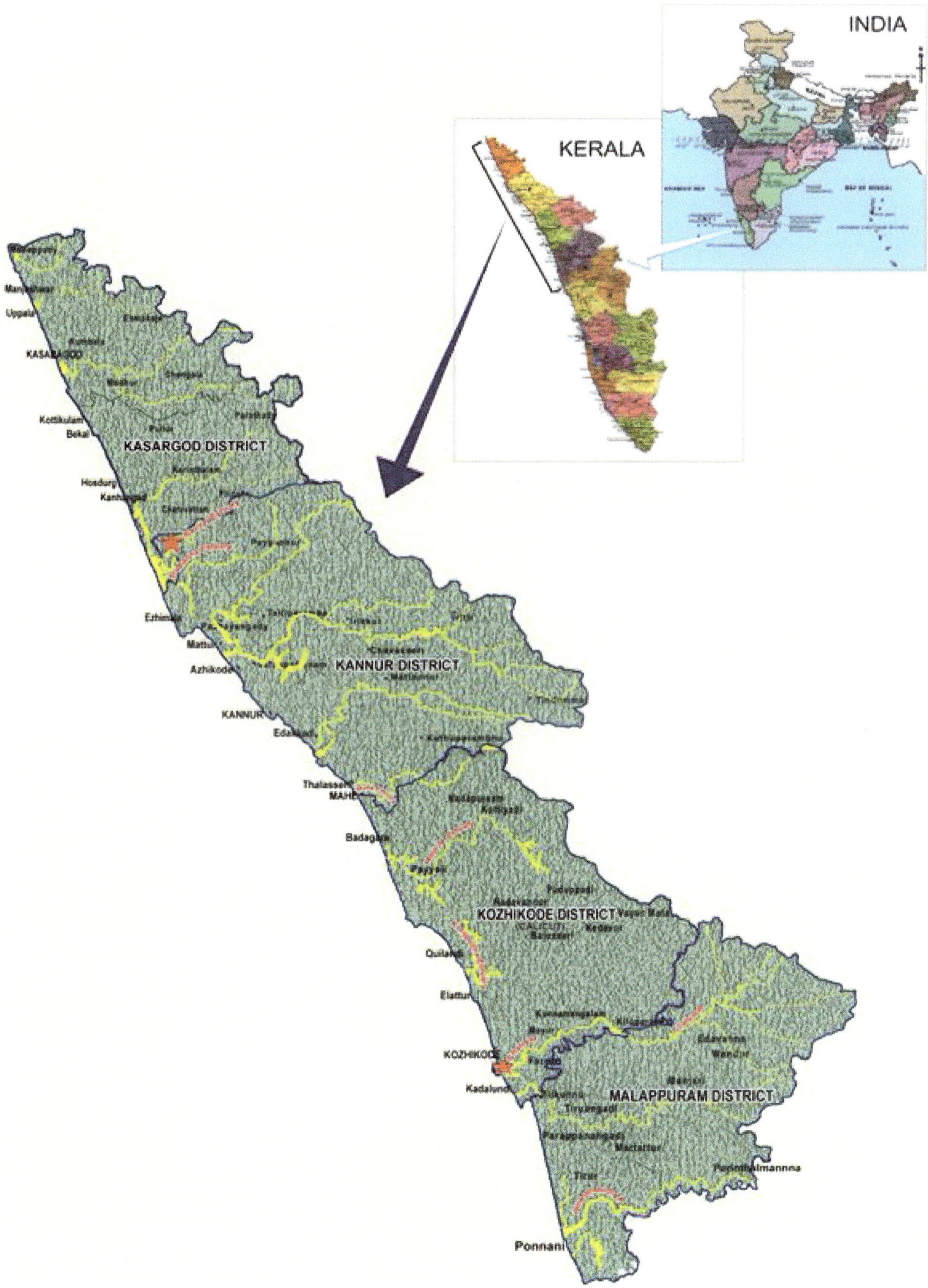

Figure 1 . ★Sampling sites of Meretrix casta in North Kerala.

Table 1. Fishery details of *M. casta* in two estuaries 2008.

Months	Chaliyar estuary			Kavvai estuary		
	Catch (T)	Effort	CPUE (Kg)	Catch (T)	Effort	CPUE (Kg)
J	322.4	1612.0	200.0	0.0	0.0	0.0
F	174.0	1392.0	125.0	0.0	0.0	0.0
M	262.3	1457.0	180.0	387.0	1480.0	261.5
A	173.4	990.0	175.2	273.6	1520.0	180.0
M	35.0	279.0	125.6	298.0	1560.0	191.0
J	30.0	150.0	200.0	198.8	1440.0	138.1
JU	0.0	0.0	0.0	173.8	1580.0	110.0
A	15.8	144.0	110.0	0.0	0.0	0.0
S	0.0	0.0	0.0	0.0	0.0	0.0
O	31.5	105.0	300.0	261.5	1480.0	176.7
N	67.2	252.0	266.7	251.0	1580.0	158.9
D	27.5	110.0	250.0	261.5	1480.0	176.7
Total	1139.1	6491.0	175.5	2105.2	12120.0	173.7
Mean	94.9	540.9	161.0	175.4	1010.0	116.1

Kavvai estuary

The size range of *M. casta* contributing to the fishery in Kavvai estuary was 11.2 to 37.8 mm. The modal classes were 18-20 (28%), 20-22 (25.4%) and 16-18 (14.6%). The mean length was 20.7 ± 3.14 mm (n = 1559) and mean weight was 4.67 ± 4.56. The sex ratio was 1:0.46. Over 34% of the females were in mature condition. The mean condition index of *M. casta* during the year was 7.0%.

Length weight relationship

Chaliyar estuary

The length of *M. casta* in the fishery ranged from 15.8 to 33.4 mm and the weight ranged from 1.17 to 14.75 g. The length weight relationship is given by Log Y = -1.259635861+ 3.090977418 Log X, r = 0.903010085. The computed growth coefficient (b) was 3.09 and the growth was isometric.

Kavvai estuary

The length of *M. casta* in the fishery ranged from 10.8 to 37.8 mm and the weight ranged from 0.26 to 32.85 g. The length weight relationship is given by Log Y = -0.98701877 + 3.054878207 Log X, r = 0.613724895. The computed growth coefficient (b) was 3.05.

Growth parameters

Chaliyar estuary

Asymptotic length (L_∞) of the VBGF was 24.24 mm and the growth coefficient (K) was 1.8^{-1} for *M. casta* from Chaliyar estuary. The growth performance index (ф') was found to be 1.91606 and t_0 was calculated at -0.04283547 years. The von Bertalanffy growth equation can be written as $L_t = 25.44\ \{1\text{-}e^{-1.8(t-\_0.0428)}\}$. Accordingly, the clam attained a size of 21.2, 24.7 and 25.3 mm by the end of 1st, 2nd and 3rd year respectively. The longevity of *M. casta* from Chaliyar estuary was estimated as 2.5 to 3 years (Figures 2 and 3).

Kavvai estuary

Asymptotic length (L_∞) of the VBGF was 26.5 mm and the growth coefficient (K) was 2.0^{-1} for *M. casta* from Kavvai estuary. The growth performance index (ф') was found to be 2.025305865 and t_0 was calculated at -0.03441665 years. The von Bertalanffy growth equation can be written as $Lt = 26.5\ \{1\text{-}e^{-2(t-\_0.0344)}\}$. Accordingly, the clam attained a size of 22.9, 26.0 and 26.4 mm by the end of 1st, 2nd and 3rd year respectively. The longevity of *M. casta* from Kavvai estuary was estimated at 3 years (Figures 2 and 3). The age and growth curve and the restructured length distribution using the growth parameters are shown in Figures 4 and 5.

Mortality, exploitation and virtual population analysis (VPA)

Chaliyar estuary

The total mortality (Z) estimates for *M. casta from* Chaliyar estuary was 3.92 $year^{-1}$. Natural mortality (M) was 2.78 $year^{-1}$ and fishing mortality (F) 2.12 $year^{-1}$ for

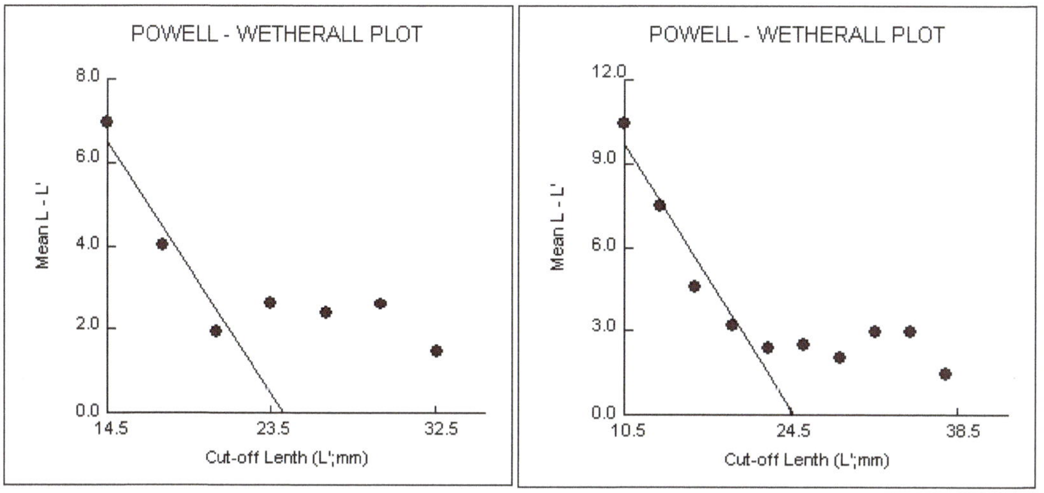

REGRESSION EQUATION

Chaliyar estuary	Kavvai estuary
Y =16.11 + (-0.665)* X, r =-0.923	Y =16.80 + (-0.678)* X, r =-0.952
Estimate of L_α = 24.24 mm, Estimate of Z/K = 0.50	Estimate of L_α = 24.77 mm, Estimate of Z/K = 0.47

Figure 2. Estimation of L_α and Z/K of *Meretrix casta* (Chaliyar and Kavvai estuary) using Powell-Wetherall plot.

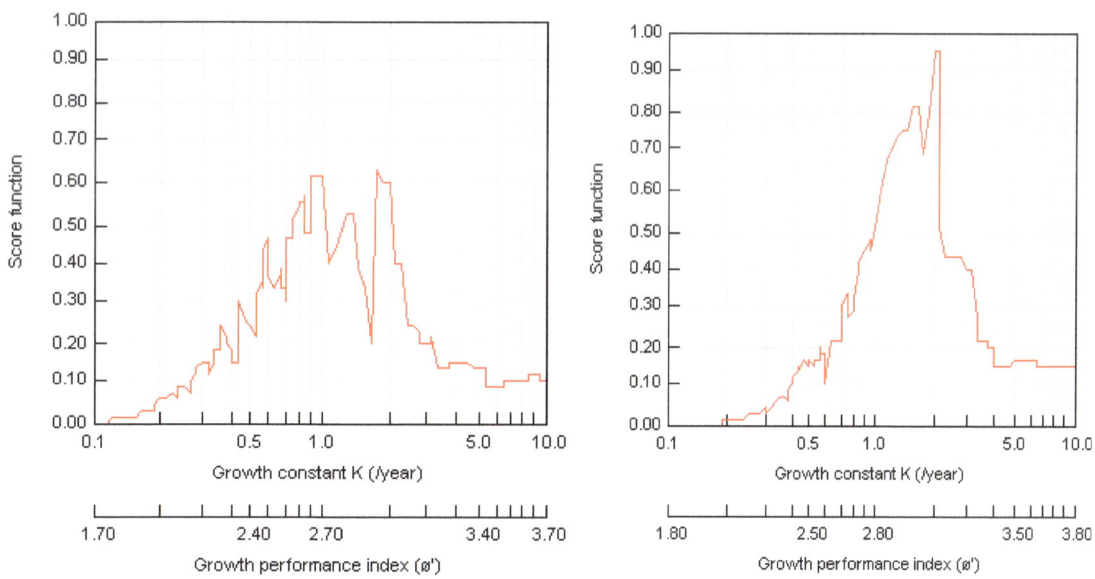

Figure 3. Estimation of K of *M. casta* and (Chaliyar estuary andKavvai estuary).

M. casta (Table 2). Exploitation level (*E)* of *M. casta* was 0.54 (Table 2).

Kavvai estuary

The total mortality (Z) estimates for *M. casta from* Kavvai estuary *was* 2.98 year $^{-1}$. Natural mortality (M) was 2.96 year $^{-1}$ and fishing mortality (F) 0.98 year $^{-1}$ for *M. casta* (Table 2). Exploitation level (*E*) of *M. casta* was 0.33 (Table 2).

Recruitment pattern

Chaliyar estuary

The recruitment pattern of *M. casta* was continuous throughout the year, however, two major peaks of recruitment was observed during April and June to August

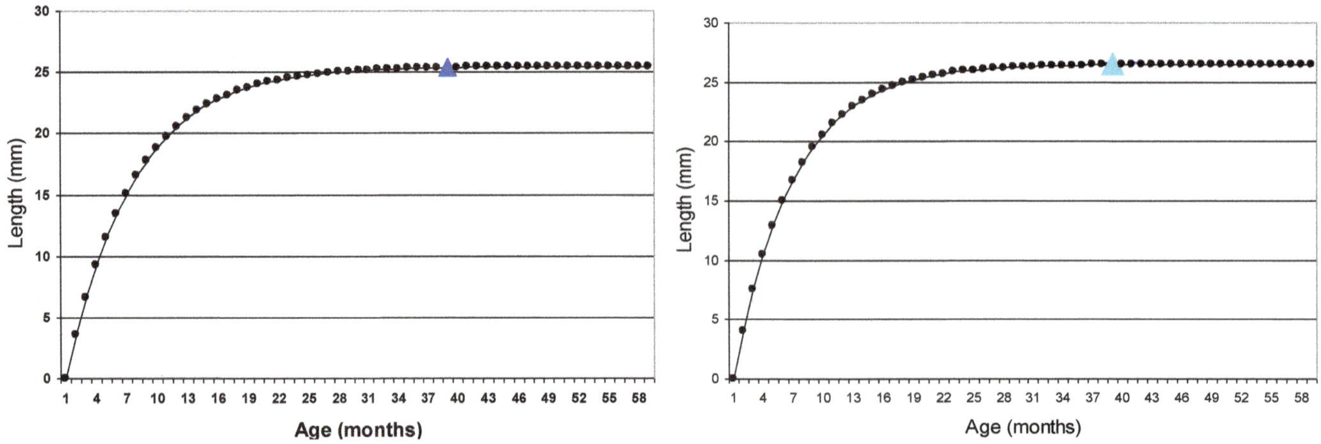

Figure 4. Plot of age and growth of *M. casta* (Chaliyar and Kavvai estuary) based on computed growth parameters.

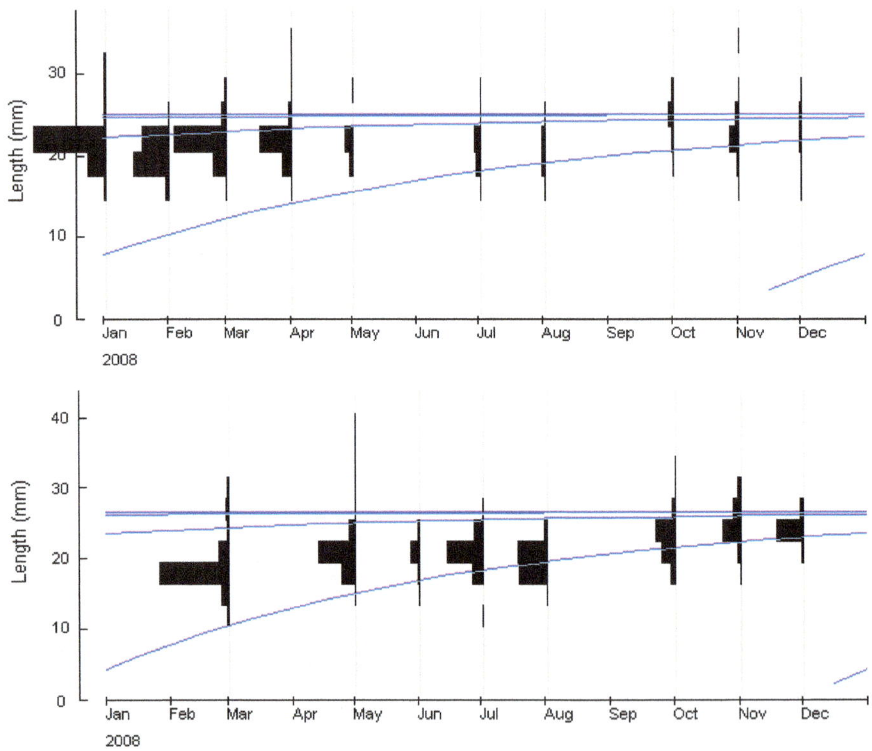

Figure 5. Restructured length frequency distribution with growth curves superimposed using ELEFAN I (L_{α} = 25.44 mm, K = 1.8 year $^{-1}$) *M. casta* (Chaliyari estuary) (L_{α} = 29.51 mm, K = 2 year $^{-1}$) *M. casta* (Kavvai estuary).

(Figure 6). The peak pulses produced over 70% of the observed recruitment during the year.

Kavvai estuary

The recruitment pattern of *M. casta* was continuous t hroughout the year, however, two major peaks of recruitment were observed during March-April and June to August (Figure 6). The peak pulses produced over 70% of the observed recruitment during the year.

Estimation of stock and MSY

Chaliyar estuary

The annual total stock (2008) of *M. casta* was estimated

Table 2. Population parameters of *Meretrix casta*, from two different estuaries, south west coast of India.

Population parameter	Chaliyar estuary	Kavvai estuary
a	0.055000184	0.10789001
b	3.0909	3.054878
Asymptotic length (L∞) in cm	2.544	2.65
Growth coefficient (K) year $^{-1}$	1.8	2.0
Natural mortality (M) year $^{-1}$	2.788	2.96
Fishing mortality (F) year $^{-1}$	2.12	0.98
Total mortality (Z) year $^{-1}$	3.92	2.98
Exploitation level (E)	0.54	0.328
Length range (cm)	1.58-3.34	1.08-3.78
Weight range (g)	1.17 – 14.75	0.26 – 32.85
Sample	1560	1559

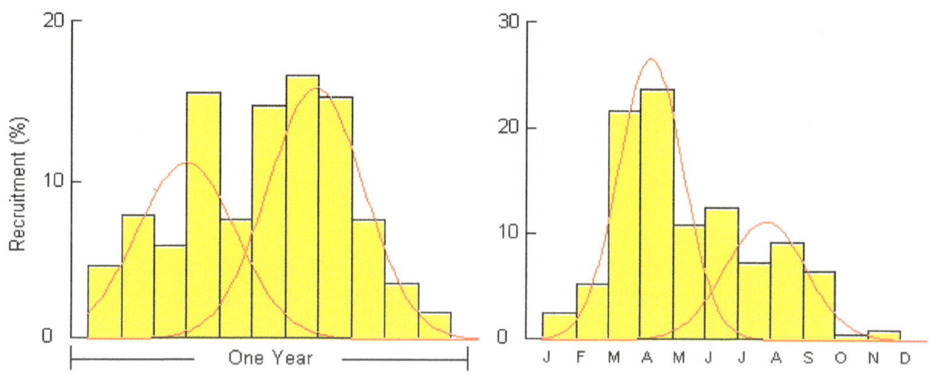

Figure 6. Recruitment pattern of *M. casta (Chaliyar* and Kavvai estuary).

at 0.0004654 t. The biomass (standing stock) was estimated at 537.3 t and maximum sustainable yield (MSY) of *M. casta* was estimated at 1053.13 t. The exploitation rate (U) was 0.53.

Kavvai estuary

The annual total stock of *M casta* (2008) was estimated at 0.0001483 t. The biomass (standing stock) was estimated at 2148.2 t and maximum sustainable yield (MSY) of *M. casta* was estimated at 3200.8 t. The exploitation rate (U) was 0.31.

DISCUSSION

M. casta forms a significant sustenance - level fishery in all major estuaries of India (Narasimham, 1991). *M. casta* is euryhaline, can tolerate low salinities and prefers high sand and silt habitat (Narasimham and Laxmilatha, 1991). The fishery of the edible clam is dependent on the spat fall and subsequent recruitment of the fishery. Salinity, temperature sediment texture and organic matter contribute to the abundance of bivalves (Lee, 1972). The spat fall is affected by the benthic state and pollution by increased pH due to coconut fibre retting and alteration of the sand bed due to sand mining. This has been reported for *M. casta* in Moorad estuary and Chettuva estuaries which has affected spat fall of *M. casta* from year to year (Laxmilatha et al., 2006a, b). The abundance of *M. casta* is also positively correlated to the organic matter content in the substrate (Fraga and Vives, 1960; Jayawardene and Wijayaratne, 2009). The salinity and total suspended solids (TSS) were critical factors influencing the abundance of estuarine clams. The pH and iron concentrations in bottom sediment water strongly related to clam densities (Jayawickrema and Wijeyaratne, 2009; Tanyaros and Tongnunui, 2011).

The mean size of *M. casta* contributing to the fishery was slightly higher from Chaliyar estuary at 21.5 mm although smaller sizes (11 mm) were being exploited from Kavvai estuary. *M. casta* attained 21.2 mm in Chaliyar and 23 mm in Kavvai estuary in 12 months. However, *M. casta* is reported to have attained 29.5 mm in 9 months in Adyar estuary (Abraham, 1963), 33.5 mm in 9 months (Salih, 1973) and 35.4 mm in 11 months in Cochin barmouth, west coast (Salih, 1973). Durve (1970) reported slower growth rate of 0.79 mm per month in the fish farm in Mandapam due to high saline conditions. Sreenivasan (1983) reported growth of 34 in 13 months in *M. casta* transplanted to Vellar estuary. Balasubramaniam and Natarajan (1987) reported of 23 mm in the first year, 38.3 mm in the second year and 50.6 mm in the third year for *M. casta* in Vellar estuary. These variations in growth are due to the variations in hydrological conditions in the estuaries (Balasubramanian and Natarajan, 1987; Jayabal and Kalyani, 1986; Sreenivasan, 1983). The mean weight of clams fished from Kavvai estuary were higher (4.7 g) than those fished from Chaliyar estuary. This implies that the hydrographical conditions and productivity of the two estuaries have an important bearing on the growth of the clams (Abraham, 1963; Seshappa, 1971). The condition indexes of the clams from both the estuaries were almost same.

M. casta from Adyar backwater has been reported to attain a size of over 56.5 mm in a period of three years. However, since over 95% of the clams are fished before they reach a length of 30 mm (Abraham, 1963). In Korapuzha and Beypore estuaries on the west coast maximum size of 31 mm was obtained during 1950-1951) (Sheshappa, 1971). In Bhimunipatnam estuary on the east coast also, *M. casta* of over 20 mm formed only 5% of the fished clams clams over 40 mm were rarely rare (Ramamohana-Rao et al., 1977). Similar is the case in Chaliyar and Kavvai estuaries, clams of over 40 mm were rarely seen. The growth parameters reported for *M. casta* from five different sites in the Dutch canal of Sri Lanka ranges from 34 to 43.1 mm and growth coefficient ranged from 0.84^{-1} to 1.44^{-1} (Jayawickrema and Wijeyaratne, 2009). The growth attained by *M. casta* in both estuaries was also same and life span estimated was also 3 years in both cases. The computed growth co efficient (b) of *M. casta* from both water bodies are nearly same and growth was isometric. Isometric growth has been reported in *M. casta* (Narasimham et al., 1988).

Despite the large scale indiscriminate fishing of the clams of smaller size (<20 mm), the stock of *M. casta* has remained resilient in almost all the major estuaries including Chaliyar and Kavvai. The factors that have enabled this resilience are the high rate of growth, especially during early stages, the abbreviated larval stages, early attainment of sexual maturity, discontinuous breeding throughout the year with at least two peaks and euryhaline tolerance to salinity (Abraham, 1963; Balasubramanian and Natarajan, 1987; Jayawickrema and Wijeyaratne, 2009; Tanyaros and Tongnunui, 2011).

The total mortality (Z) estimate of *M. casta* from Chaliyar was higher compared to that of Kavvai estuary, fishing mortality was higher for *M. casta* from Chaliyar compared to Kavvai The exploitation level of *M. casta* in Chaliyar was 0.54 while it was very low in Kavvai at 0.33. The standing stock of *M. casta* in Chaliyar estuary is much lower compared to that in Kavvai estuary.

However, the clam fishery has sustained the livelihood of the local pickers for a long period of time and in recent years spat fall has become highly variable due to the changes in the bed due to environmental changes and climatic factors. It is reported that post Tsunami of 2004, there was very poor spatfall of *M. casta* in the Chaliyar estuary and consequently very poor recruitment in the following year. Heavy mortality due to heavy flooding in the west coast and poor yield of clam in the Mumbai coast due to excessive rainfall has been reported by Hornell (1916) and Rai (1932). In Kavvai estuary, several species of clams contribute to the fishery along with fishes. *Meretrix meretrix*, *Paphia malabarica* and *Villorita cyprinoides* are landed in significant quantities apart from *M. casta* and have higher market demand than *M. casta*. Therefore, a huge stock of *M. casta* is present which can still continue to sustain the fishery. The present level of exploitation of *M. casta* is far below the MSY and therefore production of *M. casta* can be significantly enhanced on a sustainable level. In Chaliyar, the present exploitation is only 50% of the MSY.

The population characteristics of *M. casta* from the two estuaries indicate that the two stocks can still be exploited on a sustainable level. However, the major issue of concern is the changes in the habitat / sand bed of the clams. Large scale sand mining also affects the spat settlement due to high levels of total suspended solids (Laxmilatha et al., 2006a, b). *M. casta* is fished for the meat (for local consumption) and shell in most estuaries. The undersized clams are exploited during the peak season, the meat is allowed to putrefy and shells are used by the lime industry (Laxmilatha et al., 2006a, b). Conservation measures such as relaying, ranching of seed for stock enhancement, regulations to restrict exploitation of seed clams, measures to regulate sand mining are imperative to manage the stocks of *M. casta*, which is gradually on its way to the endangered status (already cited as vulnerable by IUCN) due to loss of habitat. This in turn also affects the livelihood of the fishers dependent on this important fishery for their sustenance.

ACKNOWLEDGEMENT

The authors thank the Director, Central Marine Fisheries Research Institute, Cochin for support and encouragement in carrying out this work.

REFERENCES

Abraham KC (1963). Observations on the biology of *Meretrix casta* (Chemnitz). J. Zool. Soc. India 5(2):163-190.

Alagarswami K, Meiyyappan MM (1989b). Prospects and problems of management and development of marine molluscan resources (other than cephalopods) in India. Bull. Cent. Mar. Fish. Res. Inst. 44(1):250-261.

Ananda Rao T, Molur S, Walker S (Eds.) (1998). Report of the workshop on 'Conservation assessment and Management plan for mangroves of India' Zoo outreach organization, Coimbatore, India, P. 106.

Balasubramanian K, Natarajan R (1987). Age and growth of *Meretrix casta (chemnitz)* in Vellar estuary, Parangipettai *Bull. Cent.* Mar. Fish. Res. Inst. 42(1):145-147.

Bertalanffy L von (1934). Untersuchungen uber die Gesetzlichkeiten des Wachstums1. llgemeine Grundlagen der Theorie. Roux Arch. Entwicklungsmech. Org. 131:613-653.

Beverton RJH, Holt SJ (1957). On the dynamics of exploited fish populations. Fish. Invest. Ser. II London 19:533.

Durve VS (1964). Preliminary observations on the seasonal gonadal changes and spawning in the clam Meretrix casta (Chemmitz) from the marine fish farm. J. Mar. Biol. Assoc. India 6(2):241-248:

Durve VS (1970). On the growth of the clam *Meretrix casta* (Chemnitz) from the Marine Fish Farm. J. Mar. Biol. Assoc. India. 12:125-135

Fraga F, Vives F (1960). Retention of particulate matter, In: Marine mussels: Their Ecology, and Physiology (B. L. Bayne ed.). pp.13-65, Cambridge University Press, UK.

Gayanilo FC, Jr Sparre P, Pauly D (Eds.) (1995). The FAO- ICLARM Stock Assessment Tools (FiSAT) User's Manual. *FAO Computerized Information Series (Fisheries)* No. 8: Rome, FAO. P. 185.

Gayanilo FC, Jr Sparre P, Pauly D (1997). The FAO- ICLARM Stock Assessment Tools (FiSAT) User's Manual. Reference Manual, FAO, Rome, Italy, P. 126.

Harkantra SN (1975). Some observations on the clam beds of Kali estuary, Karwar. Mahasagar. 8:101-108.

Gulland JA (1979). *Report of the* FAO/UNDP workshop on the fishery resources of the Western Indian Ocean – South Equator. FAO, Rome, 10FC/DEV/79/ 45:1-37.

Hornell J (1916). The utilization of coral and shells for lime burning in the Madras Presidency. Madras Fish. Department Bull. 8:105-126.

Jayabal R, Kalyani M (1986). Age and growth of the estuarine clam *Meretrix meretrix* (L) inhabiting the Vellar stuary. Mahasagar 19(2):141-146.

Jayawickrema EM, Wijeyaratne MJS (2009). Distribution and population dynamics of the edible bivalve species *Meretrix casta* (Chemnitz) in the Dutch canal of Sri Lanka. Sri Lanka. J. Aquatic Sci. 14:29-44.

Krishnakumari L, Rajagopal MD, Sumitra V (1977). Some aspects of biology and biochemistry of the backwater clam *Meretrix casta* (Chemnitz). Mahasagar Bull. Nat. Inst. Oceangr. 10(3/4):157-163

Laxmilatha P, Velayudhan TS, Mohamed KS, Kripa V, Radhakrishnan P, Jenny Sharma C (*2006a).* The fishery and biology of *Meretrix casta* (Chemnitz) in the Moorad estuary, Kerala Indian J. Fish. 53(1):109-113

Laxmilatha P, Velayudhan TS, Mohamed KS, Kripa V, Radhakrishnan P, Jenny Sharma C (2006b). Bivalve resources of the Chettuva estuary. Kerala Indian J. Fish. 53(4):481-486,

Lee C (1972). An ecological study on some marine bivalve beds of Korean tide lands. In: proceedings of the Indo-pacific Fisheries Council 13th Sessions, Brisbane, Australia. pp.326– 331,

Narasimham K, Muthiah AP, Sundararajan D, Vaithinathan N (1988). Biology of the great clam *Meretrix meretrix* in the Korampallam creek, Tuticorin. Indian J. Fish. 35(4):288-293.

Narasimham KA (1991). Present status of clam fisheries of India. J. Mar. Biol. Assoc. India 33:(1.2):76-88.

Parulekar AH, Dwivedi SN, Dhargalkar VK (1973). Ecology of clam beds in Mandovi, Cumbarjua Canal and Zuari esuarine system of Goa. Indian J. Mar. Sci. 2:122-126.

Pauly D (1980). On the interrelationships between natural mortality, growth parameters and mean environmental temperature in 175 fish stocks. ICES. J. Cons. 39:175-192.

Pauly D (1984). Length converted catch curves: A powerful tool for fisheries research in the tropics (Part II). Fishbyte 2(1):17-19.

Pauly D (1986). On improving operation and use of the ELEFAN programs. Part II. Improving the estimation of L. ICLARM Fishbyte, 4(1):18-20.

Pauly D, Munro JL (1984). Once more on growth comparisons in fish and invertebrates. Fishbyte 2(1):21.

Rai HS (1932). The shell fisheries of the Bombay Presidency. J. Bombay Nat. History Soc. 35:826-847.

Ramamohana-Rao V, Appa Rao T, Reuben S, Sampson, Manickam PE (1977). On the culture possibilities of *Meretrix casta* in Bhimunipatnam backwaters. Indian J. Fish. 24 (1,2):263-268.

Rao GS (1988). Biology of *Meretrix casta* (Chemnitz) and *Paphia malabarica* (Chemnitz) from Mulky estuary, Dakshina Karnataka. *In*: S. Mahadevan, KA, Narasimham, K. Satyanarayana Rao, KMS, Ameer Hamsa P Muthiah (Eds.), National seminar on shell fish resources and farming– Tuticorin. 19-21 January 1987. Bull. Cent. Mar. Fish. Res. Inst. 42(1):148-153

Ricker WE (1975). Computation and interpretation of biological statistics of fish populations. Bull. Fish. Res. Bd. Canada 191:382

Seshappa G (1971). Some observations on the backwater clam *Meretrix casta (Chemnitz)* in the Beypore and Korapuzha estuaries. Indian J. Fish. 14 (1&2):298-305

Salih MKY (1973). On the growth of the backwater clam *Meretrix casta* (Chemnitz) in the clam beds off Cochin barmouth. J. Mar. Biol. Ass. India 15(1):345-353.

Sparre P, Venema SC (1992). Introduction to tropical fish stock assessment, Part I, *annual.* FAO Fish. Technical. 306:376.

Sreenivasan PV (1983), Growth of the clam *Meretrix casta* (Chemnitz) transplanted in the Vellar estuary. Proc. Symp. Coastal Aquac. 2:564-568

Tanyaros S, Tongnunui P (2011). Influence of environmental variables o the abundance of estuarine clam *Meretrix casta* (Chemnitz, 1782) in Trang Province, Southern Thailand. Songklanakarin J. Sci. Technol. 33(1):107-115.

Thangavelu R, Poovannan P (1994). Some aspects of biology of the clam Meretrix casta (Chemnitz) in Muttukadu backwaters. J. Mar. Biol. Assoc. India. 36(1&2):77-80.

Thangavelu R, Sanjeevraj PJ (1985). Fishery and biology of the clam *Meretrix casta* (Chemnitz) in Pulicat Lake. J. Mar. Biol. Assoc. India, 27(1,2):75- 83.

5

Effect of different dosages of zinc on the growth and biomass in five marine microalgae

Dinesh Kumar S., P. Santhanam , S. Ananth, A. Shenbaga Devi, R. Nandakumar, B. Balaji Prasath, S. Jeyanthi, T. Jayalakshmi and P. Ananthi

Marine Planktonology and Aquaculture Laboratory, Department of Marine Science, School of Marine Sciences, Bharathidasan University, Tiruchirappalli-620 024, Tamil Nadu, India.

Marine environment often restrain toxic heavy metals that may enter into the food web via uptake by microalgae and eventually cause severe poisoning problems at higher tropic levels. The effects of Zinc cations upon growth of five native microalgal species, *Chlorella marina*, *Isochrysis galbana*, *Tetraselmis* sp., *Nannochloropsis* sp., and *Dunaliella salina* were evaluated. Growth inhibition of the microalgal cells were determined by exposing them to different concentrations of aqueous solutions of zinc metal for 15 days. A major reduction in cell density was observed in all the five cultures in the concentration of 50 ppm. Among the five micro algal species tested, *Tetraselmis* sp. alone showed growth up to 250 ppm concentration of zinc metal till the final day (15th day) of experiment.

Key words: Zinc, microalgae, heavy metal, *Chlorella marina*, *Tetraselmis* sp.

INTRODUCTION

Marine unicellular algae with towering metabolism rates fill the essential function in the primary production and focus of heavy metals, which are pollutants with multiple ingestion pathways to marine ecosystems. The hazard of heavy metal pollution is owing to its capability to move within marine and near-shore ecosystems for along-lasting length of time. By accumulating heavy metals in their cells, algae encourage further accumulation in the successive parts of the food chain, as well as in commercial aquatic organisms (Amiard et al., 1980). The responses to the occurrence of lethal concentrations of heavy metals vary considerably in different species. Among the marine organisms, phytoplankton resistant to heavy metals have been recommended as bioindicators for the assessment of marine pollution (Rijstenbeil et al., 1994; Kapkov and Belenikina, 2003, 2007). Based on the physical and chemical properties of phytoplankton, in addition to its reaction patterns, a chain of lethal heavy metals for phytoplankton has been proposed, in which the following chemical elements are arranged in the following order: Hg > Cu > Cd > Zn >Pb (Eichenberger, 1993). Most of the artificial methods are currently famous for removing heavy metals from aqueous solutions such as precipitation, electrolysis, ionic exchange, filtration, evaporation and others. But the main disadvantages of these methods are high cost, low efficiency, especially during removing small amounts of heavy metals and a necessity of slag committal. In order to overcome these disadvantages an alternative other than the chemical is needed for detoxication and removal of metals from aqueous solutions (Nalimova et al., 2005). Microalgae have been found to be very effective in removing heavy metals from wastewater due to their large surface area and high binding affinity (Chong et al., 2000). The toxicity

Table 1. Summary of protocol for marine microalgae species growth inhibition bioassay.

S/N	Task	Condition
1	Test type	Static
2	Temperature	23±2°C
3	Light quality	Cool white fluorescent lighting
4	Light intensity	5000 lux
5	Illumination	12:12 h light and dark cycle
6	Conical flask size	250 ml
7	Culture solution volume	100 ml
8	Renewal of test solutions	None
9	Age of test organisms	Every 3 days
10	Initial cell density in conical flasks	$1.5x10^4$ cells ml^{-1}
11	Number of replicate	3
12	Mixing rate	Constant aeration
13	Test medium	Walne medium
14	Metal concentrations	Minimum of six and a control
15	Test duration	15 days
16	Test endpoint	Growth rate

of heavy metals to different algal types varies (Fitzgerald and Faust, 1963). Some algal species demonstrated tolerance to different heavy metals possessing mechanisms to prevent their toxic effects on metabolism or damage of intracellular structures. Still now, most of the researchers studied the effect of heavy metal toxicity on fresh water algae cultivated under laboratory conditions. In this respect, study on the effects of the most dangerous metal zinc on the growth of algal species as an essential factor of natural phytocenoses seems enormously main, as it plays a determining role in near-shoremarine ecosystems. Identifying algae of different taxonomic ranks that are sensitive and resistant to heavy metals would also be needed for the development of biological markers of pollution in marine ecosystems. The aim of this study is to assess the effects of $Zinc^{2+}$ toxicity on growth, chlorophyll production and viability of the marine microalgae such as *Chlorella marina*, *Isochrysis galbana*, *Tetraselmis* sp., *Nannochloropsis* sp. and *Dunaliella salina*.

MATERIALS AND METHODS

Algal stock cultures

Experiments were performed with the algal cultures such as *C. marina*, *I. galbana*, *Tetraselmis* sp., *Nannochloropsis* sp., and *D. salina*. These strains were obtained from the Central Institute of Brackishwater Aquaculture, Chennai. The microalgae were grown in the Walne medium at pH 8 which was adjusted using 0.1 N $NaHCO_3$ and HCl (Walne, 1970). Stock cultures were incubated in 250 ml conical flasks containing 100 ml of sterilized seawater under a photon irradiance of approximately 5000 lux (using two cool white fluorescent tube lights) in 12:12 h light and dark cycle and temperature were maintained at 23±2°C.

Toxicity analysis

Growth inhibition bioassay (Table 1) for microalgal species were conducted under laboratory conditions for fifteen days. To minimize the metal contamination, all laboratory wares contact with the culture medium were soaked in 1% HNO_3 for 24 h and rinsed with Milli-Q water. Zinc solution at a stock concentration of 1000 mg/L^{-1} was prepared using analytical grade $ZnCl_2.2H_2O$ (Dinesh Kumar et al., 2013). About 4.397 g of $ZnSO_4$ $7H_2O$ salt was dissolved in 1000 ml of distilled water to make stock solution. From the stock solution, 0.1 ml was taken and dissolved in 100 ml of distilled water to make 1 ppm of zinc solution. The culture media with various zinc concentrations such as 5, 50, 250, 500, 1000 and 1500 ppm were prepared by diluting the stock according to Dinesh Kumar et al. (2013). Control medium was prepared in the same manner without adding the toxicant. A known cell density of the algal suspension with exponential growth phase was harvested from the stock cultures and inoculated in the test medium prepared with different concentrations of zinc and control fed with Walne culture medium, the initial cell density was approximately 1.5×10^4 cells ml^{-1}. The growth rates of algal cultures were monitored carefully at regular intervals such as 3, 6, 9, 12 and 15 days of incubation by measuring the optical density of algal suspension at 650 nm (Wetherell, 1961). Triplicates were maintained for all the treatments and control.

At the end of experiment chlorophyll 'a' content of the microalgal cultures was estimated by following the method of Mantoura and Llewellyn (1983). As brief 10 ml of algal culture sample was filtered using Millipore filtering system fitted with a 4.5 cm diameter GF/C filter paper by applying low suction. Before filtering the sample, a thin bed of magnesium carbonate (2 ml) was poured over the GF/C filter paper for effective filtration. After filtration, the filtrate was removed and filter paper with algal cells ground with 90% of acetone using mortar and pestle. The resulting samples were transferred to screw cap test tubes covered with black cloth and incubated in the refrigerator for 24 h. After 24 h incubation, the contents were ground with 90% of acetone and centrifuged at 3000 rpm for 10 min. Then the clear supernatant was collected and optical density was measured at different wavelengths such as 630, 645 and 665 nm using UV- visible Spectrophotometer (1800

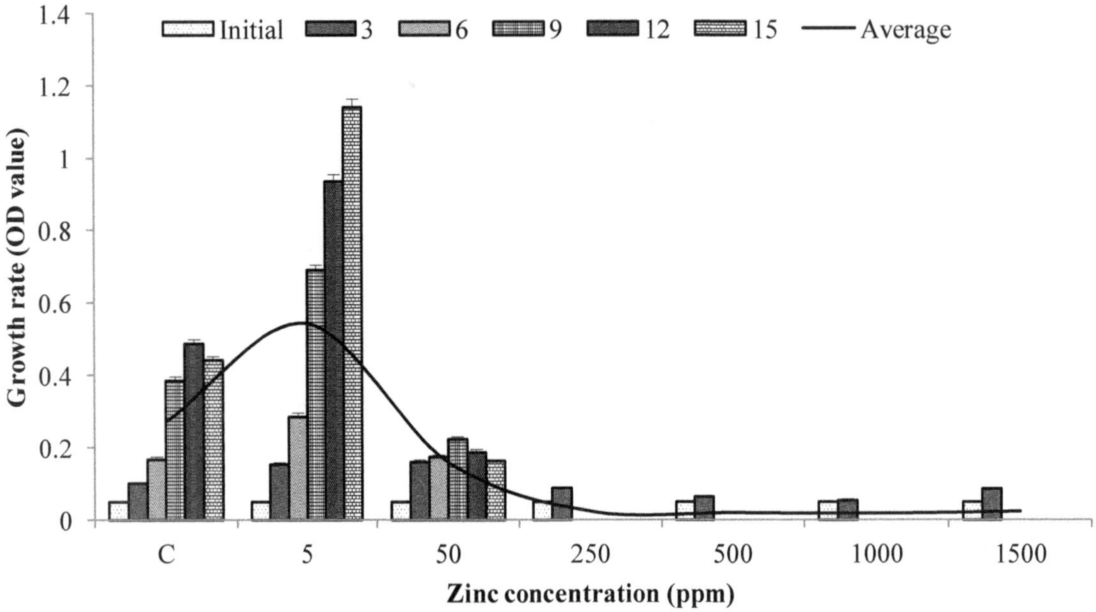

Figure 1. Effect of zinc on the growth of *C. marina*.

Shimadzu UV) for chlorophyll 'a' estimation.

RESULTS AND DISCUSSION

Effect of zinc on the growth of the microalgal cultures

In the present investigation, the effect of zinc on the growth of unicellular marine microalgae such as *C. marina*, *I. galbana*, *Tetraselmis* sp., *Nannochloropsis* sp., and *D. salina* were evaluated by using growth inhibition bioassay. The result indicates that the *Chlorella marina* showed good growth (average- 0.101±0.003 abs) in all the concentration of zinc up to 3 days of incubation. However, when the concentration of the zinc and incubation days increased, the growth of *C. marina* decreased drastically. The *Chlorella marina* displayed good growth in 5 and 50 ppm concentration of zinc compared to the control (0.153±0.005 and 0.158±0.007 abs respectively). The growth rate was significantly increased in 5 ppm of zinc concentration up to 15 days of incubation. Though, the growth rate of *C. marina* was less in 50 ppm (0.158±0.005 abs) when compared to the control (0.271±0.012 abs), it show's steady growth up to the end of the experiment (Figure 1). These results were accordance with the previous published data on other algae (Kamp-Nielsen, 1971; Nazzi, 1972; Rachlin and Farran, 1974).

The inhibitory effect of zinc on the microalgae *I. galbana* demonstrated that their growth was decreased after 3 days of incubation as *C. marina*. In 50 ppm zinc concentration, the growth (0.113±0.019 abs) was found to increase from the initial cell density (0.05±0.002 abs) till 6 days of incubation and after that a steady growth rate was observed till the end of the experiment. The increased optical density (0.278±0.035 abs) in 5 ppm concentration of zinc up to 15 days of incubation showed that the *I. galbana* had good ability to tolerate at 5 ppm of zinc concentration. In addition, comparatively the cell density of *I. galbana* was high during 12 (0.257±0.029 abs) and 15 days (0.278±0.02 abs) of incubation than the control which denotes that they have grown well in the Walne medium with the presence of 5 ppm of zinc (Figure 2). It is clear that after 9th day the algal growth was found to be reduced with reference to metal concentration. It is understand that the metal concentration have made significant adverse impact on the algal growth as agreed earlier by Davies (1974). However, Fisher et al. (1984) have stated that the surface reactivity of the *I. galbana* did not showed any change for heavy metals over an incubation period. It is also suggested that the variability of metal accumulation in algae is probably the result of competition between the metals to bind with such polysaccharides (Kohn, 1975; Hurlburet et al., 1976; Veroy et al., 1980).

The optical density results of *Tetraselmis* sp. revealed that they were able to tolerate up to the concentration of 250 ppm zinc whereas in the highest concentrations such as 500, 1000 and 1500 ppm of zinc no growth was observed after 3 days of incubation. The maximum growth in *Tetraselmis* sp. was observed in 50 and 250 ppm concentrations as 0.498±0.049 and 0.387±0.043 abs respectively on 15th day of incubation (Figure 3). Zinc can easily be rationalized since it is a micronutrient required for microalgal metabolism (Bascik-Remisiewicz et al., 2009; Vallee and Auld, 1990) as it is a part of

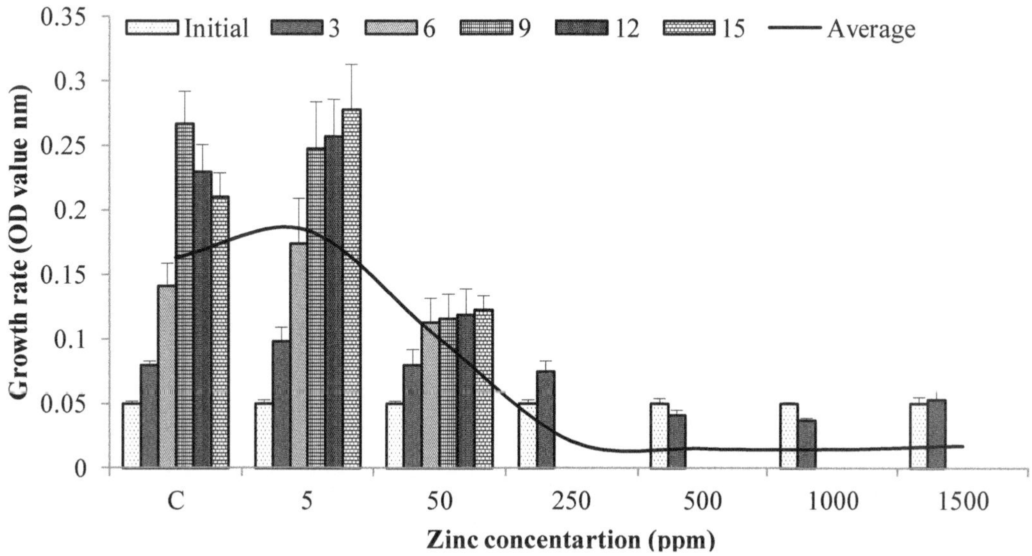

Figure 2. Effect of zinc on the growth of *I. galbana*.

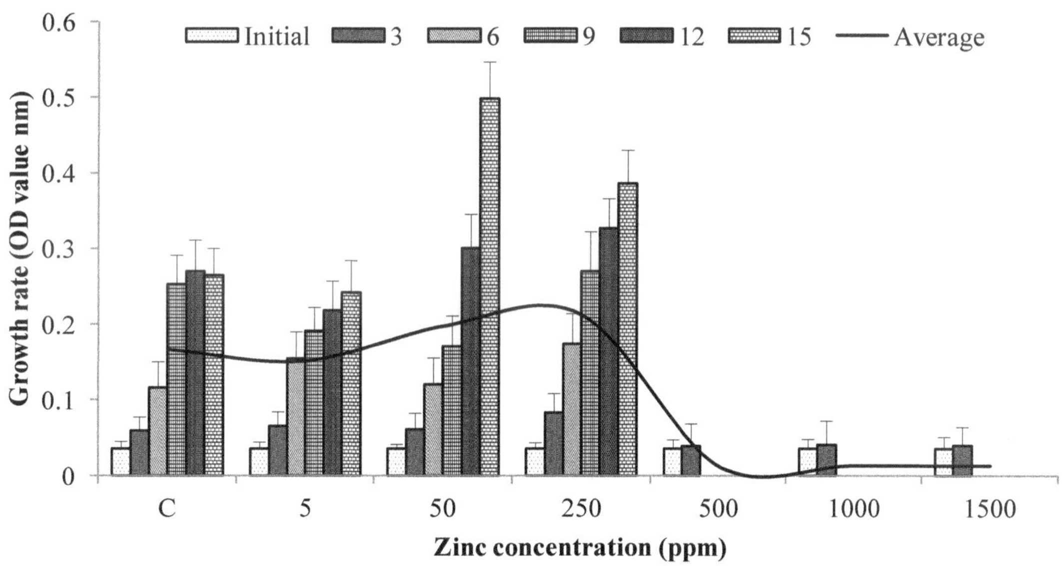

Figure 3. Effect of zinc on the growth of *Tetraselmis* sp.

prosthetic moieties of some of its relevant enzymes. In the present study, the toxicity effect of zinc was progressively decreased when the exposure time was increased which was judged by the increased growth rate of the cultures. This may be due to the accumulation of zinc onto algal cells in higher concentrations which significantly reduced the available zinc concentration and its toxicity to surviving cells in the final stage of experiment as agreed by Lim et al. (2006). Interestingly, the growth rate of the *Tetraselmis* sp., was found higher in the zinc supplemented medium than the control medium (without heavy metal) which clearly defined that the zinc act as growth stimulant of the *Tetraselmis* sp. and it can survive up to 250 ppm zinc concentration.

The effect of zinc on *Nannochloropsis* sp. was obtained by measuring the growth at 650 nm which showed that, they were relatively high in 5 (0.312±0.035 abs) and 50 ppm (0.288±0.029 abs) than the control (0.283±0.035 abs) up to 6th day of incubation. At the 9th day of incubation, the growth rate was increased in 5 ppm and decreased at 50 ppm zinc concentration. After 9th day of incubation, growth rate was decreased from an initial value of 0.36±0.037 to 0.268±0.031 abs at 5 ppm concentration whereas in highest concentrations such as 250, 500, 1000 and 1500 ppm, no growth was noticed after 3 days of incubation, which denotes that the culture

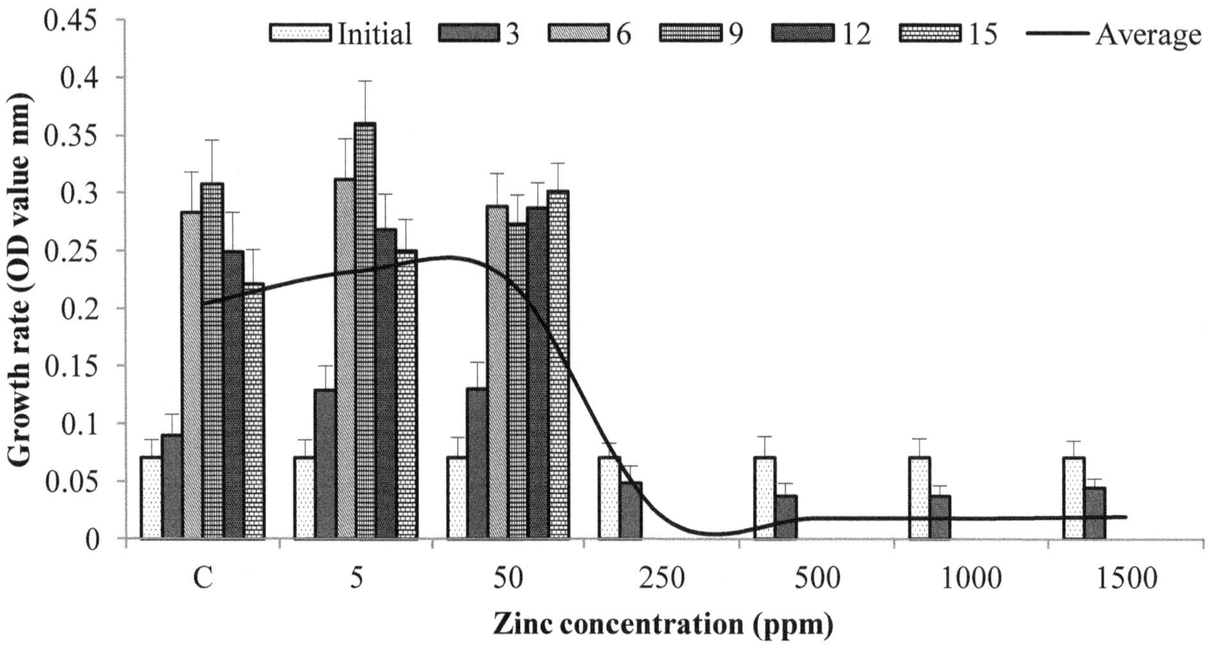

Figure 4. Effect of zinc on the growth of *Nannochloropsis* sp.

could not tolerate when the concentration of zinc increased (Figure 4). This occurs because the metal ion is interacted with the fat layer on the surface of microalgae cells (Darmono, 1995). The body surface of microalgae is covered by the cell membrane, therefore, the potential interaction between the membrane and the metal ion in the water is high (Fhencel, 1988). The surface of cells contains various functional groups, such as N-terminal of $-NH_2$, C-terminal of $-COO^-$, S-terminal of –SH and functional groups of side chains which are potential as metal binding sites. Binding of the metal ion by the active groups proceeds on the cell surface followed by the slow transport step into the membrane and then to the cytoplasm (Chu and Hashim, 2007). Accumulation of several metals (such as Zn, Co, Mo, Ca, and Mg) on algae occurs through the intracellular active biological transport. Toxic heavy metal ions can be isolated from the cytoplasm of the cell through three possible ways that is intracellular chelating by biological polymers; precipitation of heavy metals on the surface of cell wall; or surface adsorption through metal ion binding by chemical functional groups on the cell wall (Moreno-Garrido et al., 1998). The present results showed that the *Nanochloropsis* sp., can survive up to 50 ppm of zinc concentration as agreed by Hala et al. (2012).

In all the tested concentrations of zinc, the growth of *D. salina* was not completely inhibited. In high zinc concentration, slow growth was observed and maximum growth inhibition (0.329±0.039 abs) was noticed after 6 days of exposure to 50, 250, 500, 1000 and 1500 ppm (Figure 5). In control, the growth was found good after 3 days of incubation and it was gradually decreased when the incubation days prolonged. When compared to control (0.198±0.019 abs) the growth of *D. salina* was found to be higher (0.226±0.039 abs) at 5 ppm of zinc concentration till the end of the experiment. The present result suggests that *D. salina* can survive up to 5 ppm of zinc concentration could be attributed the metal binding affinity to the cell surface coat. This conclusion is in accordance with the suggestion made by Davies (1974) who stated that the cellular surface of rnicroalgae consists of a mosaic of cationic and anionic exchange sites acting as ion exchangers for anionic and cationic species in the medium. The amount of metal bound to the ce1l surface coat at equilibrium should be a function of the affinities of the binding sites for the metal, the chemical form of the metal and its concentration in the medium. So, zinc can affects the microalgal growth during the growth period as agreed by previous researchers (Rebhun and Amotz, 1986; Imani et al., 2011; Shariati and Yahyaabadi, 2006).

On an average the relative growth rates of the algae tested decreased upon increasing zinc content in the sea water. Most of the trace metals are known to promote microalgal growth at low concentrations and inhibits at high concentrations, though, the impact may vary with type of metal and the microlagal species (Gao et al., 2004). In this study the tolerance limit of zinc was 50 ppm except *Tetraselmis* sp. (250 ppm) and *D. salina* (5 ppm). Among the microalgae tested *Tetraselmis* sp. alone tolerated Zn concentrations up to 250 ppm. Our species were more tolerant to Zn concentrations than others reported (Imani et al., 2011; Tripathi and Gaur, 2006).

This may be due to the fact that this metal was an

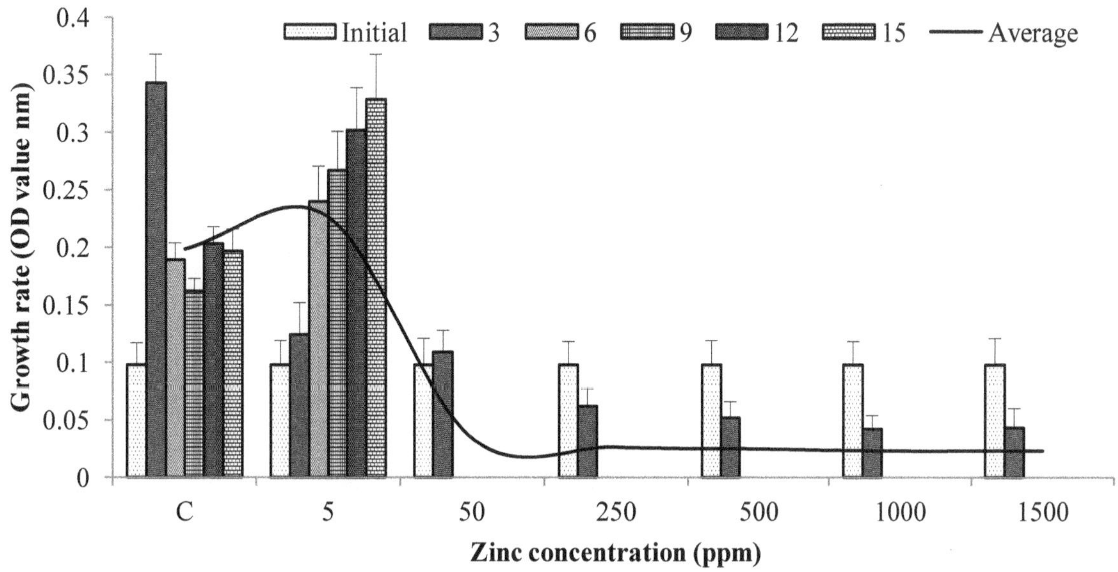

Figure 5. Effect of zinc on the growth of *D. salina*.

abundant contaminant in the original source site of our microalgae strains may account for our exceptional observations.

The effect of zinc on the algae becomes clearer when not only the growth rates in the early exponential phase, but also the total development of the cultures are considered. Many studies have proved non toxicity of zinc (in terms of cell division) at lower concentrations. Rasko and Rachin (1977) have also showed that 7.5 mgl^{-l} Zn does depress the concentration of chlorophyll a, suggesting that Zn, as well reduces algal cell division and is an aspect of its relatively mild toxicity. Thus long-term effect of the exposure to increased levels of zinc was, therefore, more serious than that expressed by the growth rate.

Effect of zinc in chlorophyll 'a' concentration

In this investigation chlorophyll 'a' content of the microalgal cultures were estimated on the final day of experiment (15th day) for zinc treated cultures as well as for the control. The results declared that the zinc significantly affects the chlorophyll 'a' concentration in all the tested marine microalgal cultures. Among the five microalgae tested, the *Tetraselmis* sp. showed maximum chlorophyll 'a' (8.21±0.97 µg/l) content than rest of the algae. Likewise the lowest chlorophyll of 0.19±0.11 µg/l was noticed in *Nannochloropsis* sp. at 50 ppm. The overall average chlorophyll 'a' concentration in all the tested zinc concentration was recorded to be higher in *Tetraselmis* sp. (4.81±0.52 µg/l) followed by *Isochrysis galbana* (1.80±0.14 µg/l) and *Chlorella marina* (1.11±0.22 µg/l) as shown in Figure 6.

Conclusion

The present study provides clear evidence that the addition of zinc in various concentrations brings about significant differences in the growth of microalgal species due to the differences in their resistances. Our study determined the intervals between the threshold and lethal concentrations of the most dangerous heavy metals for bunch species of marine microalgae and identified the variations in resistance to this metal in various species of microalgae. From the results, we conclude that *Tetraselmis* sp., having the most metal tolerating ability which survives up to 250 ppm of zinc. Except *D. salina* other microalgal species such as *C. marina, I. galbana* and *Nannochloropsis* sp., shows resistance up to 50 ppm of zinc concentration. The *D. salina* can survive only 5 ppm zinc concentration. From the above results it could be understood that the marine phytoplankton can survive up to those limits. In view of the increasing contamination of aquatic ecosystems by toxic metals and given the placement of microalgae at the basis of the food chain, ecotoxicological studies encompassing wild strains are of greater importance towards the design of bioremediation strategies and integrated environmental monitoring.

ACKNOWLEDGEMENTS

Authors express their thanks to the Head, Department of Marine Science and authorities of Bharathidasan University for the facilities provided. Authors acknowledge Dr. A. R. Thirunavukkarasu, Principal Scientist, Fish Culture Division, Central Institute of Brackishwater Aquaculture, Chennai, for providing algal

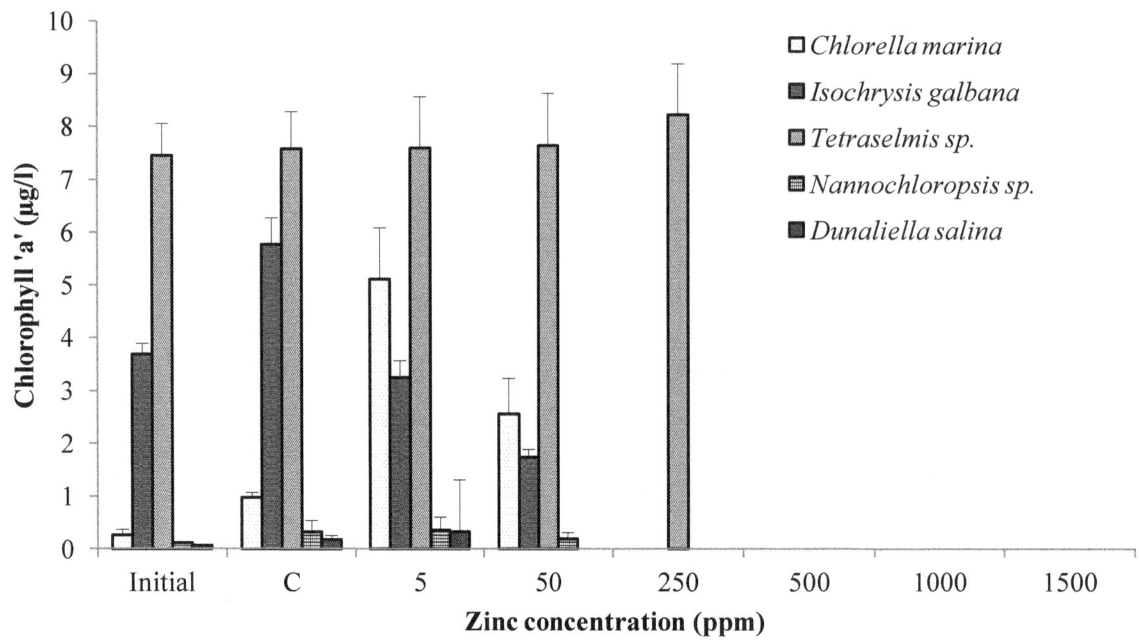

Figure 6. Effect of zinc on the chlorophyll 'a' content of five marine microalgae.

culture for our experiments.

REFERENCES

Amiard JC, Amiard-Triquet C, Metayer J, Ferre R (1980). Etude du transfert DE Cd, Pb, Cu ET Zn dans les chaines trophiques neritiques et estuariennes—I. Etat dans l'estuaire interne de la loire (France) au cours de l'ete 1978. Water Res. 14:665–673.

Báscik-Remisiewicz A, Tomaszewska E, Labuda K, Tukaj Z (2009). The effect of Zn and Mn on the toxicity of Cd to the green microalga *Desmodesmus armatus* cultured at ambient and elevated (2%) CO_2 concentrations. Polish J. Environ. Stud. 18(5):775–780.

Chong AMY, Wong YS, Tam NFY (2000). Performance of different microalgal species in removing nickel and zinc from industrial wastewater. Chemosphere 41(1-2):251-257.

Chu KH, Hashim MA (2007). Copper biosorption on immobilized seaweed biomass: Column breakthrough characteristics. J. Environ. Sci. 19(8):928-932.

Darmono (1995). Logam dalam sistem biologi makhluk hidup, UI-Press, Penerbit Universitas Indonesia, Jakarta.

Davies AG (1974). The growth kinetics of *Isochrysis galbana* in cultures containing sublethal concentrations of mercuric chloride. J. Mar. Biol. Ass. U.K. 54(1):157-169.

Dinesh Kumar S, Santhanam P, Jayalakshmi T, Nandakumar R, Ananth S, Shenbaga Devi A, Balaji Prasath B (2013). Optimization of pH and retention time on the removal of nutrients and heavy metal (zinc) using immobilized marine microalga *Chlorella marina*. J. Biol. Sci. 13(5):400-405.

Eichenberger ZE (1993). Relationship between the necessity and toxicity of metals in water ecosystems. in: K.H. Zigel and A. Zigel, eds, *Nekotorye voprosy toksichnosti ionov metallov* (Aspects of Toxicity of Metallic Ions), Moscow, Mir, pp. 62–87.

Fhencel T (1988). Marine plankton food chains. Ann. Rev. Ecol. 10(2):165-173.

Fisher NS, Bohe M, Teyssie JL (1984). Accumulation and toxicity of Cd, Zn, Ag, and Hg in four marine phytoplankters. Mar. Ecol. Progress Ser. 18:201-213.

Fitzgerald G, Faust SL (1963). Factors affecting the algicidal and algistatic properties of copper. Appl. Microbiol. 11(4):345–351.

Gao KS, Yan JI, Tanaka J (2004). Quantitative evaluation of wind effect during emersion on *Porhpyra haitanensis*(Rhodophyta), a farmed species in southern China. Fish. Sci. 70:710-712.

Hala Y, Syahrul M, Suryati E, Taba P (2012). Biosorption of zn^{2+} with *Nannochloropsis salina*. The 2nd International Seminar on New Paradigm and Innovation on Natural Sciences and its Application (INSPINSA-2), Semarang.

Hurlburet AJP, Tanaka Y, Skoryna SC (1976). Isolation and metal binding properties of fucoidan. Botanica Marina 19(5):327-328.

Imani S, Zarchi SR, Hashemi M, Borna H, Javid A, Zand AM, Abarghoue HB (2011). Hg, Cd and Pb heavy metal bioremediation by *Dunaliella* alga. J. Med. Plants Res. 5(13):2775-2780.

Kamp-Nielsen L (1971). The effect of deleterious concentrations of mercury on the photosynthesis and growth of *Chlorella pyrenoidosa*. Physiologia Plantarum 24(3):556-561.

Kapkov VI Belenikina OA (2003). Biomarkers of pollution of marine ecosystems with heavy metals. Water Ecosyst. Organ. (Moscow), 6:68–69.

Kapkov VI, Belenikina OA (2007). A Study of the resistance of mass marine algae to heavy metals. Moscow Univ. Biol. Sci. Bull. 62(1):30–33.

Kohn R (1975). Ion binding on polyuronates-alginate and pectin. Pure Appl. Chem. 42(3):371-397.

Lim CY, Yoo YH, Sidharthan M, Ma CW, Bang IC, Kim JM, Lee KS, Park NS, Shin HW (2006). Effects of copper (I) oxide on growth and biochemical compositions of two marine microalgae. J. Environ. Biol. 27(3):461-466.

Mantoura RFC, Llewellyn GA (1983). The rapid determination of algal chlorophyll and carotenoid and their breakdown products in natural waters by reverse-phase high-performance liquid chromatography. Analitica Chimica Acta 151:297-314.

Moreno-Garrido I, Blasco J, Gonzalez-Delvalle M, Lubian LM (1998). Differences in copper accumulation by the marine microalga *Nannochloropsis gaditana* Lubian, submitted to two different thermal treatments. Ecotoxicol. Environ. Restoration 1(1):43-47.

Nalimova AA, Popova VV, Tsoglin LN, Pronina NA (2005). The effects of copper and zinc on Spirulina *platensis* growth and heavy metal accumulation in its cells. Russ. J. Plant Physiol. 52(2): 229-234.

Nazzi R (1972). Toxicity of mercury to phytoplankton. Nature 237:38-40.

Rachlin JW, Farran M (1974). Growth response of the green algae

Chlorella vulgaris to selective concentrations of zinc. Water Res. 8(8):575-577.

Rebhun S, Amotz AB (1986). Effect of NaCl concentration on cadmium uptake by the halophilic alga *Dunaliella salina*. Mar. Ecol. Progress Ser. 30:215-219.

Rijstenbeil JW, Derksen JW, Gerrinda LJ (1994). Oxidative stress induced by copper: Defence and damages in the marine planktonic diatomea grow in continuous culture with high–low zinc levels. Mar. Biol. 119:583–590.

Rasko JJ, Rachlin JW (1977). Effect of Cd, Hg, Zn and Pb on cell division, growth and chlorophyll a content of the chlorophyte, *Chlorella vulgaris*. Bull. Torrey Bot. Club 104:226-233.

Shariati M, Yahyaabadi S (2006). The effects of different concentrations of cadmium on the growth rate and beta-carotene synthesis in unicellular green algae *Dunaliella salina*. Iran. J. Sci. Technol. 30:57-63

Tripathi BN, Gaur JP (2006) Physiological behavior of *Scenedesmus* sp. during exposure to elevated levels of Cu and Zn and after withdrawal of metal stress. Protoplasma 229:1-9.

Vallee BL, Auld DS (1990). Zinc coordination, function and structure of zinc enzymes and other proteins. Biochem. 29:5647–5659.

Veroy RL, Montano N, de Guzman MLB, Laserna EC, Cajipe GJB (1980). Studies on the binding of heavy metals to algal polysaccharides from Philippine seaweeds. I. Carrageenan and the binding of lead and cadmium. Botanica Marina 23:59-62.

Walne PR (1970). Studies on the food value of nineteen genera of alga e to juvenile bivalves of the genera *Ostrea, Crassostrea, Mercenaria* and *Mytilis*. Fish. Investig. 26:1-62.

Wetherell DF (1961). Culture of fresh water algae in enriched natural sea water. Physoilogia Plantarum 14(1):1-6.

6

Comparative survey of helminth parasites of *Clarias gariepinus* (Burchell, 1822) and *Clarias pachynema* (Boulenger, 1903) from the Ogun River and Asejire Dam in south-west Nigeria

Salawu M. T.[1], Morenikeji O. A.[2*], Sowunmi A. A.[2] and Odaibo A. B.[2]

[1]Department of Biological Sciences, Osun State University, Osun State, Nigeria.
[2]Department of Zoology, University of Ibadan, Oyo State, Nigeria.

A total of 250 randomly selected fish specimens consisting of 72 *Clarias gariepinus* and 51 *Clarias pachynema* from Ogun River around the cattle market, Isheri, Ogun State. 73 *C. pachynema* and 54 *C. gariepinus* from Asejire Dam (which has not been reported to be polluted) in south-west Nigeria were sampled for helminth infection between September 2008 and February 2009. All helminthic infections observed and recorded were restricted to the stomach, intestine and gill chamber. The helminths recovered included three cestodes, *Wenyonia* sp., *Polyonchobothrium* sp. and Pleurocercoid larva, a nematode, *Procamallanus* sp. and one digenean, *Clinostomum* sp. Prevalence and mean intensity of parasitic infection were higher in specimens from the Ogun River than those from the Asejire Dam. Overall, parasite prevalence of 75 and 45.1% were recorded for *C. gariepinus* and *C. pachynema* from the Ogun River, respectively, while 25.9 and 31.5% were recorded for *C. gariepinus* and *C. pachynema*, respectively, for the Asejire Dam. Male *C. gariepinus* from the Ogun River had the highest infection rate (76.5%), while those from Asejire Dam had the lowest (21.43%). No significant difference in the prevalence of parasitic infection in relation to the host sizes (P > 0.05) was recorded; however, largest size fishes had no parasitic infection. *Clinostomum* sp. was found only in *C. gariepinus* and *C. pachynema* from Ogun River.

Key words: Sharptooth catfish, *Clarias pachynema*, helminth parasites, south-west Nigeria.

INTRODUCTION

Fish has continued to be the most affordable source of animal protein to the average Nigerian family (Haruna, 2006). Parasitic infection causes production and economic losses through direct fish mortality, reduction in fish growth, fecundity and increase in the susceptibility of fish to diseases. *Clarias gariepinus* (Burchell, 1822) from the family Clariidae is one of the most important tropical catfish species for aquaculture in West Africa (Clay, 1979). It is identified by its elongated body and a large armored head enclosed by bony plates. The fish is scale-less and the body is dorsoventrally flattened (Adewumi, 2005). The fish is a general carnivorous feeder whose food consists mainly of fishes, terrestrial invertebrates, aquatic insects and zooplankton.

Clarias pachynema (Boulenger, 1903) is one of the fishes colonizing the freshwater bodies in Africa. The fish share some resemblance with *C. gariepinus* by being scale-less and elongated. The fish inhabit rainforest stagnant water bodies of streams, ponds, pools and swamps; they undergo breeding between May and July in shallow water (Idodo-Umeh, 2003). Their diets include aquatic insects, seeds, fruit, palmnuts and detritus. This study reports on helminth infection of both *C. gariepinus* and *C. pachynema* coexisting in the Ogun River and Asejire Dam experiencing contrasting impacts of abbatoir

*Corresponding author. E-mail: jumokemorenikeji@yahoo.co.uk.

and non-abbatoir wastes, respectively.

MATERIALS AND METHODS

Study areas

The Ogun River along the cattle market in Isheri, Ogun State is one of the perennial rivers found in southwest Nigeria. It rises at approximately 8°41'N, 3°28'E with an altitude of 380 m flowing for 320 km southwards into the Lagos lagoon at approximately 6°35'N, 3°25'E. It has a total drainage area of about 21,800 km^2. Detailed information on hydrology had been provided by Sydenham (1977) and Adebisi (1981). Abattoirs at Lafenwa and Isheri-Olofin along the river course process between 300 and 320 cattle and 40 and 50 goats/rams daily. The processing wastes are washed directly into the river system. This is in addition to effluents from breweries, dyeing industries, tanneries and domestic wastewater before finally discharging to the Lagos lagoon.

Asejire Dam (04°05'E; 07°21'N) was constructed in 1970 on the Oshun River. It has an impounded area of 2342 ha with gross storage capacity of 7403 million litres and located about 30 km east of Ibadan, southwest Nigeria at an altitude of 137 m (Egborge, 1979; Ekpo, 1993; Ayoade et al., 2006). The impoundment was created primarily for the provision of public water supply with fisheries development as a major ancillary benefit.

Fish collection and identification

From September 2008 to February 2009, 250 randomly selected fresh fish specimens comprising of *C. gariepinus* (72 and 54 from Ogun and Asejire, respectively) and *C. pachynema* (51 and 73 from Ogun and Asejire, respectively) were purchased live from the local fishermen. The fishes were transported live in a container containing water from the study site to the laboratory where they were sorted into different sizes and species. Identification was done based on external features as described by Idodo-Umeh (2003). Sexes of fish were determined by the presence or absence of a distinct sexual papilla conspicuously located behind the anus. The sexual papillae are absent in females. This was later confirmed after dissection by the presence of testis (male) or ovaries (female). Length and weight of the fishes were taken using a measuring board and digital weighing balance, respectively.

Examination of parasites

The fish were rendered inactive by cervical dislocation for easy handling prior to dissection. A cut was made on the ventral side from the anal opening to the lower jaw. Then, two more cuts were made on the lateral side to expose the body cavity with alimentary canal and other internal organs. The alimentary canals were removed and cut into parts (intestine and stomach) in 0.09% physiological saline for parasite recovery. Each part was further carefully slit open to aid the emergence of parasites. Gastro-intestinal parasites were further recognized by their wriggling movements on emergence. Gills and liver were examined under normal saline for presence of parasites.

The recovered helminth parasites were shaken vigorously in physiological saline to enable them to stretch for better observation. The parasites were counted and recorded before fixing in 70% alcohol.

Staining of helminth parasites for identification was done by placing in acetocarmine stain for 5 to 10 min. The stained parasites were washed in 70% alcohol, dehydrated in absolute alcohol and cleared in xylene before finally mounted in Canada balsam (MAFF, 1971).

The specimens were then viewed with the aid of a microscope and identification was done using the keys by Yamaguti (1959) and Paperna (1996).

Data analysis

The overall prevalence of the parasitic infection was calculated. Chi-square (χ^2) was used to calculate the significant difference between levels of infection in the two study sites. Parasite biometrics were calculated using standardized Shannon-weaver (1949) (H') function, mean species richness (R_{MS}), Jaccard (J) index of species overlap and percentage similarity (PS) (Hulbet, 1978; Siddall et al., 1994; Lile, 1998).

RESULTS

Parasites recovered included a nematode, *Procamallanus* sp., three cestode, *Polyonchobothrium* sp., *Wenyonia* sp., Pleurocercoid larva and one digenean *Clinostomum* sp. Concurrent infections in intestine with *Procamallanus* sp. and *Polyonchobothrium* sp were common.

A total of 487 helminths were recovered in fishes from the Ogun River when compared with 203 helminths in fishes from the Asejire Dam. 62.6% of total fish were infected in Ogun River against 28.4% from Asejire Dam (Table 1). The highest prevalence (75.0%) was recorded in *C. gariepinus* from the Ogun River. The prevalence of *C. pachynema* (31.5%) was higher than that of *C. gariepinus* (24.1%) in Asejire Dam. A higher proportion of fish species were more infected in Ogun River as compared to Asejire Dam (Table 2), although *C. gariepinus* (>70%) recorded higher infection rates when compared with *C. pachynema* (<50%) in Ogun River. The lack of significant differences (P>0.05) in infection levels between the sexes indicated absence of parasite reliability on either sex. Infection was pronounced in fishes below 18 cm in length (Tables 3 and 4) although association of host and infections were statistically different for this study. The distribution of parasite in organs showed the intestine to be the most infected. Flukes were restricted to the gills of fishes from Ogun River with preference for *C. gariepinus*. The cestodes dominated parasite abundance in all cases (Table 5). Biometrics showed that *C. gariepinus* recorded marked diversities from the two locations, this contrasted with *C. pachynema* with comparable diversities (Table 6). In contrast, species richness differed between locations for both species and parasite overlaps were also low in all cases for the species. The quantitative percentage parasite similarity was higher in *C. gariepinus* (PS = 0.99) when compared with *C. pachynema* (PS = 0.69).

However, identical PS was observed between the sampling locations.

DISCUSSION

The helminth species composition observed in this study

Table 1. Overall prevalence of Helminth infections in fishes from Ogun River and Asejire dam, Nigeria.

Fish host	Ogun River				Asejire Dam			
	Number examined	Number infected	Prevalence (%)	Total parasite	Number examined	Number infected	Prevalence (%)	Total parasite
C. gariepinus	72	54	75.00	294	54	13	24.07	111
C. pachynema	51	23	45.10	193	73	23	31.51	92
Total	123	77	62.60	487	127	36	28.35	203

Table 2. Helminth infection in relation to host sex in Ogun River and Asejire Dam, Nigeria.

Fish host	Ogun River				Asejire Dam			
	C. gariepinus		*C. pachynema*		*C. gariepinus*		*C. pachynema*	
Sex	Male	Female	Male	Female	Male	Female	Male	Female
No examined	34	38	24	27	28	26	30	43
No infected	26	28	11	12	6	7	11	12
Infection rate (%)	76.47	73.68	45.83	44.44	21.43	26.92	36.67	30.23
	χ^2=0.01, d.f =1; P>0.05		χ^2=0.004, d.f =1; P>0.05		χ^2=0.138, d.f =1; P>0.05		χ^2=0.32, d.f =1; P>0.05	

Table 3. Helminth infection in relation to size of *C. gariepinus* in Ogun River and Asejire Dam, Nigeria.

Group length (cm)	Ogun River			Asejire Dam		
	Number examined	Number infected	Percentage of infection	Number examined	Number infected	Percentage of infection
14.0-17.9	6	6	100	11	4	36.36
18.0-21.9	34	27	79.41	27	6	22.22
22.0-25.9	28	20	71.43	14	3	21.43
26.0-29.9	4	3	75.00	2	0	0
Total	72	56	77.78	54	13	24.07
	χ^2=1.563; d.f= 3; P>0.05			χ^2=1.057, d.f= 3; P>0.05		

Table 4. Helminth infection in relation to size of the *C. pachynema* in Ogun River and Asejire Dam, Nigeria.

Group length (cm)	Ogun River			Asejire Dam		
	Number examined	Number infected	Percentage of infection	Number examined	Number infected	Percentage of infection
14.0-17.9	5	4	80.00	7	3	42.86
18.0-21.9	12	5	41.67	31	10	32.26
22.0-25.9	25	11	44.00	30	9	30.00
26.0-29.9	9	3	33.33	5	0	0.00
Total	51	23	45.10	73	22	30.14
	χ^2=0.973,d.f= 3; P>0.05			χ^2=1.805, d.f= 3; P>0.05		

differed from previous reports on species other than *Clarias* (Ukoli, 1972; Okaka, 1998; Auta et al., 1999; Omoniyi and Olofintoye, 2001; Oniye et al., 2004) and *C. gariepinus* (Omoniyi and Olofintoye, 2001; Oniye et al., 2004) from Ado-Ekiti and Zaria (Nigeria), respectively.

None reported helminth infections in *C. pachynema*. The present study is the first record of *Procamallanus* sp. and *Wenyonia* sp. in *C. pachynema* from Nigeria; *Procamallanus laevionchus, Polyonchobothriium* sp. and *Wenyonia* sp. were part of previously reported helminth

Table 5. Parasite distribution in organs of hosts from the study areas.

Fish species	Parasite taxa	Ogun River			Asejire Dam	
		Stomach	Intestine	Gills	Stomach	Intestine
C. gariepinus	Cestoda	42 (68.9%)	141 (67.1%)	0	7 (63.6%)	76 (78.4%)
	Nematoda	19 (31.1%)	34 (16.2%)	0	4 (36.6%)	21 (21.6%)
	Flukes	0	0	18 (100%)	0	0
	Flatworms	0	35 (16.7%)	0	0	0
C. pachynema	Cestoda	27 (96.4%)	89 (90.8%)	0	7 (70.0%)	65 (91.5%)
	Nematoda	1(3.6%)	9 (9.2%)	0	3 (30.0%)	6 (8.5%)
	Flukes	0	0	9 (100%)	0	0

Table 6. Helminth biometrics in hosts.

Parameter	*C. pachynema*		*C.gariepinus*	
	Asejire Dam	Ogun River	Asejire Dam	Ogun River
Shannon-weaver (H') function	0.64	0.63	0.55	0.72
Jaccard index (J)	0.33	0.30	0.27	0.29
Mean species richness (R_{MS})	0.89	0.38	0.47	0.24
	0.69		0.99	
Percentage similarity (PS)			0.50	
	0.51			

parasites from *C. gariepinus* by Omoniyi and Olofintoye (2001), Oniye et al. (2004) and Sowunmi and Akinlolu (2009). The occurrence of *Clinostomum* sp. in *C. gariepinus* from South Africa has been reported by Barson et al. (2008) and Maguza-Tembo and Mfitilodze (2008); however, information on the parasite in Nigeria is scarce. *Clinostomum* sp. was found in *C. gariepinus* and *C. pachynema* from Ogun River and none was recovered from Asejire Dam. This suggested the presence of the snail intermediate host of *Clinostomum* sp. in Ogun River and possibly its ability to withstand the level of pollution and increase in number as a result of organic pollutant present in the water body. The presence of *Wenyonia* sp. extended the suggested reliability on siluroids (Ukoli, 1972; Sowunmi and Akinlolu, 2009) and as suggested by Lile (1998), which could be classified as a specialist.

Prevalence of infection in the two species of fish was higher in Ogun River, with 75% prevalence in *C. gariepinus* and 45.10% in *C. pachynema*. The highest overall prevalence of helminth parasites observed in *C. gariepinus* from Ogun River around Isheri cattle market may be attributed to pollution load from the cattle market. Oketola et al. (2006) reported that this section of Ogun River is presently being overstressed by pollution loads mainly from the cattle market and run-off from the busy Lagos-Ibadan express road. This agreed with reported high positive correlation between pollution and prevalence rate of helminth parasite by Sosanya (2002).

In both *C. gariepinus* and *C. pachynema* from Asejire Dam and Ogun River, there was no significant difference in infection rate between the sexes. Adegbaju (2007) reported that this could be attributed to random selection of the specimen from study site. There was no significant prevalence difference observed among the host size in relation to helminth parasites recorded from Ogun River and Asejire Dam fish. However, the longest fishes in both species had no infection while the shortest were most infected. Lagler et al. (1979) reported correlation between parasitic infection and fish length which also corresponds to fish age. Akinsanya et al. (2007) attributed this to the random selection and the low level of immunity in the smaller sized fish.

Parasite biometrics indicated similar host-parasite dynamics in the two water bodies for each species. However, the observed differences may be due to foraging specializations, similar to the report of Dorucu et al. (1995) on sympatric morphs, *Salvelinus alpinus*. The diet of *C. gariepinus* facilitated more exposure to helminth parasites when compared with *C. pachynema*.

The parasitic fauna of *C. pachynema* still needs further studies due to apparent lack of information when compared with other related species.

REFERENCES

Adebisi AA (1981). The physico-chemical hydrology of a tropical seasonal river-upper Ogun river. Hydrobiologia (79):157-165.
Adegbaju WA (2007). Parasitofauna impacts on *Clarias gariepinus* and

Synodontis clarias from Lekki Lagoon. M.Sc. Thesis, Univ. Ibadan. p. 72.

Adewumi (2005). The effect of heating time of soyabean for blood stock nutrition on the reproductive performance of *Clarias gariepinus* (Burchell, 1822). Ph.D. Thesis Obafemi Awolowo Univ. Nigeria. p. 160.

Akinsanya B, Hassan AA, Otubanjo OA (2007). A comparative study of the parasitic helminth fauna of *Gymnarchus niloticus* (Gymnarchidae) and *Heterotis niloticus* (Osteoglossidae) from Lekki Lagoon Lagos, Nigeria. Pakistani J. Biol. Sci. 10(3):427-432.

Auta J, Oniye SJ, Adakole JA (1999). The helminth parasies of the gastrointestinal tract of *Synodontis* species in Zaria, Nigeria. J. Pure Appl. Sci. 2(2):47-53.

Ayoade AA, Fagade SO, Adebisi AA (2006). Dynamics of limnological features of two man-made lakes in relation to fish production. Afr. J. Biotechnol. 5(10):1013-1021.

Barson M, Bray R, Ollevier F, Huyse T (2008). Taxonomy and Faunistics of the Helminth Parasites of *Clarias gariepinus* (Burchell, 1822) and *Oreochromis mossambicus* (Peters, 1852) from Temporary Pans Floodplain, Zimbabwe. Comparative Parasitol. 75(2):228-240.

Clay D (1979). Population biology, growth and feeding of the Africans catfish *Clarias gariepinus* with special reference to juveniles and their importance in fish culture. Arch. Hydrobiol. 87(4):458-482.

Dorucu M, Adams CE, Huntingford FA and Crompton DWT (1995). How fish-helminth associations arise: an example from Arctic charr in Loch Rannoch. J. Fish Biol. (47):1038-1043.

Egborge ABM (1979). The effects of impoundment on the phytoplankton of the River Osun, Nigeria. Nova Hedwiga (31):407-417.

Ekpo AO (1993). Growth, feeding and reproductive biology of *Hydrocyanus forskalii, Alestes macrolepidotus and Channa obscurus* in Asejire Reservoir. Ph.D. thesis Univ. Ibadan, Nigeria. p. 244.

Haruna AB (2006). Studies on the aspect of socio-economic factors influencing fish-farming in Adamawa State- Nigeria. J. Arid zone Fish. 2(1):8-14

Hulbet SH (1978). The measurement of niche overlap and some relatives. Ecol. 59(1):67-77.

Idodo-Umeh G (2003). Fresh-Water Fishes of Nigeria: Taxonomy, Ecological Notes, Diet & Utilization. Idodo-Umeh Publishers Ltd., Nigeria. 232 pp.

Lagler KF, Bardachi JE, Miller RR (1979). Ichthyology. John Wiley, New York. p. 882.

Lile NK (1998). Alimentary tracts helminths of four pleuronectid flatfish in relations to host phylogeny ecol. J. Fish Biol. 53:945-953.

MAFF (1971). Manual of Veterinary Parasitologia. Ministry of Agriculture, Fisheries and Food London. Technical Bulletin No. 18.

Maguza-Tembo F, Mfitilodze MW (2008). Occurrence (incidence) of parasites in three fish species (*Clarias gariepinus, Oreochromis shiranus* and *Happlochromis*) from Bunda Reservoir. Bunda J. Agric. Environ. Sci. Technol. 3:3-7.

Okaka CE (1998). Plerocercosis and other helminthic infections among freshwater fishes of Osomo and Benin rivers in Southern Nigeria. Tropical Freshwater Biol. 7:73-80.

Oketola AA, Osibanjo O, Ejelonu BC, Oladimeji YB, Damazio OA (2006). Water Quality Assessment of River Ogun Around the cattle Market of Isheri, Nigeria. J. Appl. Sci. 6(3):511-517.

Omoniyi IT, Olofintoye LK (2001). A survey of endohelminth parasites of fishes from water reservoir and Elemi River in Ado-Ekiti, Ekiti State, Nigeria. Biosci. Res. Commun. 13(1)87-94.

Oniye SJ, Adebote DA, Ayanda OI (2004). Helminth parasites of *Clarias gariepinus* (Teugels) in Zaria, Nigeria. J. Aquatic Sci. 19(2):71-75.

Paperna I (1996). Parasites infections and diseases of fishes in Africa - An update (CIFA Technical) paper 31:1-220.

Shannon CE, Weaver W (1949). The Mathematical Theory of Communication. University of Illinois Press,Urbana.

Siddall R, Pike AW, McVicar AH (1994). Parasites of flatfish in relation to sewage sludge dumping. J. Fish Biol., 45: 193-209.

Sosanya MO (2002). Fish parasite as indicators of Environment quality M.Sc Thesis, Univ. Ibadan. p. 142.

Sowunmi AA, Akinlolu DA (2009). Helmtinth parasites of fish species from Epe Lagoon Southwest Nigeria. Afr. Anim. Biomed. Sci. 4(2):38-41.

Sydenham DHJ (1977). The qualitative composition and longitudinal zonation of fish fauna of the River Ogun, Western Nigeria. Revue de Zoologie Africaine 91(4), 974-997.

Ukoli FMA (1972). Occurrence, morphology and systematics of Caryophyllaeid cestode of the genus *Weyonia* Woodland, 1923 from fishes in River Niger, Nigeria. J. West Afr. Sci. Assoc. (17):49-67.

Yamaguti S (1959). Systema Helminth. Vol. 1. Digenetic Trematodes parts 1 and 11. Inter-sci. publishers Inc. New York. p. 1575.

7

Developmental changes in intestinal brush border enzyme activity in wild, juvenile Nile perch *Lates niloticus* (Linnaeus, 1758)

Kato C. D.[1], Nyatia E.[1], Matovu E.[1], Uni Z.[2], Kedar O.[2], Hakim Y.[2], Levavi-Sivan B.[2] and Rutaisire J.[3]

[1]School of Biosecurity, Biotechnical and Laboratory Sciences, Makerere University P. O. Box 7062, Kampala Uganda.
[2]Department of Animal Sciences, Faculty of Agriculture, Food and Environment, the Hebrew University of Jerusalem, P. O. Box 12, Rehovot 76100, Israel.
[3]Aquaculture Research and Development Centre, P. O. Box 530, Kampala, Uganda.

The Nile perch, *Lates niloticus* is a carnivorous fish and a potential candidate for aquaculture. The relationship between fish size (total length) and the activity of three brush border enzymes: leucine aminopeptidase (LAP), γ-glutamyl transferase (γ-GT) and maltase was evaluated in six size groups (1 to 5, 6 to 10, 11 to 15, 16 to 20, 21 to 25 and 26 to 30 cm) of wild juvenile Nile perch. Enzymatic activity was evaluated in three intestinal sections: pyloric caeca, upper intestine and lower intestine. All the three enzymes were influenced by the fish size, intestinal section, and the interaction between the two factors ($p < 0.05$). The highest specific activity of LAP and maltase was observed in the upper intestine while that of γ-GT was highest in the lower intestine. The specific enzyme activities were significantly higher ($p < 0.05$) in the 11 to 15 cm and 16 to 20 cm size groups in all the tested enzymes. The results reveal that, the most critical stage in the nutrition of juvenile Nile perch occurs when the fish attains a total length of 11 to 20 cm. Dietary formulations for this species should hence be in cooperate substantive amount of proteins and carbohydrates at this stage so as to maximize weight gain. The presence of maltase is an indicator that carbohydrates cannot be precluded from formulated diets for this species.

Key words: Nile Perch, enzyme activity, aquaculture, brush border.

INTRODUCTION

The Nile perch, *Lates niloticus* is a carnivorous fish common to all major river basins in Africa including the Nile, Volta, Senegal, Chad and Congo. The species is also native to Lakes Albert and Turkana in Uganda and Kenya, respectively. *L. niloticus* was introduced to Lake Victoria in the late 1950s (Hughes, 1992) near Jinja, Uganda and then spreading to both Kenya and Tanzania. The population of the Nile perch grew over the years to become the most economically important species of the East African freshwater fisheries (Ogutu-Ohwayo, 1993). Fish products are Uganda's second most important export for generating export revenue, providing employment and supporting food security in the region (Benkenstein, 2011). According to the Lake Victoria Fisheries Organisation (2008) the species supported the regional fish export industry valued at USD 250 million in 2007. However, the high international demand has led to a decline in biomass and catch per unit effort (Matsuishi et al., 2006). Consequently, decreased fish catches have been reported in most areas of the lake and fish factories are operating with inadequate raw material (LVFO, 2008; Muhoozi et al., 2005; Ogutu-Ohwayo and Kirema-Mukasa, 2005). In order to revive the fish industry, *L. niloticus* is currently being explored as a potential candidate for aquaculture at the Kajjansi Aquaculture Research and Development Center, Uganda. However, before culture of any candidate species its ability to digest artificial diets has to be investigated (Kolkovski, 2001). The Nile perch is a carnivorous fish feeding on haplochromine cichlids, shrimp, dagaa and juvenile Nile perch. Attempts to feed the Nile perch on artificial commercial diets designed for the catfish (omnivorous) and *Lates calcarifer* (carnivorous) have so far been unsuccessful since the fish would not consume the diet or it would pass out half-digested compared to its natural diet of live fish. In order to understand the nutritional requirements for this species, there is need to investigate its digestive physiology. Elucidation of digestive enzyme activity is key to understanding the nutritional physiology of fish. Of special interest are enzymes located in the brush border membrane that are responsible for the final stages of degradation and assimilation of nutrients (Klein et al., 1998). The characteristics of digestive enzymes provide insights into the ability of a given species to digest and utilize dietary components (Ali et al., 2009; Debnath et al., 2007).

Fish, like other animals, are thought to exhibit plasticity in production of digestive enzymes (Caviedes-Vidal et al., 2000). For instance, carnivorous fish have been shown to have higher proteolytic enzyme activities to digest high protein diets while herbivorous fish exhibit higher carbohydrase activities (Cahu et al., 1998; Hakim et al., 2007; Harpaz et al., 2005b; Harpaz and Uni, 1999; Hidalgo et al., 1999; Sabapathy and Teo, 1993). Nevertheless, Chakrabarti et al. (1995) studied the food habits and enzymatic activity of multiple freshwater fish species and found that adult fish produce almost all digestive enzymes, irrespective of diet. Similar results were reported by Harpaz and Uni (1999) in the carnivorous hybrid striped bass, omnivorous tilapia hybrids and the silver carp fed on natural food and commercial pellets containing 30% protein and 5% fat.

The intestinal section has been shown to influence the level of enzyme activity with variations across species. A higher proteolytic enzyme activity has been reported in the pyloric caeca of *L. calcarifer* (Harpaz et al., 2005a) indicating the role of this section in protein degradation and absorption (Horn, 2005). Furthermore the carnivorous species on high protein diets have been shown to exhibit a 3-fold increase in proteolytic enzyme activity compared to omnivorous species in the pyloric caeca. However studies by Tibaldi et al. (2006) in the European sea bass (*Dicentrarchus labrax*) found the highest activity of γ-GT, a protein degrading enzyme in the lower intestinal tract. Though the role of the pyloric caeca is still unclear, it has been suggested that this section takes part in absorption of carbohydrates and fats hence supplementing the role of the stomach and intestine (Hepher, 1988).

Digestive enzyme activity is also influenced by age and diet (Péres et al., 1998). The influence of diet and age on digestive enzyme activity can best be demonstrated in species that undergo an ontogenetic change in diet (German et al., 2004; Kuz'mina, 1996). Indeed, prickleback fishes (*Cebidichths violaceus* and *Xiphister mucosus*) from the central California coast increase their carbohydrase activities with an ontogenetic shift from carnivory to herbivory (German et al., 2004). Studies based on gut content analysis have shown that juvenile Nile perch undergo an ontogenetic shift in diet (Katunzi et al., 2006; Mkumbo and Ligtvoet, 1992) depending on food availability in the different habitants (Paterson and Chapman, 2009). Nile perch of <5 cm exclusively fed on zooplankton before shifting to midge larvae and shrimp. Above 10 cm the main prey for Nile perch in most habitants was reported to be juvenile Nile perch (< 5 cm) and dagaa (*Rastrineobola argentea*) (Katunzi et al., 2006). Therefore, the aim of this paper was to investigate if brush border enzyme activities would vary with this ontogenetic shift in diet.

Although, attempts to wean candidate aquaculture species onto artificial diets typically begins with larval stages (Creswell et al., 2010; Vega-Orellana et al., 2006), induced spawning of Nile perch has not yet been successful, hence nutritional studies can only be done on wild fish. We studied the activity of three brush border enzymes in three intestinal sections: pyloric caeca, upper intestine and lower intestine in six different size groups. Two proteolytic enzymes, leucine amino peptidase (LAP, EC 3.4.11.1), gamma glutamyl transferase (γ-GT, EC 2.3.2.2) and a carbohydrase, maltase (EC 3.2.1.20) were assayed. The activity of gut enzymes like trypsin and amylase in the wild juvenile Nile perch have been investigated (Namulawa et al., 2013). The following questions were specifically addressed: Are there differences in the enzymatic activities of LAP, γ-GT and maltase across the different fish size groups? How does the activity of these enzymes vary along the different intestinal sections? Is there a correlation between the activities of these enzymes along the different intestinal sections and the different fish size groups? In order to answer these questions, juvenile *L. niloticus* at different stages of development were captured from the shores of Lake Victoria and their digestive enzyme activities at specific intestinal sites analyzed. Results from this study would provide useful information to guide the formulation of artificial diets for the Nile perch.

MATERIALS AND METHODS

Study design and sampling strategy

Wild juvenile Nile perch at different stages of development were collected from Kigungu landing site on the shores of Lake Victoria (GPS: 36" 43755 E, 0003036 N, and 1131 m above sea level). Fish with a total length ranging from 1 to 30 cm were considered juvenile (Katunzi et al., 2006). They were captured by beach seining and immediately transferred to oxygenated tanks (2 m × 2 m) containing lake water and taken to the laboratory where they were kept in the tank overnight so as to recover from handling stress. The fish were classified into six size groups of total length (1 to 5, 6 to 10, 11 to 15, 16 to 20, 21 to 25 and 26 to 30 cm) based on the observation by Katunzi et al. (2006) that wild juvenile *L. niloticus* undergo an ontogenetic dietary shift within the selected range of size groups.

Ten (10) fish from each group were randomly selected, sacrificed using a scalpel to sever the spine and their body weight taken. The abdomen was opened and the entire digestive tract removed. The intestine was dissected, adipose tissue carefully cleaned off and the total intestinal weight was measured using a weigh balance (Sartorius, Germany). The intestinal tract was then washed with distilled water in order to remove any residual prey debris and divided into the pyloric caeca, upper intestine and lower intestine. For fish between 1 to 10 cm total lengths, the whole gut was used because intestinal sectioning was not possible. The dissected sections were then washed with double distilled water and immediately stored in a -20°C freezer for evaluation on the following day.

Evaluation of brush border enzymes

Brush border enzyme activity evaluation was done following the method of Harpaz and Uni (1999). Prior to evaluation, stored sections were briefly thawed and eight milliliters of double distilled water added to 1 ml of tissue. Zirkonium glass beads (Muhlmeier, Germany) were added to each tube and the mixture homogenized using a mini-Bead Beater (Bio-Spec, U.S.A.) for 5 min. The Homogenate was then centrifuged for 5 min at a speed of 4,000 rpm using an Eppendorf Centrifuge (Eppendorf, Germany). The extract was then used to assay the specific activity of LAP, γ-GT and maltase. Specific activity was calculated as the amount of enzyme required to release 1 μmol of product per minute per milligram protein.

Leucine amino peptidase (EC 3.4.11.1) activity

The activity of leucine amino peptidase (LAP) was measured following the method of Harpaz and Uni (1999) using L-leucine-p-nitroanilide reagent (Randox, Ireland). The reagent (250 μl) was added to 15 μl of the homogenate in an ELISA plate and read at 1 min intervals for 15 min at 405 nm using a microplate reader (Tecan Sunrise, Switzerland). One unit of activity was determined as the quantity required to release 1 μmol L-leucine and p-nitro aniline per minute per milligram protein. Units of activity were determined as below

$$\text{LAP (U /mgP)} = ((\Delta A/\text{min}*TV) / (EF*SV))*(D/P)$$

Where: ΔA/min: total increment per min, TV: 265 μl: The total reaction volume (homogenate + reagent), EF: 1.167=excitation coefficient of P-nitrophenoxide at 405 nm, SV: homogenate=15 μl, D: dilution factor, P: protein (mg) in the homogenate.

Gamma glutamyl transferase (EC 2.3.2.2) activity

The activity of γ-GT was measured following the method of Harpaz and Uni (1999) using a kit manufactured by Randox, Ireland. Twenty microliters of homogenate sample was placed on an ELISA plate followed by 200 μl of reagent. The plate was immediately read at 405 nm at 1 min intervals for 15 min using a microplate reader (Tecan Sunrise, Switzerland). One unit of activity was determined as the quantity required to release 1 μmol of 5-amino-2-nitrobenzoate per minute per milligram protein. Units of activity were as for LAP except for the EF coefficient, which is 9.5 for the 5-amino-2-nitrobenzoate produced in this reaction at 405 nm.

Maltase (EC 3.2.1.20) activity

The hydrolysis of maltose by maltase was assayed following the method of Harpaz and Uni (1999) following the equation below:

$$\text{Maltose} \xrightarrow{\text{Maltase}} \text{Glucose} + \text{Glucose}$$

Twenty microliters of sample extract were mixed with 20 μl of substrate (0.056M maltose) in an ELISA plate. The plate was then incubated at 37°C for 45 min for glucose to form. The reaction was stopped by putting the plate on ice for 4 min. The presence of glucose as an indicator of enzymatic activity was then measured using the glucose oxidase reagent (Thermo electron corporation, USA). After incubation, 2 μl were taken from the reaction mixture and placed in an ELISA plate. Glucose oxidase reagent (300 μl) was then added to all the wells and the plate incubated at 37°C for 10 min. After incubation, the plate was read using a microplate reader (Tecan Sunrise, Switzerland) at 500 nm and compared to a glucose standard (Thermo Scientific, USA).

Protein determination

Protein concentration in the sample extracts was determined using a kit (BIO-RAD, USA) so as to calculate the specific activity per milligram protein. The protein concentration was measured by putting 5 μl of sample extract in an ELISA plate and adding 5 μl of reagent A to all wells followed by 200 μl of reagent B and incubation at room temperature for 10 min. The OD was read at 750 nm using a microplate reader (Tecan Sunrise, Switzerland) and compared with a 2 mg/ml BSA standard (BIO-RAD, USA).

Statistical analysis

Statistical analysis of enzyme activity was done using GraphPad 5.0 statistical package. The effect of intestinal section, size group and their interaction on the activity of Leucine amino peptidase,

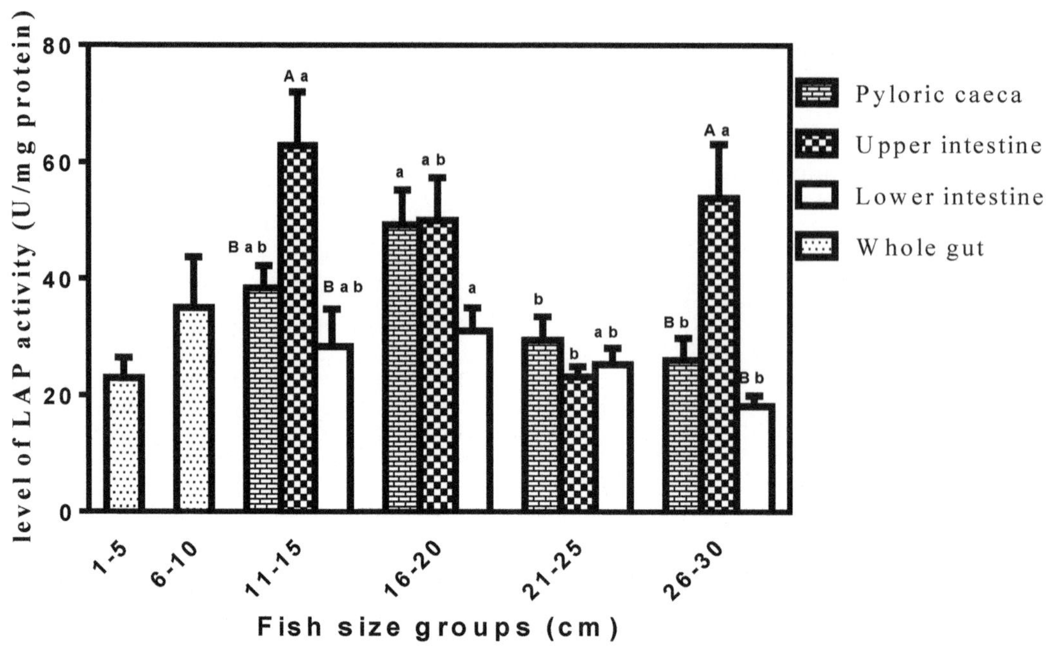

Figure 1. Specific activity of leucine amino peptidase measured in the different intestinal sections of six different fish size groups. Capital letters above the columns represent the significant differences ($p<0.05$) among intestinal sections in each size group. Small letters represent significant difference among the fish size groups within each segment (Size Groups 1 to 5 and 6 to 10 were excluded from the analysis).

Gamma glutamyl transferase and maltase was analyzed using a two-way ANOVA. The 1 to 5 and 5 to 10 cm size groups were not included in the ANOVA. Comparison between the enzymatic activity of the different intestinal segments and that between the different size groups was done using Tukey`s Multiple Comparison Test set at $P<0.05$. To determine the relationships between the activity of Leucine amino peptidase, Gamma glutamyl transferase and maltase across the different size groups, and correlation tests using Pearson correlation matrix set at a $P<0.05$ were done.

RESULTS

Leucine amino peptidase (LAP)

LAP specific activity was influenced by the intestinal section ($F_{2, 92}=14.38$, $p<0.0001$), the fish size ($F_{3, 92}=6.64$, $p<0.0004$) and the interaction between the two factors ($F_{6, 92} = 3.37$, $p<0.0048$). Significantly higher activity was detected in the upper intestine of the 11-15 cm (62.83±9.154 U/mg protein) and 26-30cm (53.81±9.27 U/mg protein) groups as compared to the pyloric caeca and lower intestine. Comparison of LAP activity in the same intestinal segment among different groups revealed significantly higher activity ($p=0.005$) in the pyloric caeca of the 16-20 cm group (49.15±6.0 U/mg protein) as compared to the 21-25 and 26-30 groups. In the upper intestine, a higher activity ($p=0.002$) was found in the 11-15 and 26-30 groups as compared to 21-25 group. In the lower intestine, a significantly higher activity ($p=0.03$) was found in the 16-20 group (31.014±4.02 U/mg protein) as compared to the 26-30 group (Figure 1).

Gamma glutamyl transferase (γ-GT)

The results show that the activity of γ-GT was influenced by the intestinal section ($F_{2, 88}=2.72$, $p<0.0247$), the fish size ($F_{3, 88}=8.25$, $p<0.0001$) and the interaction between the two factors ($F_{6, 88}=2.72$, $p< 0.018$). Comparing the activity of γ-GT in the same size group among different intestinal sections revealed a significantly higher activity in the lower intestine of fish in the 21 to 25 (814.35±152.08 U/mg protein) and 26 to 30 (703.87± 99.05 U/mg protein) cm groups as compared to the upper intestine. Comparison of the activity of γ-GT in the same intestinal tract among different groups revealed a higher activity ($p=0.01$) in the pyloric caeca of the 16 to 20 cm group (1145.68±241.5 U/mg protein) as compared to the 21 to 25 and 26 to 30 groups. In the upper section, a higher γ-GT activity ($p=0.001$) was found in the 11 to 15 group (1010.5±194 U/mg protein) as compared to the 21 to 25 and 26 to 30 groups. No significant difference was found in the activity of the enzymes in the lower intestine across the different size groups (Figure 2).

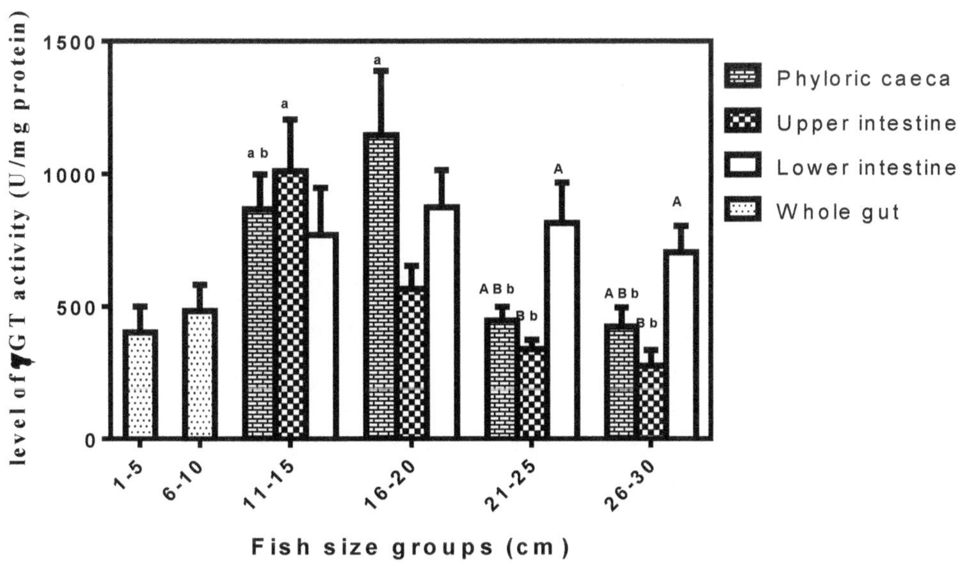

Figure 2. Specific activity of γ-glutamyl transferase measured in the different intestinal sections of six different fish size groups. Capital letters above the columns represent significant differences (p<0.05) among intestinal sections in each size group. Small letters represent significant difference among the fish size groups within each segment (Size groups 1 to 5 and 6 to 10 were excluded from the statistical analysis).

Maltase

The activity of maltase was influenced by the intestinal section ($F_{2, 103}$=5.90, p<0.0038), the fish size ($F_{3, 103}$= 21.23, p<0.0001) and the interaction between the two factors ($F_{6, 103}$=3.48, p< 0.0036). Comparing the activity of maltase in the same fish size group among different intestinal sections revealed a significantly higher activity in the upper intestine of the 11 to 15 (112.32±11.08 U/mg protein) and 26 to 30 (66.32±12 U/mg protein) groups. Comparison of maltase activity in the same intestinal tract among different groups revealed a significantly higher activity (p=0.002) in the pyloric caeca of the 11 to 15 (65.23±3.89 U/mg protein) and 16 to 20 (56.61±8.20 U/mg protein) groups. In the upper intestine, a significantly higher activity (p=0.01) was found in the 11 to 15 (112.32±11.08 U/mg protein) and 26 to 30 (66.32±12 U/mg protein) groups. In the lower intestine, a higher activity (p=0.04) was also revealed in the 11 to 15 (73.7±15.51 U/mg protein) size group (Figure 3).

Correlations between enzyme activities

The highest correlation among enzyme activities was observed between LAP and maltase in all fish size groups, followed by that between γ-GT and maltase and lowest between LAP and γ-GT. The strongest correlations were revealed in the pyloric caeca of all size groups as compared to other intestinal segments (Table 1).

DISCUSSION

The ability of fish to utilize a given diet depends on the activity of its enzymes along the digestive tract (Tengjaroenkul et al., 2000). Digestive capacity of fish relates to the age and dietary composition (Lovett and Felder, 1989; Péres et al., 1998). Chakrabarti et al. (1995), while studying 11 adult fresh water teleost fish demonstrated that at any time, fish possess a panel of enzymes with activities varying according to feeding habit. Indeed, Harpaz and Uni (1999) reported similar findings while comparing carnivorous, omnivorous and herbivorous fish species. Similar results were observed in this study in that all tested enzymes were present in the different size groups of the wild juvenile *L. niloticus* but with activity variation relating to size and intestinal section. These findings will help reveal dietary needs for juvenile Nile perch.

Maltase is a disaccharide degrading enzyme required for the breakdown of maltose produced during the hydrolysis of starch and its highest activity was observed in the upper intestine. A high activity of maltase has also been reported in other carnivorous fish like the Asian sea bass (Harpaz et al., 2005b; Sabapathy and Teo, 1993) and in birds fed on a high protein, carbohydrate free diet

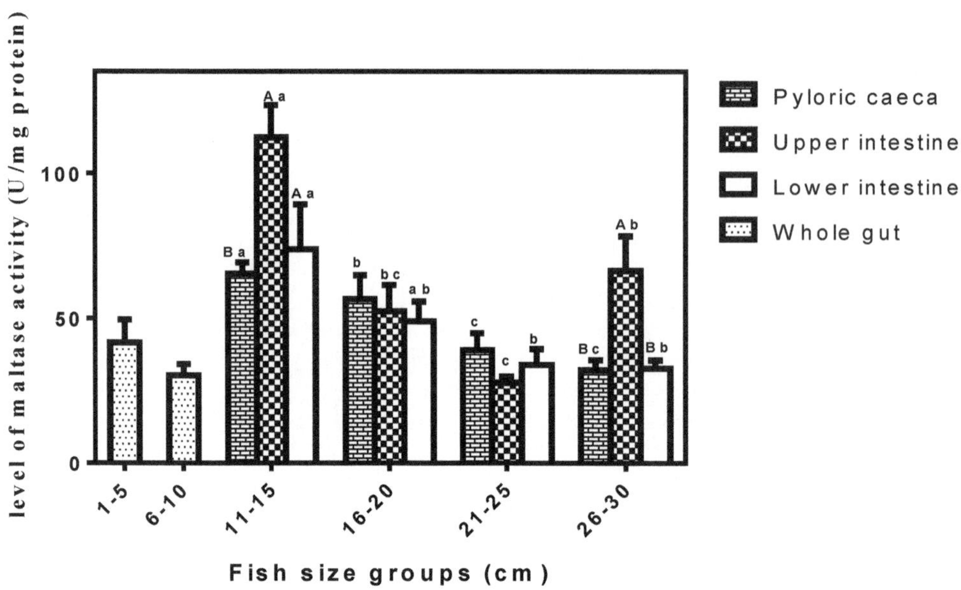

Figure 3. Specific activity of maltase measured in the different intestinal sections of six different fish size groups. Capital letters above the columns represent significant differences (p<0.05) among intestinal sections in each size group. Small letters represent significant difference among the fish size groups within each segment (Size groups 1 to 5 and 6 to 10 were excluded from the statistical analysis).

Table 1. Correlation between the specific enzyme activities of brush border enzymes among the different fish size groups and intestinal segments.

Size group (cm)	Intestinal segment	Lap/γ-GT	Lap/Maltase	γ-GT/Maltase
	Pc	0.362	0.978*	0.323
11-15	Upper	0.298	0.676*	0.610
	Lower	0.512	0.688*	0.862*
	Pc	0.831	0.914*	0.634
16 - 20	Upper	0.632	0.824*	0.654
	Lower	0.369	0.845*	0.778*
	Pc	0.790	0.920*	0.273
21 - 25	Upper	0.660	0.850*	0.570
	Lower	0.680	0.579	0.920*
	Pc	0.954*	0.935*	0.900*
26 - 30	Upper	0.234	0.342	0.965*
	Lower	0.321	0.314	0.600

The values presented are correlation coefficients (r). Figures with asterisk indicate significant values (p<0.05). Pc is the pyloric caeca.

(Sabat et al., 1998). Presence of maltase and its strong correlation to other brush border enzymes in a carnivorous fish on a natural diet suggests that the activity of brush border digestive enzymes is regulated under the same gene control mechanism. Moreover there was a strong correlation in the activity of these enzymes especially in the pyloric caeca confirming that this region is an extension of the intestine with an additive role in nutrient digestion and absorption (Horn, 2005). Similarly, Hakim et al. (2007) observed a linear correlation between mRNA expression for the different brush border enzymes and transporters in European sea bass (*Dicentrarchus labrax*). The presence of maltase in the Nile perch gives prospects for the design of artificial diets.

Our results revealed that intestinal segment had a significant effect on all assayed enzymes. As was shown previously (Hakim et al., 2007; Harpaz et al., 2005a; Tibaldi et al., 2006) the highest activity of LAP was revealed in the upper intestine. LAP is responsible for the digestion of the products of pancreatic proteases. The high activity of this enzyme in the upper intestine indicates importance of this intestinal section in protein digestion in the wild juvenile *L. niloticus*. The highest activity of γ-glutamyl transferase was revealed in the lower intestine. These results are in agreement with other studies in the hybrid striped bass (Harpaz and Uni, 1999), the Asian sea bass (Harpaz et al., 2005b) and the European sea bass (Tibaldi et al., 2006). The activity of γ-GT is indicative of active amino acid transportation across the intestinal surface (Semenza, 1986). A higher activity of this enzyme in the lower intestine indicates that absorption of amino acids is elevated with movement along the gut and might be an adaptation to efficiently absorb nutrients in order to compensate for the short intestinal tract in this species.

The results revealed a significant influence of the fish size on the activity of all the tested brush border enzymes. Several studies have shown that in the European sea bass (*D. labrax*) and Senegal sole (*Solea senegalensis*), juveniles have a higher enzymatic activity than adults (Infante and Cahu, 1994; Martinez et al., 1999; Ribeiro et al., 1999; Wang et al., 2006). Kuz'mina (1996) also observed an abrupt increase in the activity of proteolytic enzymes in carnivorous pike at an early age. Fish diet is one of the factors that influence the activity of digestive enzymes (Fernandez et al., 2001) especially in species that undergo ontogenetic shifts in diet (German et al., 2004). Different authors (Katunzi et al., 2006; Mkumbo and Ligtvoet, 1992; Ogutu-Ohwayo, 2004; Schofield and Chapman, 1999) have reported that *L. niloticus* undergo an ontogenetic dietary shift. Katunzi et al. (2006) observed that in most habitats, juvenile *L. niloticus* < 5 cm exclusively fed on zooplankton before shifting to midge larvae and shrimp. Above 10 cm the main prey for Nile perch in most habitats was reported to be juvenile Nile perch (< 5 cm) and *R. argentea*. In this study it was observed that the specific activity of both carbohydrase and proteolytic enzymes was significantly higher in the 11 to 15 and 16 to 20 cm groups, coinciding with the change to a higher protein diet. This high protein diet necessitates an increase in the activity of proteolytic enzymes in order to degrade and absorb these nutrients. Furthermore, maturation of the intestinal tract and the relative increase in intestinal length increase the total capacity of the intestine to digest and assimilate nutrients (Ferraris et al., 1986). The lower enzyme activity in Groups 1 to 5 and 6 to 10 cm relates to immaturity of the digestive tract as compared to the larger groups. This increase in enzyme activity with stage of development has been reported in other species (Kuz'mina, 1996; Kvåle et al., 2007; Zouiten et al., 2008).

Results from this study reveal the presence of proteases and carbohydrase enzymes in the gut of the Nile perch. Presence of these enzymes show that the Nile perch is capable of utilizing artificial diets. The results further reveal that juvenile Nile perch have the highest capacity to utilize dietary proteins and carbohydrates at a total length of 11 to 20 cm. This shows that proteins and carbohydrates are critical at this stage of development and hence artificial diets should contain substantive amounts at this critical stage. Moreover, this stage has been characterized by a shift to a higher protein diet in the wild further showing that it is possible to wean the juveniles at this stage since they are capable of utilizing carbohydrates that make up much of the ingredients in aquaculture feeds on the market. Presence of maltase is an indicator that carbohydrates cannot be precluded from formulated diets for this species and would hence form a cost effective venture to the aqua-culturist. Research to try and spawn Nile perch in captivity is ongoing. When this induced spawning becomes successful, it will be important to study the activity of digestive enzymes in the larval stages since this stage is critical in determining survival.

ACKNOWLEDGEMENTS

This study was supported by a grant from The World Bank under the Millennium Science Initiative administered by the National Council for Science and Technology in Uganda. Special thanks go to The National Fisheries Resources Research Institute (NaFIRRI) and Kajjansi Aquaculture Research and Development Center for running the project. Appreciation also goes to OfraKedar and Zehava. Uni from the Hebrew University of Jerusalem for all the laboratory assistance provided.

REFERENCES

Ali H, Haque M, Chowdhury M, Shariful M (2009). In vitro protein digestibility of different feed ingredients in Thai koi (*Anabas testudineus*). J. Bangl. Agric. Univ. 7(1):205-210.

Benkenstein A (2011). Governance of Africa's Resources Programme Troubled Waters' Sustaining Uganda's Lake Victoria Nile Perch Fishery (Vol. 9). South Africa: The South African Institute of International Affairs.

Cahu C, Zambonino IJ, Peres A, Quazuguel P, Le Gall M (1998). Algal addition in sea bass *Dicentrarchus labrax* larvae rearing: Effect on digestive enzymes. Aquaculture 161(1):479-489.

http://dx.doi.org/10.1016/S0044-8486(97)00295-0

Caviedes-Vidal E, Afik D, Martinez del Rio C, Karasov WH (2000). Dietary modulation of intestinal enzymes of the house sparrow Passer domesticus: Testing an adaptive hypothesis. Comp. Biochem. Physiol. Part A: Mol. Integrat. Physiol. 125(1):11-24. http://dx.doi.org/10.1016/S1095-6433(99)00163-4

Chakrabarti I, Gani MA, Chaki K, Sur R, Misra K (1995). Digestive enzymes in 11 freshwater teleost fish species in relation to food habit and niche segregation. Comp. Biochem. Physiol. Part A: Physiol. 112(1):167-177. http://dx.doi.org/10.1016/0300-9629(95)00072-F

Creswell RL, Ohs CL, Miller CL (2010). Candidate Species for Florida Aquaculture: Atlantic Croaker, Micropogonias undulatus. University of Florida, Florida FA-148: 1-4. http://edis.ifas.ufl.edu/fa148

Debnath D, Pal A, Sahu N, Yengkokpam S, Baruah K, Choudhury D, Venkateshwarlu G (2007). Digestive enzymes and metabolic profile of< i> Labeo rohita</i> fingerlings fed diets with different crude protein levels. Comp. Biochem. Physiol. Part B: Biochem. Mol. Biol. 146(1):107-114. http://dx.doi.org/10.1016/j.cbpb.2006.09.008

Fernandez I, Moyano F, Dıaz M, Martınez T (2001). Characterization of α-amylase activity in five species of Mediterranean sparid fishes (Sparidae, Teleostei). J. Exp. Mar. Biol. Ecol. 262(1):1-12. http://dx.doi.org/10.1016/S0022-0981(01)00228-3

Ferraris RP, Catacutan MR, Mabelin RL, Jazul AP (1986). Digestibility in milkfish Chanos chanos (Forsskal): Effects of protein source, fish size and salinity. Aquaculture 59(2):93-105. http://dx.doi.org/10.1016/0044-8486(86)90123-7

German DP, Horn MH, Gawlicka A (2004). Digestive enzyme activities in herbivorous and carnivorous prickleback fishes (Teleostei: Stichaeidae): Ontogenetic, dietary, and phylogenetic effects. Physiol. Biochem. Zool. 77(5):789-804. http://dx.doi.org/10.1086/422228

Hakim Y, Rowland SJ, Guy JA, Mifsud C, Uni Z, Harpaz S (2007). Effects of genetic strain and holding facility on the characteristics of alkaline phosphatase and brush border enzymes in silver perch (Bidyanus bidyanus). Aquacult. Res. 38(4):361-372. http://dx.doi.org/10.1111/j.1365-2109.2007.01674.x

Harpaz S, Hakim Y, Barki A, Karplus I, Slosman T, Tufan Eroldogan O (2005a). Effects of different feeding levels during day and/or night on growth and brush-border enzyme activity in juvenile Lates calcarifer reared in freshwater re-circulating tanks. Aquaculture 248(1):325-335. http://dx.doi.org/10.1016/j.aquaculture.2005.04.033

Harpaz S, Hakim Y, Slosman T, Eroldogan OT (2005b). Effects of adding salt to the diet of Asian sea bass Lates calcarifer reared in fresh or salt water recirculating tanks, on growth and brush border enzyme activity. Aquaculture 248(1):315-324. http://dx.doi.org/10.1016/j.aquaculture.2005.03.007

Harpaz S, Uni Z (1999). Activity of intestinal mucosal brush border membrane enzymes in relation to the feeding habits of three aquaculture fish species. Comp. Biochem. Physiol. Part A: Mol. Integrat. Physiol. 124(2):155-160. http://dx.doi.org/10.1016/S1095-6433(99)00106-3

Hepher B (1988). Nutrition of pond fishes: Cambridge University Press. pp 281-289. http://f3.tiera.ru/1/genesis/680-684/682000/7dbcab8dda1a0448f2c4eed17efcff50

Hidalgo M, Urea E, Sanz A (1999). Comparative study of digestive enzymes in fish with different nutritional habits: Proteolytic and amylase activities. Aquaculture 170(3):267-283. http://dx.doi.org/10.1016/S0044-8486(98)00413-X

Horn M (2005). Feeding and digestion (3 Ed.). Boca Raton, FL, USA: CRC press.

Hughes NF (1992). Nile perch, *Lates niloticus*, predation on the freshwater prawn, Caridina nilotica, in the Nyanza Gulf, Lake Victoria, East Africa. Environ. Biol. Fishes 33(3):307-309. http://dx.doi.org/10.1007/BF00005874

Infante JZ, Cahu C (1994). Development and response to a diet change of some digestive enzymes in sea bass (*Dicentrarchus labrax*) larvae. Fish Physiol. Biochem. 12(5):399-408. http://dx.doi.org/10.1007/BF00004304

Katunzi E, Van Densen W, Wanink J, Witte F (2006). Spatial and seasonal patterns in the feeding habits of juvenile *Lates niloticus* (L.), in the Mwanza Gulf of Lake Victoria. Hydrobiologia 568(1): 121-133. http://dx.doi.org/10.1007/s10750-006-0033-3

Klein S, Cohn S, Alpers D (1998). The alimentary tract in nutrition (9th ed.). Baltimore, USA: Lippincott Williams and Wilkins. pp. 605-629.

Kolkovski S (2001). Digestive enzymes in fish larvae and juveniles - Implications and applications to formulated diets. Aquaculture 200(1):181-201. http://dx.doi.org/10.1016/S0044-8486(01)00700-1

Kuz'mina V (1996). Influence of age on digestive enzyme activity in some freshwater teleosts. Aquaculture 148(1):25-37. http://dx.doi.org/10.1016/S0044-8486(96)01370-1

Kvåle A, Mangor-Jensen A, Moren M, Espe M, Hamre K (2007). Development and characterisation of some intestinal enzymes in Atlantic cod (*Gadus morhua* L.) and Atlantic halibut (*Hippoglossus hippoglossus* L.) larvae. Aquaculture 264(1):457-468. http://dx.doi.org/10.1016/j.aquaculture.2006.12.024

Lovett DL, Felder DL (1989). Ontogeny of gut morphology in the white shrimp *Penaeus setiferus* (Decapoda, Penaeidae). J. Morphol. 201(3):253-272. http://dx.doi.org/10.1002/jmor.1052010305

LVFO (2008). Fisheries Management Plan II for Lake Victoria 2009-2014. http://www.lvfo.org/index.php/component/docman/doc_details/40-fisheries-management-plan-ii-for-lake-victoria-2009-2014?tmpl=componentandItemid=95

Martinez I, Moyano F, Fernández-Díaz C, Yúfera M (1999). Digestive enzyme activity during larval development of the Senegal sole (Solea senegalensis). Fish Physiol. Biochem. 21(4):317-323. http://dx.doi.org/10.1023/A:1007802708459

Matsuishi T, Muhoozi L, Mkumbo O, Budeba Y, Njiru M, Asila A, Othina A, Cowx I (2006). Are the exploitation pressures on the Nile perch fisheries resources of Lake Victoria a cause for concern? Fisher. Manage. Ecol. 13(1):53-71. http://dx.doi.org/10.1111/j.1365-2400.2006.00477.x

Mkumbo O, Ligtvoet W (1992). Changes in the diet of Nile perch, *Lates niloticus* (L), in the Mwanza Gulf, Lake Victoria. Hydrobiologia, 232(1):79-83. http://dx.doi.org/10.1007/BF00014615

Muhoozi L, Nzungi P, Othina A, Lyimo E (2005). Trends in exploitation of the fisheries resources. In: the state of the fisheries resources of Lake Victoria and their Management. Paper presented at the Entebbe regional stakeholders' conference, Entebbe Uganda. http://www.lvfo.org/index.php/component/docman/cat_view/5-conference-reports?Itemid=95

Namulawa V, Kato C, Rutaisire J, Britz P, Beukes N, Pletschke B, Whiteley C (2013). Enzyme activity in the Nile perch gut: Implications to Nile perch culture. Int. J. Fisher. Aquacult. 5(9):221-228.

Ogutu-Ohwayo R, Kirema-Mukasa C (2005). The Regional Role of LVFO in promoting sustainable utilization. Paper presented at the FAO Expert Consultation on Fish Technology in Africa, Bagamoyo, Tanzania.

Ogutu-Ohwayo R (1993). The effects of predation by Nile perch, *Lates niloticus* L., on the fish of Lake Nabugabo, with suggestions for conservation of endangered endemic cichlids. Conserv. Biol. 7(3):701-711. http://dx.doi.org/10.1046/j.1523-1739.1993.07030701.x

Ogutu-Ohwayo R (2004). Management of the Nile perch, *Lates niloticus* fishery in Lake Victoria in light of the changes in its life history characteristics. Afr. J. Ecol. 42(4):306-314. http://dx.doi.org/10.1111/j.1365-2028.2004.00527.x

Paterson J, Chapman L (2009). Fishing down and fishing hard: ecological change in the Nile perch of Lake Nabugabo, Uganda. Ecol. Freshw. Fish 18(3):380-394. http://dx.doi.org/10.1111/j.1600-0633.2009.00355.x

Péres A, Infante JZ, Cahu C (1998). Dietary regulation of activities and mRNA levels of trypsin and amylase in sea bass (*Dicentrarchus labrax*) larvae. Fish Physiol. Biochem. 19(2):145-152. http://dx.doi.org/10.1023/A:1007775501340

Ribeiro L, Zambonino-Infante J, Cahu C, Dinis M (1999). Development of digestive enzymes in larvae of Solea senegalensis, Kaup 1858. Aquaculture 179(1):465-473. http://dx.doi.org/10.1016/S0044-8486(99)00180-5

Sabapathy U, Teo L (1993). A quantitative study of some digestive enzymes in the rabbitfish, *Siganus canaliculatus* and the sea bass, Lates calcarifer. J. Fish Biol. 42(4): 595-602.

http://dx.doi.org/10.1111/j.1095-8649.1993.tb00362.x

Sabat P, Novoa F, Bozinovic F, Martínez del Rio C (1998). Dietary flexibility and intestinal plasticity in birds: A field and laboratory study. Physiol. Biochem. Zool. 71(2):226-236.

Schofield PJ, Chapman LJ (1999). Interactions between Nile perch, *Lates niloticus*, and other fishes in Lake Nabugabo, Uganda. Environ. Biol. Fishes 55(4):343-358. http://dx.doi.org/10.1023/A:1007544017989

Semenza G (1986). Anchoring and biosynthesis of stalked brush border membrane proteins: Glycosidases and peptidases of enterocytes and renal tubuli. Annu. Rev. Cell Biol. 2(1):255-307. http://dx.doi.org/10.1146/annurev.cellbio.2.1.255

Tengjaroenkul B, Smith BJ, Caceci T, Smith SA (2000). Distribution of intestinal enzyme activities along the intestinal tract of cultured Nile tilapia, *Oreochromis niloticus* L. Aquaculture 182(3):317-327. http://dx.doi.org/10.1016/S0044-8486(99)00270-7

Tibaldi E, Hakim Y, Uni Z, Tulli F, de Francesco M, Luzzana U, Harpaz S (2006). Effects of the partial substitution of dietary fish meal by differently processed soybean meals on growth performance, nutrient digestibility and activity of intestinal brush border enzymes in the European sea bass (*Dicentrarchus labrax*). Aquaculture 261(1):182-193. http://dx.doi.org/10.1016/j.aquaculture.2006.06.026

Vega-Orellana OM, Fracalossi DM, Sugai JK (2006). Dourado *Salminus brasiliensis* larviculture: Weaning and ontogenetic development of digestive proteinases. Aquaculture 252(2):484-493. http://dx.doi.org/10.1016/j.aquaculture.2005.07.002

Wang C, Xie S, Zhu X, Lei W, Yang Y, Liu J (2006). Effects of age and dietary protein level on digestive enzyme activity and gene expression of Pelteobagrus fulvidraco larvae. Aquaculture 254(1):554-562. http://dx.doi.org/10.1016/j.aquaculture.2005.11.036

Zouiten D, Khemis IB, Besbes R, Cahu C (2008). Ontogeny of the digestive tract of thick lipped grey mullet (Chelon labrosus) larvae reared in "mesocosms". Aquaculture 279(1):166-172. http://dx.doi.org/10.1016/j.aquaculture.2008.03.039

Contribution of small water bodies and small-holder aquaculture towards poverty alleviation and enhancing household food security in Zambia

Confred G. Musuka and Felix F. Musonda

School of Natural Resources, Copperbelt University, P. O. Box 21692, Kitwe, Zambia.

A study was conducted between January and May 2011, through administration of questionnaires and personal interviews in 6 of the 10 provinces of Zambia (namely: Lusaka, Eastern, Central, North Western, Copperbelt, and Southern) in order to assess the contribution of small water bodies (SWBs) and small-holder aquaculture towards poverty alleviation and enhancing household food security. Using simple random sampling method, a total of 120 respondents were selected. Results indicated that, a total of 1,082 SWBs were constructed throughout the country covering a total area of 5,410 ha. Most of them were constructed for irrigation purposes and as sources of drinking water for livestock. At the same time, aquaculture adoption at household and intra household levels had increased by 4.7% in all the districts over the years. There were more than 11,327 small-holder farmers who owned 21,658 ponds, covering 578.86 ha of land. There was also a 10% increase in number of learning institutions and the hospitality industries that had adopted aquaculture activities either for recreational or learning purposes. Existing statistics indicated that, there were 11,327 small-holder farmers who owned 21,658 ponds, covering 578.86 ha of land with an annual fish production of 3,985.16 metric tonnes, while that from SWBs stood at 2,705 metric tons. It was however noted that, the majority of these farmers (65%) produced less than 0.5 tons of fish per hectare/year, which was considered to be very low. However, fish production from SWBs remained almost unchanged because extension support had remained very inadequate and the designed programme to enhance productivity in community small-water bodies was not being implemented. Most of the fish harvested comprised mainly of: *Oreochromis andersonii, Oreochromis macrochir*, and *Tilapia rendalli*, which were readily acceptable to the consumers.The study also revealed that, adoption of small-holder aquaculture helped in poverty alleviation, improved rural household food security and better nutritional status compared to non-fish farming families. Most of the small-holder farmers cultivated various agricultural crops through irrigation and were also involved in livestock rearing, from which extra income was realized.

Key words: Contribution, small-water bodies, small-holder aquaculture, poverty alleviation, food security, Zambia.

INTRODUCTION

Poverty levels in Zambian are very high especially among the rural communities. This can largely be attributed to the economic downturn of the 1980s and early 1990s when Copper prices plummeted to their lowest levels ever, causing most citizens to become poverty stricken. According to CSO (2009) and Aquaculture Development Plan (ADP) 2008 – 2011 (2009), the country's poverty levels stood at 64% of the population, of which the

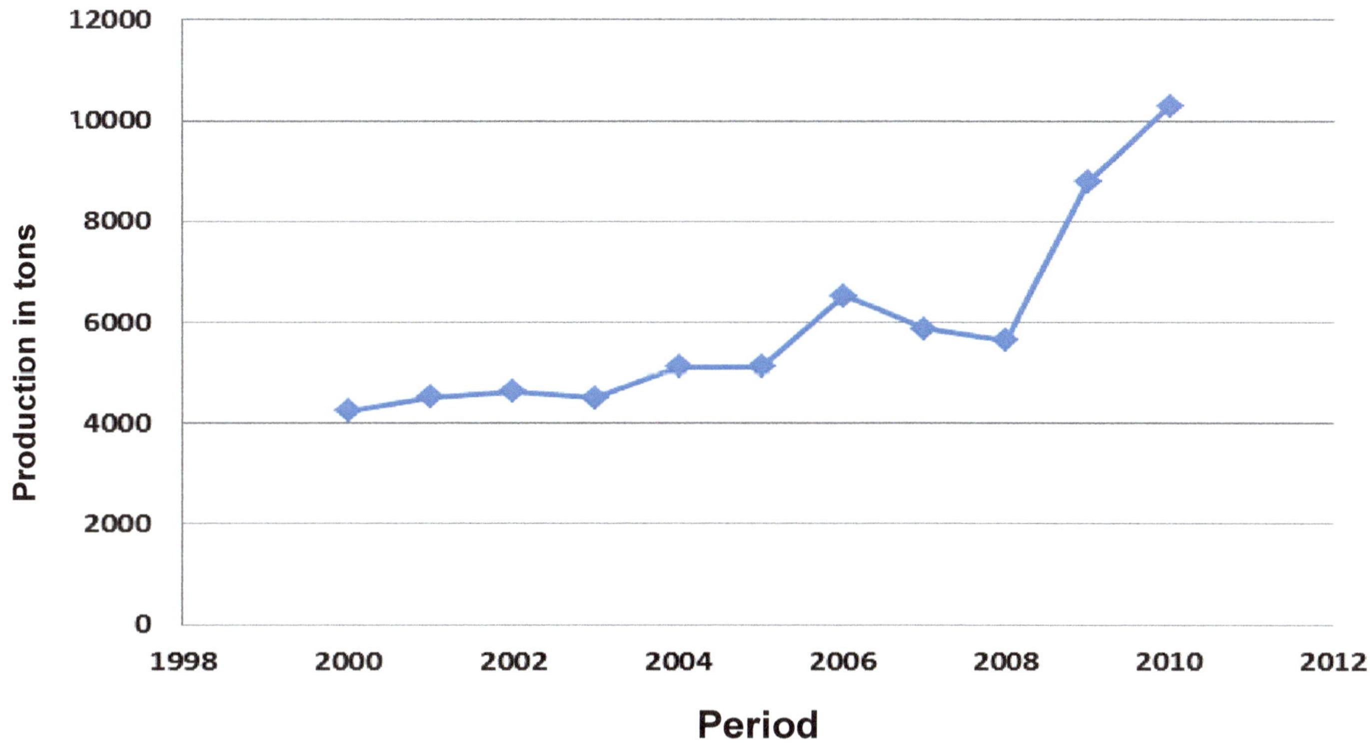

Figure 1. National Aquaculture production in Zambia (2000 to 2010) in metric tons (Source: DoF).

majority were those residing in rural areas. However, in 2010, the Civil Society for Poverty Reduction (CSPR) also observed that, poverty levels in Zambia were still on the increase, with an average of 85% of the 7,978,274 people who resided in the rural areas and 34% of the 5, 068, 234 in the urban areas were still living under the poverty datum line (Lusaka times.com 2010 and CSO report, 2010). Poverty and food insecurity in Zambia generally stem from over-reliance on rain-fed agriculture, and associated effects of frequent unfavorable climatic conditions, along with inadequate incomes, access to markets and transport facilities to enable the transfer or purchase of food, compounded by low economic diversification into sectors such as fisheries that could supplement crop production (Musumali et al., 2009).

The fisheries sector contributed between US$51 and 135 000 000 per annum to GDP over the period of 2002 to 2007, averaging around 1.24% of GDP at current prices (Musumali et al., 2009). Zambia has the potential of supplying 70 metric tons of fish from the natural fishery per annum (The New Nation, 2009). Fish makes up 40% of animal protein in the diet of Zambians (ACF/FSRP, 2009). It was undisputable that, the demand for fish in Zambia was very overwhelming and almost of all the fish produced in the country was intended for human consumption. In fact, the demand for fish by the local market was in excess of 100,000 tons. As such, there was a shortage of over 30,000 tons, which was being supplemented through fish imports from Asia and neighboring countries. Fish imports from Mozambique (Kapenta), Namibia (Horse Mackerel), Zimbabwe (Tilapia) and China (Tilapia) have gradually increased in order to meet the ever increasing demand (DoF, 2010; Musumali et al., 2009). More than 50% of fish imports, especially those from China were gutted and filleted fresh tilapias (breams), packaged in 50 kg boxes.

Aquaculture or fish farming as was popularly known in Zambia, was in the range of extensive to more intensive (commercial) and integrated systems relying on organic manure and formulated feeds (ACF/FSRP, 2009). Production from aquaculture was between 7,000 and 10,000 tons (DoF, 2010), although Aquaculture Development Plan (ADP) 2008 – 2011 (2009) estimated that, only 4,549 tons valued at US $20.9 000 000 per annum was what was produced, which was way below their potential. The sector was reported to only represent 8.8% of total fish production; a level that was not sufficient to arrest Zambia's declining national fish per capita consumption of 5.8 kg (Figure 1).

There were however, enormous opportunities and potential to increasing production 10-fold (Table 1) in existing aquaculture facilities (that is, ponds, community small water bodies (SWBs) and cages) by expanding into new areas (DoF, 2010). It was estimated that, 40 000 ha in Zambia could be brought under pond culture and further million hectares of land from lake areas was

Table 1. Potential fish pond development sites by Province.

Province	Fish pond (ha)	Location	Inlet	Water M^3/day
Central	1.400	Luswishi River	Pump	97.000
Copperbelt	1.200	Run-off Machiya River	Pump	80.000
Eastern	8.000	Luangwa River, Petauke	Pump	581.000
Luapula	4.105	Samfya and Bangweulu Lake	Gravity	273.000
Lusaka	800	Chanyanya	Gravity	53.600
Northern	2.250	Mutale-mukonge, Chandamali	Pump	150.000
Northwestern	4.140	42 dispersed small scale areas on tributaries of Kabompo River	Weir pump	275.000
Southern	15.875	Large scale development on Kafue River	Gravity	1,097.000
Western	1.790	179 Dispersed small scale areas	Seepage	124,000
Total	39.560	165 Locations		2,730.600

Source: DoF (2010).

potentially suitable for aquaculture use DoF (2010).

MATERIALS AND METHODS

A study was conducted between January and May 2011 in 6 of the 10 provinces of Zambia (namely: Lusaka, Eastern, Central, North Western, Copperbelt, and Southern) in order to assess the contribution of SWBs and small-holder aquaculture towards poverty alleviation and enhancing household food security. Using simple random sampling method, a total of 120 respondents (90 males and 30 females), thus 25% of the total population of small-holder farmers were randomly selected for the study. A representative sample for the study from each fish farming area was based on Boyd's formula as follows:

$$n/N \times 100 = C$$

Where C-represents a figure greater or equal to 5% of the fish farmer population, N-overall population (total number of Ffsh farmers), n-number of selected fish farmers (Sample size) as by Boyd et al. (1981), required sample size: $n/N \times 100 = C$, $n = 480 \times 25/100$, $n = 120$.

Data collection from the field was obtained using semi-structured questionnaires, which were administered to randomly sampled respondents with the help of fisheries officers. In addition, data was gathered from focus group discussions and participant observation. This assisted in capturing some extra information that was not captured by the questionnaires. The major data items collected included the following: aquaculture production statistics and collection methods used; fish feed/seed availability and cost and available capacity building programmes. Statistical Package for Social Sciences (SPSS) 12.0 was used to analyze data from the field. Being categorical data, descriptive statistic were done to analyze the data and summarized in tables and figures using Microsoft Excel.

RESULTS

Figures 2, 3, and 4 show the distribution, size, and production from SWBs in each of the provinces, while Table 2 gives a summary of small-holder aquaculture national statistics, and Figure 5 shows small-holder aquaculture production in metric tons.

DISCUSSION

The study observed that a total of 1,082 SWBs were constructed throughout the country covering a total area of 5,410 ha (Figure 2). Most of these SWBs were constructed for irrigation purposes and as sources of drinking water for livestock. Southern Province had the highest number of SWBs in the country with over 700 recorded (Figure 2), representing an estimated 3,500 ha (Figure 3), although a report by "The New Nation" of February, 2009, put the figure at 800 dams and slightly above 250 earthen fish ponds. The SWBs was stocked with fish, whose annual production was 2,705 metric tons, with Southern province still taking the lead at 1,750 metric tons and Luapula produced only 7.5 metric tons (Figure 4). During the study, the authors learnt that, the Department of Fisheries in Southern Province undertook fish restocking programme of Cichlids in 2010 under the Poverty Reduction Programme (PRP), where a total of 36 reservoirs were stocked with 22,299 fingerlings. According to Mudenda (2006) the stocking rate of 1.5 fish/m2 of small water body was recommended.

An earlier report by Mudenda et al. (2005) revealed that, fisheries management of community reservoirs was almost non-existent in most areas where it was introduced. The authors called for urgent measures to be put in place to redress the situation. However, during this study, it was established that, in order to sustain fishing, especially in communally owned reservoirs, committees were formed to provide a suitable management system required to control fishing activities and also to discourage the rampant use of mosquito nets that had far reaching consequences in harvesting fish stocks.

The bulk of aquaculture in Zambia was concentrated in the small-holder category, which relied on family labour

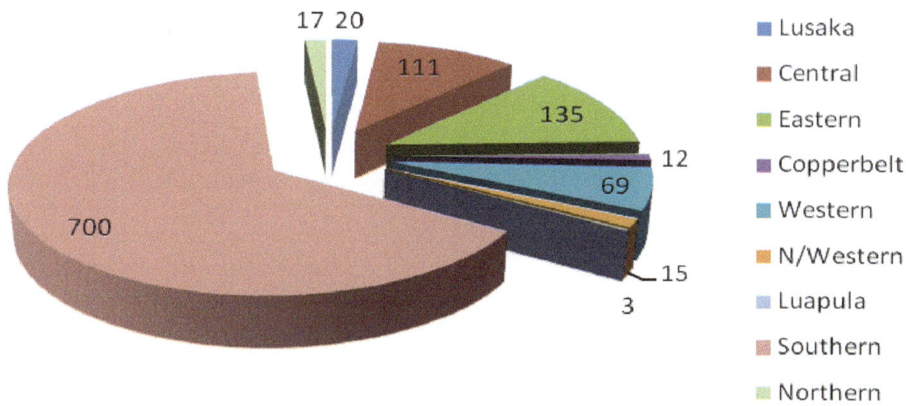

Figure 2. Distribution of SWBs per province.

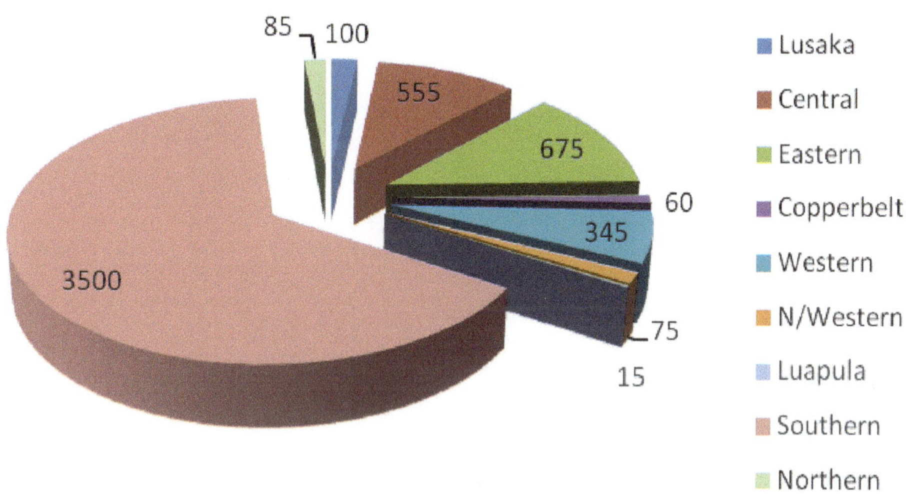

Figure 3. Coverage area in hectares (ha) of SWBs.

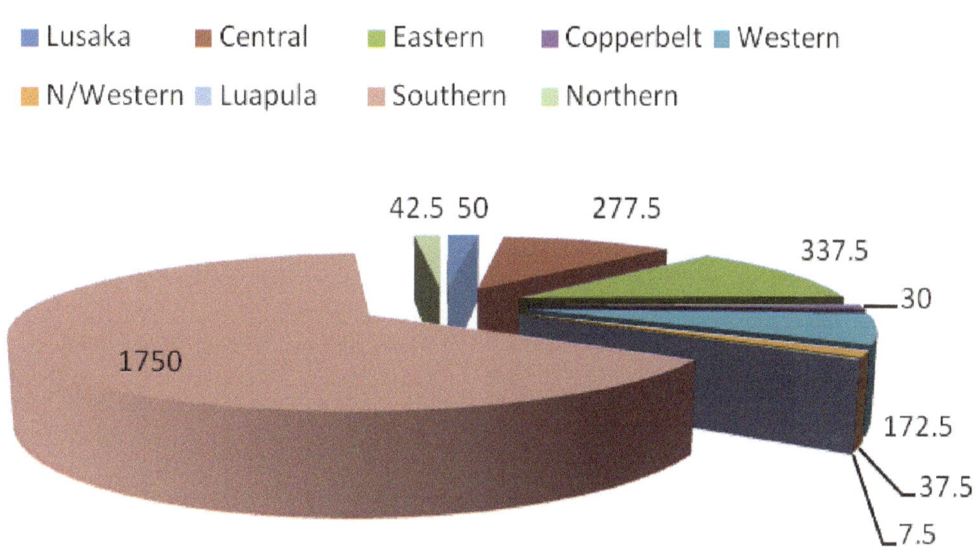

Figure 4. Fish Production in metric tons from small water bodies.

Table 2. Summary of small-holder aquaculture national statistics.

Province	Number of fish farmers	No. of ponds	Area of ponds (ha)	Prod (tons)
Central	698	925	19.69	133.3
Copper belt	597	1.436	54.07	365.96
Eastern	1.826	5.252	158.00	1125.75
Luapula	789	1.274	37.92	256.63
Lusaka	254	1.305	22.00	159.66
Northwestern	2483	4134	99.82	675.68
Northern	3771	6.457	172.07	1164.71
Southern	226	304	4.92	33.27
Western	683	571	10.37	70.2
Total	11.327	21658	578.86	3,985.16

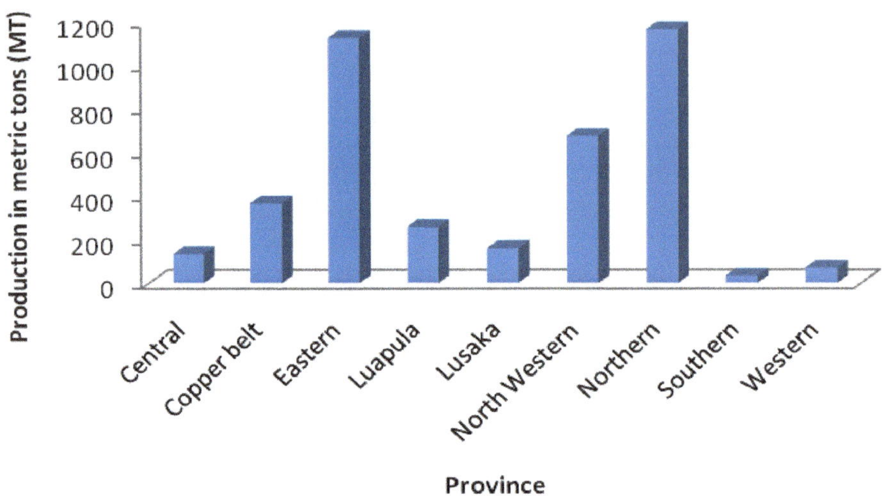

Figure 5. Small-holder aquaculture production (MT) per province.

and practised extensive culture too (ACF/FSRP, 2009). Most of them, if not all, cultivatedvarious agricultural crops and were also involved in livestock rearing. Small- holder ponds and SWBs were the main source of irrigation water for other agriculture produce, from which extra income was realized. The adoption of small-holder aquaculture helped in improving rural household food security and better nutritional status compared to non-fish farming families. Machena and Moehl (2001) further observed that, small-scale systems tended to be "rural", if not in location then in the sense that they did not rely upon urban markets for their product, with most consumed by the family or sold on the pond bank. Mudenda (2006) also reported that, small-scale farmers generally produced fish for consumption and very few of them produced for the market. In such situations farmers did not sell fish on weight basis using a balance or scale but on numbers. The author further reported that, most fish farmers did not keep performance records due to low education levels, they harvested intermittently for household consumption and sold what was produced to the neighbourhood at the end of the season while keeping some fish ponds as "banks" where they only went to harvest when need arose. The study further revealed that, with the availability of disposable income, there was a drastic reduction in poverty levels among the small-holder aquaculture communities, which in other words implied them being food secure. Food security exists when all people, at all times, have physical, social and economic access to sufficient, safe and nutritious food that meets their dietary needs and food preferences for an active and healthy life (SPFS, 2003).

In Zambia, small-holder aquaculture contributed significantly to household food security through provision of cheap fish as food, nutrition (protein), income, diversification of rural livelihood and employment generation, thus stimulating the rural economy (Mudenda, 2006; DoF, 2010). Many related institutions such as Non governmental organizations (NGOs) collaborated very well with the department of fisheries in ensuring that, the sector improved household food and nutrition security for vulnerable groups through the promotion of integrated group and livestock and fisheries production by marginalized people living under a dollar per day. Such kind of collaboration significantly contributed to the local and national economy, and provided goods and services that enhanced the social well-being for the current and future generations without compromising environmental integrity by 2030 (DoF, 2010). This was contrary to IBRD/World Bank (2006) report, which indicated that, aquaculture provided only 2% of the Sub-Saharan Africa region's supply and made only a minor contribution to economic growth, employment and foreign exchange. The authors noted that, the number of farmers taking up small-holder aquaculture in various districts had increased by 4.7% over the years; the majority of whom were retirees, youths and women's groups. The study also noted a 10% increase in number of learning institutions and the hospitality industries that had adopted aquaculture activities either for recreational or learning purposes. Existing statistics indicated that there were 11,327 small-holder aquaculturists who owned 21,658 ponds, covering 578.86 ha of land (Table 2) with an approximate total fish production of 3,985.16 metric tonnes (Figure 5). Fish harvested comprised mainly: *O. andersonii, O. macrochir*, and *T. rendalli.*

The males dominate the aquaculture sector as owners of fish farming enterprises and as workers (Mudenda et al., 2005). The study observed that, males constituted 81% while females made up a partly 19% of the existing number of small-holder farmers. This study also revealed that, 85% of the farmers were married with 8 to 10 children while the other 15% relied on either hired labour or dependants that lived with them. It can further be inferred that 52% of the existing small-holder farmers were above 50 years old. Most of the new entrants in the sector had education standards above secondary school level. For example, in Lusaka province alone, 89.5% of the respondents had attained tertiary level of education while the remainder had either primary or secondary school level of education.

It was however noted that, the majority of these farmers (65%) produced less than 0.5 tons of fish per hectare/year, which was considered to be very low. DoF (2010) also reported a reduction in production of 3.5% coming from the small-holder farmers in the same period although the overall aquaculture production and numbers of small scale fish farmers was increasing. A similar situation was observed in Malawi where 80% of farmers produced less than 18 kg of fish per annum (Britz and Hecht, 2005).

Fluctuations in production could be attributed to the attitude of farmers and government policy (Kefi et al., 2010). The fisheries law was to some extent very weak and did not in any way compel farmers to submit data on their production activities to the department of fisheries on their own resulting in disjointed or inadequate information regarding the sector. At the same time, farmers in most cases were not honest enough to disclose how much fish they were producing to avoid paying tax to the government. The situation was further compounded by shortage of qualified personnel in the department of fisheries with statistical skills.

Most small-holder producers abandoned production due to diminishing enthusiasm and in some cases produced seasonally in spite of available suitable natural conditions. The situation was also attributed to inadequate extension delivery during the year under review as a result of both reduced numbers of extension staff and zero release of funds. In most cases, small- holder farmers perceived aquaculture to be a secondary economic activity coming after crops and livestock rearing. They alleged that, aquaculture was not productive and created an impression that it was just a by-the-way activity. As a result of that notion they attended to it when they were done with their mainstream agricultural activities that were equally very involving. In some cases, aquaculture activities were relegated to other family members like women and children while men attended to what was perceived to be very serious activities.

ACF/FSRP (2009) reported that, long standing constraints in aquaculture included: inadequate extension services, lack of comprehensive training packages and materials, chronic shortages of quality fish seed and fingerlings, high cost of fish feed, and poor marketing support. Utsugi and Mazingaliwa (2002) and Mudenda et al. (2005) identified the following as the main constraints to the development of aquaculture in Zambia: inadequate quality fingerlings; insufficient animal manure, lack of affordable fish feed; lack of appropriate technology; poor rural infrastructure; lack of marketing strategy; insufficient comprehensive extension packages; insufficient extension staff; lack of data centre; inadequate operational fund for research; insufficient donor support to 10 aquaculture research station; untimely government support after project support gets discontinued; and weak research extension link. The sector's heavy reliance on hand-outs was cited as not being sustainable. Whenever there was a project aquaculture thrived but as soon as that project ended the activities declined or stopped completely. Most farmers never felt as owners of their own activities since they suffered from donor fatigue. Such successive failures discouraged potential investors and caused delayed development (DoF, 2010).

Similarly, IBRD/The World Bank (2006), Simumba (2007) and ACF/FSRP (2009) observed that, past aquaculture development efforts largely failed because of weak institutions, poor access to finance, and a heavy reliance on failing government extension services and seed production. The focus on subsistence aquaculture may have been misguided, because it often lacked the driving force of market demand and impetus provided by commercial reality (IBRD/The World Bank, 2006). Another study conducted to identify the constraints to aquaculture in Akwa Ibom State, Nigeria, Akpabio and Inyang (2007) identified lack of high yielding fingerings as one of the production constraint to fish production. Meanwhile, Kefi et al. (2010) reported that, aquaculture would be helped if those factors that derailed it were given attention by the government and by the farmers themselves.

According to FAO (2006), aquaculture management seeks to create an environment which will encourage private sector led growth so as to achieve the objectives of increased total fish production, increase per capita consumption of fish, improving nutrition and to increase income made. For example, in 2009, the government launched a five years (2010 to 2015) Sixth National Aquaculture Development Plan (SNDP) that was more precise in describing specific priority systems to promote with cluster of operators in specific high potential zones. The plan's immediate outcome was: investments in economically, socially and environmentally sound aqua businesses. These enterprises were in turn expected to contribute to a diversification of livehoods, improved nutritional status, increase income generating capacity,as well as improved employment opportunities.

The authors learnt that, the project selected fish farmers who received technical support in form of pond stocking with all-male sex-reversed indigenous *Oreochromis andersonii* produced by Revendel Tilapia

Enterprise hatchery at 5 fish/m^2. *O. andersonii* was selected on the basis of it being indigenous as well as its suitability for commercial aquaculture production in ponds because of its superior quality in terms of feeding, breeding habits, attractiveness, growth and hardiness. The fish were fed with: fry booster (1 to 4 weeks), starter crumble (at 4 weeks) and grower (from 3 months up to harvest). Lessons learnt from the project were expected to be replicated countrywide, implying that, the strategy of small-holder farming clusters was likely to impact positively in other parts of the nation.

Conclusion

The authors noted an increase in number of farmers taking up small-holder aquaculture throughout the country. Aquaculture activities were also being done at community level, thus empowering an entire community. The majority of these farmers (65%) however, produced less than 0.5 tons of fish per hectare/year, which was considered to be very low; hence the need to upscale their production in order to realize some profits from the sector.

Some producers abandoned production due to inadequate extension delivery and diminishing enthusiasm. In some cases they produced seasonally in spite of available suitable natural conditions. Others perceived aquaculture not to be very productive and viewed it as a secondary economic activity coming after crops and livestock rearing. As a result of that notion, they attended to it when they were done with their mainstream agricultural activities that were equally very involving.

Enhanced statistical data collection on aquaculture required to be improved for the better. At the same time, the government should quickly deal with land use conflicts that have been arising as farmers in rural areas grow their fish.

The adoption of small-holder aquaculture helped in improving rural household food security and better nutritional status compared to non-fish farming families. With the availability of disposable income, most families were able:

(a) To pay school fees for their children
(b) To construct decent houses with roofing sheets
(c) To buy bicycles to ease their mobility
(d) In some cases owned animals, which they used for ploughing their fields.

ACKNOWLEDGEMENTS

The authors wish to acknowledge financial and material support rendered to the team by the Copperbelt University without which this project would not have been undertaken. Special thanks to go to all members of staff in the department of fisheries throughout the country and all those who helped us in one way or another.

REFERENCES

ACF/FSRP (2009). The Status of fish population in Zambia's water bodies.

Akpabio IA, Inyang EB (2007). Major constraints affecting aquaculture development in Akwa Ibom State, Nigeria. Afr. J. Aquat. Sci. 32:45-50.

Aquaculture Development Plan (ADP) 2008 – 2011 (2009). Supporting Appropriate, Practical and Tested Business Models for Fish Farming. Lusaka, Zambia.

Britz P, Hecht T (2005). The Outlook for Aquaculture Development in Southern Africa.

Central Statistics Office (CSO) (2009). Projected mid – year population 2000 – 2009. Retrieved on 15th November 2012 from http.www.zamstats.gov.zm.

Central Statistics Office (CSO) (2010). 2010 Census of population and housing preliminary report.

DoF (2010). National Aquaculture Development Plan. Overcoming the slow growth of Aquaculture in Zambia 2010-2015.

FAO (2006). Fisheries and Aquaculture Zambia, Profiles home. Food and Agriculture Organization of the United Nations, Rome.

IBRD/The W orld Bank (2006). Aquaculture: Changing the Face of the Waters Meeting the Promise and Challenge of Sustainable Aquaculture. In Press: http://siteresources.worldbank.org/INTARD/Resources/Aquaculture_ESW _vGDP.pdf. Report No. 36622 – GLB.

Kefi AS, Chibbamulilo P, Musuka CG, Mazingaliwa K (Posthumous), Muyangali K, Mpaka C, Mbewe AG, Mutanuka E, Chilala AD, Simumba D, Nambala S, Mutale B, Masuoka M (2010). A case study on the Constraints to Aquaculture Production in Zambia. SARNISSA. www.sarnissa.org.

Lusaka Times.com (2010). Poverty levels in Zambia are still on the increase-CSPR.

Machena C, Moehl J (2001). Sub-Saharan African aquaculture: regional summary. In R.P. Subasinghe, P. Bueno, M.J. Phillips, C. Hough, S.E. McGladdery & J.R. Arthur, eds. Aquaculture in the Third Millennium. Technical Proceedings of the Conference on Aquaculture in the Third Millennium, Bangkok, Thailand, 20-25 February 2000. NACA, Bangkok and FAO, Rome. pp. 341-355.
Mudenda S, Mulenga L, Mubanga M (2005). Intesification and Commercialization of Aquaculture in ASP Areas.
Mudenda CG (2006). "Economic Perspectives of Aquaculture Development Strategy of Zambia" consultant report, Development Consultant, Lusaka Zambia.
Simumba D (2007). Assessment of Agricultural Information needs in Africa, Caribbean and Pacific (ACP). Southern Africa. Project: 4-7-41-209-6/f. (ACP-EU) Technical Centre for Agricultural and Rural Cooperation (CTA).
Musumali MM, Heck S, Husken SMC, W ishart M (2009). Fisheries in Zambia: An undervalued contributor to poverty reduction. The WorldFish Center/The world Bank. Policy Brief 1913.
Utsugi K, Mazingaliwa K (2002). Field Guide to Zambian Fishes, Planktons and Aquaculture. JICA.
SPFS (2003). The Special Programme fcr Food Security of the United Nations. FAO, Rome.
The New Nation (2009). Fish farming enhanced in Southern Province.

9

Genetic relationships between captive and wild subpopulations of *Arapaima gigas* (Schinz, in Cuvier, 1822)

Carlos Henrique dos Anjos dos Santos[1,2], Carolina Sousa de Sá Leitão[2], Maria de Nazaré Paula e Silva[2] and Vera Maria Fonseca de Almeida e Val[1,2]

[1]Universidade Nilton Lins (UNINILTONLINS), Laboratório de Genética Aplicada à Aquicultura and Biologia Molecular, Avenida Professor Nilton Lins, Flores, 69058-030, Manaus, Amazonas, Brasil.
[2]Instituto Nacional de Pesquisas da Amazônia (INPA), Laboratório de Ecofisiologia e Evolução Molecular (LEEM), Avenida André Araújo, 2936, Aleixo, 69060-001, Manaus, Amazonas, Brasil.

***Arapaima gigas*, also known as pirarucu, is endemic to the Amazon basin. There is currently considerable effort being made to cultivate this species to reduce pressure on natural populations. We characterized the diversity and genetic structure of subpopulations of wild and captive *A. gigas* based on 19 microsatellite loci. Captive subpopulations of *A. gigas* exhibited less diversity than wild individuals. We also verified the existence of outlier loci under selective pressure in both subpopulations, indicating the occurrence of local adaptation in some individuals of *A. gigas*. Furthermore, we identified a high-level genetic structure among the subpopulations, with no mixing between subpopulations. The lack of mixing between populations, the local adaptation, and the genetic structure indicate that these subpopulations should be managed and isolated to prevent captive individuals from escaping into the wild, which would reduce the diversity of the wild subpopulations. We suggest a program for the genetic management of captive subpopulations to avoid the random loss of genetic variability and the selection of characteristics that are undesirable for commercial and conservation purposes.**

Key words: Allelic diversity, osteoglossidae, conservation and management, heterezygosity, osteoglossiformes.

INTRODUCTION

The future of aquaculture largely depends upon the effective management of the genetic resources of farmed fish. Maintaining an appropriate level of genetic diversity is critical to aquaculture production and value, and is necessary to ensure the availability of affordable products for human consumption in poor areas worldwide. Unfortunately, most farmed fish species in Brazil have not been fully domesticated, and their wild relatives are preferentially used for breeding programs and related research. Furthermore, there are many species (possibly hundreds) that might have potential for aquaculture (Pullin, 2007), including *Arapaima gigas*, which is a potential candidate for Brazilian aquaculture, particularly in Northern Brazil.

A. gigas (pirarucu) is endemic to the Amazon basin and is the longest freshwater fish in the world, reaching 3 m in length and more than 200 kg in weight (Castello, 2004). Because of its high economic importance in the fisheries of Northern Brazil, *A. gigas* has recently been introduced into fish farms to facilitate its reproduction and rearing in captivity. This effort has increased in recent years due to the ability of *A. gigas* to adapt to a confined environment and improvements in facility management. Because *A. gigas* is the largest freshwater scaled fish in the world, it has high commercial value and has been overexploited for more than two decades (Viana et al., 2007). Near the principal cities of the Brazilian Amazon, *A. gigas* is considered commercially extinct, and fishermen have reported difficulties in catching *A. gigas* in certain areas (Goulding, 1980; Castello, 2008; Oliveira et al., 2005).

The International Union for the Conservation of Nature (IUCN) has included *A. gigas* on the red list of endangered species (Ortiz et al., 2007) and classified it in Appendix II of the Convention on the International Trade of Endangered Species of Wild Fauna and Flora (CITES), which lists species that are not necessarily extinct but may become extinct if their trade is not controlled (Viana et al., 2007).

To prevent overfishing and preserve the species, the Brazilian Institute of Environment and Natural Resources (IBAMA), a Brazilian government agency, established a minimum size of *A. gigas* for capture (1.5 m total length) in 1989. In 1990, the IBAMA enforced a fishing ban from December 1st to May 31st, which corresponds to the reproductive phase of the species. Since 1996, IBAMA and the Amazonas State Environmental Agency (SDS - Sustainable Development State Secretary) have prohibited the fishing and commercialization of *A. gigas*; fish from managed areas and fish farms are exempted from this prohibition (Arantes et al., 2007). These conservation actions were implemented concurrently with the creation of Sustainable Development Reserves (SDR) in Amazonas State, where fishing occurs with the direct involvement of local fishermen in the management of populations of this species. Therefore, annual portions (quotas) are permitted for the capture of *A. gigas* during a certain period of the year, according to the minimum size limitations. These actions have resulted in reduced fish extraction from the wild environment and increased the interest of fish farmers in the artificial management of breeders.

A new species of *Arapaima* was discovered in the Central Amazon of Brazil (Stewart, 2013a), and it was believed that the *Arapaima* was considered monotypic since 1868, with *A. gigas* is the only recognized species (Stewart, 2013b). This species was collected near the confluence of the Solimões and Purus Rivers, and comparisons with the pirarucus Development Mamirauá shows that there are at least two species of *Arapaima* in the Amazon State.

The cultivation of freshwater fish in Brazil is expanding rapidly due to the availability of favorable areas for cultivation, the existence of an enormous river network, and the high adaptive potential of many native species in captivity. However, aquaculture practices have reduced the genetic variability of captive animals as a result of unassisted breeding or the use of a small number of founders in fish farms (Wasko et al., 2004). Both practices alter genotype frequencies over generations and result in adaptation of the species to the confinement conditions (Santos et al., 2012).

The demand for *A. gigas* is high; therefore, large-scale data on the genetic features of wild and cultivated populations are needed. The sizes of founder populations are generally limited by farming constraints, which results in the use of a few individuals as broodstock. This practice may lead to erosion of the genetic diversity of the stocks, thereby compromising industrial performance. Due to the relatively short history of large-scale *A. gigas* commercial culture, the phenotypic effects of this erosion are not yet clear.

The present study is the first to address the genetic variability of captive *A. gigas* subpopulations. This study aimed to (i) define the genetic structure of *A. gigas* populations in a selected areas by measuring the genetic heterogeneity among wild and captive populations and (ii) estimate the degree of genetic erosion of reared stocks by comparing their genetic variability with that of geographically close wild stocks.

MATERIALS AND METHODS

Ethics committee

The group leader, Adalberto Luis Val was granted a license for capture under the umbrella project ADAPTA (SISBIO License #29837-4). Samples from wild subpopulations were collected according to the guidelines set by the IBAMA of the Ministry of Environment. All samples were collected from anesthetized animals according to the guidelines from Brazilian Council for Ethics in the use of Experimental Animals (CONCEA) and the Ministry of Science, Technology and Innovation.

Collection of samples

Wild *A. gigas* individuals were captured from the SDR during the legal period for managed fishing with the help of a harpoon, which was the only gear allowed. Local fishermen are licensed to harpoon the fish body (never the head) and, subsequently place fish in ice boxes to commercialize them. The spear catch is a traditional (cultural) way used by locals to minimize damage in fish meat. Captive individuals were captured with trawls with permission from two private farm onwers and with the permission Ceara State Aquaculture Facility. Because these fish were intended for the local

Figure 1. Geographical locations of the *Arapaima gigas* sample collection sites in the Brazilian Amazon (North) and Northeastern Brazil: (A) Departamento Nacional de Obras Contra as Secas (DNOCS) Pentecoste - Ceará; (B) Fishfarm Porto Velho (PPV), Porto Velho - Rondônia, (C) Sustainable Development Reserve Mamirauá (SDRM) Tefé - Amazonas; (D) Sustainable Development Reserve Fonte Boa (SDRFB), Fonte Boa - Amazonas and (E) Fishfarm FrigoPesca (PFP-1, -2 and -3), Manacapuru - Amazonas.

market, they were anesthetized and euthanized in ice water prior to the collection of tissue samples. A total of 10 g of white muscle was collected from each *A. gigas* sample using the tweezers process for the stilettos and then transferred to sterile eppendorf tubes and immersed in liquid nitrogen. The samples were stored at a low temperature (-80°C) until subsequent DNA analysis.

Samples of white muscle were collected from 218 wild (two subpopulations) and captive (five subpopulations) *A. gigas* individuals from Northern and Northeastern Brazil (Figure 1). The wild individuals were collected from the SDR in Mamirauá (SDRM; 25 samples) (01° 49' 00" S and 65° 42' 00" W) and Fonte Boa (SDRFB; 18 samples) (02° 30' 52" S 66° 05' 33" W) in Amazonas. All samples of captive individuals were taken from F_1 progeny and acquired from three fish farms: FrigoPesca (Amazonas), from which samples were collected three different times (every 6 months) between 2009 and 2010 (PFP-1: 47 samples, PFP-2: 18 samples, and PFP-3: 32 samples) (03° 17' 39" S and 60° 38' 04" W); Porto Velho (Rondonia State) (PPV: 48 samples) (08° 45' 43" S and 63° 54' 07" W); and Estação de Piscicultura do Departamento Nacional de Obras Contra as Secas (Ceará State) (DNOCS: 30 samples) (03° 47' 29" S and 39° 15' 58" W).

DNA extraction and PCR amplification

Total genomic DNA was extracted from the 218 tissue samples using the protocol described by Sambrook et al. (1989) with some modifications. Approximately, 100 mg of tissue was digested overnight at 37°C in 0.7 ml of lysis buffer (6 M urea, 10 mM Tris-HCl, 125 mM NaCl, 1% SDS, 10 mM EDTA, pH 7.5); 10 µl of proteinase K (Fermentas - Thermo Fisher Scientific - Suwanee, GA USA), RNase A (Invitrogen - Life Technologies- Grand Island, NY) and 1,4-dithiothreitol (Biosynth® Chemistry and Biology Rietlistr, Switzerland) were then added. After digestion, the samples were washed with phenol-chloroform-isoamyl alcohol (Invitrogen) and isopropanol on ice. The pellet was resuspended in 50 µl DNase-free H_2O. The quality of the extracted DNA was verified by electrophoresis in a 1.5% agarose gel using GelRed® (Invitrogen) and visualization using an L-PIX Molecular Image transilluminator (Loccus Biotecnologia - São Paulo, Brazil). The DNA was quantified using a NanoDrop® 2000 spectrophotometer (Thermo Scientific - Suwanee, GA USA).

The *A. gigas* samples were genotyped using the 10 microsatellite loci identified by Farias et al. (2003) (AgCTm1, AgCTm3, AgCTm4, AgCTm5, AgCTm7, AgCTm8, AgCAm4, AgCAm16, AgCAm18, and AgCAm26) and the 9 loci identified by Molecular Ecology Resources Primer Development (Consortium et al., 2011) (AG_01, AG_02, AG_03, AG_07, AG_09, AG_11, AG_12, AG_13, and AG_15). Polymerase chain reactions (PCR) were performed using a 96-well Veriti™ Thermal Cycler (Applied Biosystems - Life Technologies - Grand Island, NY) in a final reaction volume of 10 µl. Each PCR contained 1.5 µl of genomic DNA (50 ng/µl), 1.0 µl of forward primer (0.4 mM), 1.5 µl of reverse primer (0.4 mM), 1.0 µl 6-FAM fluorescent M13 primer [14] (0.4 mM), and 5.0 µl PCR Master Mix 2X (Fermentas) containing Taq DNA polymerase (0.05 U), $MgCl_2$ (1.5 mM), and dNTPs (0.4 mM). The PCR cycling conditions were as follows: 94°C for 3 min; 20 cycles at 94°C for 20 s, 56 to 58°C for 20 s, and 68°C for 30 s; 25 cycles at 94°C for 20 s, 53°C for 20 s, and 68°C for 30 s, and a final extension at 72°C for 10 min. The amplified PCR products were confirmed by electrophoresis in a 1.5% agarose gel using GelRed® (Invitrogen) and visualization using an L-PIX Molecular Image transilluminator (Loccus

Biotecnologia). The amplified products were genotyped on an ABI 3130*xl* Genetic Analyzer (Applied Biosystems) using the GeneScan Liz-500 (-250) size standard (Applied Biosystems) to determine fragment length. The alleles were scored based on the consistent pattern of their stutter peaks and the peak intensity in each individual at each locus using GeneMapper v4.0 (Applied Biosystems).

Data analysis

Genetic diversity

The genetic diversity of the *A. gigas* subpopulations was estimated based on the following parameters: the observed (H_O) and expected (H_E) heterozygosities (Fstat v2.9.3.2) (Goudet, 2001); the number of alleles per locus (Genetix v4. 5.2) (Belkhir et al., 2004); the number of private alleles or the fixation index (F_{IS}) (GDA v1.1) (Lewis and Zaykin, 2000); the polymorphism information content (PIC) (Cervus v3.0.3) (Kalinowski et al., 2007); and the effective number of alleles per locus (A_E). To calculate the effective number of alleles, we used the following mathematical formula: $A_E = 1/(1-H_E)$. Micro-Checker v2.3 software (van Oosterhout et al., 2004) was used to test for the presence of null alleles, and their frequencies were calculated using Cervus v3.0.3 software (Kalinowski et al., 2007). The linkage disequilibrium (*LD*) and Hardy-Weinberg equilibrium (*HWE*) were calculated using the Fstat software, with sequential Bonferroni corrections to minimize the errors of type I microsatellite markers (Rice, 1989). The significance level was α = 0.05.

Genetic differentiation

The levels of genetic differentiation among sampling sites were analyzed using the *F* statistics of Wright (F_{IT}, F_{IS} and F_{ST}) (Weir and Cockerham, 1984) using the software Fstat and Arlequin v3.5 (Excoffier and Lischer, 2010). The indices of genetic differentiation, F_{ST} (Weir and Cockerham, 1984) and R_{ST} (Slatkin, 1995), were applied to better address the mutational models for the microsatellite data (e.g., IAM for F_{ST} (Kimura and Crow, 1964) and SMM for R_{ST} (Kimura and Otha, 1978). We also applied 3 estimators of genetic differentiation, G_{STest} (Nei and Chesser, 1983), G'_{STest} (Hedrick, 2005), and D_{est} (Jost, 2008). These estimators were calculated using Smogd v1.2.4 software (Crawford, 2010). Analysis of molecular variance (AMOVA) (Excoffier and Slatkin, 1995) was performed for the 7 *A. gigas* subpopulations using Arlequin software.

Analysis of population structure

Structure v2.3.1 software (Pritchard et al., 2000; Falush et al., 2003) was used to verify the level of genetic structure among the *A. gigas* subpopulations under the ancestral admixture model, which correlates gene frequencies between subpopulations. Subsequently, the data generated a burn-in of 50,000 steps per 200,000 using the Markov chain Monte Carlo (MCMC) test for inference. We used the Structure Harvest v0.6.8 software (Earl and Holdt, 2012) to check the value of ΔK (according to Evanno et al., 2005), which represents the highest level of hierarchy of the genetic divisions. Furthermore, we conducted a run for each test with K = 1 and K = the number of sampled populations +2, with 3 replicates for each K. To complement the analysis of genetic structure, we performed a factorial correspondence analysis (FCA) (Benzecri, 1973) of the genetic distances between subpopulations using Genetix software, allowing the program to identify potential outlier individuals. The values of F_{ST} (Weir and Cockerham, 1984) and the genetic distance (D_A) (Nei, 1978; Cavalli-Sforza and Edwards, 1967) in pairs were constructed after comparing the subpopulations of *A. gigas*. A neighbor-joining tree (NJ) was constructed using GDA software based on genetic distance according to Nei (1978).

Bottlenecks and detection of outlier loci (F_{ST}-outlier method)

We analyzed the occurrence of bottleneck populations (with an excess of heterozygotes) in the wild and captive *A. gigas* subpopulations using Bottleneck software (Cornuet and Luikart, 1996; Piry et al., 1999). This analysis was used to detect potential bottleneck populations of no more than $2N_E$-$4N_E$ for past generations, depending on the severity of the bottleneck and the rate of change of the loci analyzed. In this study, we applied 2 mutational models [Stepwise mutation model (SMM) and Two phase model (TPM)]. These models are the most appropriate for evaluating microsatellite data (di Rienzo et al., 1994) and were implemented at 95% (SMM) and 5% (TPM) for the various stages of mutation, ranging 12% among several steps, as recommended by Piry et al. (1999).

Outlier loci were detected using 2 different approaches. First, a coalescent-based simulation approach was used to identify outlier loci with unusually high and low F_{ST} values by comparing the observed F_{ST} values with the values expected under neutrality (Beaumont and Nichols, 1996), as implemented in the Lositan selection workbench (Antao et al., 2008). We performed an initial run with 100,000 simulations on all loci using the mean neutral F_{ST} as the preliminary value. A more accurate estimate of the mean neutral F_{ST} was obtained after the first run by excluding all loci outside the 95% confidence interval because their distribution could result from selection rather than neutral evolution. This refined estimate was used for a final set of 100,000 simulations of all loci. In addition, we used the method described by Vitalis et al. (2001), which investigates outliers in a pairwise fashion based on subpopulation-specific *F*-statistics. The coalescent simulations were performed with Detsel v1.0 (Vitalis et al., 2003). The null distributions were generated using the following parameters: population size before the split N_0 = 500; mutation rate μ = 0.0001 and 0.00001; ancestral population size N_e = 500, 1000 and 10000; time since the bottleneck T_0 = 50, 100 and 1000; and time since the population split *t* = 100. The outliers were determined based on an empirical *p* value for each locus array of 50 × 50 square cells (Vitalis et al., 2001).

RESULTS

Genetic diversity of wild and captive subpopulations

Summaries of the genetic and allelic diversities of the *A. gigas* subpopulations are presented in Tables 1 and S1 (Supporting information). We observed the presence of null alleles for loci AG_03 (PPV) and AG_15 (PPV, RDSFB and RDSM). A total of 146 alleles were observed for the 19 microsatellites markers, with an average of 7.68 alleles per locus. The total number of alleles ranged from 49 (2.58 alleles/loci; DNOCS) to 88 (4.63 alleles/loci; SDRM). We also detected the occurrence of 49 private alleles (Table 2), 23 of which had a frequency of greater than 5%. The SDRM (wild) subpopulation showed higher values for the mean number of alleles (MNA = 4.632) and allelic richness (AR = 4.258), whereas the group from DNOCS (captive) showed the

Table 1. Genetic diversity indices for wild and captive subpopulations of *Arapaima gigas* based on 19 microsatellite loci.

Origin	Subpopulations	Allelic diversity					Genetic diversity			
		N	*TNA*	*MNA*	*NEA*	*NPA*	*AR*	H_E	H_O	F_{IS}
Captive	PFP-1	47	51	2.684	1.621	5	2.587	0.383	0.473	-0.225**
	PFP-2	18	59	3.105	2.066	9	3.060	0.516	0.695	-0.317**
	PFP-3	32	52	2.737	1.852	4	2.680	0.460	0.551	-0.179**
	DNOCS	30	49	2.632	1.730	3	2.562	0.422	0.542	-0.266**
	PPV	48	50	2.684	1.684	5	2.567	0.406	0.482	-0.177**
Wild	SDRFB	18	75	3.947	2.004	7	3.815	0.501	0.559	-0.085$^{(1)}$**
	SDRM	28	88	4.632	2.268	16	4.258	0.559	0.591	-0.035$^{(1)}$**

N, sample size; *TNA*, total number of alleles; *MNA*, mean number of alleles; *NEA*, number of effective alleles; *NPA*, number of private alleles; *AR*, allelic richness; H_E, genic diversity (expected heterozygosity in Hardy-Weinberg equilibrium); H_O, observed heterozygosity; F_{IS}, inbreeding coefficient of population; (n), number of loci that deviated from Hardy-Weinberg equilibrium (HWE); level of significance, $^{**}p < 0.05$.

Table 2. Values of the genetic differentiation indices for the 7 subpopulations of *Arapaima gigas* based on 19 microsatellite loci.

Locos	*PIC*	F_{IT}	F_{IS}	F_{ST}	R_{ST}	G_{Stest}	G'_{Stest}	D_{est}
AgCTm1	0.454	0.381*	-0.106	0.441*	0.544*	0.358*	0.534*	0.275*
AgCTm3	0.657	0.157*	-0.155	0.270*	0.430*	0.204*	0.526*	0.405*
AgCTm4	0.707	-0.214	-0.425	0.148*	0.855*	0.143*	0.495*	0.410*
AgCTm5	0.823	0.347*	-0.064	0.387*	0.440*	0.306*	0.828*	0.752*
AgCTm7	0.235	0.214*	0.097	0.130*	0.160*	0.113*	0.148*	0.039*
AgCTm8	0.557	0.225*	-0.430	0.458*	0.411*	0.397*	0.645*	0.411*
AgCAm4	0.299	-0.210	-0.572	0.230*	0.230*	0.192*	0.285*	0.115*
AgCAm16	0.848	0.420*	0.091	0.362*	0.347*	0.305*	0.860*	0.798*
AgCAm18	0.198	-0.008	-0.080	0.067*	0.083*	0.052*	0.068*	0.017*
AgCAm26	0.623	-0.080	-0.174	0.216*	0.152*	0.162*	0.426*	0.315*
AG_01	0.756	0.277*	-0.110	0.349*	0.335*	0.306*	0.754*	0.646*
AG_02	0.447	0.179*	0.012	0.169*	0.156*	0.174*	0.343*	0.205*
AG_03	0.628	0.036	-0.298	0.257*	0.106*	0.190*	0.463*	0.337*
AG_07	0.653	0.128*	-0.192	0.268*	0.192*	0.219*	0.545*	0.418*
AG_09	0.600	0.297	-0.034	0.320*	0.433*	0.265*	0.577*	0.424*
AG_11	0.462	0,040	-0.238	0.225*	0.326*	0.203*	0.370*	0.210*
AG_12	0.733	0.243*	-0.224	0.382*	0.366*	0.330*	0.745*	0.620*
AG_13	0.726	-0.213	-0.501	0.192*	0.135*	0.162*	0.486*	0.387*
AG_15	0.452	0.337*	0.110*	0.255*	0.408*	0.257*	0.470*	0.286*
All	0.571	0.156	-0.185	0.288	0.327	0.228	0.504	0.277

PIC, Polymorphism information content; F_{IT}, fixation index of the total population; F_{IS}, fixation index of the subpopulations; F_{ST}, fixation index result for the comparison of the subpopulations to the total populations; R_{ST}, analogous to F_{ST} of microsatellite markers; G_{Sest}, estimator of relative differentiation; G'_{Sest}, standardized measure of genetic differentiation; D_{est}, current estimate of differentiation; level of significance, $^{*}p < 0.05$.

lowest values (MNA = 2.632 and AR = 2.562).

The indices of genetic diversity in this study ranged from 0.473 to 0.695 (H_O), 0.383 to 0.559 (H_E), 0.198 to 0.848 *PIC*, and -0.317 to -0.035 (F_{IS}). Of the 19 studied loci, 13 were highly informative (Table 2). However, 6 loci were monomorphic in some subpopulations of *A. gigas* (Table S1). Only the wild subpopulations had F_{IS} values close to 0 (Table 1), indicating that these subpopulations tended to approach Hardy-Weinberg equilibrium ($F_{IS} \sim 0$; $H_O \sim H_E$). The loci AG_09 (RDSFB) and AG_15 (RDSM) significantly deviated from Hardy-Weinberg equilibrium (HWE) ($P < 0.05$, Bonferroni correction 0.0004). However, we did not detect linkage disequilibrium (LD) for the analyzed pairs of loci ($p > 0.05$).

Table 3. F_{ST} (above) and genetic distance values (below) for *A. gigas* subpopulations based on 19 microsatellite loci.

Variables	PFP-1	PFP2	PFP3	DNOCS	PPV	SDRFB	SDRM
PFP1	-	0.278	0.255	0.338	0.364	0.318	0.299
PFP2	0.341(0.341)	-	0.270	0.265	0.284	0.223	0.230
PFP3	0.281(0.000)	0.439(0.281)	-	0.280	0.341	0.242	0.241
DNOCS	0.422(0.000)	0.377(0.433)	0.374(0.396)	-	0.312	0.298	0.266
PPV	0.477(0.422)	0.390(0.000)	0.502(0.000)	0.397(0.000)	-	0.261	0.299
SDRFB	0.433(0.000)	0.371(0.440)	0.360(0.000)	0.458(0.377)	0.332(0.000)	-	0.135
SDRM	0.440(0.477)	0.458(0.000)	0.408(0.439)	0.429(0.000)	0.481(0.371)	0.207(0.458)	-

The genetic distances calculated according to Nei (1978) are shown; the distances calculated according to Cavalli-Sforza and Edwards (1967) are shown within parentheses.

Table 4. Analysis of molecular variance (AMOVA) for the 7 subpopulations of *Arapaima gigas* based on 19 microsatellite loci.

Source of variation	% var	F-stat	F-value	Standard error	p-value
Within individuals	0.866	F_{IT}	0.134	0.044	-
Between individuals	-0.081	F_{IS}	-0.103	0.043	1.000
Between populations	0.215	F_{ST}	0.215	0.016	0.001*

*Indicates a significant difference at $p < 0.05$.

Genetic differentiation and population structure

The indices of genetic differentiation (F_{ST}, R_{ST}, G_{Stest}, G'_{Stest}, and D_{est}) were significant for all examined loci ($p < 0.05$). The locus AgCAm18 was the only marker that showed low (D_{est} = 0.017) and moderate (R_{ST} = 0.083) genetic differentiation. The average value of F_{ST} (0.288) was very close to the G_{Stest} (0.228) and D_{est} (0.277) indices (Table 2). The F_{ST} values between pairs of *A. gigas* subpopulations ranged from 0.135 to 0.364, revealing that the level of genetic differentiation between the analyzed pairs varied from high to extremely high (Table 3). The AMOVA revealed that the distribution of genetic variability was 21.5% among the subpopulations and 86.6% among individuals of the subpopulations (Table 4).

We verified the occurrence of genetic structure among the *A. gigas* subpopulations with distinct groups (Figure 2); the clusters were based on the data inferred from differences in allelic frequencies. The average association coefficient (*q*) for each subpopulation (K = 7) showed that the individuals of *A. gigas* are distributed (Table S2) in specific agglomerations (q > 0.9), resulting in a high-resolution *q* value and indicating that no subpopulation was mixed. The FCA yielded values of 6.61 and 6.32% for Axes 1 and 2, respectively; the dispersion graph of those parts is shown in Figure 3. Notably, there was a clear separation of subpopulations in the wild, with Axis 1 > 0 and Axis 2 < 0, and the formation of two *A. gigas* populations: one group consisting of the subpopulations PFP-1, PFP-2, PFP-3, and DNOCS (group 1; Axis 1 < 0), and a second group consisting of the PPV, SDRM, and SDRFB subpopulations (group 2, Axis 1 > 0) (Figure 3). The separation of the subpopulations into these two distinct groups by FCA was consistent with the data obtained for the NJ tree based on the genetic distances between the subpopulations (Figure 4).

Bottlenecks and detection of outlier loci (F_{ST}-outlier method)

To detect bottlenecks, we eliminated the loci that drifted from Hardy-Weinberg equilibrium because the presence of these loci could strongly influence the final results and because of the effect of selection on the over-dominant loci with an excess of heterozygotes (Luikart and Cornuet, 1998). Furthermore, we verified the minimum number of microsatellite markers needed for an analysis of recent genetic bottlenecks, as the model for mutational markers could influence the final results if an inappropriate number of SSR loci were used.

We detected a recent population bottleneck in all analyzed subpopulations of *A. gigas*. Both the Sign and Wilcoxon tests showed a significant excess of heterozygotes for the TPM mutational model. However, only the captive subpopulations showed an excess of heterozygotes using the Sign test. The mutational model SMM was significant only for the subpopulations PFP-2 and PFP-3 using the Wilcoxon test.

We observed 10 outlier loci in the *A. gigas* subpopulations;

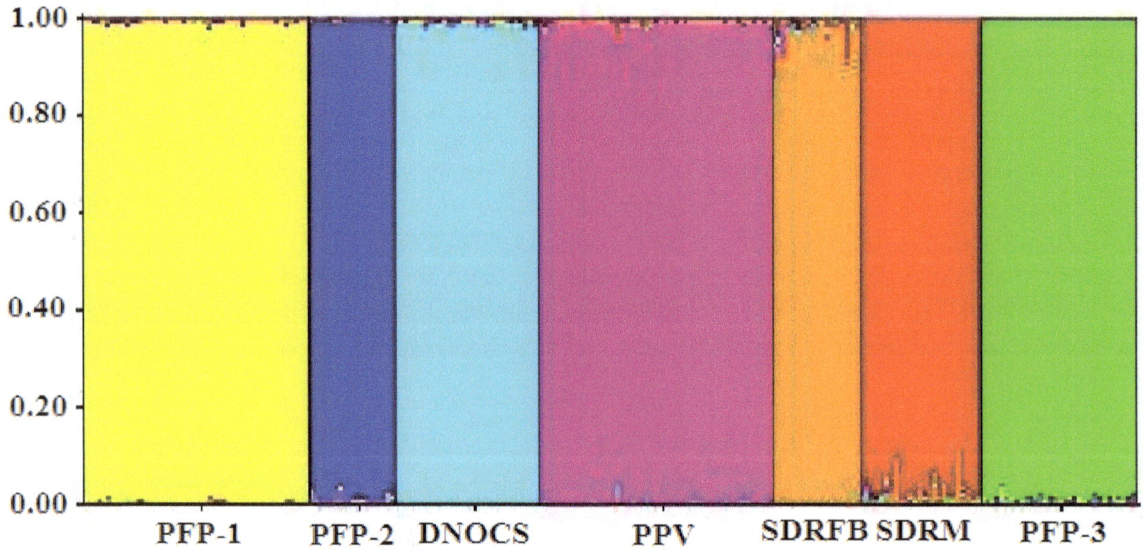

Figure 2. Structural analysis results for the 7 subpopulations based on 19 microsatellite loci. K = 7.

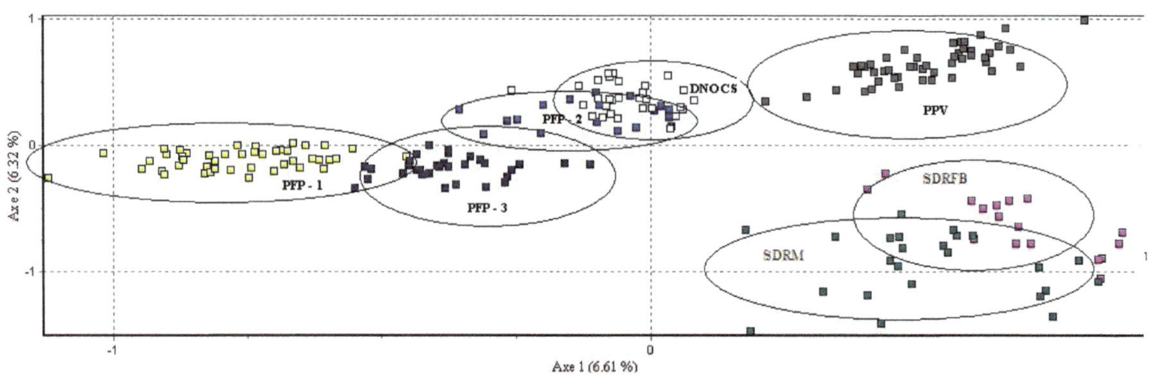

Figure 3. Factorial correspondence analysis (FCA) of the 7 subpopulations based on 19 microsatellite loci.

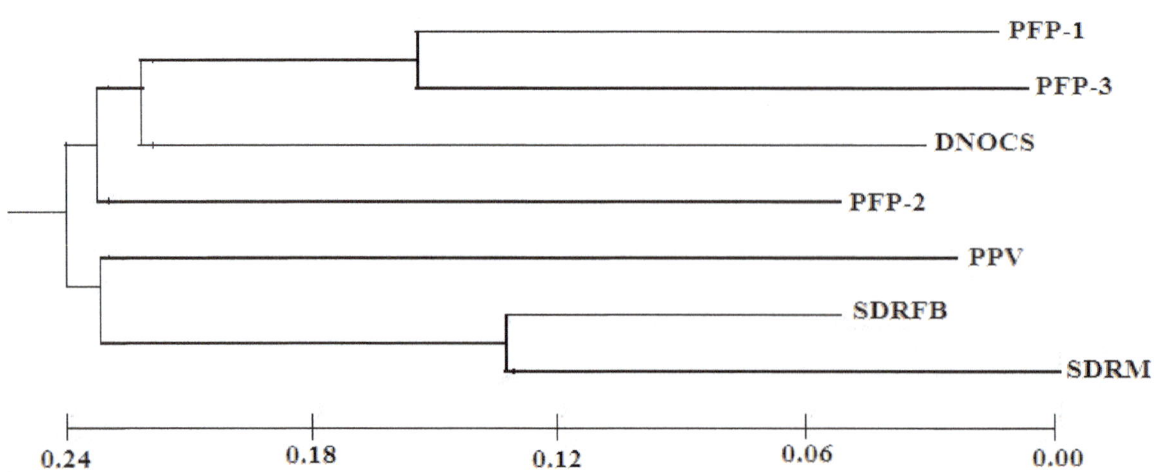

Figure 4. Neighbor-joining dendrogram for wild and captive subpopulations of *Arapaima gigas* based on 19 microsatellite loci.

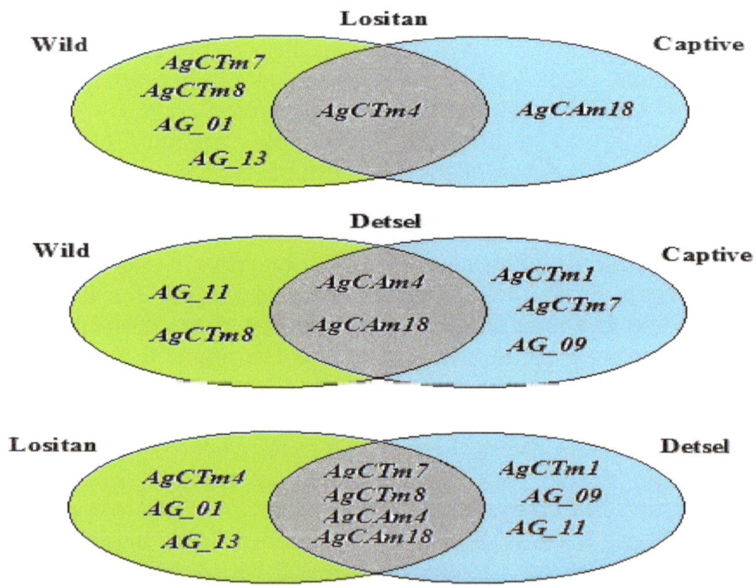

Figure 5. Outlier loci detected among the 7 studied subpopulations according to Beaumont and Nichols (1996) and Vitalis et al. (2001) using Lositan (Antao et al. 2008) and Detsel v1.0 (Vitalis et al. 2003), respectively.

four of these loci were identified as common to all subpopulations by Lositan and Detsel software (Figure 5), which suggests that these four loci may be related to the local adaptation of *A. gigas*. The Lositan program identified four outlier loci with balanced selective pressure for the 2 subpopulations of wild and captive specimens. Furthermore, we detected a locus with positive selective pressure for wild individuals, while the remaining loci were considered neutral markers (Figure S1).

DISCUSSION

Genetic differences between wild and captive subpopulations

According to Ellegren and Sheldon (2008), populations are usually in evolutionary equilibrium; however, evolutionary responses may be hidden due to deterioration or loss of the environment. In addition, the connectivity between populations plays a crucial role in the maintenance of genetic variation, particularly in small populations strongly affected by genetic drift (Broquet et al., 2010). Genetic analyses of 7 subpopulations of *A. gigas* using 19 microsatellite markers revealed that most of the markers were highly polymorphic, with 13 loci showing values of PIC > 0.5 (Table 2). The results also revealed the occurrence of moderate genetic diversity in *A. gigas* subpopulations, indicating that the evolutionary potential of the wild individuals from the SDR is greater than that of the captive individuals, as expected (Table 1). Hedrick (2001) reported that the pattern of variation within a given variable population generally consists of neutral alleles, that is, the equilibrium is predicted by a reduction of variation through genetic drift and an increase in variation by random mutation. Therefore, even when a population is small, selection may act on the variation in a particular gene, but genetic drift acts more strongly than selection on the allele frequencies.

An excess of heterozygotes was observed in all *A. gigas* populations (Table 1). We also observed decreased heterozygosity at the AgCAm16 locus in all the domesticated and wild populations, which suggests a recent evolutionary history because this locus is predicted to be under intensive selective pressure. According to Barroso et al. (2005), deficiency in heterozygous loci may occur due to positive selection, improvement of the relatedness between individuals, and population structure. We observed higher allelic richness in wild subpopulations (SDRFB and SDRM) compared to the captive subpopulations; the observed values are consistent with those reported by Silva (2008) for individuals of *A. gigas* in the Jarauá and Maraã lakes (SDRM). Araripe et al. (2013) investigated the wild subpopulations of *A. gigas* in a large geographical area; based on analyses of 7 informative microsatellite regions, they proposed that the genetic differentiation between subpopulations should increase with increasing geographical distance between them and that the differences among populations separated by geographical distances greater than 1300 km could be caused by historical bottlenecks in population size, in light of the sedentary behavior of the species.

Here, we observed that the average genetic differentiation among all subpopulations (F_{ST}) was 21.5% ($p < 0.05$), demonstrating the high discrimination of the seven studied subpopulations (Figure 2) and the absence of mixtures (Table S2). This genetic differentiation suggests that the captive subpopulations are increasingly isolated due to lack of gene flow between fish farms, which merits further consideration. However, gene flow is also limited between individuals of the wild subpopulations of the SDR. According to Teixeira (2008), *A. gigas* individuals caught in the wild may not represent the total population because a group may have descended from one pair of individuals. According to Glover et al. (2011), genetic differences have been observed between Atlantic salmon fry in fish sites populated for commercial reproduction from farms of the same company, indicating that the captive environment accelerates genetic differentiation. However, Hedrick (1999) reported that the number of highly polymorphic microsatellite F_{ST} might be underestimated and that such an underestimate could influence the observed proportion of heterozygotes in subpopulations compared to the total population. In this context, Hartl and Clark (2010) reported that the estimated F statistic must correct the effects of sampling and the limited number of subpopulations, and suggested that new indices would provide a correction of the values of F_{ST}.

The recent bottleneck in captive subpopulations is certainly due to the founder effect, which results from the use of a small number of individuals to establish reproductive stocks in fish farms. Furthermore, the lack of allelic richness in captive subpopulations may have contributed to the genetic signature of a new bottleneck (Broquet et al., 2010). These results reinforce the idea that the small number of breeding stations in fish farms may contribute to the genetic deterioration of captive populations.

The bottleneck detected in wild subpopulations is likely due to intense fishing activity in recent years. Furthermore, the practice of selective fishing in the SDR has likely decreased breeding inventories and, subsequently, the effective population size. A decrease in wild fish stocks was reported in the mid-1990s, when the *A. gigas* production reached 207.5 tons, compared with the previous decade, when production was 1,751 tons (Venturiere and Bernardino, 1999). Luikart and Cornut (1998) reported that natural populations show changes in the frequency of heterozygotes when a stock experiences a recent bottleneck.

We observed that the *A. gigas* subpopulations are undergoing selection, leading to local adaptation to the new environmental conditions that the species has been recently exposed to (Figure 5). We observed four loci with balanced and positive selection in the wild groups. In the captive groups, we identified two loci with balanced selection. However, when all groups were considered, we observed six loci with positive selection, indicating the existence of local adaptive processes in the wild and captive subpopulations. According to Orr (1998), genotyping many loci is likely important to identify complex adaptive differences between populations. The adaptive complexity provided by genotypic differentiation is directly related to balanced selection, which is common in confined populations but rare in wild populations. Furthermore, balanced selection, which favors heterozygotes, is more frequent than selection favoring homozygotes, thereby maintaining this variation (Amos and Balmford, 2001). The balanced selection of wild *A. gigas* populations observed in this study is likely due to a recent bottleneck experienced by these subpopulations due to over-fishing.

Local adaptations might be lost through intraspecific hybridization; therefore, actions that increase the amount of genetic exchange among locally adapted populations could be harmful. According to Utter (2004) and Einum and Fleming (1997), the release of individuals from hatcheries or their escape from aquaculture ponds could harm wild populations through intraspecific hybridization and the loss of local adaptations. Nunes et al. (2011) suggested that the identification of loci with adaptive importance is a key step toward understanding the process of speciation in natural populations because these loci are responsible for phenotypic variations that affect the fitness of the species in different environments. According to Sønstebø et al. (2007) and Vasemägi et al. (2005), analysis of the genetic differences between the captive broodstocks and wild populations is also important because the specific genes of each population reflect local adaptability; consequently, the survival of juveniles will be affected if the populations are mixed. Our results indicate that this species requires management and conservation.

Implications for management and conservation

There is a global need to preserve genetic resources by maintaining species and their genetic diversities. The identification of actions that will minimize the effects of genetic changes (or losses) within biological populations is one of the primary mechanisms for maintaining variable and genetically healthy populations (Boettcher et al., 2010). According to He et al. (2012), microsatellite markers, in combination with recent statistical methodologies, represent a useful tool for the conservation and management of endangered species. For *A. gigas* subpopulations, allelic richness should be maintained and conserved, with a particular focus on preserving the existing rare alleles within populations, as suggested by Pérez-Ruzafa et al. (2006). Genetic information is also needed to aid the understanding of the current level of conservation of endangered species (Laikre et al., 2009).

Conservation projects focusing on the management and conservation of the genetic diversity of endangered species must consider the likely environmental changes (Boettcher et al., 2010) caused by stochastic events or human activity. Our results suggest that subpopulations of *A. gigas* have had independent evolutionary trajectories, and thus effective conservation measures should focus on preserving the genetic integrity of each subpopulation separately. Therefore, we suggest a more detailed management of breeding groups and periodic monitoring of wild populations.

Hrbek et al. (2005) proposed the use of a large group of matrices of *A. gigas* (founders) in fish farms, regardless of their geographical origin, thereby minimizing the potential loss of genetic diversity due to founder effects. Furthermore, it has been suggested that confined populations be supplemented with wild individuals from selected areas to maintain high genetic diversity in confinement systems. However, before we consider maintaining or restoring the genetic diversity, we should be alert to individuals collected from nature that served as new breeding on fish farms, as these new breeding, can they bring in parasites from fish farms that might jeopardize the survival of older breeding and consequently reduce the effective size and negatively affect the genetic diversity of these populations.

In addition, we suggest the rotation of *A. gigas* groups within and between fish farms to maintain the evolutionary potential of the species and avoid greater adaptation to captive systems. *A. gigas* fingerlings from several wild and captive breeders could also be distributed to fish farmers. However, the formation of closely related couples should be avoided to reduce consanguinity, which could compromise the genetic health of *A. gigas* lots for future distribution.

ACKNOWLEDGEMENTS

The authors thank Prof. Izeni Pires Farias from the Laboratory of Animal Genetics and Evolution (LEGAL/UFAM) for contributions to the original PhD thesis, and LEGAL and the Departamento Nacional de Obras Contra as Secas (DNOCS, CE) for providing *A. gigas* samples for this work. This work is part of INCT ADAPTA, coordinated by Adalberto Luis Val.

Conflict of Interest

The author(s) have not declared any conflict of interest.

REFERENCES

Agata K, Alassad S, Almeida-Val VMF, Álvarez-Dios JA, Barbisan F, Beadell JS, Beltrán JF, Benítez M, Bino G, Bleay C, Bloor P, Bohlmann J, Booth W, Boscari E, Caccone A, Campos T, Carvalho BM, Clímaco GT, Clobert J, Congiu L, Cowger C, Dias G, Doadrio I, Farias IP, Ferrand N, Freitas FD, Fusco G, Galetti PM, Gallardo-Escárate C, Gaunt MW, Ocampo ZG, Gonçalves H, Gonzales EG, Haye P, Honnay O, Hyseni C, Jacquemyn H, Jowers MJ, Kakezawa A, Kawagushi E, Keeling C, Kwan Y-S, la Spina M, Lee W-O, Lesniewska M, Li Y, Liu H, Liu X, Lopes S, Martínez P, Meeus S, Murray BW, Nunes AG, Okedi LM, Ouma JO, Pardo BG, Parks R, Paula-Silva MN, Pedraza-Lara C, Perera OP, Pino-Querido A, Richard M, Rossini BC, Samarasekera NG, A Sánchez, Sanchez JA, Santos CHA, Shinohara W, Soriguer RC, Sousa ACB, Sousa CFS, Steves VM, Tejedo M, Valenzuela-Bustamante M, van de Vliet MS, Vandepitte K, Vera M, Wandeler P, Wang W, Won Y-J, Yamashiro A, Yamashiro T, Zhu C. Molecular Ecology Resources Primer Development Consortium (2011). Permanent genetic resources added to molecular ecology resources database 1 December 2010 – 31 January 2011. Mol. Ecol. Resour. 11:586-589. http://dx.doi.org/10.1111/j.1755-0998.2011.03004.x

Amos W, Balmford A (2001). When does conservation genetics matter? Heredity 87:257-265. http://dx.doi.org/10.1046/j.1365-2540.2001.00940.x

Antao T, Lopes A, Lopes RJ, Beja-Pereira A, G Luikart (2008). LOSITAN: a workbench to detect molecular adaptation based on a FST-outlier method. BMC Bioinformatics 9:323. http://dx.doi.org/10.1186/1471-2105-9-323

Arantes CC, Castello L, Garcez DS (2007). Variações entre contagem de Arapaima gigas (Schinz) (Osteoglossomorpha, Osteoglossidae) feitas por pescadores individualmente em Mamirauá, Brasil. Pan-Am. J. Aquat. Sci. 2(3):263-269.

Araripe J, Rêgo PS, Queiroz H, Sampaio I, Schneider H (2013). Dispersal capacity and genetic structure of Arapaima gigas on different geographic scales using microsatellite markers. Plos One 8(1):e54470. http://dx.doi.org/10.1371/journal.pone.0054470

Barroso RM, Hilsdorf AWS, Moreira HLM, Cabello PH, Traub-Cseko YM (2005). Genetic diversity of wild and cultured populations of Brycon opalinus (Cuvier, 1819) (Characiforme, Characidae, Bryconiae) using microsatellites. Aquaculture 247:51-65. http://dx.doi.org/10.1016/j.aquaculture.2005.02.004

Beaumont MA, Nichols RA (1996). Evaluating loci for use in the genetic analysis of population structure. Proc. R. Soc. B 363:1619-1626. http://dx.doi.org/10.1098/rspb.1996.0237

Belkhir K, Borsa P, Chikhi L, Raufaste N, Bonhomme F (2004). GENETIX 4.05, logiciel sous Windows TM pour la génétique des populations. Montpellier, France: Laboratoire Génome, Populations, Interactions, CNRS UMR 5171, Université de Montpellier II.

Benzecri JP (1973). L'Analyse des donnees. Vol 2: L'Analyse des Correspondances. Paris: Dunod.

Boettcher PJ, Tixier-Boichard M, Toro MA, Simianer H, Eding H, Gandini G, Joost S, Garcia D, Coli L, Ajmone-Marsan P, GLOBALDIV Consortium (2010). Objectives, criteria and methods for using molecular genetic data in priority setting for conservation of animal genetic resources. Anim. Genet. 41:64-77. http://dx.doi.org/10.1111/j.1365-2052.2010.02050.x

Broquet T, Angelone S, Jaquiery J, Joly P, Lena JP, Lengagne T, Plenet S, Luquet E, Perrin N (2010). Genetic bottlenecks driven by population disconnection. Conserv. Biol. 24(6):1596-1605. http://dx.doi.org/10.1111/j.1523-1739.2010.01556.x

Castello L (2008). Lateral migration of Arapaima gigas in floodplains of the Amazon. Ecol. Freshw. Fish 17:38-46. http://dx.doi.org/10.1111/j.1600-0633.2007.00255.x

Castello L (2004). A method to count pirarucu: fishers, assessment and management. N. Am. J. Fish Manage 24:379-389. http://dx.doi.org/10.1577/M02-024.1

Cavalli-Sforza LL, Edwards AWF (1967). Phylogenetic analysis: models and estimation procedures. Am. J. Hum. Genet. 19:233-257.

Cornuet JM, Luikart G (1996). Description and power analysis of two tests for detecting recent population bottlenecks from allele frequency data. Genetics 144:2001-2014.

Crawford NG (2010). SMOGD: software for the measurement of genetic diversity. Mol. Ecol. Res. 10:556-557. http://dx.doi.org/10.1111/j.1755-0998.2009.02801.x

di Rienzo A, Peterson AC, Garza JC, Valdes AM, Slatkin M, Freimer NB (1994). Mutational processes of simple sequence repeat loci in human populations. Proc. Natl. Acad. Sci. USA 91:3166-3170. http://dx.doi.org/10.1073/pnas.91.8.3166

Earl DA, von Holdt BM (2012). Structure Harvester: a website and program for visualizing structure output and implementing the Evanno method. Conserv. Gen. Res. 4(2):359-361. http://dx.doi.org/10.1007/s12686-011-9548-7

Einum S, Fleming IA (1997). Genetic divergence and interactions in the wild among native, farmed and hybrid Atlantic salmon. J. Fish Biol. 50:634-651. http://dx.doi.org/10.1111/j.1095-8649.1997.tb01955.x

Ellegren H, Sheldon BC (2008). Genetic basic of fitness differences in natural populations. Nature 452:169-175. http://dx.doi.org/10.1038/nature06737

Evanno G, Regnaut S, Goudet J (2005). Detecting the number of clusters of individuals using the software structure: a simulation study. Mol. Ecol. 14(8):2611-2620. http://dx.doi.org/10.1111/j.1365-294X.2005.02553.x

Excoffier L, Lischer HEL (2010). Arlequin suite ver 3.5: a new series of programs to performance populations genetics analyses under Linux and Windows. Mol. Ecol. Res. 10:564-567. http://dx.doi.org/10.1111/j.1755-0998.2010.02847.x

Excoffier L, Slatkin M (1995). Maximum-likelihood estimation of molecular haplotype frequencies in a diploid population. Mol. Biol. Evol. 12:921-927.

Falush D, Stephens M, Pritchard JK (2003). Inference of population structure using multilocus genotype data: linked loci and correlated allele frequencies. Genetics 164:1567-1587.

Farias IP, Hrbek T, Brinkmann H, Sampaio I, Meyer A (2003). Characterization and isolation of DNA microsatellite primers for Arapaima gigas, an economically important but severely over-exploited fish species of the Amazon basin. Mol. Ecol. Notes 3:128-130.
http://dx.doi.org/10.1046/j.1471-8286.2003.00375.x

Goudet J (2001). FSTAT, a program to estimate and test gene diversities and fixation indices (version 2.9.3.2). Updated from Goudet J (1995) FSTAT version 1.2: a computer program to calculate F-statistics. J. Hered. 86:485-486.

Goulding M (1980). The fishes and the forest. Los Angeles, CA, USA: University of California Press. P. 200.

Glover KA, Skaala Ø, Søvik AGE, TA Helle (2011). Genetic differentiation among Atlantic salmon reared in sea-cages reveals a non-random distribution of genetic material from a breeding programme to commercial production. Aquac. Res. 42:1323-1331. http://dx.doi.org/10.1111/j.1365-2109.2010.02720.x

Hartl DL, Clark AG (2010). Princípios de genética de populações, 4th Edition. Porto Alegre: Artmed. P. 660.

He L, Xie Y-N, Lu W, Wang Y, Chen L-L, Mather PB, Zhao Y-L, Wang Y-P, Wang Q (2012). Genetic diversity in three redclaw crayfish (Cherax quadricarinatus, von Martens) lines developed in culture in China. Aquac. Res. 43:75-83. http://dx.doi.org/10.1111/j.1365-2109.2011.02802.x

Hedrick PW (2005). A standardized genetic differentiation measure. Evolution 59:1633-1638.
http://dx.doi.org/10.1111/j.0014-3820.2005.tb01814.x

Hedrick PW (2001). Conservation genetics: where are we now? Trends Ecol. Evol. 16(11):629-636. http://dx.doi.org/10.1016/S0169-5347(01)02282-0

Hedrick PW (1999). Perspective: highly variable loci and their interpretation in evolution and conservation. Evolution 53:313-318. http://dx.doi.org/10.2307/2640768

Hrbek T, Farias IP, Crossa M, Sampaio I, Porto JIR, Meyer A (2005). Population genetic analysis of Arapaima gigas, one of the largest freshwater fishes of the Amazon basic: implications for its conservation. Anim. Conserv. 8:297-308. http://dx.doi.org/10.1017/S1367943005002210

Jost L (2008). GST and its relatives do not measure differentiation. Mol. Ecol. 17(18):4015-4026.
http://dx.doi.org/10.1111/j.1365-294X.2008.03887.x

Kalinowski ST, Taper ML, Marshall TC (2007). Revising how the computer program CERVUS accommodates genotyping error increases success in paternity assignment. Mol. Ecol. 16:1099-1106. http://dx.doi.org/10.1111/j.1365-294X.2007.03089.x

Kimura M, Crow J (1964). The number of alleles that can be maintained in a finite population. Genetics 49:725-738.

Kimura M, Otha T (1978). Stepwise mutation model and distribution of allelic frequencies in a finite population. Proc. Natl. Acad. Sci. USA 75:2868-2872. http://dx.doi.org/10.1073/pnas.75.6.2868

Laikre L, Nilsson T, Primmer CR, Ryman N, Allendorf FW (2009). Importance of genetics in the interpretation of favourable conservation status. Conserv. Biol. 23(6):1378-1381. http://dx.doi.org/10.1111/j.1523-1739.2009.01360.x

Lewis PO, Zaykin D (2000). Genetic Data Analysis: computer program for the analysis of allelic Data, Version 1.0. (d15). Storrs, CT: University of Connecticut.

Luikart G, Cornut JM (1998). Empirical evaluation of a test for identifying recently bottlenecked populations from allele frequency data. Conserv. Biol. 12:228-237. http://dx.doi.org/10.1111/j.1523-1739.1998.96388.x

Nei M (1978). Estimation of average heterozigosity and genetic distance from a small number of individuals. Genetics 89: 583-590.

Nei M, Chesser RK (1983). Estimation of fixation indices and gene diversities. Ann. Hum. Genet. 47(3):253-259. http://dx.doi.org/10.1111/j.1469-1809.1983.tb00993.x

Nunes VL, Beaumont MA, Butlin RK, Paulo OS (2011). Multiple approaches to detect outlier in a genome scan for selection in ocellated lizards (Lacerta lepida) along an environmental gradient. Mol. Ecol. 20:193-205. http://dx.doi.org/10.1111/j.1365-294X.2010.04936.x

Oliveira V, Poleto SL, Venere PC (2005). Feeding of juvenile pirarucu (Arapaima gigas, Arapaimidae) in their natural environment, lago Quarto Bocas, Araguaiana – MT, Brazil. Neotrop Ichthyol 3(2):312-314. http://dx.doi.org/10.1590/S1679-62252005000200010

Orr HA (1998). The population genetics of adaptation: the distribuition of factors fixed during adaptive evolution. Evolution 52:935-949. http://dx.doi.org/10.2307/2411226

Ortiz WS, Salcedo RDL, Ruíz LC, Macías JNL (2007). Potencial acuícola de pirarucu (Arapaima gigas) em la cueca Amazónica. Rev. Elect. Ing. Prod. Acuíc. 2:75-83.

Pérez-Ruzafa A, González-Wangüemert M, Lenfant P, Marcos C, García-Charton JA (2006). Effects of fishing protection on the genetic structure of fish populations. Biol. Conserv. 129:244-255. http://dx.doi.org/10.1016/j.biocon.2005.10.040

Piry S, Luikart G, Cornuet JM (1999). Bottleneck: a computer program for detecting recent reductions in the effective population size using allele frequency data. J. Hered. 90:502-503. http://dx.doi.org/10.1093/jhered/90.4.502

Pritchard JK, Stephens M, Donnelly P (2000). Inference of population structure using multilocus genotype data. Genetics 155:945-959.

Pullin RSV (2007). Genetic resources for aquaculture: status and trends. In: Bartley DM, Harvey BJ, Pullin RSV (editors). Workshop on status and trends in aquatic genetic resources: A basic for international policy: 8-10 May 2006, Victoria, British Columbia, Canada. Rome: Food & Agriculture Org.

Rice WR (1989). Analyzing tables of statistical tests. Evolution 43:223-225. http://dx.doi.org/10.2307/2409177

Sambrook J, Fritsch EF, Maniatis T (1989). Molecular cloning: a laboratory manual, 2nd ed. Cold Spring Harbor, NY: Cold Spring Harbor Laboratory.

Santos CHA, Leitão MAB, Sousa CFS, Santana GX, Paula-Silva MN, Almeida-Val VMF (2012). Genetic variability of wild and captivity populations of Colossoma macropomum (Cuvier, 1818). Acta Sci. Biol. Sci. 34(2):191-197. http://dx.doi.org/10.4025/actascibiolsci.v34i2.7149

Silva JAG (2008). Genética de populações de pirarucus (Arapaima gigas) da Reserva de Mamirauá e considerações sobre a estrutura genética para a espécie. Tese (Doutorado em Zoologia) 101. Belém, Brasil: Museu Paraense Emilio Goeldi, Universidade Federal do Pará.

Slatkin MA (1995). A measure of population subdivision based on microsatellite allele frequencies. Genetics 130:457-462.

Sønstebø JH, Borgstrøm R, Heun M (2007). Genetic structure of brwn trout (Salmo trutta L.) from the Hardangervidda mountain plateau (Norway) analyzed by microsatellite DNA: a basic for conservation guidelines. Conserv. Gen. 8:33-44. http://dx.doi.org/10.1007/s10592-006-9145-6

Stewart DJ (2013a). A new species of Arapaima (Osteoglossomorpha: Osteoglossidae) from the Solimões river, Amazonas State, Brazil. Copeia 3:470-476. http://dx.doi.org/10.1643/CI-12-017

Stewart DJ (2013b). Re-escription of Arapaima agassizzi (Valenciennes), a rare fish from Brazil (Osteoglossomorpha: Osteoglossidae). Copeia 1:38-51. http://dx.doi.org/10.1643/CI-12-013
Teixeira AS (2008). The monotony of transferrin and esterase electrophoretic patterns in pirarucu, Arapaima gigas (Schinz, 1822) from Santa Cruz lake, Tefé River, Amazonas, Brazil. Genet. Mol. Res. 7(2):407-416. http://dx.doi.org/10.4238/vol7-2gmr425
Utter F (2004). Population genetics, conservation and evolution in salmonids and other widely cultured fish; some perspectives over six decades. Rev. Fish. Biol. Fisheries 14:125-144. http://dx.doi.org/10.1007/s11160-004-3768-9
van Oosterhout C, Hutchinson WF, Wills DPM, Shipley P (2004). Micro-Checker: software for identifying and correcting genotyping errors in microsatellite data. Mol. Ecol. Notes 4:535-538. http://dx.doi.org/10.1111/j.1471-8286.2004.00684.x
Vasemägi A, Cross R, Paaver T, Koljonen ML, Nilsson J (2005). Extensive immigration from compensatory hatchery releases into wild Atlantic salmon population in the Baltic sea: spatio-temporal analysis over 18 year. Heredity 95:76-83. http://dx.doi.org/10.1038/sj.hdy.6800693
Venturiere R, Bernardino G (1999). Pirarucu, espécie ameaçada pode ser salva através do cultivo. Rev. Panor. Aquic. 9:13-21.
Viana JP, Castello L, Damasceno JMB, Amaral ESR, Estupi-án GMB, Arantes C, Batista GS, Garces DS, Barbosa S (2007). Manejo comunitário do pirarucu Arapaima gigas na Reserva de Desenvolvimento Sustentável de Mamirauá, Amazonas, Brasil. In: Áreas protegidas como instrumento de gestão pesqueira. Brasília MMA. pp. 239-261.
Vitalis R, Dawson K, Boursot P, Belkhir K (2003). DetSel 1.0: a computer program to detect markers responding to selection. J. Hered. 94(5):429-431. http://dx.doi.org/10.1093/jhered/esg083
Vitalis R, Dawson K, Boursot P (2001). Interpretation the heterogeneity of F values: a suggestion and a correction. Genetics 84:399-401.
Wasko AP, Martins C, Oliveira C, Senhorini JA, Foresti F (2004). Genetic monitoring of the Amazonian fish matrinchã (Brycon cephalus) using RAPD markers: insights into supporting and conservation programmes. J. Appl. Ichthyol. 20:48-52. http://dx.doi.org/10.1111/j.1439-0426.2004.00479.x
Weir BS, Cockerham CC (1984). Estimating F statistics for the analysis of population structure. Evolution 38:1358-1370. http://dx.doi.org/10.2307/2408641

SUPPLEMENTARY

Table S1. Basic parameters for genetic diversity associated with subpopulations of *Arapaima gigas* based on 19 microsatellite loci.

Loci	Subpopulations						
	Captive					Wild	
	PFP-1	PFP-2	PFP-3	DNOCS	PPV	SDRFB	SDRM
AgCTm1							
H_O	0.540	0.944	0.040	0.000	0.381	0.133	0.300
H_E	0.470	0.577	0.039	0.000	0.363	0.124	0.460
A	2.000	5.000	2.000	1.000	2.000	2.000	3.000
A_E	1.887	2.364	1.040	1.000	1.570	1.142	1.852
F_{IS}	-0.136	-0.619	0.000	-	-0.038	-0.037	0.370
AgCTm3							
H_O	0.667	0.714	0.875	0.654	0.512	0.786	0.348
H_E	0.552	0.766	0.650	0.587	0.414	0.566	0.401
A	3.000	5.000	4.000	3.000	3.000	4.000	4.000
A_E	2.232	4.274	2.857	2.421	1.706	2.304	1.669
F_{IS}	-0.195	0.103	-0.327	-0.095	-0.226	0.355	0.154
AgCTm4							
H_O	1.000	1.000	0.926	1.000	0.775	0.875	0.947
H_E	0.636	0.611	0.593	0.680	0.537	0.793	0.813
A	4.000	4.000	3.000	4.000	3.000	7.000	7.000
A_E	2.747	2.571	2.457	3.125	2.160	4.831	5.348
F_{IS}	-0.563	-0.615	-0.625	-0.454	-0.434	-0.071	-0.139
AgCTm5							
H_O	0.667	0.722	0.478	0.654	0.244	0.611	0.920
H_E	0.554	0.619	0.593	0.499	0.219	0.775	0.818
A	3.000	3.000	4.000	3.000	3.000	6.000	10.000
A_E	2.242	2.625	2.457	1.996	1.280	4.444	5.494
F_{IS}	-0.193	-0.139	0.214	-0.242	-0.102	0.238	-0.104
AgCTm7							
H_O	0.000	0.000	0.000	0.083	0.364	0.556	0.417
H_E	0.000	0.000	0.000	0.079	0.469	0.438	0.448
A	1.000	1.000	1.000	2.000	3.000	3.000	3.000
A_E	1.000	1.000	1.000	1.086	1.883	1.779	1.812
F_{IS}	-	-	-	-0.022	0.236	-0.241	0.091
AgCTm8							
H_O	0.526	0.875	0.577	0.040	0.650	1.000	0.000
H_E	0.410	0.492	0.482	0.039	0.466	0.500	0.000
A	3.000	2.000	2.000	2.000	2.000	2.000	1.000
A_E	1.695	1.968	1.930	1.040	1.873	2.000	1.000
F_{IS}	-0.271	-0.765	-0.179	0.000	-0.383	-1.000	-
AgCAm4							
H_O	0.000	0.500	0.222	0.923	0.674	0.000	0.840
H_E	0.000	0.375	0.198	0.497	0.447	0.000	0.487
A	1.000	2.000	2.000	2.000	2.000	1.000	2.000
A_E	1.000	1.600	1.247	1.988	1.808	1.000	1.949

Table S1. Contd.

F_{IS}	-	-0.300	-0.106	-0.852	-0.500	-	-0.714
AgCAm16							
H_O	0.425	0.533	0.556	0.458	0.585	0.706	0.550
H_E	0.429	0.560	0.578	0.478	0.608	0.808	0.714
A	3.000	3.000	3.000	3.000	3.000	8.000	5.000
A_E	1.751	2.273	2.370	1.916	2.551	5.208	3.496
F_{IS}	0.021	0.082	0.058	0.063	0.050	0.156	0.254
AgCAm18							
H_O	0.000	0.357	0.259	0.423	0.275	0.000	0.250
H_E	0.000	0.304	0.226	0.429	0.237	0.000	0.224
A	1.000	3.000	2.000	3.000	2.000	1.000	3.000
A_E	1.000	1.437	1.292	1.751	1.311	1.000	1.289
F_{IS}	-	-0.140	-0.130	0.033	-0.147	-	-0.092
AgCAm26							
H_O	0.610	0.938	0.731	0.625	0.462	0.824	0.636
H_E	0.495	0.623	0.659	0.640	0.380	0.644	0.548
A	3.000	3.000	3.000	3.000	3.000	6.000	4.000
A_E	1.980	2.652	2.932	2.778	1.613	2.809	2.212
F_{IS}	-0.220	-0.480	-0.089	0.044	-0.201	-0.251	-0.140
AG_01							
H_O	0.456	0.929	0.581	0.440	0.727	0.733	0.591
H_E	0.416	0.600	0.564	0.343	0.664	0.624	0.606
A	3.000	3.000	3.000	2.000	3.000	3.000	3.000
A_E	1.712	2.500	2.293	1.522	2.976	2.660	2.538
F_{IS}	-0.087	-0.523	-0.013	-0.263	-0.083	-0.141	0.049
AG_02							
H_O	0.372	0.533	0.370	0.440	0.349	0.438	0.524
H_E	0.313	0.580	0.417	0.385	0.318	0.482	0.622
A	3.000	3.000	2.000	2.000	2.000	2.000	3.000
A_E	1.456	2.381	1.715	1.626	1.466	1.930	2.646
F_{IS}	-0.176	0.115	-0.123	-0.088	0.125	0.182	0.130
AG_03							
H_O	0.575	1.000	0.967	0.720	0.405	0.938	0.579
H_E	0.434	0.635	0.499	0.518	0.544	0.691	0.460
A	3.000	4.000	2.000	3.000	3.000	6.000	4.000
A_E	1.767	2.740	1.996	2.075	2.193	3.236	1.852
F_{IS}	-0.314	-0.553	-0.933	0.371	0.267	-0.327	-0.234
AG_07							
H_O	0.537	1.000	0.500	0.800	0.540	0.706	0.550
H_E	0.429	0.653	0.530	0.540	0.422	0.600	0.671
A	3.000	4.000	3.000	2.000	3.000	4.000	5.000
A_E	1.751	2.882	2.128	2.174	1.730	2.500	3.040
F_{IS}	-0.240	-0.511	0.077	-0.466	-0.269	-0.146	0.205
AG_09							
H_O	0.619	0.750	0.692	0.375	0.024	0.125	0.909
H_E	0.520	0.553	0.644	0.520	0.047	0.602	0.549

Table S1. Contd.

A	3.000	3.000	3.000	3.000	3.000	4.000	3.000
A_E	2.083	2.237	2.809	2.083	1.049	2.512	2.217
F_{IS}	-0.178	-0.328	-0.055	0.298	0.500	0.804*	-0.644
AG_11							
H_O	0.286	0.600	0.875	0.733	0.200	0.000	0.920
H_E	0.272	0.460	0.655	0.518	0.180	0.000	0.755
A	3.000	3.000	3.000	3.000	2.000	1.000	6.000
A_E	1.374	1.852	2.898	2.075	1.220	1.000	4.082
F_{IS}	-0.039	-0.273	-0.316	-0.402	-0.099	-	-0.199
AG_12							
H_O	0.415	0.429	0.577	0.593	0.780	0.750	0.783
H_E	0.362	0.400	0.467	0.450	0.535	0.589	0.751
A	3.000	3.000	3.000	3.000	3.000	6.000	11.000
A_E	1.567	1.667	1.876	1.818	2.150	2.433	4.016
F_{IS}	-0.133	-0.033	-0.218	-0.300	-0.449	-0.241	-0.019
AG_13							
H_O	1.000	0.882	1.000	0.962	0.909	1.000	0.920
H_E	0.725	0.493	0.627	0.499	0.598	0.650	0.748
A	4.000	2.000	4.000	2.000	3.000	4.000	6.000
A_E	3.636	1.972	2.681	1.996	2.488	2.857	3.968
F_{IS}	-0.370	-0.778	-0.583	-0.923	-0.511	-0.519	-0.211
AG_15							
H_O	0.300	0.500	0.240	0.370	0.310	0.444	0.250
H_E	0.255	0.506	0.339	0.308	0.262	0.633	0.539
A	2.000	3.000	3.000	3.000	2.000	5.000	5.000
A_E	1.342	2.024	1.513	1.445	1.355	2.725	2.169
F_{IS}	-0.164	0.044	0.311	-0.185	-0.171	0.323	0.554*
Total							
H_O	0.473	0.695	0.551	0.542	0.482	0.559	0.591
H_E	0.383	0.516	0.460	0.422	0.406	0.501	0.559
A	2.684	3.105	2.737	2.632	2.684	3.947	4.632
A_E	1.621	2.066	1.852	1.730	1.684	2.004	2.268
F_{IS}	-0.225	-0.317	-0.179	-0.266	-0.177	-0.085	-0.035

*Locos with deviation for Hardy-Weinberg equilibrium (HWE), $p < 0.05$ (Bonferroni correction; 0.0004).

Table S2. Member proportions of each population according to the seven clusters determined with Structure Software.

Populations/cluster	1	2	3	4	5	6	7
PFP-1	0.002	0.003	0.003	**0.984**	0.002	0.003	0.002
PFP-2	0.003	0.003	**0.978**	0.004	0.005	0.006	0.003
PFP-3	0.002	**0.980**	0.003	0.004	0.003	0.004	0.003
DNOCS	0.002	0.003	0.003	0.003	0.003	**0.983**	0.002
PPV	0.003	0.003	0.003	0.003	**0.981**	0.004	0.003
SDRFB	0.012	0.005	0.004	0.003	0.011	0.004	**0.961**
SDRM	**0.962**	0.004	0.004	0.004	0.004	0.004	0.019

(Contribution > 0.9 in bold).

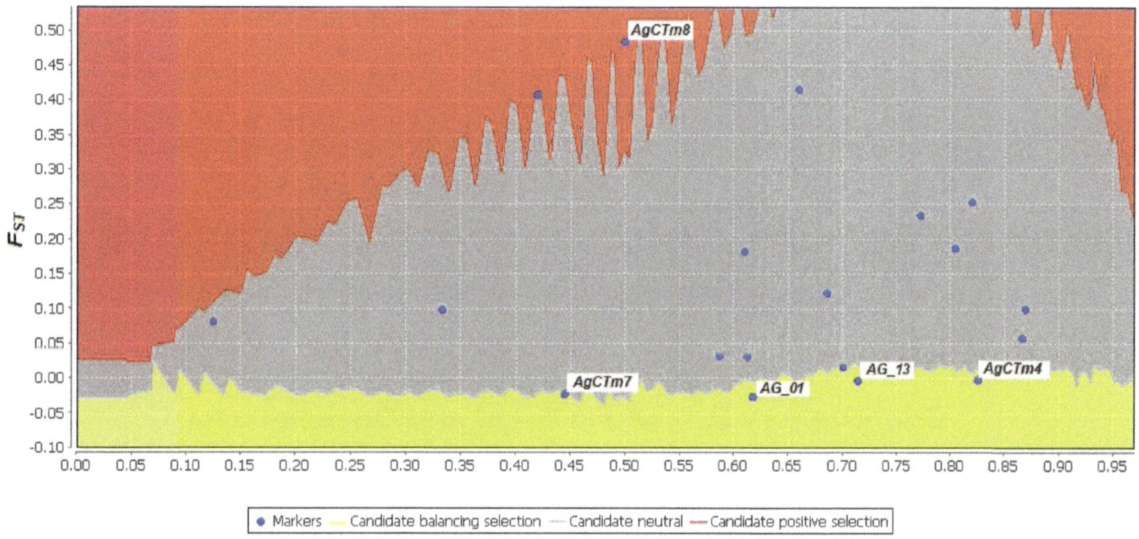

Figure S1. Outlier loci with positive selective pressure for wild individuals.

10

The biology of *Parachanna obscura* (Osteichthyes: Channidae) in Anambra River, Nigeria

Odo, Gregory E.[1]*, Onoja S. U.[2] and Onyishi Grace C.[1]

[1]Department of Zoology, University of Nigeria, Nsukka, Nigeria.
[2]Department of Home Science and Nutrition, University of Nigeria, Nsukka, Nigeria.

The biology of a commercially important teleost, *Parachanna obscura* was studied for a period of twelve months. Monthly samples of the species were randomly collected from fishermen in the lower reaches of Anambra River, Southeastern Nigeria for the twelve months. The haematological profile established were mean haematocrit of 22.5%, mean blood haemoglobin concentration of 7.23 g/dl, mean red blood cell sedimentation rate 31.25 µm, mean red blood cell count 3.72 (10^6/L), white blood cell count 2.81×10^4/L, mean cell volume 60.75 μm^3 and mean corpuscular haemoglobin 19.5 pg. The blood groups are O^+ (10%), O^- (90%), genotype AA (88%) and AS (12%). Digestive enzymes assays in the different gut regions of *P. obscura* revealed an array of glycosidase, proteases and lipases. The pattern of distribution and relative activity of the enzymes is consistent with its predatory diet. Total length (TL) ranged from 23.4 to 28.5 cm and weight from 86.0 to 140 g; males were not longer or heavier than females. Maturity occurred earlier in males than females; 50% of both sexes matured at 24.7 cm TL. Fecundity ranged from 126 to 1580 oocytes (mean 896 ± 477 oocytes). Total length was a better predictor of fecundity than ovary weight. Index of food significance (IFS) showed that insect (51.67%) was the dominant food group, followed by fish (15.66%) and frog eggs (8.02%). Food items of primary importance were ephemeroptera nymphs, chiromomidae, unidentified insects and fishes.

Key words: Anambra River, *Parachanna obscura*, digestive enzymes, haematological profile, blood groups.

INTRODUCTION

The snakeheads (Baily, 1994) are members of the freshwater perciform fish family Channidae, native to Africa and Asia. In parts of Africa, the snake head is considered a valuable food fish and is produced in aquaculture (Teugels et al., 1984). It is commonly available in local fish markets and it is one of the cheapest sources of proteins. One of the difficulties in assessing the state of health of natural fish population has been the paucity of reliable references of the normal condition. In pursuant of this goal, many fish physiologists have turned to the studies of hematology, probably because it has proved a valuable diagnostic tool in evaluating human health. Although fish hematology continues to offer the potentials of a valuable tool, progress in establishing normal range values for blood parameters has been slow and literature in this area is isolated and often incomplete (Mawdesley-Thomas, 1971). Perhaps, further confounding these data are variables such as age, sex, dietary state, and stress, all of which may alter blood values (Barnhart, 1969; McCarthy et al., 1973).

The hematological profile of some tropical African fish species have been reported, namely, *Clarias isheriensis* (Kori–Siakpere, 1985), *Clarias gariepinus, Heterobranchus longifilis* and *Clarias nigrodigitatus* (Erondu et al., 1993; Sowunmi et al., 2003) *Oreochromis niloticus* (Fagbenro et al., 1999), *Hemichromis fasciatus* and *Tilapia zillii* (Egwuyenga et al., 1999), *Sarotherodon*

*Corresponding author. E-mail: odogreg@yahoo.com.

melanotheron (Gabriel et al., 2007) and *Heterobranchus bidorsalis* (Fagbenro et al., 1993). Meanwhile, there are no reports on the haematological profile of *Parachanna obscura* from Anambra River Basin. Hence there is need to study the hematological profile to provide some useful information on aspect of its biology. Moreover, the ability of an organism to digest a given material is dependent on the presence of appropriate enzymes. No information is available on the quantitative and qualitative assays of digestive enzymes in the gut of *P. obscura*, from Anambra River Basin in contrast to other African fresh-water fishes whose digestive enzymes assays have been established (Olatunde et al., 1988, Fagbenro, 1990; Fagbenro et al., 1993). Considerable biological studies have been undertaken and documented on some economically important tropical fish families in Anambra River Basin; for instance, Clariidae et al. (1992), Mgbenka and Eyo (1992), Ezenwaji and Inyang (1998), Distochontidae (1998) and Clupeidae et al. (2003). Nevertheless, studies of the *P. obscura* feeding biology are scanty. To be able to manage a resource of such commercial importance, knowledge of its biology is imperative.

The study of dietary habits of fish based on stomach content analysis is widely used in fish ecology as an important means of investigating trophic relationships in aquatic communities Thus, this study, which forms part of a larger and on-going investigation on the fish and fisheries of the river basin, addresses aspects of the biology of the species viz. relative activities of enzymes in the different gut regions, the haematological and sero-logical profiles of *P. obscura* in the Anambra River Basin.

MATERIALS AND METHODS

Study area

The Anambra River (Figure 1) has its source in Ankpa highlands of Kogi state of Nigeria, about 100 km North of Nsukka (Azugo, 1978). It lies between latitudes 6°10' and longitude 7°15' East of the River Niger. Essentially, the river has a southward course crossing the Kogi / Enugu state boundary, and then meanders through Ogurugu to Otuocha, from where it flows down to its confluence with the River Niger at Onitsha. The main river channel, which has a total length of about 207.40 km (Azugo, 1978), has its bank covered by such plants like *Echinochloa* spp., *Salvinia nymnellula, Ludwigia decurrens, Imperata cylindrica, Andropogon* spp., *Jussica* spp., *Pennisetum* spp. and *Cynodon* spp. There is a rainy season (April - September/ October) and a dry season (October/ November-March). From December to January/February, the basin is influenced by the harmattan but their effect is not well marked. Agricultural activities are very high and crops such as yam, cassava, rice, millet, vegetables, groundnuts, potatoes, banana and plantain are produced in large quantities. Fishing methods in the river basin include bailing out of water or pumping out water from ponds with water pumps, construction of fish fences, the use of "atalla", hooks and line, set lines, lift nets, dragnets, beach seines, cast nets, among others (Awachie and Ezenwaji, 1981; Eyo and Akpati, 1993). Species of fish found in the river include *Channa (Parachanna obscura*), *Labeo, Distichodus, Alestes, Mormyrids, Clarias,* and *Heterobranchus* among others.

Fish collection, morphometrics and stomach content analysis

Monthly samples of *P obscura* were randomly collected from fishermen in the lower reaches of Anambra River, Southeastern Nigeria, between May 2007 and March 2008 inclusive. Total length (TL) to the nearest 0.1 cm and body weight (g) to the nearest 0.1 g of each *P. obscura were* measured and the sex determined. The length-weight relationships (LWR) were determined using the power curve: $W= aTL^b$ (Ricker, 1973). The parameters of the LWR were determined for the same species collected at different periods. These different estimates were considered separate 'populations'. Fulton's condition factor, K, was calculated as: $K = W/L^3 \times 100$. To determine gonad maturation, gonads were evaluated macro-scopically and four maturation stages recognized: I-immature, II-mature, III-ripe and IV – spent. Size at maturity was determined as the length at which 50% of individuals were in gonad stage II.

Fecundity is defined as the number of spawnable oocytes in both ovaries of the species, The relationship between fecundity and gonad weight and total length was determined by the least squares method. The gonad somatic index (GSI) was determined as: $GSI = W_1/W_2 \times 100$, where W_1 = gonad weight (g) and W_2 = fish weight (g) (less gonad weight). The stomach of each *P. obscura* was dissected and silt open, and its degree of fullness estimated by an arbitrary 0 - 20 point scale: thus 0, 2.5, 5, 10, 15 and 20 points were allotted to empty, trace, quarter-full, half-full, three quarter-full and fully distended stomachs. Stomach contents were sorted into categories and analyzed using relative frequency (RF) and point's method (Hynes, 1950; Hyslop, 1980). In the RF, the frequency of a particular food item in all stomachs was expressed as a percentage of the frequencies of all food items. For the point's scheme, each stomach was allotted 20 points regardless of the fish size and these were shared amongst the various contents, taking account of their relative proportion by volume. The points gained by each food item in all stomachs examined were computed and expressed as a percentage of the total points of all food items. The point scheme gave an indication of bulk contribution of each food category to the diet composition. % RF and % PP (percentage of total points) were then used to determine the index of food significance IFS as follows:

$$IFS = \frac{\%RF \times \%PP}{\sum(\%RF \times \%PP)} \times 100$$

Food with IFS ≥3% was regarded as primary, ≥ 0.1 to <3% as secondary, whereas food with <0.1% was regarded as incidental. The IFS of each were used to compute diet breadth based of Shannon-Wiener function $\overline{H}$ percentage of total points percentage of total points as follows:

$$\overline{H}_{(IFS)} = \sum_{i=1}^{s}\left(\frac{n_i}{N}\right)\log_e\left(\frac{n_i}{N}\right)$$

Where ni = IFS of each food item, N=total IFS of all food items.

Haematological profile

Twenty-three live specimens (TL 27 - 41 cm) were kept undisturbed in large glass aquarium (120 liter capacity) supplied with filtered and aerated tap water for 2 weeks of acclimation to laboratory conditions (pH 7 - 8: dissolved oxygen concentration, >6 mg/L: water, temperature, 28 - 30.5° prior to blood sampling. During this period, the fish were fed to satiation twice daily with 380 g/kg body weight of crude protein pellets. All fish were considered healthy on the basis of their appearance and absence of obvious signs of

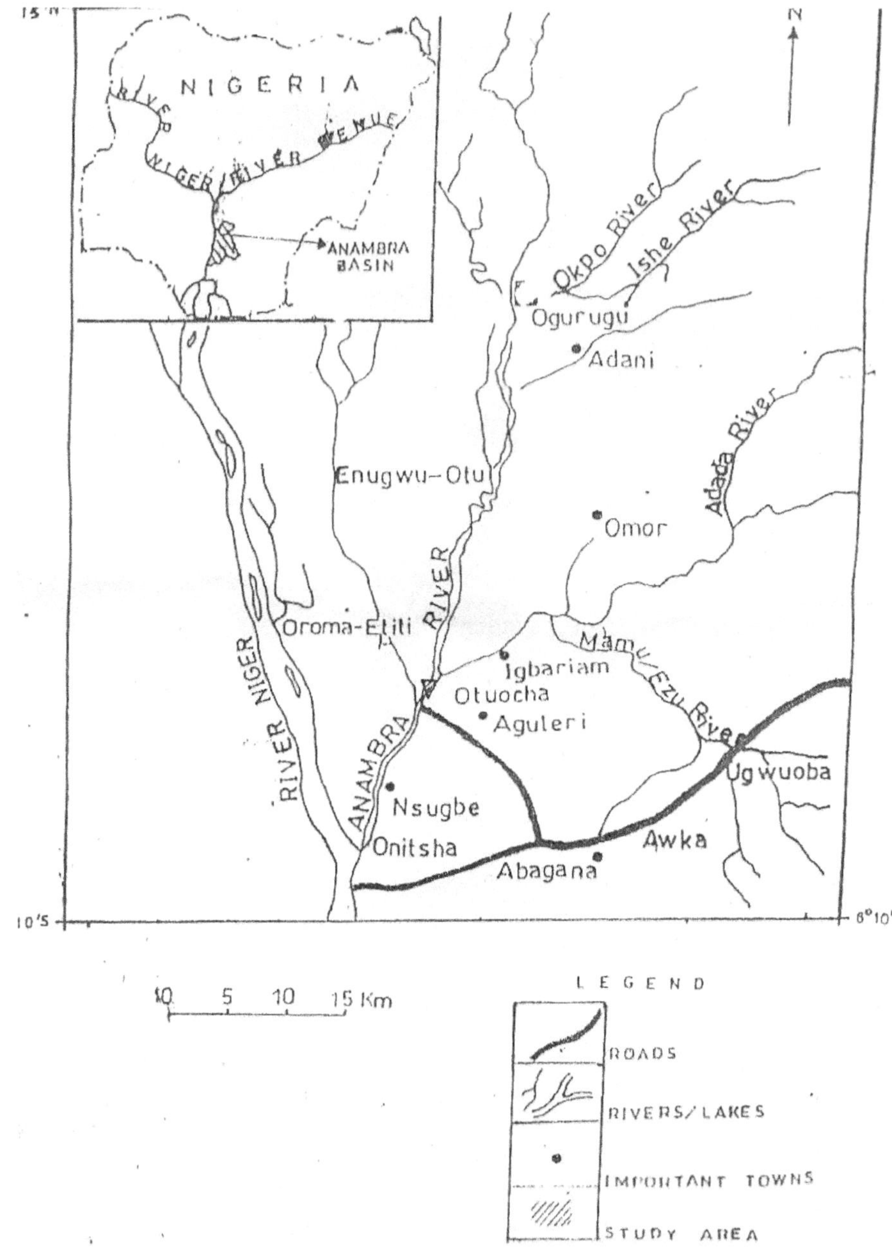

Figure 1. Map of Anambra River.

disease. No sex selection was made. Blood was collected from the caudal vein of each fish using separate heparinized disposable syringes and hypodermic needles. Haematocrit packed cell volume (PCV) was measured after centrifugation at 15000 rpm (MSE Microcentrifuge). Preparations were made for the determination of haemoglobin (Hb) content, leukocyte count (WBC), and erythrocyte count (RBC).

Serological profile

Blood grouping was performed by the test tube techniques based on agglutination tests, while the genotype was determined by haemoglobin electrophoresis (Delany and Garratty, 1969).

Digestive enzyme assays

Twenty-three adult *P. obscura* specimens (TL 29 - 37 cm) were kept unfed for 72 h inside out door concrete cisterns in order to bring them to a similar physiological state as well as to ensure the emptiness of the entire gut. They were anaesthetized with benzocaine and dissected to remove the entire gut, and later separated into the anatomically distinct regions. The different gut regions were pooled and homogenized; the homogenates were then centrifuged at 1200 rpm (Fisher Scientific Company, USA) at 4°C. The supernatants were used as crude extracts without further purification. Benedict's qualitative reagents were used for the qualitative assays of glycosidase following the method of Olatunde et al. (1988), while quantitative assays were conducted using the dinitrosalicylate (DNS) method (Plumber, 1978). Qualitative and quantitative assays of proteases followed the method of Balogun and Fisher (1970).

Data analysis

Data for corresponding months were pooled together for analysis. Food richness was defined as the number of food items in the diet with IFS ≥0.1%. Abundance data were analyzed by two-way analysis of variance (ANOVA). Food composition and sex ratio were analyzed by students-test and "χ^2" test, respectively (Bailey, 1994). Differences were considered significant at 5% level of probability. The methods described by Ogunbiyi and Okon (1976) were used to determine lip activity both qualitatively and quantitatively. Controls were run simultaneously for all assays. Regression analysis was carried out between the various haematological parameters and standard length. The coefficient of regression (r) was then analyzed for statistical significance by Student's t-test. Results were presented as mean with standard deviation (SD). The results were also analyzed using students t-test. The level of significance was $p < 0.05$ at 95% confidence limit.

RESULTS

Distribution and abundance

Of the 40 species of fin fish and three shell fish caught in Anambra River fishery totaling 2435 individuals and weighing 36.81 kg, *P. obscura* was the most abundant in both number (860, 35.32%) and weight (8.52 kg, 23.14%). The *P. obscura* was significantly more numerous in the rainy season months of June to October than in the dry months of November to March ($P<0.05$).

Size range

The length of *P. obscura* (n = 550) ranged from 26 - 37 cm total length (mean 28 ± 88 cm TL). Males ranging from 23 – 28 cm TL (25 ± 0.84) were not longer than females that range from 27 - 39 cm TL (mean 28 ± 0.84 cm TL). Weight ranged from 86 - 140 g (mean 110 ± 0.76 g). The weight of the males ranging from 87 - 151 g (mean 89 ± 0.72 g) was not significantly heavier than the weight of females that range from 45 - 69 g, (mean 54 ± 0.76 g)

Population structure

The population structure is presented in Figure 2. From this figure, it can be seen that both sexes had one mode (>53% individuals) at 26 cm TL. The 25 - 27 cm TL size classes (represented by 3-5) constitute over 90% of the total sample (Figure 2). The overall monthly sex ratio was significantly different from 1:1 (χ^2 = 5.21, df=1; $P<0.05$, (Table 1; Figure 3). Within the months, females dominated in March (χ^2 = 9.68, df = 1, $P<0.05$) and May (χ^2 = 5.33, df = 1, $P<0.05$). There is seasonal variation in the number of individuals depending on the sexes. For example, during the dry season the sexes (male = 178, 42.5%; female = 211, 57.5%) showed significant difference in the number collected (χ^2 = 5.4, df = 1, p = 0.05). However, during the rainy season the difference was not significant as the number of males collected (141, 47.2%) did not differ markedly from the number of females collected (158, 52.8%) (χ^2 = 0.96, $P >0.05$).

Morphometric parameters

The LWR analyses of the 12 populations are presented in Table 2. The analyses further indicated that the calculated corrections was significant ($P<0.05$) with coefficient of determination ranging from 48 to 98.5%. These values showed that the intercept 'a' exhibited moderate variation among the populations (CV = 52.23%) and ranged from $a_{min} = 8.6 \times 10^{-3}$ (overall) to $a_{max} = 31.7 \times 10^{-3}$ (March). Conversely, the exponent 'b' showed low variations among the population (CV = 9.64%) and varied from b_{min} = 2.192 (June) to b_{max} = 2.939 (overall). The mean exponent (b = 2.66; SD. = 0.257) was significantly less than 3 (t-test, df = 11, $P <0.05$), indicating negative allometric function for the populations. Table 2 also showed that LWR exhibited in both the rainy and dry seasons were isometric.

Fulton's conditions factor, K

The mean monthly K was 0.78 ± 0.11 and varied from 0.61 ± 0.16 in June to 0.92 ± 0.10 in March, (Figure 4). The fish species were in very good condition in September, October, February and March, whereas the poorest conditions occurred in January, June and July.

Reproductive biology

The mean monthly gonad somatic index (GSI) of 94 female *P. obscura* varied from 1.1 - 3.3% (mean 2.05 ±0.72%). There were only slight variations in the mean monthly GSI indicating an all year round reproductive pulse, although three peaks in June, September and January were evident (Figures 4 and 5). The monthly dynamics in the percentage male and female at each maturation stage (Table 3) showed that immature, mature, ripe and spent gonads were present throughout the year indicating an all year round gonad recrudescence, breeding period and recruitment. There were significantly more females than males in immature stage (χ^2 = 4.4, df = 1, $P<0.05$) and spent stage (χ^2 = 19.6, df = $P<0.05$). Males and female did not depart from a 1:1 sex ratio in mature and ripe *P. obscura* (Table 4). The size at maturity in females and males differed. The smallest female matured at 24.3 cm TL, whereas in the males it was at 23.4 cm TL. Over 50% of both sexes matured a

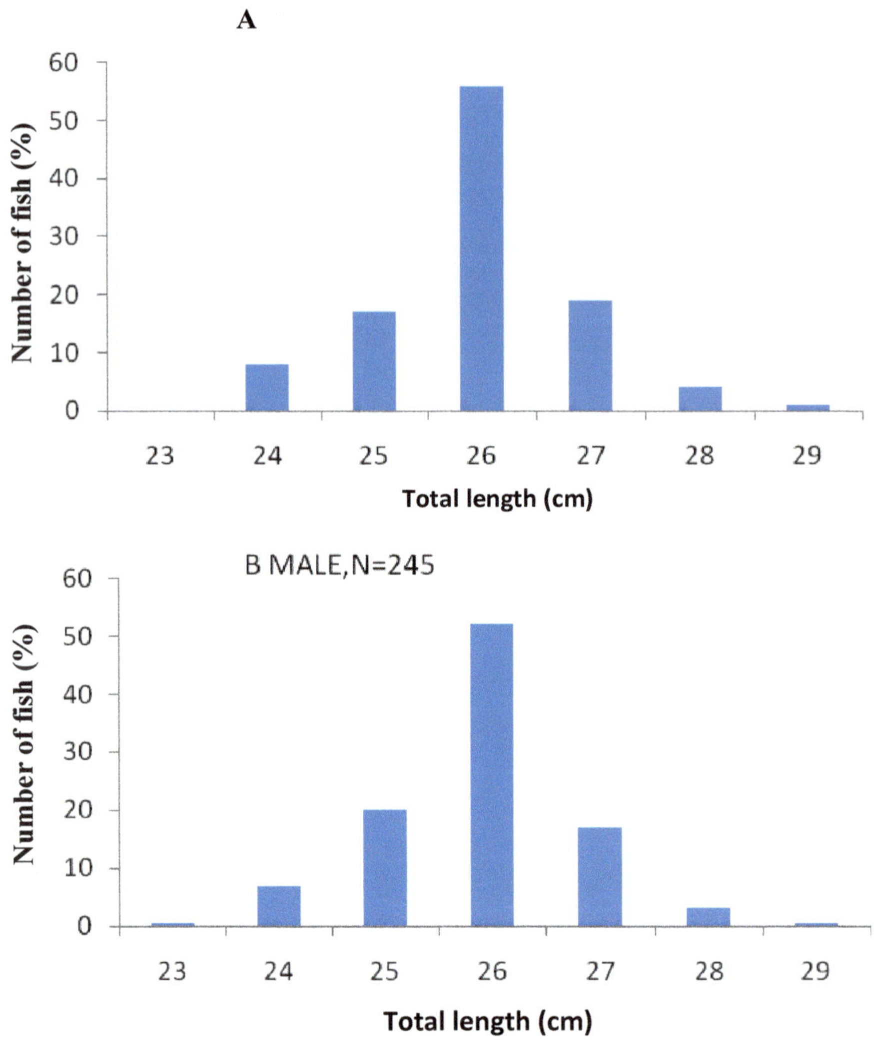

Figure 2. Length-frequency distribution of (A) female and (B) male *P. obscura* in Anambra River.

Table 1. The overall monthly sex ratio of *P. obscura* in Anambra River.

Month	Number collected		Sex ratio (M : F)
	M	F	
M	30	54	1:2.6
J	39	18	1:1.2
J	46	36	1:0.7
A	40	40	1:1.2
S	44	37	1:0.8
O	39	44	1:1.1
N	38	39	1:1.1
D	34	46	1:1.5
J	37	43	1:1.3
F	38	37	1:1.1
M	1631	46	1:2.0
Total	420	440	1:1.3

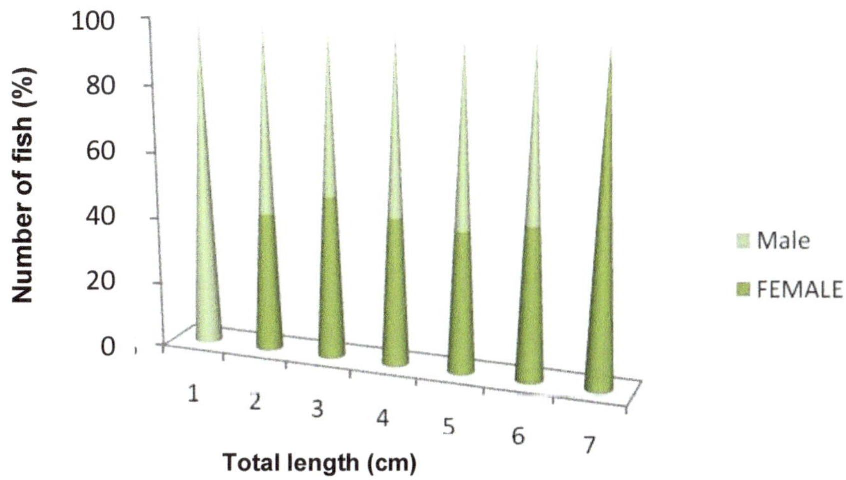

Figure 3. Variation in sex ratio of *P. obscura* in Anambra River.

Table 2. Length-weight relationships and related statistics in *P. obscura* (May 2007 - March 2008).

Month	Total length (cm)				Length-weight relationships			
	Mean	S.D	Min	Max	A	B	r	N
M	26.00	0.67	25.0	28.4	0.0099	2.821	0.946	50
J	26.37	0.73	23.4	26.7	0.0227	2.192	0.735	50
J	26.56	0.56	25.9	28.5	0.0087	2.890	0.992	50
A	26.80	0.46	26.1	28.2	0.0112	2.906	0.977	50
S	25.45	0.59	24.6	27.3	0.00153	2.704	0.972	50
O	26.48	0.65	25.6	28.3	0.0307	2.348	0.949	50
N	26.04	0.43	25.3	27.3	0.0113	2.801	0.933	50
D	24.53	0.51	23.8	26.0	0.0135	2.612	0.968	50
J	25.70	0.28	25.1	26.3	0.0133	2.564	0.692	50
F	25.78	0.81	24.7	28.0	0.0116	2.859	0.967	50
M	26.20	0.50	25.4	27.9	0.0317	2.317	0.884	50
Overall	25.99	0.63	23.4	28.4	0.0086	2.939	0.941	550
Rainy season	26.3	0.57	23.4	28.4	0.0081	2.942	0.945	300
Dry season	25.6	0.62	23.8	28.3	0.0060	3.169	0.941	250

24.7 cm TL. Oocyte count of 15 females ranged from 126 to 1580 (mean 896 ± 477 oocytes). The regression equations for the relationships between fecundity and total length, and ovary weight ($F = aX^b$, where X stands for either total length or ovary weight) were: $F = 21.93TL^{1.82}$, r = 0.37; $F = 847.65OW^{0.04}$, r = 0.02. The correlations were positive but low; total length had a better predictive value of 0.37 than ovary weight (0.02).

Diet

Out of 550 stomachs examined, 56(10.2%) had full stomachs (FS), 396 (72%) had partially-filled stomachs (PS), whereas 97(17.6%) had empty stomachs (ES). Full stomachs were highest in May (6.46%) and lowest in September, January and March (0.0%). Among the partially-filled stomachs, 41(l0.35%) were ¼ full, 120 (30.30%) ½ full, 193 (48.74%) $^3/_4$ full and 42 (10.61%) obtained traces of food. Twenty-four different food items were ingested (Table 5). Insects (51.67%) were the dominant food group followed by frog eggs (15.66%) and fish (8.02%). Foods of primary importance were *Ephemeroptera* nymph, Chiromomidae, unidentified insects and fishes. The most important food item was Chiromomidae (IFS = 36.95%); which occurred more regularly in the stomach and constituted a large proportion of the food taken. Of all the food items ingested, only *Ephemeroptera* nymphs and chironomids were consumed in all the months. Food of primary

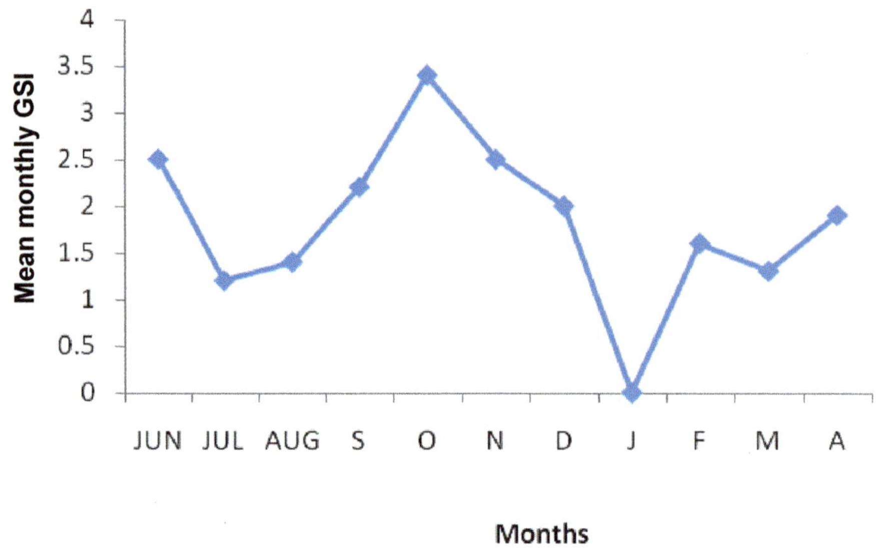

Figure 4. The mean monthly GSI of 94 mature female of *P. obscura*.

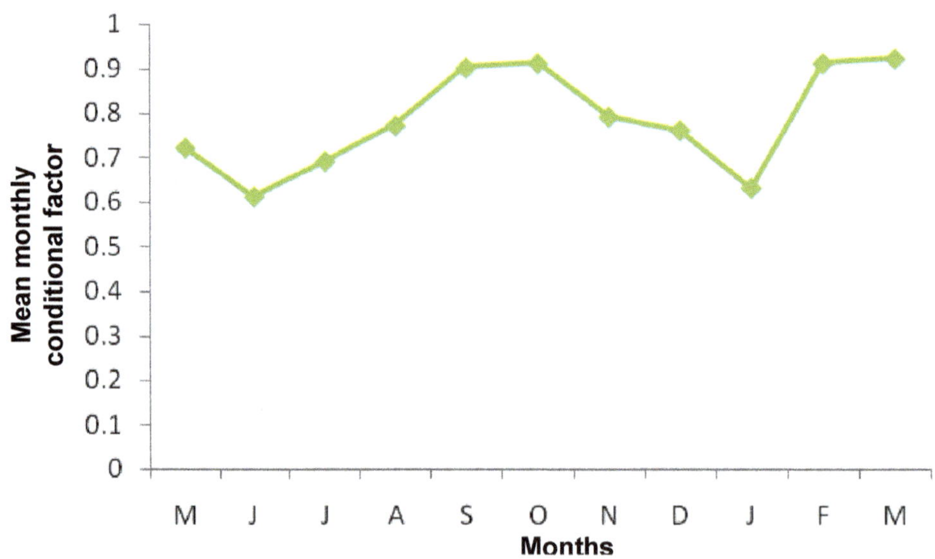

Figure 5. Mean monthly conditional factor of *P. obscura* from Anambra River.

importance varied May 1st to October 6th (Table 6). The total mean and range were also estimated

Hematological parameters

The general mean values with the standard deviation of the blood parameters of P. *obscura* are presented in Tables 7. The haematological profile established were mean haematocrit of 22.5%, mean blood haemoglobin concentration of 7.23g/dl, mean red blood cell sedimentation rate 31.25 µm, mean red blood cell count 3.72 (10^6/L), white blood cell count 2.81 × 10^4/L, mean cell volume 60.75 µm^3 and mean corpuscular haemoglobin 19.5 pg. Table 10 shows the correlation analysis of the blood parameters. The Linear regression and prediction equation of blood parameters as functions standard length is presented in Table 11, while the regression and prediction equation of blood parameters as functions of standard length and body weight are presented in Table 12. From the correlation analysis, RBC has a weak positive relationship with WBC and MCHC

Table 3. The monthly variation in the percentage female (N=440) and male (N = 420) *P. obscura* at each maturation stage.

Month	No. of females examined	Maturation stage								No. of males examined
		Female				Male				
		I	II	III	IV	I	II	III	IV	
M	36	61.1	36.1	2.8	-	28.6	71.4	-	-	14
J	27	37.0	51.9	11.1	-	21.7	56.5	21.7	-	23
J	20	25.0	25.0	45.0	5.0	16.7	46.7	33.3	3.3	30
A	27	14.8	25.9	51.9	7.4	34.8	43.5	13.0	8.7	23
S	22	22.7	31.8	27.3	18.2	46.4	28.6	21.4	3.6	28
O	26	46.2	34.6	7.7	11.5	30.4	39.1	30.4	-	23
N	24	41.7	16.7	33.3	8.3	54.5	27.3	18.2	-	22
D	28	50.0	25.0	21.4	3.6	36.8	36.8	26.3	-	19
J	28	42.9	42.9	7.1	7.1	27.3	40.9	31.8	-	22
F	26	26.9	19.2	15.4	38.5	17.4	26.1	47.8	8.7	23
M	32	3.1	9.4	59.4	28.1	12.5	6.3	-81.3	-	16
Total	296									

Table 4. The dynamics of female and male *P. obscura* in maturation stages.

Maturation stage	Female	Male	Sex ratio (M.F)
Immature	102	73	1: 0.7
Mature	86	93	1: 1.1
Ripe	74	71	1: 1
Spent	34	6	1: 0.2
Total	296	243	1: 0.8

(P<0.05). There is high positive relationship between RBC and PCV at P < 0.05 level of significance but there was a negative correlation with ESR. Moreover, there is a negative correlation between RBC and ESR. RBC has a negative correlation with MCV but has no significant correlation with MCH. WBC has a significance correlation with PCV (P<0.05 WBC has a significance correlation with HB (P<0.05) but no significance correlation with ESR, MCV, MCH and MCHC P > 0.05.

PCV has a high positive correlation with Hb at the (P<0.01), but has a negative correlation with ESR. On the other hand, PCV has no significant correlation with MCV. It has significant correlation with MCH and MCHC (P<0.01). Hb had a negative correlation with ESR but no significant correlation with MCV. PCV had a significant high positive correlation with MCH and MCHC (P<0.01); ESR had a significant correlation with MCV at P< 0.01 level of significance but no significant correlation with MCH and MCHC. MCV had a high positive relationship with MCH and MCHC at the 0.01 level of significance. MCH also had a positive correlation with MCHC which was also very high at (P<0.01; P<0.01), while weight has little or no significance with it.

Serological parameters

The blood group and genotype are presented in Tables 8 and 9, respectively. Among the 64 samples of *P. obscura*, 6 fishes were O+ with 10% prevalence, while 58 were 0^- with 90% prevalence. There is no group A, group B or group AB; the prevalent blood group is O+ genotype. Among the fish samples, 56 had AA genotype with 88% prevalence, while eight fish had AS genotype with 12% prevalence. Moreover, none expressed SS genotype.

Digestive enzyme assays

Table 12 shows the various enzymes detected in the different regions of the *P. obscura* gut, their distribution and activity varying along the gut length. A variety of glycosidase was detected (Table 14). Cellulase activity was recorded only in the pyloric caeca (Table 12). The protein hydrolyzing enzymes found in the stomach are pepsin-like while those in the pyloric caeca are alkaline proteases, possibly trypsin and /or chymotrypsin. Lipase activity occurred along the entire gut length with peaks in

Table 5. Trophic spectrum of the diet of all sizes of *P. obscura* in Anambra River.

Food species/group	%RF	%PP	IFS
Crustacean			
Ostracoda	2.31	2.48	0.53
Macrobrachium felicium	2.31	2.73	0.58
Insecta			
Ephemeroptera nymph	6.45	6.66	3.97
Coenagrion sp.	0.10	0.11	+a
Water bugs	1.15	0.89	0.09
Dystiscidae	0.48	0.55	0.02
Formicoid Hymenoptera	1.06	0.99	0.10
Mosquito larvae	0.29	0.31	36.95
Chironomid larvae and pupae	17.90	22.33	36.95
Chaoborus larvae	0.29	0.29	0.01
Terrestrial Orthoptera	2.89	3.22	0.86
Neuropteran larva	1.25	1.03	0.12
Unidentified insect	10.78	9.57	9.54
Fish			
Alestes.	0.29	0.32	0.01
Barbus callipterus	0.10	0.11	+
Epiplastys	0.38	0.60	0.02
Chrysichthys sp.	0.19	0.29	0.01
Fish remains	7.31	7.80	5.27
Fish fry	11.93	13.82	15.24
Fish egg	1.54	1.68	0.24
Nematodes	1.35	1.47	0.18
Macrobrachium felicium	10.39	8.35	8.02
Tadpole	2.02	2.29	0.43
Frog eggs	16.75	11.48	17.77

[a]+ <0.01

generally survival strategies and adaptations aimed at perpetuating the species in response to high fishing and/or natural mortality. The survivors prey on the rich variety of food available in the river; they grow fast and become recruited into the fishery. It seems probable that it is in this way that large numbers of *P. obscura* are maintained in the Anambra River. Ezenwaji (1999, 2002) has recorded similar situations in *Clarias albounclatus* and *C. ebriensis* in the same river system.

While a fairly good knowledge of the breeding biology of *P. obscura* is beginning to emerge in Nigerian lentic and lotic habitats, we still need as Marshal (1993) noted in *Limnothrissa miodon* to ascertain the environmental factors determining reproductive success, the effect of fishing on the sexually mature individuals and the relationship between stock and recruitment. Furthermore, knowledge of other demographic characteristics (such as growth and mortality) of *P. obscura* is also important in order to gain an overall understanding of factors determining its abundance in the Anambra River. The size range *P. obscura* is a good reflection of its population structure. The preponderance of the fish between 25 and 27 cm TL is probably because the shallow water habitat of fish less than 25 cm TL may not have been adequately sampled, and fish up to 9 cm TL would have been exploited as a result of the high fishing and natural mortality.

In fishes, generally, the annual production of oocytes accounts for a large proportion of the body weight. *P. obscura* of the Anambra River appears to be a notable exception. With an average of 2.05 ± 0.72%, the oocyte weight/body weight ratio of this clupeid appears to be one of the lowest in fishes.

This is slightly different from the average ovary weight/body weight ratio; 4.6% in *P. afzeius* and 7.4% in *Siermtllrissa leonensis* in Kainji lake (Otobo, 1978), and 6.1 ± 2.5% (range 2:2 -1: 2.2%) in *C. ebreiensis* in the Anambra River system (Ezenwaji,

Table 6. The monthly IFS of *P obscura* in Anambra River.

Food species/group	M	J	J	A	S	O	N	D	J	F	M
Crustacean											
Ostracoda	-	-	-	-	-	9.29	3.17	0.50	1.78	0.64	0.05
Macrobrachium felicium	-	-	-	0.01	-	11.29	0.87	0.12	-	0.28	10.10
Insecta											
Ephemeroptera nymph	0.43	41.38	0.49	3.75	0.78	0.29	0.29	0.56	6.12	5.28	6.13
Coenagrion sp	-	-	-	-	-	1.10	-	-	-	-	-
Water bugs	-	-	0.11	0.01	0.05	-	-	9.78	0.44	0.05	-
Dystiscidae	-	-	-	-	-	0.86	-	-	-	-	0.34
Formicoid Hymenoptera	-	-	-	-	-	0.18	0.04	-	-	0.08	-
Mosquito larvae	-	-	-	-	-	0.86	-	0.08	-	-	-
Chironomid larvae & pupae	99.57	55.19	8.26	35.22	0.31	1.62	0.76	1.66	0.69	1.77	20.15
Chaoborus larvae	-	-	-	-	-	0.18	0.09	-	-	-	-
Terrestrial Orthoptera	-	-	-	-	-	-	-	-	0.55	0.53	1.79
Neuropteran larvae	-	-	-	-	-	2.02	0.55		0.52	0.08	0.46
Unidentified insect	-	-	-	1.11	35.37	45.70	57.24	28.42	36.01	3.04	9.50
Fish											
Brycinus sp	-	-	0.22	0.03	-	-	0.09	-	-	-	-
Alestes	-	-	2.32	0.05	-	-	-	-	-	-	-
Chrysichthys sp	-	-	0.54	0.09	-	-	-	-	-	-	-
Barbus callipterus	-	-	-	0.03	-	-	-	-	-	-	-
Fish remains	-	1.52	36.56	8.30	3.45	1.84	0.02	1.48	13.46	3.39	-
Nematodes	-	1.27	34.31	40.34	22.78	5.60	5.09	6.79	7.58	30.41	11.62
Frog eggs	-	-	-	-	-	-	-	0.15	0.07	-	0.05
Prawns	-	-	-	-	-	1.47	-	0.32	1.18	0.05	0.25
Cray fish	-	0.07	1.88	0.59	-	7.04	3.68	28.67	11.66	41.70	4.07
Mud	-	-	-	-	-	4.86	15.65	0.08	0.25	0.25	0.20
Sand	-	0.57	15.30	10.46	37.26	5.80	12.37	20.37	19.69	12.37	35.29
Food richness	2	6	10	13	7	17	15	15	14	16	14
Diet breadth	0.03	0.85	1.47	1.40	1.25	1.94	1.40	1.71	1.83	1.55	1.85

Table 7. Summary of estimated haematological characteristics of *P. obscura.*

Haematological parameters	Mean ± SD	Range
Standard length (SL) (cm)	25 .13 ± 2.18	22.00 – 29.00
Total length (TL) (cm)	29.88 ± 2.78	26.00 – 37.00
Body weight (wt) (g)	110.00 ± 16.11	86.00 – 140.00
RBC (10^6)	3.72 ± .74	1.50 – 5.10
WBC (10^4)	2.81 ± .22	25800.00 – 31800.00
PCV (%)	22.50 ± 4.17	16.00 -32.00
Hb (g/dl)	7.23 ± 1.72	3.70 - 12.20
ESR (%)	31.25 ± 11.22	12.50 – 49.00
MCV (μm^3)	60.75 ± 3.18	57.00 - 68.00
MCH (pg)	19.50 ± 2.83	15.00 – 26.00
MHCH (%)	32.00 ± 2.89	26.00 – 40.00

Table 8. The haematological characteristics of blood group of *P. obscura*.

Blood group	No of fish positive	Prevalence
A	None	None
B	None	None
B	None	None
O^{+}	6	10%
O^{-}	58	90%

Table 9. The haematological characteristics of the genotype of *P. obscura*.

Genotype	No. of positive Fishes	Prevalence (%)
AA	56	88
AS	8	12
SS	0	-

2002). A similar situation obtained for *L. miodon* in the northwestern waters or Lake Tanganyika (Muhmbua, 1993). Thus, this phenomenon appears typical of fishes that mature early and breed all the year round.

The length (24.27 cm TL) at which 50% of the sample matured ensures that the largest number of the most fish (those in the 24 - 28 cm TL bracket) is in reproductive state. This agrees with the size at maturity in *P. obscura.* As size at maturity of fish depends on the growth rate, it does seem that there are no significant differences in the size of this teleost in Anambra River and in Kainji lake environments. A large number of autochthonous and allochthonous insects constitute important proportion of the food or many fishes inhabiting the Anambra river system (Ezenwaji and Inyang, 1998; Ezenwaji, 1999). Thus, *P. obscura* of this study, although omnivorous, subsists mainly on insects (Table 6). As ephemeropterans, chironomids, algae, fish, terrestrial Orthoptera and formicoid Hymenoptera are largely ingested, it implies that *P. obscura* feeds in or on the substratum in mid-water and in the air-water interface. Apart from eating other fish (e.g. 8 m-*bus callipterus),* the exhibition of the phenomenon of cannibalism by *P. obscura,* even in the face of a rich variety of other high quality food items, is attributed to light attraction employed by fishers which Gliwicz (1984) points out may lead to this kind of abnormal behavior in fish. Thus, *L. miodon* inhabiting lake Kariba, Zimbabwe, is omnivorous, feeding predominantly on plankton, followed by insects (particularly chironomids, ephemerterans, trichopterans and hemipterans) and fish, and exhibits cannibalism (Chifamba, 1993). *Microtllrissa acutirostris* is omnivorous in lake Mweru-Wantipa, Zambia (Mubamba, 1993); it feeds mainly on zoo- and phyto-plankton and rotifer. It appears, therefore, that those teleosts are able to exploit all food niches (bottom, mid-water and water surface) in their habitats. Thus, they exhibit wide plasticity in their feeding, primarily consuming a combination of two or more of insects, crustaceans, plankton, fish algae, plant detritus sand rotifers, depending on availability and abundance of these food items.

Clinical haematological studies are of great importance in the epidemiology, management and husbandry practices, veterinary diagnosis; treatment and prognosis in many disease conditions (Benjamin, 1961; Cole, 1967). Undoubtedly, these haematological characteristics of fish are affected by age, sex, nutrition, seasons as well as other physiological conditions of blood such as composition of serum protein and pathological state. The ranges of blood parameters determined for *P. obscura* are generally similar to those determined for African fresh water fishes reported. The mean haematocrit value of *P. obscura* is comparable to those of African Clariid fishes *C. isheriensis*, (Kori-Siakpere, 1985), *C. garienpinus* (Adeyemo, 2007) *H. longifilis* (Valenciennes) (Erondu et al., 1993) and *Sarotherodon melanotheron* (Gabriel et al., 2007). Badawi and Said (1971) listed some blood parameters of Nile Tilapia, *Tilapia zillii* and *Tilapia aurerurs* and concluded that blood picture is not a dull and static phenomenon, but It depends on what is happening to the fish at any particular locality and time and so it is very dynamic. Lagler et al. (1977) stated that whereas red blood cells range from 20,000 to 3,000, 000 per mm^3 of blood; white blood cells vary between 20,000 and 50,000 per mm in different group of fishes. The values obtained for *P. obscura* fall between these parameters. The number of circulating red blood cells is a response to both internal and external stress and in the relation to oxygen demand (Maclean, 1978). The permeability of the red blood cells is a function of the lipids and proteins of the membranes (Ness and Stengle, 1974). It is natural for the body of animals to fight infections of parasites and diseases through the immune system made up of immunoglobulin or antibodies in the blood,

Table 10. Correlations of the blood parameters.

Blood parameter	Source of variation	RBC	WBC	PCV	Hb	ESR	MCV	MCH	MCHC
RBC	Pearson Correlation	1							
	Sig. (2-tailed)								
	N	64							
WBC	Pearson Correlation	0.317(*)	1						
	Sig. (2-tailed)	0.011							
	N	64	64						
PCV	Pearson Correlation	0.900(**)	0.328(**)	1					
	Sig. (2-tailed)	0.000	0.008						
	N	64	64	64					
Hb	Pearson Correlation	0.826(**)	0.253(*)	0.894(**)	1				
	Sig. (2-tailed)	0.000	0.043	0.000					
	N	64	64	64	64				
ESR	Pearson Correlation	-0.648(**)	0.171	-0.634(**)	-0.475(**)	1			
	Sig. (2-tailed)	0.000	0.177	0.000	0.000				
	N	64	64	64	64	64			
MCV	Pearson Correlation	-0.088	0.005	0.086	0.214	0.442(**)	1		
	Sig. (2-tailed)	0.491	0.971	0.501	0.089	0.000			
	N	64	64	64	64	64	64		
MCH	Pearson Correlation	0.206	0.170	0.413(**)	0.502(**)	0.215	0.842(**)	1	
	Sig. (2-tailed)	0.102	0.179	0.001	0.000	0.088	0.000		
	N	64	64	64	64	64	64	64	
MCHC	Pearson Correlation	0.297(*)	0.069	0.447(**)	0.526(**)	0.093	0.655(**)	0.840(**)	1
	Sig. (2-tailed)	0.017	0.586	0.000	0.000	0.464	0.000	0.000	
	N	64	64	64	64	64	64	64	64

** Correlation is significant at the 0.01 level (2-tailed); * Correlation is significant at the 0.05 level (2-tailed).

the white blood cells or leucocytes play this all-important defensive role.

Weinreb (1958) used leukocytes count as a means of assessing the systematic response of the rainbow trout, *Salmo gairdneri* (R), to various injections. Leukocyte counts seem to have wide range of variation 20000 – 63000/mm^3 for brown trout, *Salmo trutta* (L). In the present work, the range of Leukocyte counts was 2.58 – 3.18 × 10^4/mm^3. There is variation between these results and those obtained by some other workers like Kori-Siakpere et al. (2005). However, the variation has no statistical significance. The PCV of P. *obscura* has a high significant positive correlation with RBC, Hb, MCH and MCHC. This is expected since PCV of blood depends on the number of RBC, shape and size of cells and the haemoglobin concentration. Consideration of the intra-species relation between blood parameter as indicated by the correlation coefficient (r) in Table 11 and the intraspecies variations as indicated by the wide range of some of the parameters in this study have been attributed to any of the many factors such as season, spawning, migration, sex and genetic variation that affect the blood picture of fish. Stress factors due to capture, handling and sampling procedures of which capture is most important are factors, which can cause intra-species haematological variation (Bouck and Ball, 1966). Stress of handling has been shown to produce a low haemo-concentration. It has also been shown that the haemoglobin concentration and haematocrit of fish blood decreases after the stress of capture and transportation (Hattingh and Van Pletzer, 1974). Positive and useful correlation between haematocrit, haemoglobin and red

Table 11. Linear regression blood parameter with standard length (SL) and body weight (WT).

Blood parameters	Regression analysis
RBC	= 2.703 + 0.044 sl (cm) - 0.001 wt (g) r^2 = 0.015 ($p < 0.05$) ($P > 0.05$)
WBC	= 1.446 + 0.040 sl (cm) - .003 wt (g) r^2 = 0.318 ($p < 0.05$) ($p > 0.05$)
PCV	= 13.602 + 0.376 sl (cm) - 0.005 wt (g) r^2 = 0.035 ($p > 0.05$) ($p > 0.05$)
Hb	= 2.246 + 0.200 sl (cm) + 0.000 wt (g) r^2 = 0.063 ($p > 0.05$) ($p > 0.05$)
ESR	= -37.720 + 2.090 sl (cm) + 0.149 wt (g) r^2 = 0.303 (p< 0 05) ($p > 0.05$)
MCV	= 41.151 + 0.898 sl (cm) - 0.027 wt (g) r^2 = 0.308 ($p < 0.05$) ($p < 0.05$)
MCH	= 1.130 + 0.729 sl (cm) + 0.000078 wt (g) r^2 = 0.316 ($p < 0.05$) ($p > 0.05$)
MCHC	= 15.608 + 0.637 sl (cm)+ 0.003 wt (g) r^2 = 0.245 ($p < 0.05$) ($p > 0.05$)

Table 12. Summary of qualitative and quantitative assays of digestive enzymes in the gut of *P. obscura* (n = 23).

Enzymes	Oesophagus	Stomach	Pyloric caeca	Duodenum	Ileum	Rectum	SE
Glycosidases[1]							
Amylase	7.1	9.4	112.2	140.1	87.3	ND	2.34
Sucrase	11.1	11.6	16	32.6	29	ND	0.42
Maltase	19	18.7	49.1	40.6	48	ND	0.59
Lactase	ND	26.2	34.1	37	31.2	ND	0.64
Cellulase	11	9.1	21.8	ND	ND	ND	-
Proteases[2]	ND	125.8	240.8	161	102.7	N	2.29
Lipase	ND	40.5	184.4	131	101.4	ND	2.67

ND = Not detected; [1]mg glucose/min/mg protein at 37°C; [2]change in optical density at 595 nm/h/mg of L-tyrosine/h at 37°C; SE=standard error.

cell count used in the computation of the haematological indices, have been demonstrated for several species (Eisler, 1965; Summerfelt et al., 1967; Houston and Dewilde, 1968).

One of the fundamental needs of a fish, like other animals, is to have an adequate supply of oxygen in the tissues so that oxidation can occur and provide the necessary energy for life. Oxygen transport in blood depends on the compound, haemoglobin, which is the respiratory pigment of the blood. According to Lagler et al. (1977), the haemoglobin content of the fish blood varies with the number of erythrocytes present, expressed as percentage of dry weight of erythrocytes. They found the values of 14 to 19% in the smooth dogfish, *Mustelus canis* and 37 to 79% in a number of marine and fresh water teleosts. The mean values of P. *obscura* studied seem to deviate from the above relationship since PCV has a negative correlation with ESR. Guyton (1991) postulates that when Hb formation is deficient in the bone marrow, the percentage of haemoglobin in the cells may fall considerably below the normal value and the volume of red blood cell decreases as well because of diminished haemoglobin to fill the cell, which also affects the MCV; this could explain this deviation. The mean value for Hb conc. of P. *obscura* was 7.23 g/dl. This value falls within the range reported for other fishes like *Labeo rohita*; 6.0 g/dl for males and 6.3g/dl for females (Seddiqui and Mishira, 1979), *Hypophthalmichthys molitrix,* 6.11 g/dl (Pieterse et al., 1981) and *C. gariepinus,* (Adeyemo, 2007; Kori-sakpere et al., 2005). The Hb value of *C. obscura* is higher than that obtained for *P. obscura* (Fagbenro et al., 1998). In fish blood, oxygen is carried in physical solution and in combination with haemoglobin. So physiologically, haemoglobin is crucial to the survival of the fish as its role is directly related to the oxygen building capacity of blood. Despite this importance, *Chaenocephalus aceratus* has been reported as having no haemoglobin (Rund, 1954), indicating that oxygen in such species is transported by blood in physical solutions only.

Blaxhall and Daisely (1973) have reported the possibility of using haematocrit as a tool in aquaculture and fishery management for checking anaemia condition. Reported values for fish haematocrit are usually between 20 - 35% and scarcely attain values greater than 50% (Clarks et al., 1979). The mean haematocrit value in this work was within this range (22.5% ± 4:7). Das (1965) stated that both the red cell number and haemoglobin concentration tend to increase with length and age. Increase haematocrit have been observed for male fishes approaching spawning, but this increase seems to be of a limited and of transient nature (Poston, 1966; Summerfelt 1967). A snake head (*Parachanna*) can live out of water for more than 24 h and be restored to normal life even after the skin has become dry (Lagler et al., 1977). Thus, the high Hb value in *P. obscura* is indicative of its air breathing character and activity. The predominant blood group in *P. obscura* is 90% prevalence. The prevalence of this blood group is advantageous to the fish since the risk of death especially of the fingerlings is minimized in case of fertilization with the positive group; the genotypes are AA (88%) and AS (12%). This is also advantageous among the species since there will be little probability of generating sickle cell off springs. It can also be deduced that this genotype group is favored by natural selection being able to withstand environmental stress.

The linear regression analysis has shown that standard length affects the blood parameters, while body weight has little or nothing to offer. In conclusion, therefore, it should be realized that blood picture is not a dull and static thing; It depends on what is happening to the fish at any particular locality and time and so it is very dynamic. For this and other reasons there is a great need to monitor with broad limits the blood values in normal healthy fish in order to improve the management and treatment practice since these variables affect the total yield and ultimately the productivity of these fishes.

Except for the occurrence of amylase in the oesophagus, Akintunde (1985) observed a similar general pattern of enzyme distribution in *Sarotherodon galilaeus* (Linnaeus, 1758), which like *P. obscura* has a microphagous dietary. Worthy of note is the occurrence of enzyme secretion in the oesophagus of *H. niloticus* (Table 12) which is rare, having been reported in only a few species (Kawai and Ikeda, 1971). The variety of glycosidase indicates the ability of *P. obscura* to digest a wide range of carbohy-drate food components. Cellulase activity was recorded only in the pyloric cacca and its origin is attributed to gut micro flora ingested along with the detritus which featured prominently in the diet. The relatively high activity levels of proteases, particularly in the duodenum (Table 12), are not surprising in view of the large proportion of protein components (mainly zooplankton) in the diet. Pepsin would hardly be expected to occur in the two distal gut regions since it is active only in strongly acid media found in the stomach. Lipase distribution and activity along the entire gut was also reported in *C. isheriensis* by Fagbebro (1980). From the foregoing, it is evident that *P. obscura* is well equipped to digest the carbohydrate, protein and lipid components of its diet.

The conclusion may be reached therefore that plasticity and resilience of biological attributes of *P. obscura* enables it to survive and thrive on both remote sahelian and coastal environments. Early sexual maturity, rapid growth and high recruitment compensate the large mortality of the species resulting from predation and fishing. Thus, the high abundance of *P. obscura* in the Anambra River is sustained.

REFERENCES

Adeyemo O K (2007). Haematological profile of *Clarias garcenpinus* (Burchell, 1822) exposed to lead. Turk. J. Fish. Aquat. Sci. 7:163-169.

Akintunde EA (1985). Digestive enzymes in the gut of *Sarotherodon galilaeus* (syn. *Tilapia galilaea*, family Cichlidae) of Lake Kainji. Nigeria. J. Sci. pp. 1822-1825.

Awachie JBE, Ezenwaji HMG (1981). The importance of *Clarias* spp. in the Fisheries development of the Anambra river basin, Nigeria. CIFA Tech. pap. 8:212-233.

Azugo WI (1978). Ecological studies of the helminth parasites of the fish of Anambra river system. M. Phil Thesis, University of Nigeria, Nsukka. p. 178.

Badawi HK, Said MM (1971). A Comparative Study of the Tulip species (pisces). Marine Biol. 8:202-204.

Bailey NTJ (1994). Statistical Methods in Biology, 3rd Edition. Cambridge University Press Cambridge, p. 255.

Baloggun RA, Fisher O (1970). Studies on the digestive enzymes of the common African toad, *Bufo regularis* Boulger. Comp. Biochem. Physiol. 33:813-820.

Barnhart RA (1969). Effects of certain variables on haematological characteristics of rainbow trout, *Salmo gairdneri* (Richardson). Trans. Am. Fish Soc. 98:411-418.

Benjamin MM (1961). Outline of veterinary clinical pathology. 2nd Ed. IOWA State University Press, America. IOWA, p. 250.

Blaxhall PC, Baibley KW (1973). Routine haematological methods for use with fish blood. J. Fish Biol. 5:771-774.

Bouck RG, Ball RC (1966). Influence of Capture methods on blood characteristics and mortality in rainbow trout (*Salmo gairdneri*). Trans. Am. Fish Soc. 95:167-176.

Chifamba PC (1993). The life history style of *Limnothrissa miodon* in Lake Kariba. CIFA Occas. Pap. No. 19:75-86.

Clarks S, Whitemore DH, Mcmahan RF (1979). Consideration of blood parameters of large mouthbass,(*Micropterus salmooides*). J. Fish. Biol. 14:147-154.

Cole EH (1967). Veterinary Clinical pathology. W. B. Saunders Company, Philadelphia, p. 235.

Das BC (1965). Age-related trends in the blood chemistry and haematology of the Indian carp (*Catla catla*). Gerontologia 10:47-64.

Delany JW, Garratty G (1969). Handbook of heamatological and blood transfusion techniques. London: Buttterworths, p. 165.

Egwunyenga AO, Nmorsi OPG Igbinosun (1999). Haemotological changes in Cichlidae of Ethiope river (Niger Delta, Nigeria) due to intestinal helminthiasis. Biol. Sci. Res. Commun. 11:361-365.

Eisler R (1965). Erythrocyte count and haemoglobin content of nine species of Marine teleosts. Chesapeake Sci. 6:116-120.

Erondu E S, Nnubia C, Nwadukwe O (1993). Haematological studies on four catfish species raised in fresh water ponds in Nigeria. J. Appl. Ichthyol. 9:250-256.

Eyo JE, Akpati CI (1993). Fishing gears and fishing methods. pp. 143-167. In: Ezenwaji, HMG Inyang NM, Orji EC (Eds). Proceeding of the UNDP assisted Agricultural and rural development programme (Ministry of Architecture Awka, Anambra State). Training Course on Artisanal Fisheries development, held at University of Nigeria Nsukka, October 29, November 12, 1995.

Eyo JE, Mgbenka BO (1992). Aspect of the Biology of *Clarias gariepinus* in Anambra river basin. 1. Oocyte diameter, fecundity and Sex ratio. J. Agric. Sci. Technol. 2(1):47-51.

Ezenwaji HMG (1999). The abundance and trophic biology of *Clarias albopunctatus* Nichols and La Monte, 1953 (Osteichthyes: Clariidae) in the Anambra flood river system. Nig. Ecol. Freshw. 8:22-25.

Ezenwaji HMG (2002). The biology of *Clarias ebriensis* Pellegrin, 1920 (Osteichthyes: Clariidae) in an African rain forest river basin. Fish. Res. 54:235-252

Ezenwaji HMG, Inyang NM (1998). Observations of the biology of *Clarias agboyiensis* Sydenham. 1980 (Osteichthyes: Clariidae) in the Anambra river flood system. Nig. Fish. Res. 36:47-60.

Ezenwaji HMG, Inyang NM (1998). Observations on the biology of *Clarias agboyiensis* Sydenham, 1980 (Osteichthyes: Clariidae) in the Anambra flood river system. Nig. Fish. Res. 36:47-60.

Fagbenro AO (1990). Food composition and digestive enzymes in the gut of pond-cultured *Clarias isheriensis* (Sydenham 1980). (Siluriformes: Clariidae). J. Appl. Ichthyol. 6:91-98.

Fagbenro OA, Adedire CO, Owoseeni E A Ayotunde EO (1993). Studies on the Biology and aquacultural potential of feral catfish, *Heterobranchus bidosalis.*(Geoffrey Sst Hilaire 1809) *(Clariidae).* Trop. Zool. 6:67-79.

Fagbenro O, Adedire CO, Ayotunde EO, Famino EO (1999). Haematological profile, food composition and digestive enzyme assay in the gut of African bony-tongue fish, *Heterotis* (clupisudis) *nitroticus* (Cuvier, 1829) (osteoglossidae). Trop. Zool. 13:1-9.

Gabriel UU, Anyanwu PE, Akinrotimi AO (2007). Haematological profile of Black chinned tilapia (*Sarotherodon melanotheron*) from Buguma creek, Niger Delta. J. Fish Biol. 2(2):186-189.

Gliwicz ZM (1984). Limnological study of Cahora Bassa reservoir with special regard to sardine fishery expansion. FAO/GCP/MOZ/OO6/SWE Field Document, 8: 71.

Gourene G, Teugels GG (1991). Revision du genre *Pellonula* des eaux deuces africaine (Pisces: Clupeidae). Ichthyol. Explor. Freshw. 131:213-225.

Guyton MD (1991). Textbook of medical physiology. W. B. Suanders Company, Philadelphia, p. 1014.

Hattingh J, Van Pletzer JJ (1974). The influence of capture and transportation on some blood parameters of fresh water fish. Comp. Biochem. Physiol. 49:607-609.

Hickman CP, Roberts LS, Larson A (2001). Integrated Principles of Zoology. McGraw Hill, New York, pp. 899.

Hora SL (1994). Hints of the Culture of Murrel Industrial Farming. Bombay Nat. Hist. Soc. 6:8.

Houston AH, Dewilde MA (1969). Haematology and blood volume of thermally acclimated brook trout, *Salvelinus fontalis* (Mitchill). Comp. Biochem. Physiol. 28:877-885.

Hynes HBN (1950). The food of freshwater sticklebacks, *Gasterosteus aculeatus* and *Pygosteus pungitius.* with a review of methods used in studies of the food of fishes. J. Anim. Ecol. 19:36-58.

Hyslop EJ (1980). Stomach content analysis-a review of methods and their applications. J. Fish Biol. 17:411-429.

Idodo GU (2003). Fresh water fishes of Nigeria (Taxonomy, Ecological Notes, Diet and Utilization). Idodo Umeh publishers Limited, Benin City, Nigeria, p. 232.

Kewai SI, Ikeda S (1971). Studies on digestive enzymes of fishes.I. Carbohydrates in digestive organs of several fishes. Bull. Jpn. Soc. Sci. Fish. 37:333-337.

Kori Siakpere O, Ake JEG, Idoge E (2005). Haematological characteristics of the African snake head, *Parachanna obscura*. Afr. J. Biotechnol. 4(6):527-530.

Kori-Siakpere O (1985). Haematological characteristics of *Clarias isheriensis*. Syndenham J. Fish Biol. 27:259-263.

Lagler KF, Bardach EJ, Miller RR, May Passino DR (1977). Ichthyology. John Wiley and sons, New York. p. 506.

Le Danois E (1955). Fishes of the world. The Countryman Press, Paris. p. 350.

Lelek A (1973). Sequences of changes in fish population in the New Tropical man-made lake Kainji, Nigeria, West Africa. Archiv fur Hydrobiologie, 71:381-420

Linnaeus C (1758). System Nature. Regnum Animale. 10th ed. L. Salvii. Holminae.

Lishev MH (1950). The food of Amur Predatory fish. Trudy Amurskoi ikhtiologychskoi ekspeditsu. 1:1945-1949.

Maclean N (1978). Haemoglobin. The institute of Biology. Series in Biology. Edward Arnold ltd, p. 69.

Malcolm J (1995). Environmental Biology of fishes. Chapman & Hall, London, pp. 455.

Marshall BE (1987). On the relationship between tigerfish and sardines in lake kariba. J. Limnol. Soc. South Africa 13:78-79.

Marshall BE (1993). *Limnothrissa* in man-made lakes: do we understand the implications of their small size? CIFA Occas. Pap. No. 19:1-17.

Mawdesley-Thomas LE (1971). Toxic chemicals-the risk to fish. New Scientist. 49:74-75.

McCarthy DH, Stevenson, JP, Roberts, MS (1973). Some blood parameters of the rainbow trout (*Salmo gaidneri* Richardson). 1. The Kamloops variety. J. Fish Biol. 5:1-8.

Mgbenka BO Eyo JE (1992). Aspects of the biology of *Clarias gariepinus* in Anambra river basin. 2. Maturation and condition factor of Aquaculture. Sci. Technol. 2(1):52-55.

Miale JB (1982). Laboratory Medicine Haematology. The C.V Mosby company, London. pp. 1083.

Ness PH, Stengle JM (1974). Historical Introduction: The Red Blood cell. (2nd ed) vol. 1 Surgenor DM (ed), Academic Press Inc, pp. 23.

Norman JR (1947). A History of Fishes. Ernest Benn Limited, London. 463 pp.

Nwani CD (1998). Aspects of the Biology of *Distichondus* sp in Anambra River, Nigeria. M.Sc Thesis Dept. of Zoology, University of Nigeria Nsukka, p. 133.

Ogunbiyi AO, Okon EE (1976). Studies on the digestive enzymes of the African fruit bat, *Eilodon helve* (Kerr.) Comp. Biochem. Physiol. 55:359-361.

Ogutu-Ohwayo R (1993). *Lates* spp. and other predators as a cause of mortality in small pelagic fish species in the African great lakes. CIFA Occas. Pap. No. 19:18-23.

Olatunde AA, Ukoha AI, Oguntayo BF (1988). Digestive enzymes in the alimentary tract of *Clarias lazera* cuvier and valenciennes family clariidae (Osteichthyes: Siluriformes). Arch. Far Hydrobiol. 112(1):107-113.

Otobo FO (1978). The reproductive biology of *Pellonula afzeliusi* Johnels and *Sierrathrissa leonensis* Thys Audernaerde in lake kainji, Nigeria. Hydrobiologia 61:99-122.

Oyowale JO (1987). Haematological and plasma Biochemical values of two breeds of domestic fowl in tropical environment. Anim. Technol.

38(1):49-52.

Petr T, Kapetsky JM (1983). Pelagic fish and fisheries of tropical and subtropical natural lake and reservoir. ICLARM Newslett. 6:9-11.

Pieterse JJ, Smith GL, Van vilet K, Schoonbe HJ, Hattingh J (1981). Observation on the blood of the Chinese silver carp, *Hypophthalmicthys molitrix* (valenciennes). J. Fish Biol. 18:455-459.

Plummer DT (1978). An introduction to practical Biochemistry. 2nd edition. London: McGraw Hill Book company, p. 283.

Poston HA (1966). Effects of sex and reproduction stage on haemoglobin levels in brown trout. Fish Res. Bull. 29:28-29.

Reynolds I (1969). The biology of the clupeids in the new Volta lake. In: *Man-made lakes: the Accra symposium,* Ed. by Obeng LE. Ghana University Press, Accra, pp. 195-203.

Reynolds JD (1970). Biology of the small pelagic fishes in the new Volta in Ghana. 1. The lake and the fish: feeding habits. Hydrobiologia 35:569-603.

Ricker WE (1973). Linear regressions in fishery research. Fish. Res. 30:409-434.

Rund JJ (1954). Vertebrates without erythrocytes and blood pigment. Nature 173:848-850.

Seddiqui AK, Mishra S (1979). Blood dyscrasia in a teleost, *Colisa fasciatus* after acute exposure to sub lethal concentrations of lead. J. Fish Biol. 14: 199-204.

Sowunmic AA (2003). Haemotology of the African catfish, *Clarias gariepinus* from Eleiyele Reservoir Ibadan South-West, Nigeria. Zoologist 2:85-91.

Summerfelt RC, Lewis WM, Ulrich MG (1967). Measurement of some haematological characteristics of the gold fish. Prog. Fish Cult. 29:13-20.

Teugels GG, Daget J (1984). *Parachanna* nom. Nov. for the African snakeheads and rehabilitation of *Parachanna insignis* (Sauvage, 1884) (Pisces, Channidae). Cybium. 8(4):1-7.

Weinreb EL (1958). Studies on the histology and histopathology of the rainbow trout, *Salmo gairdneri* (Richardson). 1. Haematology. Under normal and experimental conditions of inflammations. Zoologica, New York, pp. 46-153.

11

Toxic effect of potassium permanganate on *Oreochromis niloticus* based on hematological parameters and biomarkers of oxidative stress

Jakeline G. França[1]*, Maria J. T. Ranzani-Paiva[2], Julio V. Lombardi[2], Solange de Carvalho[1], Francisco Filipak-Neto[3] and Ciro A. Oliveira-Ribeiro[3]

[1]Aquaculture Center, UNESP, Via de Acesso Prof. Paulo Donato Castellane, 14884-900 Jaboticabal, SP, Brazil.
[2]Fisheries Institute/APTA/SAA SP, Avenida Francisco Matarazzo, 05001-900 São Paulo, SP, Brazil.
[3]Departament of Celular Biology, University Federal do Paraná, 81.531-990 Curitiba, PR, Brazil.

Potassium permanganate ($KMnO_4$) is an oxidizing agent used in aquaculture for many years. Despite its known efficacy for the treatment and prevention of fish diseases, information about their potential toxicity in non-target organisms is still very limited. The objective of this study was to evaluate the sublethal effects of potassium permanganate (1.0 and 4.0 mg L^{-1}) on Nile tilapia, *Oreochromis niloticus*, using hematological analysis and oxidative stress (reduced glutathione concentration, glutathione *S*-transferase activity, catalase and lipid peroxidation) as biomarkers. Fish exposed to a concentration at 1.0 mg L^{-1} had no significant variations in blood parameters. At 4.0 mg L^{-1}, the changes in hematological parameters of the exposed animals indicated hemolysis as a result of the oxidizing action of potassium permanganate. Regarding the oxidative stress analyses, only the reduced glutathione presented significant increase in fish exposed to potassium permanganate, indicating adaptive and protection responses against the oxidative stress. This study demonstrated that concentrations of potassium permanganate (1.0 to 4.0 mg L^{-1}) usually recommended for the treatment of fish diseases, can be toxic to Nile tilapia (*O. niloticus*).

Key words: Potassium permanganate, Nile tilapia, hematology, oxidative stress.

INTRODUCTION

Potassium permanganate is an oxidizing agent used for many years in aquaculture for the treatment or prevention of diseases in fish, removal of parasites, disinfection of fish tanks and aquariums, reduction of products toxic to fish as rotenone and antimycin, and for the control of fungi and algae for managing problems when temporary reduction of oxygen in tanks occurs (Lay, 1971; Duncan, 1974). Despite its widespread use for the treatment of diseases in cultivated fish, this chemical compound is not approved by the U.S. Food and Drug Administration as a therapeutic agent in aquaculture (FDA, 2007).

Potassium permanganate is potentially toxic to humans and other organisms (Kegley et al., 2010), including several fish species (Marking and Bills, 1975; Cruz and Tamse, 1989; Bills et al., 1993; Straus, 2004; Silva et al., 2006) but is not included in regulatory categories. Therefore, there is no pattern or established limit to this compound and it may be used freely and without restriction in aquaculture. According to the FDA (2007), the inclusion of this compound in a regulatory class still depends on further studies. Several studies on the implementation and effectiveness of potassium permanganate in aquaculture are available in the literature, but information

*Corresponding author. E-mail: jkgfranca@gmail.com.

about its toxic potential in organs and tissues in fish is still scanty.

The impact of contaminants on the aquatic ecosystem can be measured through a variety of parameters, from low levels of biological organization (molecular and biochemical responses) to high organization levels (population and community response) (Au, 2004). In this context, the use of biomarkers at cellular and molecular levels is extremely important as a sensitive tool to measure the biological effects during the assessment of environmental quality (Sarkar et al., 2006), since they are more specific, sensitive, reproducible and easy to determine (Au, 2004).

The Nile tilapia, *O. niloticus* (Linnaeus, 1758) was selected as the test organism in this study for its great aquaculture and commercial value. Tilapia is the third most important cultured fish group in the world, after carps and salmonids. Their hardiness and adaptability to a wide range of culture systems has led to the commercialization of tilapia production in more than 100 countries. Nile tilapia represents approximately 84% of total global tilapia production. In 2008, Nile tilapia culture alone was ranked fifth among the most cultured species in the world, with a total aquaculture production of 2.3 million tonnes (FAO, 2009). In 2010, it is anticipated that global Nile tilapia production will reach nearly 2.5 million tonnes (FAO, 2010).

The objective of this study was to evaluate the sub-lethal effects of potassium permanganate for tilapia, *O. niloticus*, using the hematological analyses and oxidative stress as biomarkers.

MATERIALS AND METHODS

This study was carried out at the Laboratory of Pathology of Aquatic Organisms (Fisheries Institute, Sao Paulo State, Brazil) under a controlled environment.

The fish used as test organisms were juveniles of tilapia, *O. niloticus*, purchased in commercial fish farming, with mean weight of 44.52 ± 11.07 g and mean length of 13.87 ± 1.14 cm. Fish were previously acclimated for 15 days in the same environmental conditions of testing containers, that is, they were randomly distributed into three 300 L tanks equipped with artificial aeration and biological filters previously matured, at the density of 70 fish per tank (7.6 g L^{-1}). The water quality variables were measured daily during the period of acclimation and they were appropriate for the specie in the initial conditions: temperature 22 ± 0.7°C; pH 7.7 ± 0.2; dissolved oxygen 6.9 ± 0.3 mg L^{-1} and total ammonia between 0.33 and 0.40 mg NH_4 L^{-1}. Test solutions were set up by dissolving potassium permanganate ($KMnO_4$) Merck^{TM} in dechlorinated tap water. Nominal test concentrations of potassium permanganate (1.0 and 4.0 mg L^{-1}) were used in the assays, and it was chosen based on minimum and maximum doses of this compound reported in the literature to treat diseases in fish (Francis-Floyd and Klinger, 1997). A control group with no addition of potassium permanganate was also carried out in the same conditions of testing containers.

Test solutions were partially replaced every three days at the proportion of 1/3 of total volume (renewal test system). Fish were fed with commercial ration 32% crude protein, and food residues were cleaned up at every operation of water replacement.

The assay was carried out for 30 days, with sampling of 10 individuals per group for the hematological analyses and biochemical after 7, 15 and 30 days of exposure time.

For hematological analyses, the animals were anesthetized with benzocaine (3%); blood was drawn by caudal puncture, with the help of a needle and syringe previously heparinized. The blood specimens were assayed for: number of erythrocytes (RBC), counted in a Neubauer chamber; hematocrit (Ht) by the microhematocrit technique; and hemoglobin level (Hb) by the cyanomethemoglobin method. After these procedures, the following RBC indices were calculated: mean corpuscular volume (MCV), mean corpuscular hemoglobin (MCH) and mean corpuscular hemoglobin concentration (MCHC). Smears were made using the same blood aliquots, and the slides were later stained with May-Grunwald-Giemsa, which were utilized for total leukocyte number (WBC) and differential count for each type of leukocyte (lymphocytes, neutrophils, basophils, eosinophils and monocytes) and total thrombocytes.

The biochemical analyses were performed at the Laboratory of Cell Toxicology, in the Federal University of Parana (UFPR), Curitiba, Parana State, Brazil. Fish were sacrificed by deep sedation with benzocaine. Portions of the liver were sampled and maintained at −80°C freezer before performing analysis. Then, samples were homogenized in phosphate buffered saline (PBS, pH 7.4) and frozen. Further, aliquots were separated from samples to the determinations of total protein concentration, glutathione *S*-transferase (GST) and catalase (CAT) activities, as well as the concentrations of reduced glutathione (GSH) and lipid peroxidation (LPO). Total protein concentration was determined by the Bradford method (Bradford, 1976). The activity of glutathione *S*-transferase was monitored by the reaction of the substrate 1-chloro-2,4-dinitrobenzene with GSH, generating a thioether that absorbs light at 340 nm (Habig et al., 1974; Keen et al., 1976). The concentration of GSH was determined according to Sedlak and Lindsay (1968) after protein precipitation with 10% of trichloroacetic acid, and the activity of CAT was determined according to Aebi (1984). Lipid peroxidation was determined by the FOX method (Ferrous Oxidation/ Xylenol Orange Method) according to Jiang et al. (1991, 1992).

In order to assure that the probable toxic effects observed were attributed to potassium permanganate, the following physical and chemical variables of water were measured at the beginning of the experiment and then at every 24 h: temperature (°C) and dissolved oxygen (mg L^{-1} and % saturation) were measured using the oximeter YSI - Mod.55; pH values were obtained using potentiometer and the electrical conductivity (μScm^{-1}) using conductivimeter. The analysis of hardness was done by the titration method with EDTA, alkalinity (mg $CaCO_3$ L^{-1}) by titration. The following analyses were performed weekly: the hardness, alkalinity (mg $CaCO_3$ L^{-1}) and total ammonia (mg NH_4L^{-1}) according to standardized methodology in APHA et al. (2005). Non-ionized ammonia (NH_3, mg L^{-1}) was calculated according to Bower and Bidwell (1978).

For data analyses we used the statistical program TOXSTAT 3.5 (West and Gulley, 1996). Data normality was tested by Shapiro Wilk's test and homoscedasticity of variances by the Bartlett's test. Then, data was analyzed using the statistical test ANOVA. Alterations were considered significant when $P < 0.05$.

RESULTS

Survival rate of 100% was observed in experiments and all fish appeared healthy. Physical and chemical variables

Table 1. Mean values and standard error of the mean of red blood series of *Oreochromis niloticus* exposed to potassium permanganate.

Treatments	RBC (10^6 µL^{-1})	Ht (%)	Hb (g dL^{-1})	MCH (pg)	MCV (fL)	MCHC (%)
7 days						
Control	2.03 ± 0.05[a]	30.10 ± 1.10[a]	6.43 ± 0.25[a]	32.01 ± 1.83[a]	149.92 ± 08.33[a]	21.39 ± 0.48[a]
1 mg L^{-1}	2.11 ± 0.07[a]	27.20 ± 0.36[a]	5.92 ± 0.36[a]	28.47 ± 2.13[a]	130.29 ± 07.75[a]	21.86 ± 1.04[a]
4 mg L^{-1}	1.52 ± 0.06[b]	26.85 ± 1.82[a]	6.41 ± 0.54[a]	42.19 ± 3.36[a]	177.78 ± 12.55[a]	23.60 ± 0.82[a]
15 days						
Control	1.98 ± 0.06[a]	28.00 ± 1.08[a]	6.67 ± 0.36[a]	33.58 ± 1.39[a]	141.39 ± 08.33[a]	24.09 ± 1.42[a]
1 mg L^{-1}	1.86 ± 0.06[ab]	29.80 ± 0.26[a]	5.92 ± 0.36[a]	39.49 ± 1.48[a]	161.15 ± 07.75[a]	24.69 ± 0.96[a]
4 mg L^{-1}	1.71 ± 0.06[bc]	26.45 ± 0.78[a]	6.41 ± 0.54[a]	35.20 ± 2.44[a]	156.03 ± 12.55[a]	22.70 ± 1.45[a]
30 days						
Control	1.79 ± 0.13[a]	28.25 ± 1.07[a]	6.58 ± 0.32[a]	37.58 ± 2.07[a]	162.08 ± 08.53[a]	23.32 ± 0.83[a]
1 mg L^{-1}	1.48 ± 0.09[a]	26.50 ± 0.92[ab]	6.19 ± 0.37[a]	42.62 ± 2.48[a]	183.71 ± 09.16[a]	23.27 ± 0.95[a]
4 mg L^{-1}	1.64 ± 0.11[a]	24.15 ± 0.87[bc]	6.09 ± 0.18[a]	38.18 ± 2.03[a]	152.68 ± 11.56[a]	25.42 ± 0.92[a]

[abc] Different letters indicate significant different ($P < 0.05$) mean values between treatments. RBC, Number of erythrocytes; Ht: hematocrit; Hb, hemoglobin; MCH, mean corpuscular hemoglobin; MCV, mean corpuscular volume; MCHC, mean corpuscular hemoglobin concentration.

of water used in the bioassays were similar among different treatments (P=0.05). The mean data recorded were: temperature (°C): 22.54 ± 1.30; pH: 7.77 ± 0.10; dissolved oxygen (% saturation): 80.67 ± 1.39; dissolved oxygen (mg L^{-1}): 7.00 ± 0.41; electrical conductivity (µs cm^{-1}): 248.16 ± 31.11; alkalinity (mg $CaCO_3$ L^{-1}): 76.56 ± 13.51; total ammonia (mg NH_4 L^{-1}): 0.40 ± 0.20 and un-ionized ammonia (mg NH_3 L^{-1}): 0.009 ± 0.006. Fish kept at a concentration of 1 mg L^{-1} $KMnO_4$ did not present significant variations in blood variables herein analyzed (RBC, Ht and Hb) and in hematimetric indices (MCV, MCH and MCHC) (Table 1).

The animals exposed to a concentration at 4 mg L^{-1} had a significant reduction (P<0.05) in red blood cells on the seventh and 15th day after collection, and a reduction of hematocrit on the 30th day (Table 1). On the seventh day of collection, a significant increase of MCH (P<0.05) was detected in fish exposed to a concentration of 4 mg L^{-1}.

Thrombocytes were quite numerous in fish exposed to potassium permanganate on the seventh collection day. However, only at 1 mg L^{-1} a significant difference was registered (Table 2). The total count of leukocytes and lymphocytes in tilapia exposed to 4.0 mg L^{-1} $KMnO_4$ was significantly lower on the seventh day of sampling (Table 2). On the 30th day, a significant decrease was observed in the total count of leukocytes and lymphocytes in both treatments with potassium permanganate (1.0 and 4.0 mg L^{-1}). Eosinophils were not verified in any treatment.

But basophils were found in two samples of the control and one sample of the treatment at 1 mg L^{-1}.

On the 15th day of collection, significant increase of GSH was observed in the two tested concentrations of potassium permanganate (Figure 1A). No significant change occurred in the activity of GST, CAT and LPO levels in fish exposed to potassium permanganate (Figure 1B to D).

DISCUSSION

The reduced values of hematimetric values found in this present study suggest the induction of anemia caused by the toxicity of potassium permanganate. The anemia may be due to the inhibition of erythropoiesis and hemo-synthesis and to an increase in the rate of erythrocyte destruction in hemopoietic organs, resulting in a reduced RBC (Adhikari et al., 2004). Other xenobiotics also induce changes that give evidence of decreased hematopoiesis followed by anemia induction in fish, for example, changes in erythrocyte profile induced by sublethal exposure to cypermethrin and carbofuran in *Labeo rohita* (Adhikari et al., 2004), diazinon in *Cyprinus carpio* (Banaee et al., 2008), mercuric chloride in *O. niloticus* (Ishikawa et al., 2007). deltamethrin in *Cyprinus carpio* (Svoboda et al., 2003) and chromium in *Labeo rohita* (Vutukuru, 2005).

Another observed response was the significant increase of MHC in fish exposed to potassium permanganate, indicating the presence of cells with larger volume and greater amount of hemoglobin. Even with the inverse variation of MHC in relation to the number of erythrocytes, the level of hematocrit, hemoglobin and MCHC had no significant alteration (P > 0.05), with values close to control (Table 1). Moreover, the mean value

Table 2. Mean values and standard error of the mean of leukocytes and thrombocytes of *O. niloticus* exposed to potassium permanganate.

Treatments	Lc (10^4 µL^{-1})	Lf (10^4 µL^{-1})	Nt (10^4 µL^{-1})	Mn (10^4 µL^{-1})	Tr (10^4 µL^{-1})
7 days					
Control	4.45 ± 0.66^a	4.27 ± 0.64^a	0.07 ± 0.01^a	0.08 ± 0.03^a	0.44 ± 0.14^a
1 mg L^{-1}	4.86 ± 0.67^a	4.65 ± 0.69^a	0.12 ± 0.05^a	0.07 ± 0.02^a	1.68 ± 0.30^b
4 mg L^{-1}	2.78 ± 0.38^b	2.65 ± 0.39^b	0.08 ± 0.03^a	0.04 ± 0.01^a	1.22 ± 0.38^a
15 days					
Control	4.24 ± 0.45^a	4.13 ± 0.45^a	0.05 ± 0.01^a	0.06 ± 0.01^a	0.65 ± 0.19^a
1 mg L^{-1}	2.84 ± 0.44^a	2.75 ± 0.43^a	0.05 ± 0.02^a	0.03 ± 0.01^a	1.05 ± 0.25^a
4 mg L^{-1}	2.81 ± 0.51^a	2.77 ± 0.51^a	0.03 ± 0.01^a	0.05 ± 0.01^a	0.94 ± 0.23^a
30 days					
Control	3.70 ± 0.22^a	3.53 ± 0.20^a	0.07 ± 0.02^a	0.07 ± 0.01^a	0.49 ± 0.14^a
1 mg L^{-1}	2.00 ± 0.25^b	1.92 ± 0.25^b	0.03 ± 0.01^a	0.04 ± 0.01^a	0.76 ± 0.22^a
4 mg L^{-1}	2.06 ± 0.31^b	1.99 ± 0.31^b	0.02 ± 0.01^a	0.04 ± 0.01^a	0.51 ± 0.12^a

^{ab} Different letters indicate significant different (P < 0.05) mean values between treatments. Lc, Leukocytes; Lf, lymphocytes; Nt, neutrophils; Mn, monocytes; Tr, thrombocytes.

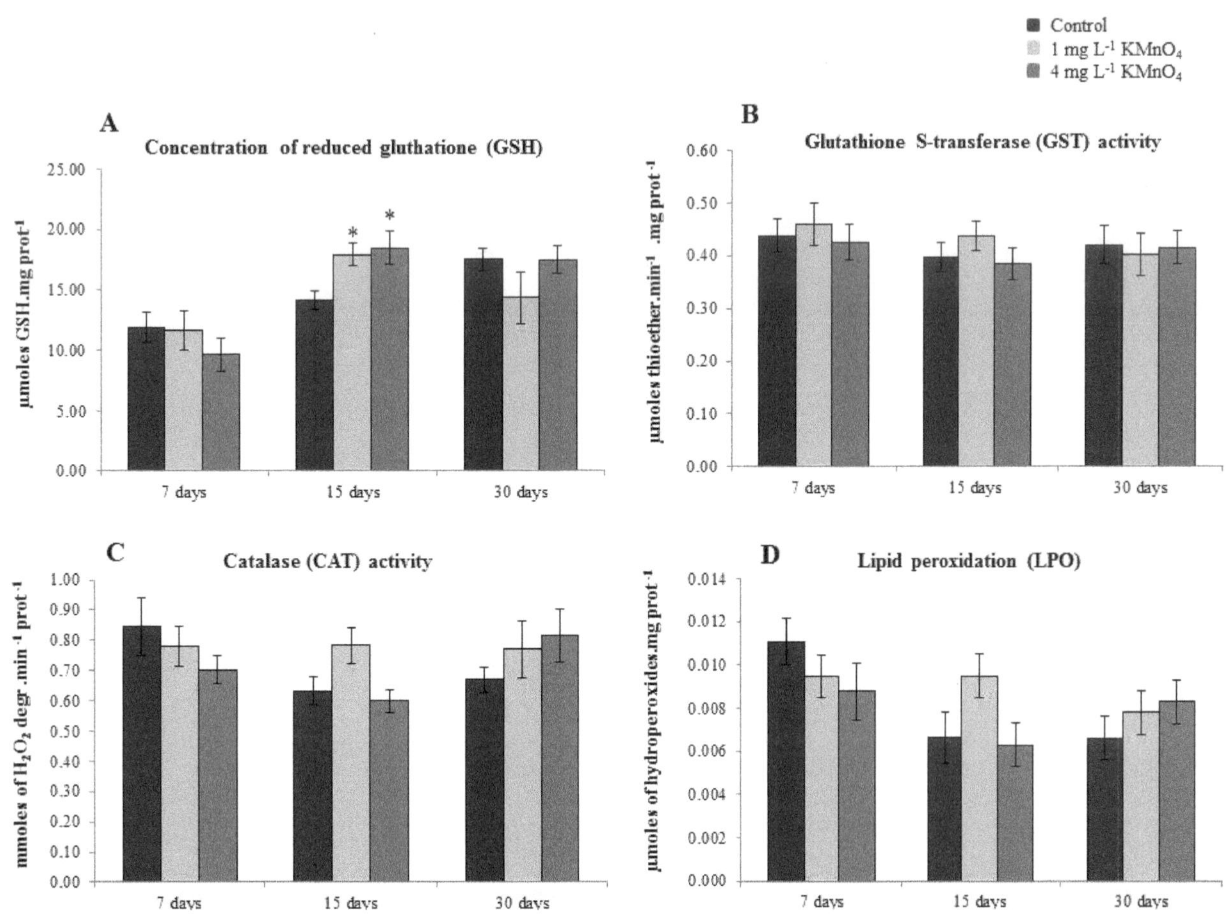

Figure 1. Concentration of GSH (A); activity of GST (B); activity of CAT (C); Lipid peroxidation (D), in liver of *O. niloticus*, exposed to potassium permanganate. Data are reported as mean ± standard error (n=10). *, Significant difference from control group (P < 0.05).

of MCV was higher than other groups, but without significant difference this indicates that the same amount of hemoglobin has occupied a larger volume in the cell. These results suggested the presence of a great amount of old and large erythrocytes, probably owing the prolonged mechanism or failure to replace these cells (Oliveira-Ribeiro et al., 2006), inducing a macrocytic anemia. Changes in the differential count of leukocytes provide an evidence for a decrease in non-specific immunity in fish after exposure to toxic substances (Svoboda et al., 2003; Oliveira-Ribeiro et al., 2006; El-Sayed et al., 2007). Thus, the leukopenia observed in fish exposed to potassium permanganate is primarily assigned to the reduction in the number of lymphocytes probably due to a reduced production and rapid destruction, leading to an increased rate of removal of these cells. Svoboda et al. (2001) and Banaee et al. (2008) observed a decrease in the percentage of lymphocytes and monocytes in *Cyprinus carpio* after exposure to diazinon, indicating a decrease in immunity, in response to stress caused by the toxic substance, corroborating the results of this present study.

The increased number of thrombocytes registered in stressed fish after being exposed to pollutants is related to damage to organs and tissues (Mazon et al., 2002; Gabriel et al., 2007). According to Francis-Floyd and Klinger (1997), the potassium permanganate is a compound capable of reacting indiscriminately with any organic material in the water, including fish tissues like skin and gills. Since it is a strong oxidizing agent, the high values of thrombocytes found in fish of both treatments (1.0 and 4.0 mg L^{-1} $KMnO_4$) can be associated with injuries in tissues such as skin and gills, once these organs are directly exposed to this chemical in the water.

According to Zhang et al. (2004), under severe stress, suppression in the levels of GSH may occur, due to the reduction capacity of synthesis, regeneration or oxidation of GSH. Moreover, the increase in GSH observed in this study reflects the stimulation of the detoxification metabolism, suggesting an adaptive response and a protection response of this biomolecule against a rapid oxidative stress, induced by contaminants (Pandey et al., 2003; Zhang et al., 2004; Farombi et al., 2007).

As observed in this study, Ventura-Lima et al. (2009) found no significant change in the levels of GST in *Danio rerio* exposed to arsenic, indicating that the tested concentrations were not deleterious enough to interfere in the activity of this enzyme. Sanchez et al. (2005) reported that the exposure to copper promoted antioxidant response in fish, revealed by the rapid increase of SOD (superoxide dismutase), catalase and GPx (glutathione peroxidase), but significant change was not observed for GST.

However, as observed for the GST, CAT and LPO levels did not vary significantly (Figures 1B to D). This may be associated with the increased levels of GSH observed, to balance the oxidative stress and as adaptive response to avoid the increase of reactive oxygen species (ROS) and oxidative damage. Similar results were reported by Ventura-Lima et al. (2009), where the absence of changes in the activity of GST, CAT and LPO levels in fish exposed to arsenic was due to the effectiveness of the increased levels of GSH for preventing additional oxidative disturbances. Thus, the results of this present study confirm that GSH is often the first defense against oxidative stress.

Conclusion

Information on the physiological changes in fish by acute and subchronic exposure to potassium permanganate is limited. Study performed by França et al. (2011) in *O. niloticus*, Kori-Siakpere (2008) in *Clarias gariepinus* and Silva et al. (2006) in *Colossoma macropomum*, revealed that the compost has high toxicity to fish in a short term exposure.

The results obtained in this present study have demonstrated that the prolonged use of concentrations considered safe by literature for prophylaxis and treatment of diseases in fish farming (1.0 and 4.0 mg L^{-1} $KMnO_4$) may cause considerable changes to health of exposed fish, evidencing the toxic potential of potassium permanganate in non-target organisms. Such alterations, especially in hematological variables, have pointed out that potassium permanganate induced the destruction of red blood cells in a short period of exposure to the chemical, indicating the direct action of its toxicity. Therefore, the use of different biomarkers become important as they can reflect more accurately the toxicity of contaminant substances studied and their effects impacting the aquatic ecosystem.

ACKNOWLEDGMENT

The authors would like to thank the São Paulo Research Foundation (FAPESP) for financial support (Process 0756481-6R).

REFERENCES

Adhikari S, Sarkar B, Chatterjee A, Mahapatra CT Ayyappan S (2004). Effects of cypermethrin and carbofuran on certain hematological parameters and prediction of their recovery in a freshwater teleost, *Labeo rohita* (Hamilton). Ecotox. Environ. Safe. 58:220-226.

Aebi H (1984). Catalase *in vitro*. Methods Enzymol. 105:121-126.

American Public Health Association (APHA) (2005). Standard Methods for the Examination of Water and Wastewater, 21th ed. American Public Health Association, Washington, D.C. p. 1325.

Au DWT (2004). The application of histo-cytopathological biomarkers in marine pollution monitoring: a review. Mar. Pollut. Bull. 48:817-834.

Banaee M, Mirvagefei AR, Rafei GR, Majazi AB (2008). Effect of sub-lethal diazinon concentrations on blood plasma biochemistry. Int. J. Environ. Res. 22:189-198.

Bills TD, Marking LL, Howe GE (1993). Sensitivity of juvenile striped bass to chemicals used in aquaculture, Resource publication 192, U.S. Department of Interior, Fish and Wildlife Service, Washington, D.C. pp. 1-11.

Bower CE, Bidwell JP (1978). Ionization of ammonia in seawater: effects of temperature, pH, and salinity. J. Fish. Res. Board. Can. 35:1012-1016.

Bradford M (1976). A rapid and sensitive method for the quantification of microgram quantities of protein utilizing the principle of protein-dye binding. Anal. Biochem. 72:248-254.

Cruz ER, Tamse CT (1989). Acute toxicity of potassium permanganate to milkfish fingerlings, Chanos chanos. Bull. Environ. Contam. Toxicol. 43:785-788.

Duncan TO (1974). A review of literature on the use of potassium permanganate ($KMnO_4$) in fisheries. US Fish and Wildlife Service, Report FWS-LR-74-14. p. 61.

El-Sayed YS, Saad TT, El-Bahr SM (2007). Acute intoxication of deltamethrin in monosex Nile tilapia, *Oreochromis niloticus* with special reference to the clinical, biochemical and haematological effects. Environ. Toxicol. Pharm. 24:212-217.

Farombi EO, Adelowo OA, Ajimoko YR (2007). Biomarkers of oxidative stress and heavy metal levels as indicators of environmental pollution in African Cat Fish (*Clarias gariepinus)* from Nigeria Ogun River. Int. J. Environ. Res. Public Health 4:158-165.

Food and Agriculture Organization of the United Nations (FAO). 2009. FAO yearbook. Fishery and aquaculture statistics. http://www.fao.org/fishery/publications/yearbooks/en.

Food and Agriculture Organization of the United Nations (FAO). 2010. Cultured Aquatic Species Information Programme, Oreochromis niloticus (Linnaeus, 1758). http://www.fao.org/fishery/culturedspecies/Oreochromis_niloticus/en.

França JG, Paiva MJTR, Carvalho S, Miashiro L, Lombardi JV (2011). Toxicity of the therapeutic potassium permanganate to tilapia *Oreochromis niloticus* and to non-target organisms *Ceriodaphnia dubia* (microcrustacean cladocera) and *Pseudokirchneriella subcapitata* (green microalgae). Aquaculture 322-323:249–254.

Francis-Floyd R, Klinger R (1997). Use of potassium permanganate to control external infections of ornamental fish. University of Florida Institute of Food and Agricultural Sciences Extension Fact Sheet FA-37. p. 4.

Gabriel UU, Amakiriand EU, Ezeri GNO (2007). Haematology and gill pathology of *Clarias gariepinus* exposed to refined petroleum oil, kerosene under laboratory conditions. J. Anim. Vet. Adv. 63:461-465.

Habig WH, Pabst MT, Jacoby WB (1974). Glutathione S-transferases: The first enzyme step in mercapturic acid formation. J. Biol. Chem. 249:7130-7139.

Ishikawa NM, Ranzani-Paiva MJT, Lombardi JV, Ferreira, CM (2007). Hematological challenge with tilapia, *Oreochromis niloticus* exposed to sub-letal mercury concentrations. Braz. Arch. Biol. Technol. 50:619-626.

Jiang ZY, Hunt JV, Wolff SP (1992). Ferrous ion oxidation in the presence of xylenol orange for detection of lipid hydroperoxide in low density lipoprotein. Anal. Biochem. 202:384-389.

Jiang ZY, Woollard ACS, Wolff SP (1991). Lipid hydroperoxides measurement by oxidation of Fe^{2+} in the presence of xylenol orange. Comparison with the TBA assay and an iodometric method. Lipids 26:853-856.

Keen JH, Habig WH, Jakoby WB (1976). Mechanism for several activities of the glutathione-S-transferases. J. Biol. Chem. 251:6183-6188.

Kegley SE, Hill BR, Orme S, Choi AH (2010). Pesticide Database, Pesticide Action Network, North America, San Francisco. Available at: http:www.pesticideinfo.org.

Kori-Siakpere O (2008). Acute toxicity of potassium permanganate to fingerlings of the African catfish, *Clarias gariepinus* (Burchell, 1822). Afr. J. Biotechnol. 7:2514–2520.

Lay BA (1971). Applications for potassium permanganate in fish culture. Trans. Am. Fish. Soc. 100:813-816.

Marking LL, Bills TD (1975). Toxicity of potassium permanganate to fish and its effectiveness for detoxifying antimycin. Trans. Am. Fish. Soc. 104:579-581.

Mazon AF, Monteiro EAS, Pinheiro GHD, Fernandes MN (2002). Hematological and physiological changes induced by short-term exposure to copper in the freshwater fish, *Prochilodus scrofa*. Braz. J. Biol. 62:621-631.

Oliveira-Ribeiro CA, Filipak-Neto F, Mela M, Silva PH, Randi MAF, Rabitto IS, Alves-Costa JRM, Pelletier E (2006). Hematological finding in neotropical fish *Hoplias malabaricus* exposed to subcrhonic and dietary doses of methylmercury, inorganic lead, and tributyltin chloride. Environ. Res. 101:74-80.

Pandey S, Parvez S, Sayeed I, Haque R, Bin-Hafeez B, Raisuddin S (2003). Biomarkers of oxidative stress: a comparative study of river Yamuna fish *Wallago attu* (Bl. & Schn.). Sci. Total Environ. 309:105-115.

Sanchez W, Palluel O, Meunier L, Coquery M, Porcher JM, Aït-Aïssa S (2005). Copper-induced oxidative stress in three-spined stickleback: relationship with hepatic metal levels. Environ. Toxicol. Pharmacol. 19:177-183.

Sarkar A, Ray D, Shrivastava AN, Sarker S (2006). Molecular Biomarkers: Their significance and application in marine pollution monitoring. Ecotoxicology 15:333-340.

Sedlak J, Lindsay RH (1968). Estimation of total protein bound and nonprotein sulphydril groups in tissues with Ellman's reagent. Anal. Biochem. 25:192-205.

Silva ALF, Chagas EC, Gomes LC, Araujo LD, Silva CR and Brandão FR (2006). Toxicity and sublethal effects of potassium permanganate in Tambaqui (*Colossoma macropomum*). J. World Aquacult. Soc. 37:318-321.

Straus DL (2004). Comparison of the acute toxicity of potassium permanganate to Hybrid Striped Bass in well water and diluted well water. J. World Aquacult. Soc. 5:55-60.

Svoboda M, Lusková V, Drastichová J, Zlábek V (2001). The effect of diazinon on hematological indices of common carp (*Cyprinus carpio* L.). Acta Vet. Brno.70:457-465.

Svoboda Z, Lusková V, Drastichová J, Svoboda M, Zlábek V (2003). Effect of Deltamethrin on Haematological Indices of Common Carp (*Cyprinus carpio* L.). Acta Vet. Brno. 72:79-85.

US Food and Drug Administration (FDA) (2007). Enforcement priorities for drug use in aquaculture, part B and C. In CVMs policy and procedures manual 1240.4200. Departament of Health and Human Services, Center for Veterinary Medice. http://www.fda.gov/downloads/AnimalVeterinary/GuidanceComplianceEnforcement/PoliciesProceduresManual/ucm046931.pdf

Ventura-Lima J, Castro MR, Acosta D, Fattorini D, Regoli F, Carvalho LM, Bohrer D, Geracitano LA, Barros DM, Marins LFF, Silva RS, Bonan CD, Bogo MR, Monserrat JM (2009). Effects of arsenic (As) exposure on the antioxidant status of gills of the zebrafish *Danio rerio* (Cyprinidae). Comp. Biochem. Physiol. C. 149:538-543.

Vutukuru SS (2005). Acute effects of hexavalent chromium on survival, oxygen consumption, hematological parameters and some biochemical profiles of the Indian Major Carp, *Labeo rohita*. Int. J. Environ. Res. Public Health 2:456-462.

West INC, Gulley D (1996). TOXTAT 3.5. University of Wyoming.

Zhang J, Shen H, Wang X, Wu J, Xue Y (2004). Effects of chronic exposure of 2,4-dichlorophenol on the antioxidant system in liver of freshwater fish *Carassius auratus*. Chemosphere 55:167-174.

12

Bacterial population in intensive tilapia (*Oreochromis niloticus*) culture pond sediment in Hai Duong province, Vietnam

Huong N. T. Thanh[1], Hien L. Thuy[2], Wenresti G. Gallardo[3] and Hai N. Thanh[4,5]

[1]Department of Science Technology and International Cooperation, Vietnam Directorate of Fisheries, Vietnam.
[2]Department of Petroleum Microbiology, Institute of Biotechnology (IBT), Vietnamese Academy of Science and Technology, Vietnam.
[3]Aquaculture and Aquatic Resources Management, Asian Institute of Technology, Thailand.
[4]Shanghai Ocean University, Shanghai, China.
[5]Vietnam Institute of Fisheries Economics and Planning, Ha Noi, Vietnam.

Three intensive tilapia ponds were studied to determine the bacterial population during the culture period in Hai Duong province, Vietnam. Bacteria were identified to genus or species level. Total bacteria count in sediment ranged from 6.49×10^5 to 1.29×10^6 cfu g^{-1}. In total, 11 bacterial genera and 13 species were identified from pond sediments, in which both gram-negative and gram-positive rod-shaped bacteria dominated the population accounting for 52 and 48%, respectively. Both beneficial and harmful bacteria were found in sediments, in which the harmful bacterial species dominated (62.5%) and appeared more at the beginning of culture. The harmful bacteria were *Pseudomonas fluorescens, Pseudomonas* sp., *Vibrio pectenicida, Enterococcus*, sp.HD31, *Enterococcus* sp.HD1, *Vibrio damsella, Aeromonas* sp.HD5, *Vibrio* sp. HD7, *Vibrio* sp.HD331, *Aeromonas* sp.HD361, *Enterococcus* sp.HD371, *Vibrio* sp.V31, *Aeromonas hydrophyla* and *Pseudomonas putida*. In contrast, beneficial bacteria accounted for 20.8% only of total isolates. The beneficial bacteria were *Carnobacterium inhibens, Lactobacillus plantarum, Bacillus cereus, Lactobacillus delbrueckii* and *Bacillus pumilus*; they appeared more in the later stage of culture period.

Key words: Bacteria, *Oreochromis niloticus*, intensive culture pond, sediment, beneficial bacteria, harmful bacteria.

INTRODUCTION

Tilapia is an important cultured species in most of tropical, subtropical and temperate regions and currently second to carps in total production; production of tilapia is 2,505,465 tonnes in 2007 (FAO, 2009). However, the success of intensive culture of this species depends on water quality in pond which is influenced by native microflora. Intensive production usually leads to an increase in disease due to poor water quality and high stocking densities of species. Tilapia is susceptible to many infectious agents such as bacteria and parasites (Shoemaker et al., 2006). The important pathogen of tilapia is *Streptococcus iniae*, a gram-positive bacteria responsible for significant losses in intensive culture (Perera et al., 1994; Shoemaker et al., 2006).

The organic matter content in pond is normally due to the feed input and fertilizer (Boyd, 1990). The uneaten feed will be broken down into ammonia, nitrite, and nitrate through the processes of mineralization, nitrification and denitrification by different microorganisms.

The activities of algae and microorganism govern the dissolved oxygen (DO) in water. Photosysnthesis of algae produces DO using sunlight, while bacteria consume oxygen for respiration. The microbial processes including aerobic and anaerobic, also effect on the ammonia content and water quality factor such as pH, DO, nitrite (Moriarty, 1997).

In intensive tilapia pond, the bacterial community may contribute to food web. They may be eaten directly by tilapia. Moreover, the nitrogen and phosphorus may be recycled to stimulate primary production through the activity of decomposed microorganism (Moriarty, 1997).

Recently, there are some studies of Hien et al. (2005) about bacterial composition in shrimp pond in Thanh Hoa province, Vietnam. These studies applied the useful native bacteria to produce probiotic to improve water quality in shrimp pond. Bacterial population in intensive aquaculture pond has an important role in improving water quality because of decomposing organic matter and reducing harmful bacteria in sediments.

Studies on bacterial population in aquaculture are important in predicting the possibility of disease outbreaks and providing information for developing an optimal strategy for effective management to improve production and environmental condition. However, no study has been reported on bacterial variation in freshwater tilapia pond sediments, particularly in the north of Vietnam. Recent studies have only focused on the variation of bacteria in tilapia brackish water pond in Saudi Arabia (Al-Harbi and Uddin, 2003, 2005). This study determined the variation of bacterial population in intensive freshwater tilapia pond during the culture period. Research findings can be useful as basis to implement optimal management strategies to improve production.

MATERIALS AND METHODS

Pond information

Three intensive tilapia ponds were selected for this study in Hai Duong province, Vietnam. The area of ponds ranges from 3000 to 5000 m^2 with an average depth of 2.0 m. Ponds were constructed 7 years ago. All ponds are used for intensive tilapia culture using commercial feed. Tilapia stocking density is 6 fish per m^2. Pond water is exchanged once or twice per month. The culture period is 6 months. Probiotic was applied only once a month in Pond 1.

Sample collection

Water and sediment samples were collected from the three intensive tilapia ponds at different sites such as near inlet, outlet, center of each pond and every month starting from the third month of culture. Water samples were collected in sterile glass bottles (250 ml) sub-surface and sediment samples (20 to 25 g) were collected from the top 10 cm of pond bottom using grab and kept in polyethylene bags during transport to the laboratory. The samples were stored with ice and transferred to the laboratory for analysis at the Institute of Biotechnology (Hanoi, Vietnam).

Analysis of samples

Viable count for enumeration of cell

Sediment samples in each pond were pooled, weighed and pulverized with a mortar and pestle.

Dilution method: 1 g sediment sample mixed with 9 ml of sterile distilled water and shaken well. It was assumed that the bacteria were evenly distributed between solid and liquid. Use the pipette to remove 1 ml and delivered into the next dilution tube. Discarded the pipette, and continued for the required number of dilutions (10-, 100-, 1000-fold dilution). Pipetted 0.1 ml of each dilution of sediment samples into center of a plate agar. A separate plate for each dilution was used. Spread each inoculum using sterile bent glass rod over the plate and incubated for 24 to 48 h and then observed the growth. The number of colonies on a plate were counted and calculated.

Cultivation and isolation of bacteria

Sediment samples from three ponds were analyzed to isolate the bacterial strains. The bacteria were divided into different types based on the morphological characteristics such as colony form, size and color. After recording the above characteristics, 3 to 5 representatives of each colony type were streaked on MPA plates to purify the isolates.

Identification of bacteria

All the purified isolates were observed for cell shape, motility, spores, and encapsulation through gram staining and microscopy after incubating at 28 to 30°C for 24 to 48 h. Simultaneously, the commercial API 50CHB and API 50CHL system (Biomerieux, France) method were used to classify the gram-positive bacteria; API 20NE was used for identifying gram-negative bacteria.

Identification of beneficial and harmful bacteria in sediment

Bacteria were classified into beneficial and harmful bacteria based on the following papers:

1) Role and functions of beneficial microorganisms in sustainable aquaculture (Zhou et al., 2009).
2) Bacteria from fish and other aquatic animals (Buller, 2004).

Physiochemical environmental factors

Water temperature and DO were measured by thermometer and DO test kits *in situ* during water sampling.

Table 1. Mean ± standard error of physicochemical parameters of tilapia pond water at different culture month.

Culture month	Temperature (°C)	pH	DO (mg/L)
3rd	25.5 ± 0.29^{a}	6.5 ± 0.15^{a}	4.9 ± 0.35^{a}
4th	22.5 ± 0.28^{b}	6.63 ± 0.14^{a}	5.43 ± 0.29^{a}
5th	18.8 ± 0.43^{c}	7.06 ± 0.17^{a}	4.96 ± 0.08^{a}
6th	14.7 ± 0.44^{d}	7.1 ± 0.2^{a}	4.46 ± 0.14^{a}

Table 2. Bacterial quantity, mean ± standard error (cfu/g) in tilapia pond sediment.

Pond (cfu/g)	Quantity of bacteria
Pond 1	1.29 ± 0.68 × 10^{6a}
Pond 2	6.49 ± 1.33 × 10^{5a}
Pond 3	1.18 ± 0.39 × 10^{6a}

Data analysis

Statistical analysis was performed using Excel and SPSS 13.0. Data are presented as the mean ± S.E. The bacterial population in sediments at different culture months was tested for significant differences ($p < 0.05$) using one-way and two-way analysis of variance (ANOVA), followed by multiple comparisons of means using SPSS version 13.0.

RESULTS

Physicochemical parameters

Mean values for physicochemical parameters of water for all the three ponds at different culture months are shown in Table 1. The difference in pH and DO was not significant ($p > 0.05$) during the culture months, however, water temperature was significantly different ($p < 0.05$) among the culture months; water temperature was 25.5°C in the third culture month which declined to 14.7°C in the sixth month because it was already winter. Pond water pH was slightly acidic (varied from 6.5 to 7.1). DO range from 4.46 to 5.43 mg L^{-1} during the culture period.

Bacterial population in pond sediment

Quantitative data

The quantity of aerobic bacteria in freshwater tilapia pond sediment in Hai Duong is shown in Table 2. Each number is the average value of viable colonies which grew on duplicate agar plates for each sample. The quantity of bacteria in sediments range from 6.49 × 10^5 to 1.29 × 10^6 cfu g^{-1} in the three ponds but bacteria count were not significantly different among the ponds ($p > 0.05$). Although probiotic was used in Pond 1 to improve water quality during the culture period, the quantity of bacteria did not show any significant difference. This is probably due to the ineffectiveness of the kind of probiotic or the frequency of its application (only once a month).

Bacterial composition in pond sediments

Bacteria in tilapia pond sediment were isolated and identified at genus or species level. The composition of bacterial flora is given and showing the beneficial and harmful bacterial flora which appeared in three studied ponds in Table 3.

In this study, 11 bacterial genera and 13 species were identified from sediment, in which, both gram-negative and gram-positive rod-shaped bacteria dominated all the populations (accounting for 52 and 48%, respectively). Ten bacterial species were common in all three studied ponds: *Pseudomonas fluorescens, Carnobacterium inhibens, Vibrio damsella, Lactobacillus plantarum, Pseudomonas putida, Aeromonas hydrophyla, Granulicatella* sp.V34, *Lactobacillus delbrueckii, Enterococcus* sp.HD371, *Pseudomonas* sp., comprising 41.6% of the total isolates. The bacterial species appeared in two of three ponds: *Vagococcus fluvialis, Granulicatella* sp.V1, *Aeromonas* sp.HD5, *Vibrio* sp.V31, *Bacillus cereus*, and *Bacillus pumilus* (account for 25% of total isolates). Eight bacterial species which were present in only one of the three ponds were *Aeromonas popoffii, Aeromonas* sp.HD361, *Enterococcus* sp.HD31, *Enterococcus* sp.HD1, *Gemella haemolysans, Vibrio* sp.HD7, *Vibrio* sp.HD331, *Vibrio pectenicida*.

Beneficial and harmful bacteria in pond sediments

Bacterial species in this study were classified into

Table 3. Bacterial population in intensive tilapia pond sediments in Hai Duong province.

Pond	Bacteria	Name of species	Gram	Presence during culture month				Beneficial/Harmful
				3rd	4th	5th	6th	
1,2,3	HD1.1	*Pseudomonas fluorescens*	-	x				H
1,3	HD1.2	*Vagococcus fluvialis*	+	x				Environmental organism
1	HD2.1	*Aeromonas popoffii*	-	x				Environmental organism
1,2, 3	HD 2.2, HD (2), HD3.5.1	*Pseudomonas* sp.	-	x	x	x		H
1	HD2.3	*Vibrio pectenicida*	-	x				H
2	HD3.1	*Enterococcus sp.HD31*	+	x				H
2,3	V1	*Granulicatella sp.V1*	+	x				H
1,2, 3	V4	*Camobacterium inhibens*	+	x	x			B
1	HD (1)	*Enterococcus sp.HD1*	+		x			H
3	HD (3)	*Gemella haemolysans*	+		x			H
1,2, 3	HD (4)	*Vibrio damsella*	-		x	x		H
1,3	HD (5)	*Aeromonas* sp.HD5	-		x			H
2	HD (7)	*Vibrio* sp.HD7	-		x			H
1	HD3.3.1	*Vibrio* sp.HD331	-			x		H
1,2, 3	HD3.4.1	*Lactobacillus plantarum*	-			x		B
2	HD 3.6.1	*Aeromonas* sp.HD361	-			x		H
1,2,3	HD3.7.1	*Enterococcus* sp.HD371	+			x		H
1,2	V3.1	*Vibrio* sp.V31	-			x		H
1,2,3	V3.3	*Aeromonas hydrophyla*	-			x		H
1,2,3	V3.4	*Granulicatella* sp.V34	+			x		H
2,3	HD4.2.1	*Bacillus cereus*	+				x	B
1,2,3	HD4.4.1	*Lactobacillus delbrueckii*	+				x	B
1,2,3	HD4.5.1	*Pseudomonas putida*	-				x	H
2,3	HD4.6.1	*Bacillus pumilus*	+				x	B
		Unidentified rod bacteria	+				x	
			-				x	

beneficial and harmful bacteria based on several papers reviewed for this study (Table 3). Zhou et al. (2009) classified bacteria into beneficial, harmless, harmful and pathogenic bacteria. The bacterial population in the intestines of animals will be normal when the health of animals is well, and beneficial bacteria will maintain the balance of intestinal microecosystem.

In all populations, the harmful bacterial species dominated (66.6%) including *P. fluorescens, Pseudomonas* sp., *V. pectenicida, Enterococcus* sp. HD31, *Enterococcus* sp.HD1, *V. damsella, Aeromonas* sp.HD5, *Vibrio* sp.HD7, *Vibrio* sp. HD331, *Aeromonas* sp. HD361, *Enterococcus* sp.HD371, *Vibrio* sp.V31, *A. hydrophyla*, and *P. putida.* The beneficial bacterial species accounted for 20.8% only of total isolates. The beneficial bacteria were *C. inhibens, L. plantarum, B. cereus, L. delbrueckii*, and *B. pumilus.* Others were environmental organisms such as *A. popoffii*, and *V. fluvialis.*

DISCUSSION

Physicochemical parameters

The above measured values are suitable for tilapia culture, except temperature in the sixth month of culture because it was in winter. Low temperature possibly affects fish growth and microbial decomposition in pond sediment. Boyd (1990) indicated that a temperature increase of 10°C often doubles the rate of decomposition and oxygen consumption. Uddin and Al-Harbi (2004) reported that water temperature range from 14.5 ± 1.5 to 33.0 ± 2.3°C, pH slightly alkaline and DO from 3.8 ± 0.2 to 6.2 ± 0.3 ppm for tilapia culture system in Saudi Arabia. Slightly alkaline ponds were suitable for fish culture particularly tilapia (Ntengma and Edema, 2008).

Bacterial population in pond sediment

Bacteria counts observed in the present study are lower than the results reported by Al-Harbi and Uddin (2003). They reported that the quantity of bacteria in hybrid tilapia pond sediment ranged from $9.3 \pm 1.1 \times 10^6$ to $1.9 \pm 1.5 \times 10^8$ cfu g^{-1}. Another study of Al-Harbi and Uddin (2005) showed that the bacteria loads in brackish water tilapia pond sediment ranged from $1.2 \pm 3.1 \times 10^6$ to $7.3 \pm 1.1 \times 10^7$ cfu g^{-1}.

Only 17 bacterial species, of which 87% were gram-negative rod-shaped bacteria were found in tilapia pond sediment in the study of Al-Harbi and Uddin (2005), including *Vibrio parahaemolyticus, Vibrio carchariae, Vibrio alginolyticus, Vibrio vulnificus, Chryseomonas* sp., *Streptococcus* sp., *Shewanella putrefaciens, Pseudomonas* sp., and *Bacillus* sp.,. These species differed from the identified species in the present study, especially, some pathogenic bacteria such as *V. parahaemolyticus, V. carchariae, V. alginolyticus, V. vulnificus, Chryseomonas* sp., and *S. putrefaciens* were not found in the present study due to these species is only common in brackish water environment. Furthermore, *A. hydrophyla* was identified in this study, however, it was not found in sediment in their study. Uddin and Al-Harbi (2004) indicated that *A. hydrophyla* was the dominant bacteria in pond water for all the different seasons. In another study, Al-Harbi and Uddin (2003) reported that the dominant bacteria in pond sediment were *Corynebacterium urealyticum, S. putrefaciens* and *Escherichia coli*, these species were not found in present study, however, *A. hydrophyla* and *P. fluorescens* were also found in this study.

The beneficial bacteria found in this study mainly belong to *Lactobacillus, Bacillus* and *Carnobacterium* genera. These genera are beneficial because they are used in many microbial products to improve water quality to reduce prevalence of fish disease because beneficial bacteria can hinder the growth of harmful bacteria due to the survival competition or increase the health of fish. *Bacillus* sp. could improve water quality because these gram-positive bacteria are better converters of organic matter back to CO_2 than gram-negative bacteria. The high levels of gram-positive bacteria can minimize the buildup of dissolved and particulate organic carbon during the production cycle. Zhou et al. (2009) reported the dominant species in the intestine (>99%) are *Bacillus, Lactobacillus*, and *Bifidobacterium*. The beneficial bacteria in this study can be used to produce probiotic (including beneficial bacteria and *Bacillus* spp.) to restrict the development of pathogens and improve water quality. Dalmin et al. (2001) used probiotic to reduce the *Vibrio* pathogen and enhance beneficial bacilli in the culture leading to improved water quality, survival and growth rate and increase health status of juvenile shrimp *Penaeus monodon*. Qi et al. (2009) used *Bacillus* strains isolated from fish gut or shrimp ponds as an antagonist to *A. hydrophyla* and decrease ammonia level in culture water.

L. plantarum and *L. delbrueckii,* which belong to *Lactobacillus* genus, are important lactic bacteria because they can produce lactic and acetic acids through metabolism and thus lower pH in the intestine and enable beneficial bacteria to be dominant species (Zhou et al., 2009). They can be found in intestine of fish. Many applications of these species are reported due to their fermentative ability as well as their health and nutritional benefits (Rossland et al., 2003). These bacteria were used as microbial feed supplement to improve intestinal microbial balance of fish. *Lactobacillus* genus is also applied to improve water quality and sediment condition.

Beside beneficial bacteria, many harmful bacteria were also found in pond sediment in this study including *A. hydrophyla, P. fluorescens, V. pectenicida, Enterococcus* sp., *V. damsella, Vibrio* sp., and *P. putida*. *A. hydrophyla* is an opportunistic organism, however, pathogenesis and virulence of this species remains unclear. This species can be found in fresh, brackish and coastal waters (Kuijper et al., 1989). Many toxins were produced such as haemolysins and cytotoxins (Chopra et al., 1993). A dominant *A. hydrophyla* was found which is associated with the outbreak of disease in aquaculture of China (Nielsen et al., 2001). Many reports showed that this species was found in tilapia, scallop larvae (*Argopecten purpuratus*) (Riquelme et al., 1996), channel catfish (*Ictalurus punctatus* Rafinesque) (Buller, 2004), European seabass (*Dicentrarchus labrax* Linnaeus), seabass (*Puntazzo puntazzo* Cuvier) (Doukas et al, 1998), blue discus fish (*Symphysodon aequifasciatus* Pellegrin), and walking catfish (*Clarias batrachus* Linnaeus) (Angka et al., 1995) to cause mortality. Moreover, *A. hydrophyla* was also found in the lung and liver of grey seal (*Halichoerus grypus*) (Krovacek et al., 1998) and faeces of human (Brew et al., 1999).

P. fluorescens found in this study is an opportunistic pathogen. It has been found in freshwater ornamental fish, carp, tilapia, goldfish, seabream, rainbow trout (Kusuda and Yamaoka 1974) and appeared in the gills of cultured salmonids (Rainbow trout, Chinook salmon, Atlantic salmon) (Shotts and Waltman, 1990). Vibriosis is one of the oldest recognized infectious diseases. *V. pectenicida* was identified in this study. It is a pathogenic bacterium and was found in scallop (*Pecten maximus*) larvae, shrimp and prawn.

Results of this study also show the presence of more beneficial bacteria in the later stage of culture period. This demonstrates that these beneficial bacteria can decompose rapidly the organic matter in pond sediment and thus improve water quality in tilapia pond. On the other hand, these bacteria can restrict the development of harmful bacteria which appear mostly at the earlier stage of culture period so that in the later stage, less harmful bacteria were found. From these results, it suggests that culturist should apply some beneficial microorganism in the earlier stage of culture not only to improve water quality but also control the development of pathogenic bacteria to improve the health of fish. However, the beneficial bacteria found in this study also appeared in Ponds 2 and 3 which did not use any probiotic. This suggests that the application of probiotic did not increase the beneficial bacteria in tilapia ponds. The presence of these beneficial bacteria may be due to the frequent water exchange in ponds, resulting in good water quality suitable for beneficial bacteria and cultured fish.

These findings will not only lead to the better control of water quality but also apply the right kind of bacteria or microbial products in ponds. The information on bacterial population may be useful to improve management of intensive tilapia pond. The harmful bacteria appeared in the early stage of culture, thus, more care should be given at this stage and proper pond management such as more water exchange and aeration should be done to enhance the beneficial bacteria.

Conflict of Interest

The author(s) have not declared any conflict of interest.

ACKNOWLEDGEMENTS

The authors thank the tilapia farmers from Hai Duong province, Vietnam for allowing the collection of sediment samples from their ponds and to the Department of Petroleum Microbiology for the laboratory facilities. Ms. Huong is grateful to ASEAN Foundation Scholarship for the support to study in AIT.

REFERENCES

Al-Harbi AH, Uddin N (2003). Quantitative and qualitative studies on bacterial flora of hybrid tilapia (*Oreochromis niloticus* x *O. aureus*) cultured in earthern ponds in Saudi Arabia. Aquac. Res. 34:43-48.http://dx.doi.org/10.1046/j.1365-2109.2003.00791.x

Al-Harbi AH, Uddin N (2005). Bacterial diversity of tilapia (*Oreochromis niloticus*) cultured in brackish water in Saudi Arabia. Aquac. 250:566-572. http://dx.doi.org/10.1016/j.aquaculture.2005.01.026

Angka SL, Lam TJ, Sin YM (1995). Some virulence characteristics of Aeromonas hydrophyla in walking catfish (*Clarias gariepinus*). Aquaculture 130:103-112. http://dx.doi.org/10.1016/0044-8486(94)00216-B

Boyd CE (1990). Water quality in ponds for aquaculture. Auburn, Al: Alabama Agriculture Experiment Station.

Brew SD, Perrett LL, Stack JA, MacMillan AP, Staunton NJ (1999). Human exposure to *Brucella* recovered from a sea mammal. Vet. Record 144:483.PMid:10358880

Buller NB (2004). Bacteria from fish and other aquatic animals: A practical identification manual. CABI Publishing.

Chopra A, Houston C, Peterson J, Jin GF (1993). Cloning, expression, and sequence analysis of a cytolytic enterotoxin gene from *Aeromonas hydrophyla*. Canad. J. Microbiol. 39:513-523. http://dx.doi.org/10.1139/m93-073PMid:8330262

Dalmin G, Kathiresan K, Purushothaman A (2001). Effect of probiotics on bacterial population and health status of shrimp in culture pond ecosystem. Indian J. Exp. Biol. 39:939-942.PMid:11831382

Doukas V, Athanassopoulou F, Karagouni E, Dotsika E (1998). Aeromonas hydrophyla infection in cultured sea bass, *Dicentrarchus labrax* L., and *Puntazzo puntazzo* Cuvier from Aegean Sea. J. Fish Dis. 21:317-320.http://dx.doi.org/10.1046/j.1365-2761.1998.00105.x

Food and Agriculture Organization of the United Nations (FAO) (2009). FAO FishStat plus. Aquaculture Production 1970-2000. Rome, Italy.

Krovacek K, Huang K, Sternberg S, Svenson SB (1998). *Aeromonas ydrophyla* sapticaemia in a grey seal (*Halichoerus grypus*) from the Baltic Sea: A case study. Comparat. Immunol. Microbiol. Infect. Dis. 21:43-49.http://dx.doi.org/10.1016/S0147-9571(97)00015-5

Kuijper EJ, Steigerwalt AG, Schoenmakers BSCIM, Peeters MF, Zanen HC, Brenner DJ (1989). Phenotypic characterization and DNA relatedness in human fecal isolates of *Aeromonas* spp. J. Clin. Microbiol. 27:132-138. PMid:2913025 PMCid:PMC267248

Kusuda R, Yamaoka M (1972). Etiological studies on bacterial pseudotuberculosis in cultured yellow tail with *Pasteurella piscicida* as the causative agent. I. On the morphological and biochemical properties. Bull. Japan. Soc. Sci. Fisher. 38:1325-1332. http://dx.doi.org/10.2331/suisan.38.1325

Hien LT, Phuong DT, Anh VP, Nga DP, Hang PT, Nga VT, Cong LTN, Yen NT, Tu NB (2005). Application of probiotic for bioremediation of industrial shrimp farming area in Hoang Hoa, Thanh Hoa. Regional symposium on chemical engineering, Hanoi Dec. 2005

Moriarty DJW (1997). The role of microorganisms in aquaculture ponds. Aquaculture 151:333-349. http://dx.doi.org/10.1016/S0044-8486(96)01487-1

Nielsen ME, Hoi L, Schmidt A, Qian D, Shimada T, Shen J, Larsen J (2001). Is *Aeromonas hydrophyla* the dominant motile *Aeromonas* species that causes disease outbreaks in aquaculture production in the Zhejiang province in China. Dis. Aquat. Organ. 46:23-29. http://dx.doi.org/10.3354/dao046023PMid:11592699

Ntengma FW, Edema MO (2008). Physico-chemical and microbiological characteristics of water for fish production using small ponds.

Perera RP, Johnson SK, Collins MD, Lewis DH (1994). *Streptococcus iniae* associated with mortality of *Tilapia nilotica* × *T. aurea* hybrids. J. Aquat. Anim. Health 6:335–340. http://dx.doi.org/10.1577/1548-8667(1994)006<0335:SIAWMO>2.3.CO;2

Qi Z, Zhang X, Boon N, Bossier P (2009). Probiotics in aquaculture of China-Current state, problems and prospect. Aquaculture 290:15-21.http://dx.doi.org/10.1016/j.aquaculture.2009.02.012

Rossland E, Borge GIA, Langsrud T, SØrhaug T (2003). Inhibition of Bacillus cereus by strains of Lactobacillus and Lactococcus in milk. Int. J. Food Microbiol. 89:205-212. http://dx.doi.org/10.1016/S0168-1605(03)00149-1

Shoemaker CA, De-Hai X, Evans JJ, Klesius PH (2006). Parasites and Diseases. In C. Lim and C.D. Webster (Eds) Tilapia: Biology, Culture and Nutrition (pp. 561-582). Binghamton, NY: The Haworth Press, Inc.

Shotts EB, Waltman WDII (1990). A medium for the selective isolation of Edwardsiella ictaluri. J. Wildl. Dis. 26:214-218. http://dx.doi.org/10.7589/0090-3558-26.2.214PMid:2338726

Uddin MN, Al-Harbi AH (2004). Seasonal variation of bacterial flora in ponds in Saudi Arabia used for tilapia aquaculture. J. Appl. Aquac. 16:53-61. http://dx.doi.org/10.1300/J028v16n01_04

Zhou Q, Li K, Jun X, Bo L (2009). Role and functions of beneficial microorganisms in sustainable aquaculture. Bioresou. Technol. 100:3780-3786. http://dx.doi.org/10.1016/j.biortech.2008.12.037 PMid:19261470

Evaluation of physicochemical parameters of waste water from rainbow trout fish farms and their impacts on water quality of Koohrang stream – Iran

Firooz Fadaeifard[1]*, Mehdi Raissy[1], Mostafa Faghani[2], Alireza Majlesi[3] and Gholamhossein Nodeh Farahani[4]

[1]Department of Aquatic Animal Disease and Health, Islamic Azad University, Shahrekord Branch, Shahrekord – Iran.
[2]Department of Animal Science, Islamic Azad University, Shahrekord Branch, Shahrekord – Iran.
[3]Faculty of Veterinary Medicine, Islamic Azad University, Shahrekord Branch, Shahrekord – Iran.
[4]Department of Chemistry, Islamic Azad University, Shahrekord Branch, Shahrekord – Iran.

Physicochemical parameters of rainbow trout farm water were evaluated to assess the potential of fish farm effluents on stream water quality. Seven fish farms were selected at Rostamabad region located on 70 km at south-west of Shahrekord-Iran. Water samples were collected from fish farms inlet and outlet at first and second sampling point. Significant differences were observed in some water factors such as total hardness, total dissolved solid, total suspended solid, COD, BOD_5, dissolved oxygen, phosphate, nitrite, nitrate and total ammonia between inlet and outlet water of fish farms where as there were no significant changes in pH, sodium chloride and water temperatures. All of the water factors because of self-purification potential of the stream were in desirable amounts in third sampling point. Feasibility study carried out to assess the distance for making consecutive fish farm along the stream. It was observed that it is possible to make consecutive fish farms at around 1500 m distance but it is depended on stream water discharge and self-purification of the stream.

Key words: Water pollution, trout farm effluents, environmental impact, Koohrang stream.

INTRODUCTION

Nowadays, water pollution from discharging waste water of fish farms is real concern in the world (Boyd, 2003). Changing on river and bank of river ecosystems by effluents of consecutive fish farms is caused to revise on managing of fish farms waste water control (Mumpton and Fishman, 1997). Fish farms could have different types of environmental effects on rivers such as changing on river hydrology, introducing non-native species in natural water and water pollution (Read et al., 2001). In flow-through fish farms system, effluents discharges into the environment with high concentration of nutrients and solid materials. If fish farms effluents discharge into the environment without treatment, it will have undesirable and harmful effects on the environment (Schulz et al., 2003; Miller and Semmens, 2002; Forenshell, 2001). Determination of physicochemical parameter of fish farms effluents and their environmental effects will help fish farmers to manage and develop their waste water systems (Pulatsu et al., 2004).

Generally salmonids fatten by consuming lots of nutrients. Different types of metabolites will release in their body which is known as unpleasant impact on aquatic environments (Papatryphon et al., 2005; True et al., 2004; Bureau and Cho, 1999). Recently, according to sustainable development in fish farming, aquaculture industry is moving toward reducing resources of pollutions by proper managing of fish feeding and waste water treatment. Although competent authorities must

*Corresponding author. E-mail: fadaeifard@gmail.com.

also have supervision and control on water resources in order to minimize environmental impacts of fish farms (Forenshell, 2001; Read et al., 2001; Midlen and Redding, 1998). Fish farms pollutants are fish metabolic products and fish feces as well as not-used nutrients residue (Kajimura et al., 2004). The amounts of pollutants are related to food chemical composition and its stability as well as fish feeding methods. Fish food ingredient or composition which is precipitated in fish farms can be very broad (Teodorowicz et al., 2006). Some of physicochemical parameters of waste waters which can impact on the aquatic environments are including total suspended solids, total dissolved solids, total hardness, BOD_5, $N\text{-}NO_2$, $N\text{-}NO_3$, $N\text{-}NH_3$ and pH.

It is common to see a certain amount of turbidity in fish farms due to existing suspended clay, soil sediments, artificial and natural sludge coming from constructional activities, mining or separating sediments from fish farms bed. Solids suspended materials cause asphyxia during incubation period on growing eggs. They can also have injure or scratch and cover fish gills physically. There are no standards for suspended solids particles such as volcanic ashes with irregular and sharp edges which potentially can injure fish gills (Wedemeyer, 1996).

Water pH could be affected by the type of water resources, geological conditions as well as aquatic plants. Discharging of acidic and basic waste water and also sewage with different types of minerals can either have effect on water pH. Total hardness and water alkalinity are usually known as touchstone for buffer capacity. Soft and hard waters are usually acidic and basic respectively. Soft water has got low calcium and other minerals, therefore fishes can endure this situation if the amounts of these materials increase in fish diets (Wedemeyer, 1996).

Warm water fishes are able to endure low dissolved oxygen in comparison with cold water fishes (Lawson, 2001). Nitrogenated compounds as well as metabolites which are produced by metabolism in fish body are harmful for aquatic. Total ammonia in water depends on water temperature and salinity. Non-ionic nitrogen compounds are toxic for fishes, although most of excreted nitrogen compounds from metabolism in fish body are in shape of ammonia. High amount of ammonia in water causes different problems in fishes such as toxicity which is harmful to the gills and blood hemoglobin. It reduces capability of blood for transferring oxygen as well as increasing oxygen demand in fish tissues (Lloyd and Orr, 1969). Nitrite is a toxic nitrogenated compound accumulated in anaerobic sludge which is able to combine with hemoglobin to form methemoglobinemia. This compound cannot carry oxygen, thus fishes will suffer from anoxia. Nitrate has less toxicity in comparison with other inorganic nitrogenated compound with LC50, 1000 – 3000 mg/L in 96 h (Forenshell, 2001). Phosphate is found in different shapes of water and waste water. Main sources of phosphate are detergents, fertilizers and industrial sewages which are harmful for fish's life.

Objectives of the study are (1) to monitor water quality through determination of total suspended solids, total dissolved solids, total hardness, total phosphate, NaCl, COD, BOD_5, $N\text{-}NO_2$, $N\text{-}NO_3$, $N\text{-}NH_3$, electrical conductivity and pH and (2) to assess required distance between two consecutive fish farms along the stream. In this study, it will be shown whether the self-purification potential of stream could have effect on improvement of physicochemical properties of the water in stream at a distance of 1,500 m far from trout farms. Therefore it will be investigated if the self-purification of stream allows reusing water for aquaculture.

MATERIALS AND METHODS

Study area

The study area where is known to be Rostam Abad is located on distance of 70 km at south-west of Shahrekord-Iran. There are seven trout farms in this region. The water supply of the farms is a mountain spring which is called Sardab Rostamabad spring with average discharge of 1500-2000 L/s. Fish farm effluents discharge into Koohrang stream, which is one of the branches of Karoon River (the biggest river in west of Iran). The annual production of these farms is almost 700 tons. Figure 1 shows three stations as sampling points where called; S-1(farms inlet), S-2 (farms outlet) and S-3 (1500 m far from the farms outlets along the Koohrang stream). The times of samplings were at 10 am before fish feeding in the morning at three sampling points. Daily diets data was also recorded during sampling.

Sampling and water sample analysis

Water samples were collected from seven rainbow trout farms from July to December 2009 (summer and autumn). Water samples were kept in plastic and glass screw capped bottles. Glass bottles were used to determine dissolved oxygen (DO). The water samples in plastic bottles used for determination of twelve physicochemical parameters of water samples such as total suspended solids, total dissolved solids, total hardness, total phosphate, NaCl, COD, BOD_5, $N\text{-}NO_2$, $N\text{-}NO_3$, $N\text{-}NH_3$, electrical conductivity and pH which were determined by classic and new methods. Labeled bottles after sampling were transferred to the laboratory within 1-2 h. Testo(R) 240 – Germany, a portable measuring instrument, was used to measure temperature, electrical conductivity (EC) and the amount of sodium chloride. Water pH was measured by using portable pH meter (HACH-EC30, USA). The rest of the water factors were determined by classical methods. Total hardness (TH) was measured by ASTM-2340C method (EDTA titration method). ASTM – D5907 method was used to determine total dissolved solid (TDS) and total suspended solid (TSS). ASTM WK4052 method was used to determine total phosphate. COD and BOD_5 were determined by ASTM D1252-06 and ASTM D6238-98(2003) respectively. Nitrite and nitrate were measured by ASTM D3867-09 method and ammonia by ASTM D1426-08 method.

Statistical analysis

All data were expressed as mean ± SD and means between different months compared with statistical tests

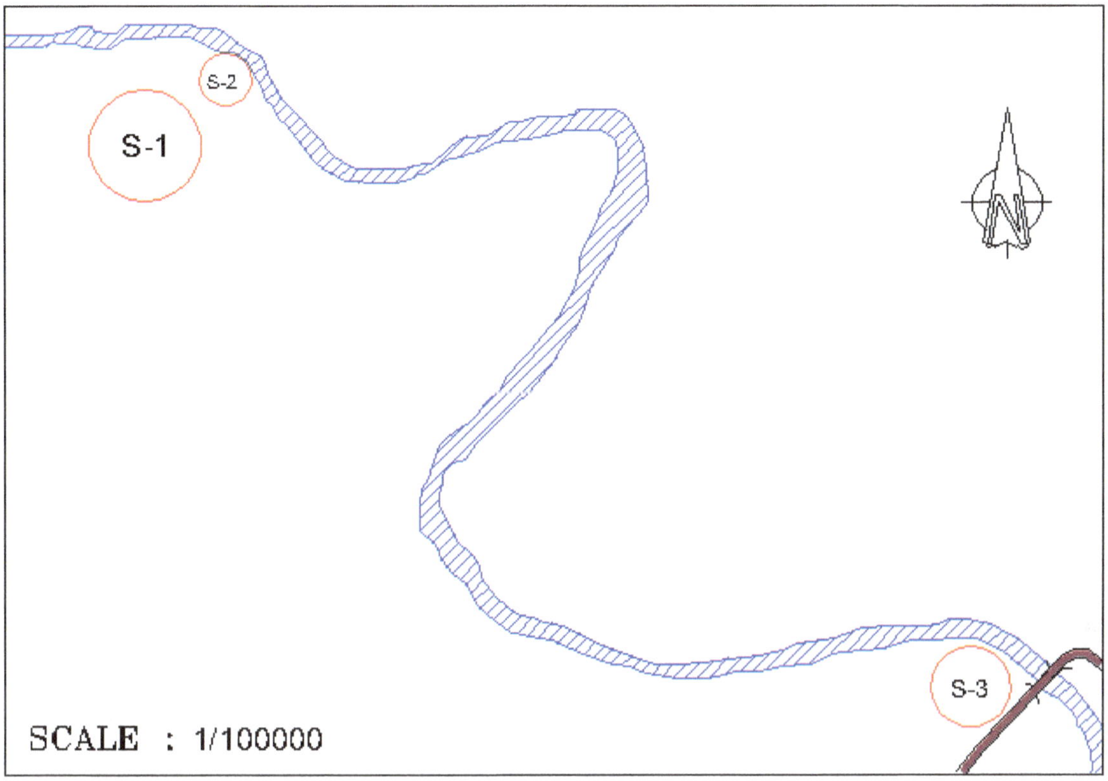

Figure 1. The location of sampling points at different stations along Koohrang stream-Iran.

(One Way ANOVA) analysis and Duncan's multiple at 95% confidence interval. Also SPSS software (version 19) was used for statistical analysis.

RESULTS AND DISCUSSION

Physicochemical parameters of water samples are given in Tables 1, 2 and 3. Physicochemical parameters of water inlets and outlets of seven fish farms were shown in Tables 1 and 2 respectively. Table 3 shows physico-chemical factors of water from the last sampling point where located at 1500 m far from the fish farm waste water discharge in Koohrang stream.

It was observed that there is no significant difference in electrical conductivity of the water inlet and outlet of fish farm (Tables 1 and 2). However there is a significant difference in water EC at third sampling point (Table 3). It is due to higher salinity, hardness, total dissolved solid, stream bed and the corrosion on stream bed. Significant difference was observed in third sampling point for sodium chloride which was more than others. It is due to the source of this compound in stream bed which could affect on water EC. A same trend was observed for total dissolved solid and total solid. It is due to presence of some metal carbonate, bicarbonate, sulfate and etc. Fish foods also have effect on TDS between inlet and outlet water, however it is not that significant because samplings were done before fish feeding.

Tables 1, 2 and 3 show the water pH which is almost same in all water samples in 3 sampling points at different sampling time. The impact of trout farm effluents on the pH of the receiving water was not significant. Water pH is affected by rainfall, snow melting and stream bed and sediments. Water hardness and alkalinity play a main roll to prohibit changes in water pH (Stoskopf, 1992). Water pH at range 6.5 – 9 is suitable for fish farming. If pH ranges were between 4 – 6 or 9 – 10, although fishes will be alive but fish growth rate will be reduced. Regeneration will also decrease if the water pH becomes less than 6. Fish mortality will start in pH less than 4 (acidic) or more than 11 (basic) in fish farms (Lawson, 2001).

Significant differences were shown between dissolved oxygen (DO) in inlet and outlet waters. Reduction of DO level, because of consumption of oxygen in trout farms, was found statistically significant. Fish biomass increases the level of excretory products of fish lead to a decrease in effluent DO level and an increase in BOD5, COD, and total nitrogen in the effluent (Maillard et al., 2005). There was a remarkable difference in DO between the third sampling point compared with inlet and outlet water (Figure 2). It was significantly higher than DO in inlet and effluent waters. However increase in third station is probably due to high river flows.

There is a direct relationship between stream water

Table 1. Comparison of all parameters (means±SD) of fish farm's inflows in the different months.

Month / Parameter	July	August	September	October	November	December
Temp (°C)	9.43± 0.59^{B}	10.33± 0.31^{C}	9.29± 0.27^{AB}	9.29± 0.27^{AB}	9.00± 0.0^{A}	9.00± 0.0^{A}
pH	7.96 ± 0.08^{A}	7.99± 0.13^{A}	7.96 ± 0.17^{A}	8.10 ± 0.37^{A}	8.09 ± 0.07^{A}	8.08 ± 0.08^{A}
NaCl (mg/L)	137.21± 52.09^{A}	150.91± 88.34^{A}	189.57± 119.27^{A}	147.86± 5.49^{A}	154.40± 39.53^{A}	158.43± 36.59^{A}
EC(µs/cm)	288.57 ± 106.29^{A}	314.57± 71.63^{A}	392 ± 241.65^{A}	307.86 ± 12.06^{A}	321.86 ± 80.35^{A}	330 ± 74.36^{A}
TDS(mg/L)	163.57 ±52.19^{A}	174.43 ±40.71^{A}	222.29 ±117.86^{A}	173.29 ±11.07^{A}	184.81 ±53.05^{A}	172.57 ±39.54^{A}
TSS(mg/L)	28.29 ± 14.80^{A}	15.29 ± 8.22^{B}	13.00 ± 4.90^{B}	13.14 ± 3.19^{B}	12.57 ± 3.69^{B}	10.14 ± 1.35^{B}
TS(mg/L)	191 ±52.07^{A}	189.71 ±47.55^{A}	233.86 ± 117.18^{A}	186.43 ±10.01^{A}	197.39±58.54^{A}	182.71± 40.39^{A}
TH(mg/L)	145.43±11.65^{D}	193.14±18.51^{BC}	244.29±20.77^{ABC}	230 ±39.16^{A}	191.29±17.51^{C}	226.86±17.04^{AB}
DO(mg/L)	8.33± 1.01^{A}	8.97± 0.52^{A}	8.56± 0.32^{A}	8.61± 0.50^{A}	8.72± 0.28^{A}	8.17± 0.30^{A}
BOD_5(mg/L)	3.64 ± 3.20^{A}	1.86±0.63^{AB}	1.64± 0.48^{AB}	2.21 ±0.95^{AB}	1.50 ±0.41^{AB}	1.30± 0.48^{B}
COD(mg/L)	25.14±13.38^{A}	10.29 ± 3.04^{B}	10.86 ± 5.87^{AB}	23.66 ± 11.55^{AB}	20 ± 3.83^{AB}	19.26 ± 11.44^{AB}
NH4(mg/L)	0.24± 0.17^{A}	0.17± 0.17^{A}	0.14± 0.18^{A}	0.05± 0.06^{A}	0.09± 0.06^{A}	0.04± 0.04^{A}
N-NO_2(mg/L)	0.02± 0.01^{A}	0.01± 0.01^{B}	0.01± 0.01^{B}	0.01± 0.01^{B}	0.01± 0.01^{B}	0.01± 0.01^{B}
N-NO_3(mg/L)	50.90± 1.50^{B}	2.51±1.23^{B}	1.61±0.80^{B}	5.06±1.46^{B}	6.37±3.79^{B}	14.57±10.18^{A}
PO_4(mg/L)	0.24±0.14^{A}	0.16±0.08^{ABC}	0.21±0.10^{AB}	0.10±0.05^{BC}	0.06±0.05^{C}	0.06±0.04^{C}

*Means within rows with different superscript letters are significant statistical difference (P≤0/01).

Table 2. Comparison of all parameters (means±SD) in outflow of fish farms in the different months.

Month / Parameter	July	August	September	October	November	December
Temp (°C)	10.77± 1.22^{B}	12.33± 1.07^{C}	10.00± 0.50^{AB}	9.29± 0.27^{AB}	9.50± 0.00^{A}	9.50± 0.00^{A}
pH	7.81± 0.11^{A}	7.70± 0.22^{A}	7.82± 0.26^{A}	8.10 ± 0.37^{A}	7.87± 0.14^{A}	7.81± 0.04^{A}
NaCl (mg/L)	139.22 ±52.94^{A}	157.50 ±42.03^{A}	172.14 ±51.53^{A}	147.86± 5.49^{A}	161.47 ±44.78^{A}	165 ±45.03^{A}
EC(µs/cm)	292.57 ± 108.30^{A}	311.29 ± 94.21^{A}	348.43 ± 108.85^{A}	307.86 ± 12.06^{A}	336 ± 91.29^{A}	343.71 ± 91.71^{A}
TDS(mg/L)	163.43 ± 43.91^{A}	181.43 ± 48.83^{A}	203.43 ± 44.21^{A}	173.29 ±11.07^{A}	203.14 ± 57.49^{A}	200.86 ± 55.92^{A}
TSS(mg/L)	28.86 ±16.32^{A}	22.71 ± 12.58^{A}	27.29 ± 11.57^{A}	13.14 ± 3.19^{B}	23.57 ± 6.55^{A}	22.86 ± 6.18^{A}
TS(mg/L)	192.29±50.81^{A}	204.14±59.92^{A}	230.71±52.13^{A}	186.43 ±10.01^{A}	226.71±61.60^{A}	223.71±58.80^{A}
TH(mg/L)	167.43± 18.61^{A}	216.76± 27.20^{A}	246.29 ± 24.78^{A}	230 ±39.16^{A}	210.14 ±22.21^{A}	241.14 ±21.41^{A}
DO(mg/L)	7.06^{A}±1.37	7.06^{A}±0.78	7.27^{A}±0.35	8.61± 0.50^{A}	7.76^{A}±1.19	6.90 ± 0.58^{A}
BOD_5(mg/L)	6.64±4.10^{A}	5.36±3.28^{AB}	5.50± 1.66^{B}	2.21 ±0.95^{AB}	2.86±0.63^{B}	2.63± 0.97^{B}
COD(mg/L)	47.20± 21.50^{A}	28.70± 15.50^{AB}	22.86± 7.80^{B}	23.66 ± 11.55^{AB}	37.14± 3.24^{AB}	35± 16.94^{AB}
NH_4(mg/L)	0.58± 0.33^{A}	0.31± 0.31^{A}	0.32± 0.17^{A}	0.05± 0.06^{A}	0.22± 0.13^{A}	0.25± 0.26^{A}
N-NO_2(mg/L)	0.06± 0.06^{A}	0.04± 0.06^{A}	0.02± 0.04^{A}	0.01± 0.01^{B}	0.01± 0.01^{A}	0.03^{A}±0.02
N-NO_3(mg/L)	4.76± 2.03^{C}	6.76± 3.52^{BC}	6.14± 3.36^{BC}	5.06±1.46^{B}	10.20 ±50.08^{BC}	20.31± 12.23^{A}
PO_4(mg/L)	0.14^{A}±0.31	0.20± 0.09^{AB}	0.24± 0.15^{AB}	0.10±0.05^{BC}	0.15± 0.10^{AB}	0.14± 0.03^{B}

*Means within rows with different superscript letters are significant statistical difference (P≤ 0/01).

discharge and total dissolved solid. Rainbow trout fishes are able to adjust themselves to 30 g/L salinity, but the highest amount of salinity which is suitable for non euryhaline fishes are 10 g/L. They are able to endure the osmotic pressure (Wedemeyer, 1996). Total suspended solid in effluent water was more than inlet water due to colloid particles, organometallic compounds, natural sludge, fish feces and left over foods. This causes to increase turbidity too. TSS more than 80 - 100 mg/L will cause injure fish gills (Teodorowicz et al., 2006). TSS, in all water samples, was less than 35 mg/L which does not have undesirable effect on fish farming. It was found that TSS in third sampling points has significant fluctuation (Figure 3). It was increased because of amounts of foods used in these months and changes in water discharge of the stream which is decreased in cold season. Stewart et al. (2006) treated waste water of rainbow trout fish farms by using artificial substrates for coagulation and baffled sedimentation basin simultaneously for TSS treatment. They reported that using these two techniques is useful

Table 3. Comparisons of all parameters at third sampling point (1500 m after waste water discharge of fish farms along the stream) in different months.

Month / Parameter	July	August	September	October	November	December
Temp (°C)	18.5	18.5	14	14	10	9.5
pH	8.35	8.25	8.20	8.22	8.14	6.55
NaCl (mg/L)	238	235	280	260	299	301
EC(μs/cm)	492	487	577	531	614	616
TDS(mg/L)	350	345	303	203	420	388
TSS(mg/L)	21	23	12	35	24	36
TS(mg/L)	371	371	315	238	444	424
TH(mg/L)	206	209	190	230	250	260
DO(mg/L)	10.6	10.1	7	8.5	9.8	8.76
BOD_5(mg/L)	4	4	4	4.8	3	3
COD(mg/L)	8	7	22	58	40	40
NH_4(mg/L)	0.15	0.14	0.10	0.15	0.03	0.11
N-NO_2(mg/L)	0.001	0.002	0.001	0.013	0.006	0.004
N-NO_3(mg/L)	4	4	7.7	11.5	10	24
PO_4(mg/L)	0.1	0.1	0.1	0.15	0.09	0.05

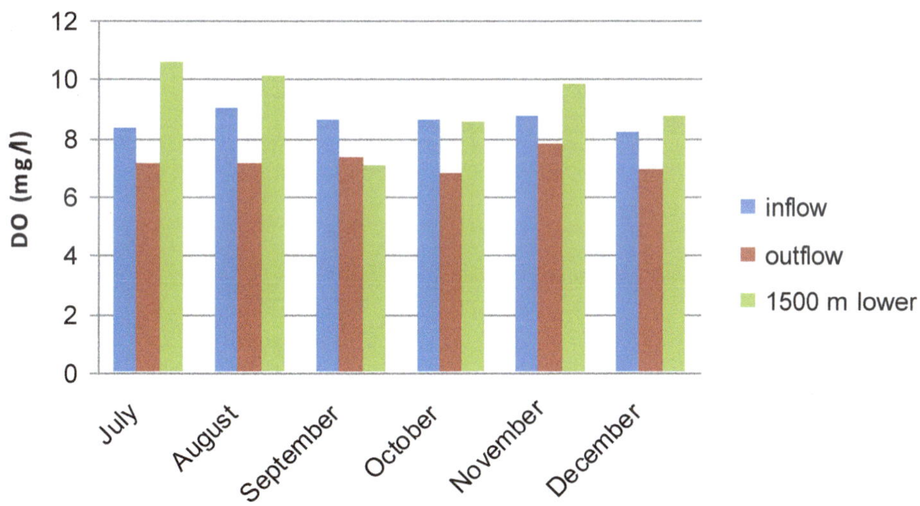

Figure 2. Change of dissolved oxygen in three sampling points; farms inflow, farms outflow and 1500 m far from the farms outflow along the Koohrang stream.

for TSS treatment.

Total hardness in effluent water was more than inlet water and in the third sampling point was more than inlet and outlet waters (Tables 1, 2 and 3). The minimum amount of hardness which is proper for fish growing in fish farming is 100 mg/L (Stoskopf, 1992). Hard waters is better than other types of water for trout farming because it can provide required calcium for fish growing. It also reduces required energy to regulate osmotic pressure for substitution of blood electrolytes (Wedemeyer, 1996). In fresh water, with hardness of more than 200 mg/L, fishes use methabolic energy to regulate osmotic pressure whereas in soft water with hardness of less than 30 mg/L, this energy is used for fish growing (Stoskopf, 1992).

According to the sampling time that was from summer to autumn, water temperature was gradually reduced. It has effect on water physicochemical factors and it is very important aspect in fish farming. Higher feeding rates increase the output of organic matter from farms either as left over food or feces and result in marked elevation in the BOD of receiving water (Miller and Semmens, 2002). BOD_5 in effluent water were more than inlet water, other researchers have also confirmed that the impact of trout farm effluents on the BOD_5 of the receiving water

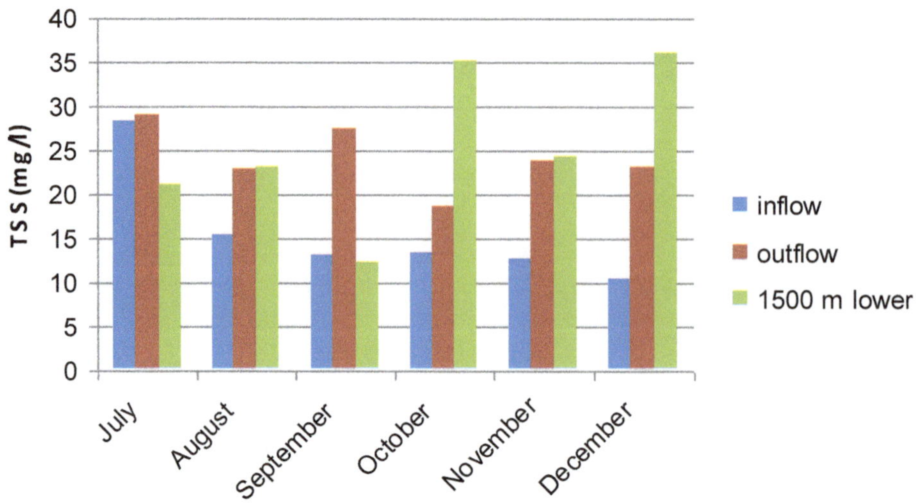

Figure 3. Total suspended solids in three sampling points: farms inflow, farms outflow and 1500 m far from the farms outflow along the Koohrang stream.

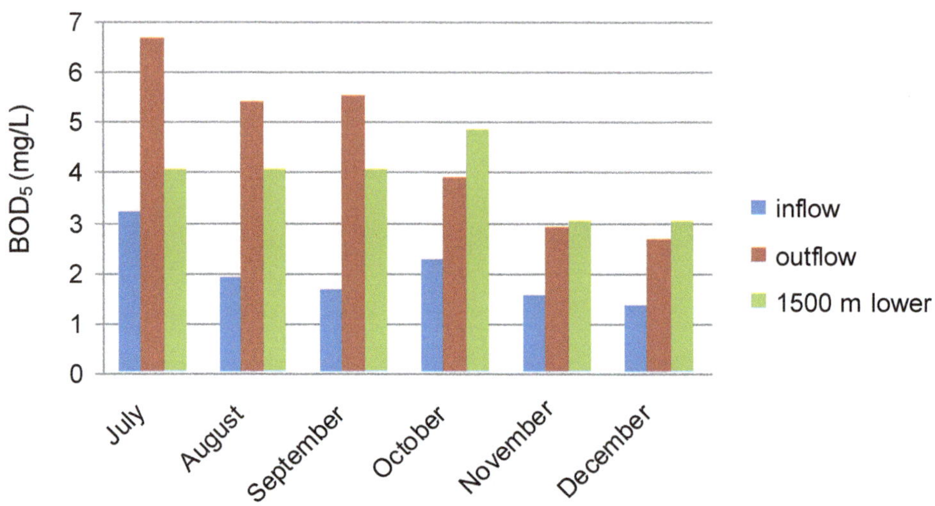

Figure 4. Levels of biological oxygen demand in three sampling points; farms inflow, farms outflow and 1500 m far from the farms outflow along the Koohrang stream.

was significant (Pulatsu et al., 2004; Maillard et al., 2005). It was observed that, in the third sampling point, BOD_5 was almost less than effluent waters (Figure 4). It should be due to stream's self-purification ability. A same trend was observed for COD. COD values in the effluent increase due to aquaculture activities (Pillay, 2004). Boaventura et al. (1997) has reported that BOD returned to the acceptable value for feeding water 2–3 km downstream from the point of effluent discharge in the Fornelo and Inha rivers in Portugal.

Same as BOD_5 and COD, total ammonia values increased in the discharged water. Figure 5 shows significant differences between amount of ammonia in inlet and effluent water which is due to the farm production capacity, feeding level and food composition. It was observed that, total ammonia at third sampling point was more than that in inlet water but it was less than that in effluent due to high flow of stream. It has a good consistency with previous studies (Sawyer and Mc Carty, 1978). The NH_3-N concentration of Koohrang stream remained less than the maximum allowable level of 1.0 mg/L recommended by the EEC for protection and improvement of fish in freshwater (Boaventura et al., 1997). It was found that the level of total ammonia decreased gradually from hot season to cold season due to reduction of metabolism of organic materials in aquatic ecosystems (Figure 5). A same trend was observed for nitrite and nitrate which is because of biological

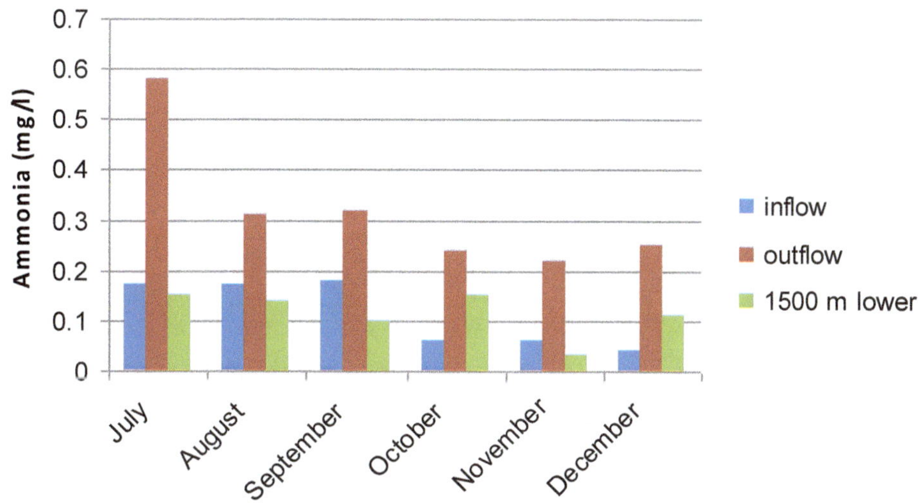

Figure 5. Changes of ammonia levels in three sampling points; farms inflow, farms outflow and 1500 m far from the farms outflow along the Koohrang stream.

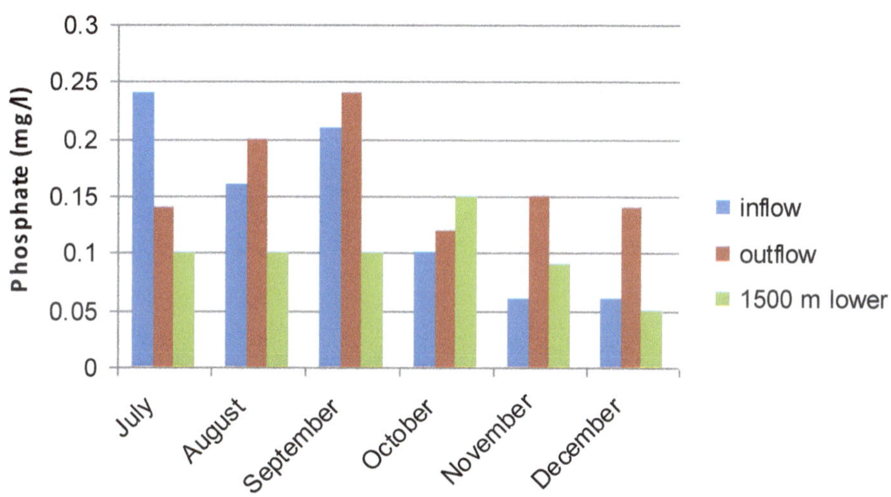

Figure 6. Amount of phosphate in three sampling points; farms inflow, farms outflow and 1500 m far from the farms outflow along the Koohrang stream.

degradation of proteins in water (Russo and Thuston, 1991).

It was found that the mean concentrations of NO_2-N and NO_3-N decreased significantly in the third sampling point (Tables 1, 2 and 3). The amount of nitrite in inlet water of fish farms was less than the standard amount which is 0.1 mg/L. The concentration level of NO_2-N and NO_3-N were below the recommended values for farm effluents (Schwartz and Boyd, 1994). Variation of phosphate in inlet, effluent and water samples from third sampling point were similar to total ammonia. It has got declining rate from hot season to cold season due to reduction of metabolic activities in aquatic animals. Maximum amount of phosphate which is harmful for fishes is more than 0.7 mg/L (Boaventura et al., 1997). Figure 6 shows that the amounts of phosphate in all water samples were less than 0.7 mg/L. Bergheim et al. (1982) studied on four fish farms in Norway. They found that each fish farm are able to make pollution 30 to 1260 times more than its biomass which is affected by fishes sizes, type of fish feeding, water flow rate and operation of fish farm cleaning. Metabolic products of fishes and left over fish's foods from fish farms make pollution in water environment. The pollution level depends on fish foods composition and its persistence in water as well as feeding practices. Amount and composition of precipitated fish foods in water could be very widespread (Teodorowicz et al., 2006).

It is possible to treat waste water by using zeolites which is able to keep ammonium ions in itself. If water flow rate in fish farms was maximum 10 L/s per 1 metric ton biomass, zeolite filters can treat waste water of fish farms, otherwise water should be diluted (Bergero et al., 1997). Sindilariu et al. (2009) studied on influence of pollution on thirteen rainbow trout fish farms which have water discharge with flow rate of 30 to 800 L/s. They found that inlet water quality, type of fish farming, amount of fish foods and the technique using for water treatment are affected on waste water treatment of fish farms. They also found that total phosphate, COD, BOD5 and TSS could be affected between 29 to 53% on waste water treatment. Mcmillan et al. (2003) found that precipitation of particles and materials in fish farm outlet water is the best way to reduce total solids in fish farms which flow-through system. Finally it is necessary to mention that the velocity and flow of stream water, seasons and effluent rate of farms could have great impacts on self-purification potential of the streams.

Conclusions

The results of this study were obtained from determination of twelve water physicochemical factors of three sampling point; (1) inlet water, (2) outlet water and (3) the last sampling point at 1500 m far from water outlet along the Koohrang stream. It was observed that trout farm effluents had a significant impact on the water quality of the Koohrang stream. There were significant differences in most of water physicochemical factors between inlet and effluent water, where as this variation were modified at third sampling point. The flow and velocity rate of Koohrang stream caused to reduce the level of some potentially harmful physicochemical water factors. Some of these factors have undesirable effects on stream water environment, although they can be treated by self-purification potential of the stream regarding to the water discharge of the stream. It was found that the level of all of the studied physicochemical factors of the water reduced to less than standard amounts for trout farming. It was concluded that there is a possibility to make other consecutive fish farms around 1500 m or nearer along the stream, but further studies are necessary to consider other environmental aspects of fish farm waste waters.

ACKNOWLEDGEMENT

Authors would like to show appreciation to the deputy of research of Islamic Azad University, Shahrekord branch because of its financial support for this project.

REFERENCES

Bergero D, Boccignone M, Di Natale F, Forneris G, Palmegiano GB, Zoccarato I (1997). Ammonium removal capacity of natural zeolites in rainbow trout farming. In: Kirov, G., Filizova, L., Petrov O (Eds), Natural Zeolites-Sofia'95. Pensoft, Sofia, Bulgaria, pp.75-82.

Bergheim A, Sivertsena A,Selmer-Olsen AR (1982). Estimated pollution loading from Norwegian Fish Farms .I. Investigations 1978-1979, Aquaculture, 28(3-4):347-361.

Boaventura R, Pedro AM, Coimbra J, Lencastre E (1997). Trout farm effluents: characterization and impacts on the receiving streams. Environ. Pollut. 95(3):379-387.

Boyd CE (2003). Guidelines for aquaculture effluent management at farm-level. Aquaculture 226:101-112.

Bureau DP, Cho CY (1999). Phosphorus utilization by rainbow trout (*Oncorhynchus mykiss*): estimation of dissolved phosphorus waste output. Aquaculture 179:127-140.

Forenshell G (2001). Setting basin design. Western Regional Aquaculture Center, WRAC-106. USA p. 6.

Kajimura M, Croke SJ, Glover CN, Wood CM (2004). Dogmats and controversies in the handling of nitrogenous wastes: The effect of feeding fasting on the execretion of ammonia, urea and other nitrogenous waste products in rainbow trout. J. Exp. Biol. 207:1993-2002.

Lawson TB (2001). Fundamentals of Aquacultural Engineering. Chapman and Hall, New York p. 364.

Lloyd R, Orr D (1969). The diuretic response of rainbow trout to sublethal concentrations of ammonia. Water Res. 3:335-344.

Maillard VM, Boardman GD, Nyland JE, Kuhn DD (2005). Water quality and sludge characterization at raceway-system trout farms. Aquacult. Eng. 33:271-284.

McMillan JD, Wheaton FW, Hochheimer JN, Soares J (2003). Pumping effect on particle sizes in a recirculating aquaculture system. Aquacult. Eng. 27:53-59.

Midlen A, Redding TA (1998). Environmental Management for Aquaculture. Kluwer Academic Publishers, London p. 215.

Miller D, Semmens K (2002). Waste Management in Aquaculture. West Virginia University Extension Service Publication No. AQ02-1. USA p. 8.

Mumpton FA , Fishman PH (1977). The application of natural zeolites in animal science and aquaculture. J. Anim. Sci. 45:1188-1203.

Papatryphon E, Petit J, Hayo V, Kaushik SJ, Claver K (2005). Nutrient balance modelling as a tool for environmental management in aquaculture: The case of trout farming in France. J. Environ. Manag. 35:161-174.

Pillay TVR (2004). Aquaculture and the environment. 2nd Ed. Blackwell Publishing, Oxford.

Pulatsu S, Rad F, Aksal G, Aydin F, Benil AC, Topcu A (2004). The Impact of Rainbow Trout Farm Effluents on Water Quality of Karasu. Turk. J. Fish. Aquat. Sci. 4:09-15.

Read PA, Fernandes TF, Miller KL (2001). The derivation of scientific guidelines for best environmental practice for the monitoring and regulation of marine aquaculture in Europe. J. Appl. Ichthyol. 17:146-152.

Russo RC, Thurston RV (1991). Toxicity of ammonia, nitrite, and nitrate to fishes. In: Brune DE, Tomasso JR (eds) Aquaculture and Water Quality. World Aquaculture Soc, Baton Rouge pp. 58-89.

Sawyer CN, McCarty PL (1978). Chemistry for Environmental Engineering. Third Edition. McGraw-Hill, New York.

Schulz C, Gelbrecht J, Rennert B (2003). Treatment of rainbow trout farm effluents in constructed wetland with emergent plants and subsurface horizontal water flow. Aquaculture 217:207-221.

Schwartz MF, Boyd CE (1994). Channel catfish pond effluents. Prog. Fish. Cult. 56:273-281.

Sindilariu PD, Reiter R, Wedekind H (2009). Impact of trout aquaculture on water quality and farm effluent treatment options. Aquat. Living Resour. 22(1):93-103.

Stewart NT, Boardman GD, Helfrich LA (2006). Treatment of rainbow trout (*Oncorhynchus mykiss*) raceway effluent using baffled sedimentation and artificial substrates. Aquacult. Eng. 35(2):166-178.

Stoskopf M (1992). Fish medicine. W.B. Saunders Company. pp. 333-337.

Teodorowicz M, Gawrońska H, Lossow K, Łopata M (2006). Impact of Trout farms on water quality in the MARÓZKA stream (Mazurian Lakeland, Poland), Arch. Pol. Fish. 14(2):243-255.

True B, Johnson W, Chen S (2004). Reducing phosphorus discharge from flow-through aquaculture III: assessing high-rate filtration media for effluent solids and phosphorus removal. Aquacult. Eng. 32:161-170.

Wedemeyer GA (1996). Physiology of Fish in Intensive Culture Systems. Chapman and Hall, New York p. 300.

Selectivity and fish catches of gillnets in stratum VII (Yeji sector) of Lake Volta for sustainable management

Daniel N. Akongyuure*, Patrick K. Ofori-Danson and Francis K. E. Nunoo

Department of Oceanography and Fisheries, University of Ghana, P. O. Box LG 99, Legon, Accra – Ghana.

The study was conducted in Stratum VII (Yeji sector) of Lake Volta where commercial fishermen arbitrarily use gillnets, from June, 2009 to March, 2010. The catches of monofilament gillnets, the most dominant gear used in Stratum VII were investigated for dominant fish species composition and selectivity parameters. The study revealed the dominance of two fish species (*Synodontis membranaceus* and *Schilbe mystus*) which together contributed about 45% by number of gillnet catches. The optimum selection length (total length) of the 5 cm mesh size gillnet for *S. membranaceus* was 19.7 cm and *S. mystus* was 18.2 cm. The result showed that using monofilament gillnets with mesh sizes 3 and 5 cm could have increase pressure on juvenile population. The mean monthly catch for the study period was estimated to be in the range of 0.3961 × 10^{-3} – 3.2427 × 10^{-3} (standard deviation, δ=1003.9) metric tonnes. With reference to precautionary approach to fisheries management, a legal minimum mesh size of gillnets of 8 cm may be appropriate for conservation and sustainable exploitation in Stratum VII.

Key words: Catch, gillnet, management, selectivity, species.

INTRODUCTION

Ghana has a long tradition of a very active fishing industry. Ghana's fishing industry has made tremendous strides over the years, developing from a predominantly traditional canoe fleet to a mix of traditional and modern fleet (Agyenin-Boateng, 1989). Fish provides the Ghanaian consumer with about 60% of animal protein needs (FAO, 2008). According to Ghana's 2007 Budget Statement, the country's total annual fish requirement is estimated at 720,000 metric tonnes (mt), while annual production averages 400,000 mt (Atta-Kesson and Atugba, 2008). This leaves an annual deficit of 320,000 mt which is made up through the importation of US$200 million worth of fish into the country yearly. Lake Volta contributes between 80 and 90% of fish catches from the inland fisheries of Ghana (Ofori-Danson, 2002). In order to meet the needs of consumers, maintenance of the quality, diversity and availability of fish in sufficient quantities must be promoted.

The use of inappropriate fishing methods and techniques accompanied with free access of fishermen into Lake Volta pose inverse effects on fish catches (Braimah, 1998). It is observed that over a decade now no significant research work has been undertaken on changes of the fish stocks at Yeji of Lake Volta. As a result, information on the status of the present fish stocks is lacking. To obtain an insight into the potential for increased production and management purposes of the lake, a reliable estimate of the total annual fish catch of the major fishing gear would be necessary. In the light of the above, this study primarily sought to assess fish catch and effects of gillnet selection at Stratum VII (Yeji Sector) of Lake Volta.

MATERIALS AND METHODS

Stratum VII (Figure 1) of Lake Volta lies between longitudes 0°10' and 1°05'W and latitudes 8° 8' and 8° 20'N and extends for about 60 km south 50 km north of Yeji . Stratum VII is chosen because it is currently one of the areas with the highest fishing activities and boasts of the largest fish market centre at Yeji.

For the purpose of sampling, two landing sites namely Jaklai and Site Area as shown in Figure 1 were selected. Jaklai and Site Area

*Corresponding author. E-mail: bayotitoloju@yahoo.com.

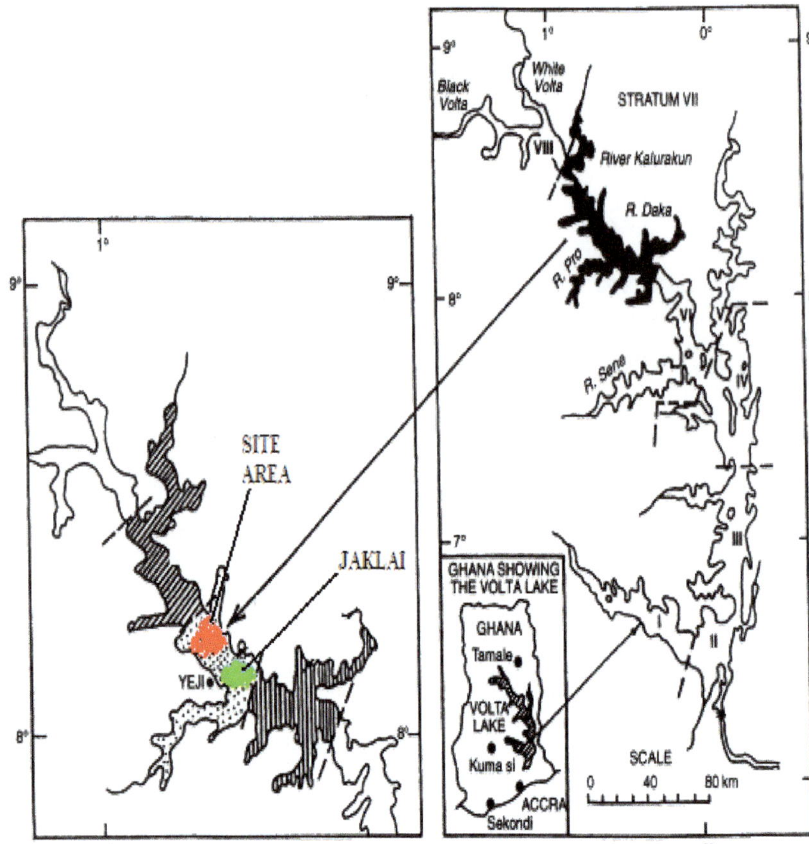

Figure 1. Map of Volta Lake showing Stratum VII and sampling sites (Source: Directorate of Fisheries, Yeji).

are located at the central part of Stratum VII. These landing sites were selected because they form the largest and most active among the five major landing sites of Stratum VII.

The data collected were in two folds namely primary data and secondary data. The primary data included gillnet catch, selectivity and physico-chemical data. The secondary data comprised monthly catches and mean rainfall values compiled from the records of Fisheries Directorate of the Ministry of Food and Agriculture, Yeji and Ghana Meteorological Agency, Accra respectively. Sampling for catch and determination of fishing effort were carried out at the selected sites once every week for 10 months (from June, 2009 to March, 2010). At each landing site, about 2.5 to 5.0 kg of fish was weighed using a balance scale and a bowl after sorting from each landed canoe and the following data collected: (i) fish species composition (ii) total body length (from snout to anal fin) and weight of fish in kg (iii) mesh size of gillnets in cm (iv) number of fishermen per canoe and (v) total number of operational and non-operational canoes.

Fish samples obtained from commercial gillnet catches were identified up to species using freshwater fish identification guide (Dankwa et al., 1999). Monthly fish catches (C) were estimated using C = Catch per unit effort (CPUE) × Number of fishing days (Fd).The CPUE was estimated as catch per canoe per day (kg/canoe/day). On selectivity parameters, two monofilament gillnets of stretched mesh sizes 3 and 5 cm representing small and large sizes respectively with equal surface area (100 m2) and hanging depth of between 1.5 and 3.0 m were set for a period of about 16 h before removal once every month (from June, 2009 to March, 2010). The catches then were sorted for species composition. The weight of the fish sample and then the overall catch was recorded. *In situ* measurements of temperature, pH, dissolved oxygen and transparency were taken from the surface to the bottom of the lake by the help of a Hyro-Bio water sampler of 1.7 L monthly.

In order to show the distribution of dominant species and total lengths of fish sampled, pie chart and bar graph of Microsoft Office Excel 2007 were used. Again, the distribution and trends of fish catch were established using percentage and graph (curves) of Microsoft Office Excel 2007. Data on gillnet selectivity was analysed using Fish Stock Assessment Tool II, Version 1.1.3 (FISAT II). The selectivity parameters, optimum selection length and selection factor were estimated using Holt (1963) model. The model assumes that when the catches (CA, CB) of two mesh sizes (ma, mb) overlap, their common standard deviation and estimate of optimum selection length for catch of the mesh sizes can be obtained by a linear regression of the form:

$$\ln (C_A/C_B) = a + b\,L$$

Where 'L' is length interval midpoints, 'a' the intercept and 'b' the slope of the regression line, C_A = catch of mesh size m_a and C_B = catch of mesh size m_b. The optimum lengths (L_{ma} and L_{mb}) for the 2 nets was then estimated from equations of Sparre and Venema (1992) and Pauly (1984).

$$L_{ma} = -2\,[(a \times m_a) / (b)\,(m_a + m_b)]$$

Table 1. Some physico-chemical parameters in Stratum VII of Lake Volta.

Parameter	Depth (m)				Average
	0	2	6	10	
Temperature (°C)	31.6	30.8	29.4	29.0	30.2 ± 0.0738
pH (units)	7.0	6.9	6.8	6.5	6.8 ± 0.0738
Dissolved oxygen (mg/L)	9.2	8.6	7.6	6.8	8.1 ± 0.0738
Transparency (cm)	60.5	-	-	-	60.5 ± 0.0738

$L_{mh} = -2[(a \times m_h) / (b) (m_a + m_b)]$

The common standard deviation is given by:

$SD = \sqrt{2a} (m_a - m_b) / b^2 (m_a + m_b)$

The selection factor (SF) of the 2 nets was estimated from

$SF = 2a / b (m_a + m_b)$

The probability of capture (P) at any given length for meshes m_a and m_b was estimated from:

$P_{ma} = \exp [- (L - L_{ma})^2 / 2 S.D^2]$
$P_{mb} = \exp [- (L - L_{mb})^2 / 2 S.D^2]$

A plot of the natural logarithms of the probability of capture against length interval midpoints provided estimates of L_{ma}, L_{mb} and the selection curves of the two mesh sizes. To be able to determine the physico-chemical parameters that influenced the CPUE (catch per unit effort) greatly, multi-linear regression of Statistical Package for Social Sciences (SPSS), Version 14.0 for windows was performed. Matrix Laboratory 2008a (MATLAB) was also used in depicting the relationship between precipitation and CPUE.

RESULTS

Physico-chemical condition of Stratum VII during the period of study

The vertical profile measurements of temperature, pH, dissolved oxygen and transparency showed decreasing values with increasing depth. The average values however for the parameters were 30.2 °C (temperature), 6.8 (pH), and 8.1 mg/l (dissolved oxygen) and 60.5 cm (transparency) with a standard deviation of ± 0.0738 as shown in Table 1.

Fish stock composition of commercial gillnet catches in Stratum VII of Lake Volta

Gillnet catches were dominated in terms of numbers by *Synodontis membranaceus* representing 24% of the total catch for the study period. Following closely was *Schilbe mystus* (21%). 'Others' (*Auchenoglanis occidentalis, Distichodus rostratus Hydrocynus brevis, Sarotherodon melantheron, Mormyrus macrophthalmus, Hippopotamyrus psittacus*, etc.) contributed the least in terms of numbers as these together represented 2% as shown in Figure 2. On the hand, *Labeo coubie* (28%) recorded the highest percent by weight followed by *Oreochromis niloticus* (16%) and thirdly *Bagrus bajad* (11%) as shown in Figure 3.

Length distribution of fish stock caught in gillnets during the period of study

Fish caught by gillnets ranged between 10 and 40 cm total length as shown in Figure 4. Fifty percent of fish caught measured around 20 cm total length, 37.5% of recorded 30 cm total length and 13.5% for 40 cm total length.

Mean monthly gillnet catches of fish recorded during the study period

In Figure 5, the greatest catch, 3.2427×10^{-3} metric tonnes (3242.7 kg) occurred around November, 2009. There was a continuous decline in catch after this peak. The lowest catch, 0.3961×10^{-3} metric tonnes (396.1 kg) however was recorded in March, 2010.

Fish catches for 2008 and 2009 in Stratum VII of Lake Volta recorded by the fisheries Directorate, Yeji

Figure 6 shows the fish catches by all gears for 2008 and 2009 based on secondary data obtained by courtesy of the Fisheries Directorate, Yeji. It can be observed that the total catch of fish for 2009 (1.066084×10^{-1} metric tonnes with a standard deviation, $\sigma = 4804.6$) was greater than that of 2008 (8.28521×10^{-1} metric tonnes, $\sigma = 3259.5$). Also, the peak catches were recorded around July and August for 2009 and 2008 respectively. There was a general decline in catches for both years after August.

The annual catches of dominant fish caught by diverse gears at Stratum VII of Lake Volta are shown in Table 2. Mochokidae (*S. membranaceus S. schall, S. ocellifer S. nigrita*) and Schilbeidae (*Auchenoglandis occidentalis Schilbe mystus, S. intermedius*) recorded13.6 and 5.05% by weight in 2008 respectively. In 2009, Mochokidae recorded 12.59% and Schilbeidae recorded 7.43% by

Table 2. Fresh fish catches for 2008 and 2009 in Stratum VII recorded by the department of fisheries, Yeji.

Family	Species	Catch (kg/year)		Percentage equivalent (%)	
		2008	2009	2008	2009
Characidae	*Hydrocynus brevis H. forskalii Alestes baremoze, A. dentex, A. nurse*	2783.9	1944.5	3.36	1.82
Bagridae	*Bagrus bajad, B. docmac, Chrysichthys auratus, C.nigrodigitatus Auchenoglandis occidentalis*	24094.1	29679.0	29.08	27.84
Citharinidae	*Citharinus citharus*	214	656	0.26	0.62
Claridae	*Clarias anguillaris*	2533.8	2479.1	2.65	2.33
Clupeidae	*Odaxothrissa mento, Pellonula leonensis, Sierrathrissa leonensis*	7356.4	10387.7	8.88	9.74
Distichodontidae	*Distichodus rostratus*	603.2	1086.9	0.73	1.02
Osteoglossidae	*Heterotis niloticus*	16.5	9.1	0.02	0.01
Cyprinidae	*Labeo coubie, L. senegalensis*	6225.6	7972.7	7.51	7.48
Malapteruridae	*Malapterurus electricus*	9	12	0.01	0.01
Mormyridae	*Mormyrus rume, M. macropthalmus*	2000.7	3131.5	2.41	2.94
Schilbedae	*Schilbe mystus, S. intermedius*	4178.8	7921.2	5.05	7.43
Mochokidae	*Synodontis membranaceus, S. schall, S. ocellifer, S. nigrita*	11276.2	13418.8	13.61	12.59
Cichlidae	*Tilapia dageti, T. zillii, T. guineansis, Oreochromis niloticus, Sarotherodon galilaeus, Hemichromis fasciatus, H. bimaculatus*	21560	27549.7	26.02	25.84
Total		82852.1	106608.4	100	100

weight in Stratum VII, Yeji. The family Bagridae recorded the highest percentage by weight for both 2008 (29.08%) and 2009 (27.84%). The families with the least percentage presentation were Osteoglossidae and Malapteruridae each of which had 0.01%.

Monofilament gillnet selectivity

Catch and length records of S. membranaceus and S. mystus

The total number of *S. membranaceus* and *S. mystus* caught by both 3 cm stretched mesh size and 5 cm stretched mesh size during the study period was 1222 with 35 kg total weight. The 3 cm stretched mesh size net contributed 527 individual fish while 695 individual fish was recorded for the 5 cm stretched mesh size net. *S. membranaceus* caught in mesh size 3 cm ranged between 5.4 to 24.3 cm. Those caught in mesh size 5 cm ranged between 7.5 and 25.5 cm total length. The mean modal lengths for the 3 cm mesh and 5 cm mesh size gillnets were 18.0 and 22.2 cm, respectively. On the other hand, a total body length range of 5.2 to 24.3 cm and 7.5 to 26.6 cm were recorded for the 3 cm mesh size gillnets and 5 cm mesh, respectively for *S. mystus.* Its mean modal lengths were 15.9 cm (3 cm mesh size) and 18.0 cm (5 cm mesh size) as shown in Figure 11.

Gillnet selectivity plot

The optimum length for capture of *S. membranaceus* increased from 11.8 to 19.7 cm (Total length) for the 3 and 5 cm mesh sizes (Figure 7). The plot of the regression of the natural logarithms of the catch ratios of

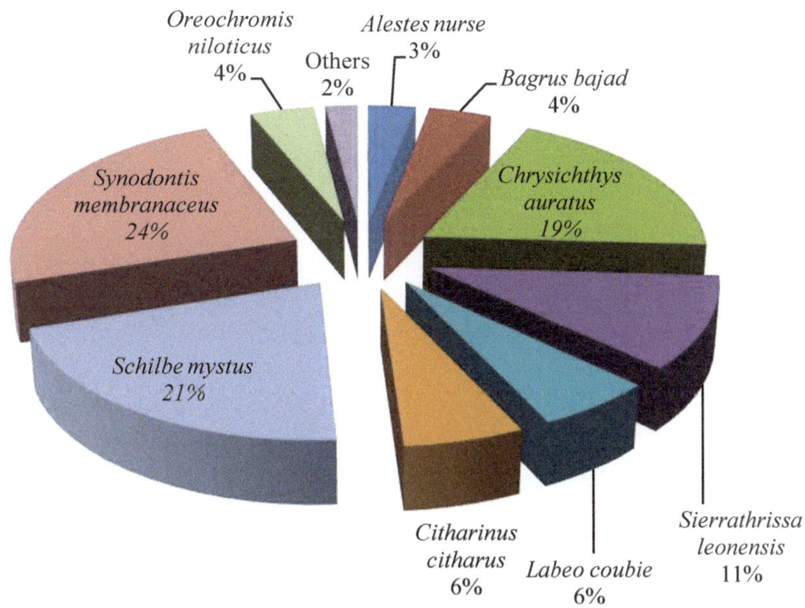

Figure 2. Percentage composition by number of main fish species caught by gillnets during the study period (June, 2009 - March, 2010).

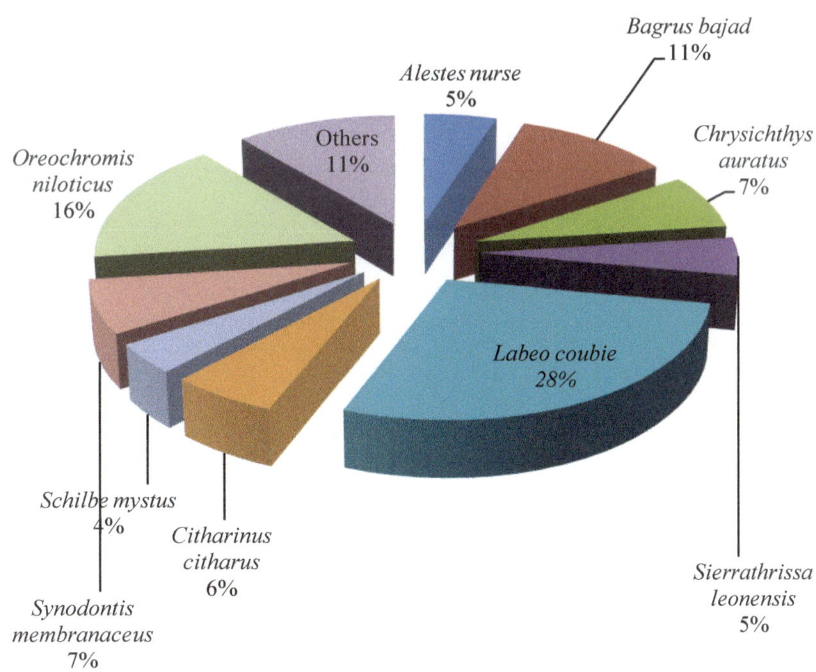

Figure 3. Percentage composition by weight of the main commercial fish species caught by gillnets in Stratum VII from June, 2009 to March, 2010.

mesh sizes 3 and 5 cm on the length interval midpoints for *S. membranaceus* (Figure 8). Figure 9 shows new optimum length for capture of *S. membranaceus* in a log-normal plot of the same variables. The selection ranges for *S. membranaceus* were 5.4 to 24.3 cm and 7.5 to 28.5 cm for mesh sizes 3 and 5 cm respectively. The log-normal plot of the regression of the natural logarithms of the catch ratios of mesh sizes 3 cm and 5 cm on the length interval midpoints for *S. membranaceus* (Figure 10) indicated that the optimum length for capture in the

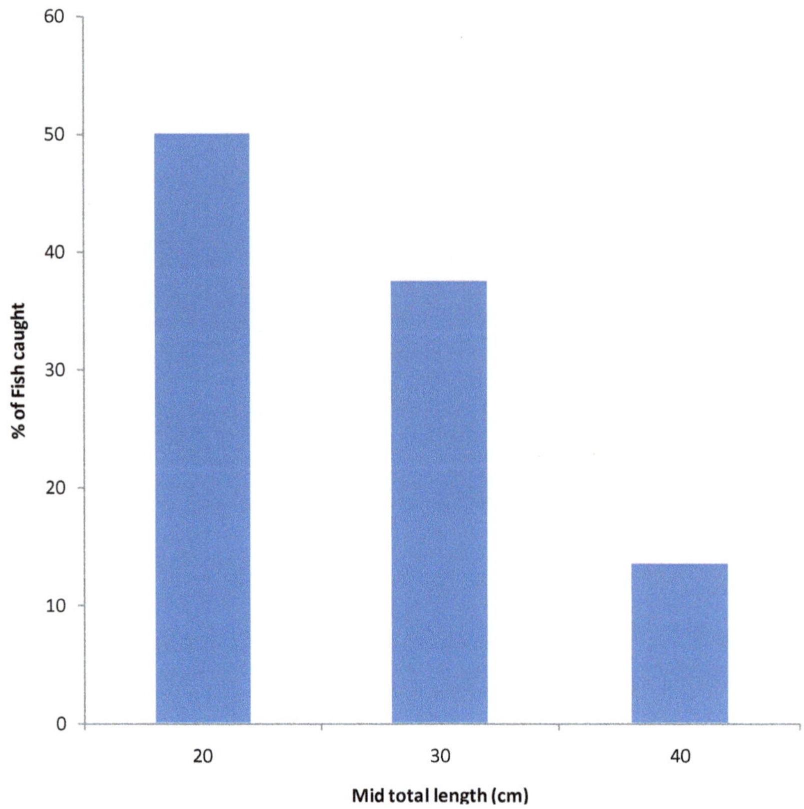

Figure 4. Length distribution of major fish stock caught by gillnets in Stratum VII of Lake Volta from June, 2009 to March, 2010.

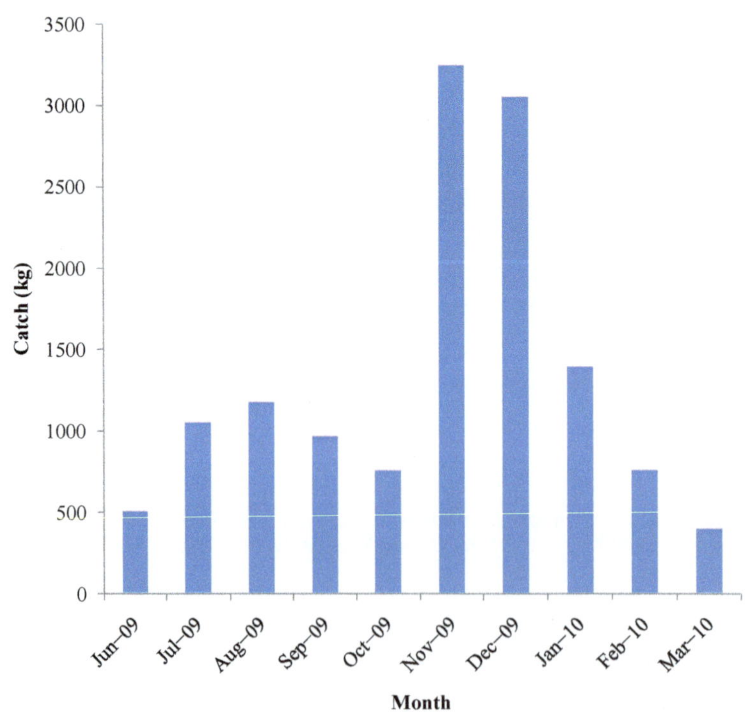

Figure 5. Mean monthly gillnet catches of fish recorded during the study in Stratum VII of Lake Volta.

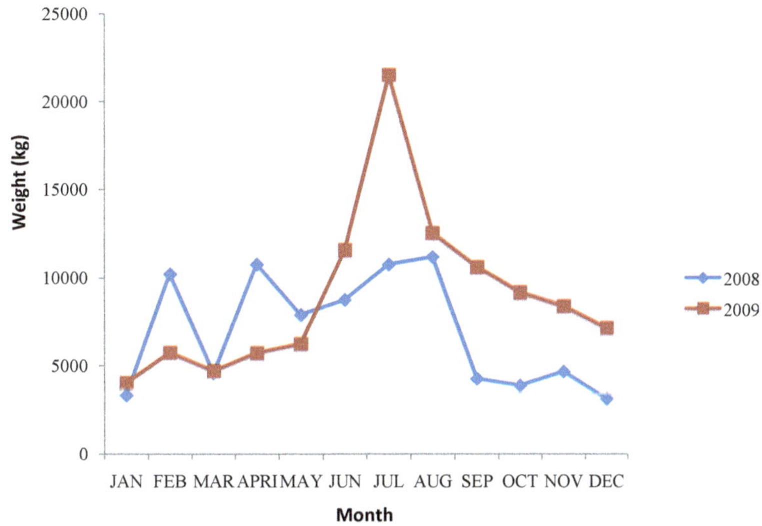

Figure 6. Monthly fish catches recorded by the Fisheries Directorate, Yeji for 2008 and 2009 by all gears in Stratum VII of Lake Volta.

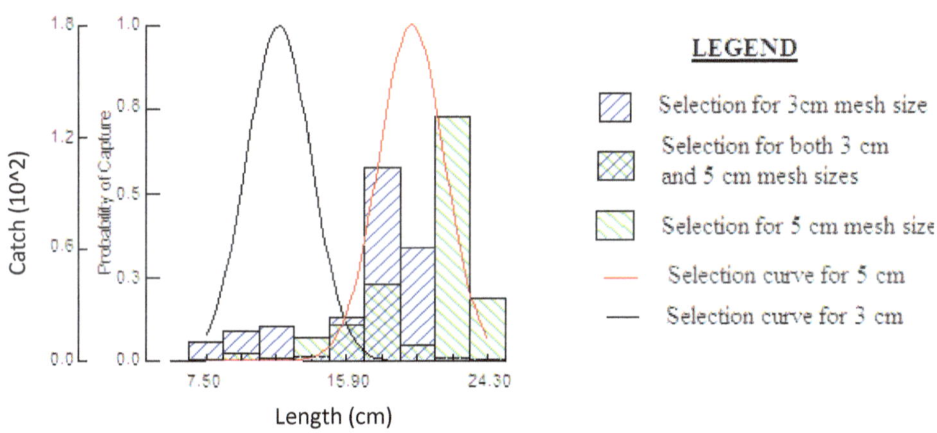

Figure 7. Gillnet (3 and 5 cm mesh sizes) selectivity curves obtained with FISAT for *S. membranaceus* in Stratum VII.

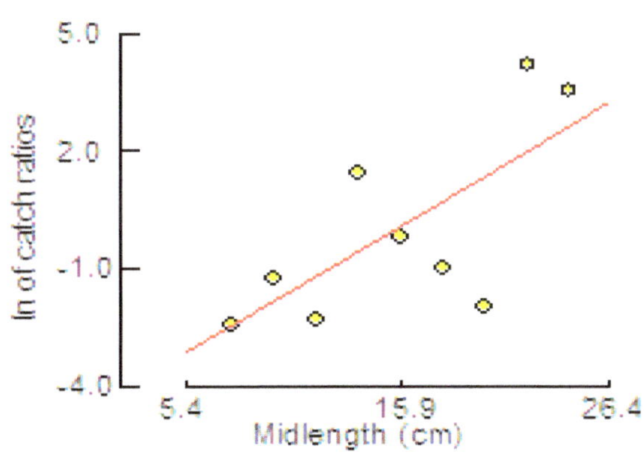

Figure 8. FISAT output of ln of catch ratios versus mid-length for *S. membranaceus* caught with gillnets of mesh sizes 3 and 5 cm.

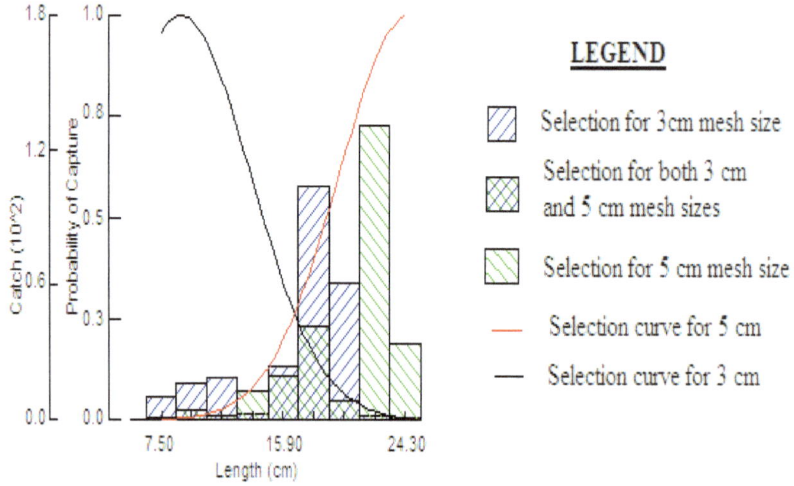

Figure 9. FISAT Log-normal plot of catch and probability of capture versus total length for *S. membranaceus* caught by gillnets of 3 cm and 5 cm mesh sizes in Stratum VII.

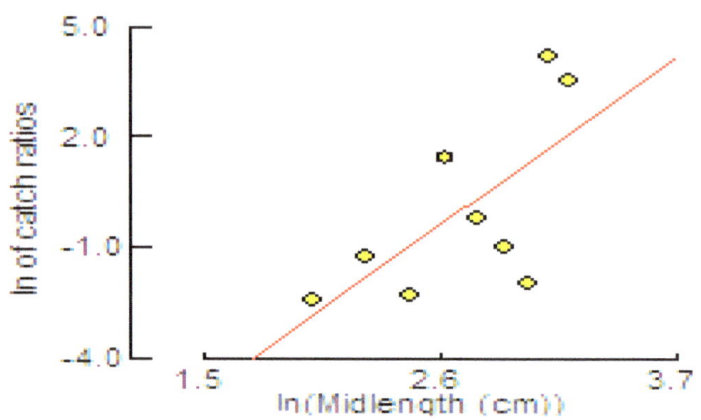

Figure 10. FISAT Log-normal plot of catch ratio versus mid-length for *S. membranaceus* caught by gillnets of 3 and 5 cm mesh sizes in Stratum VII.

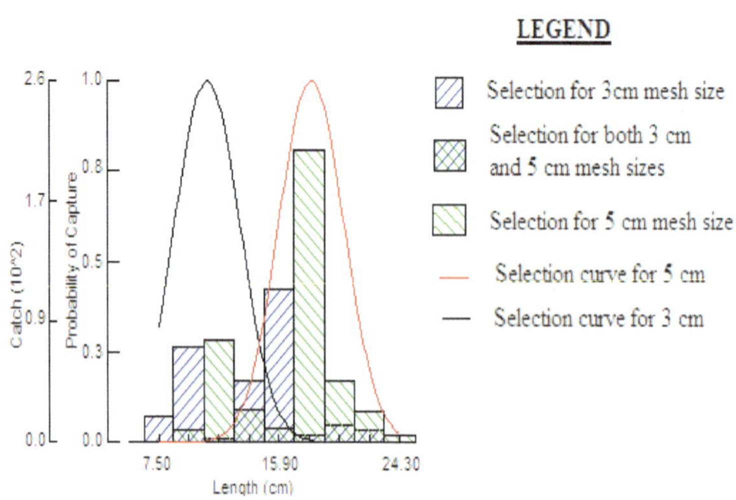

Figure 11. Gillnet (3 and 5 cm mesh sizes) selectivity curves obtained with FISAT for *S. mystus* in Stratum VII.

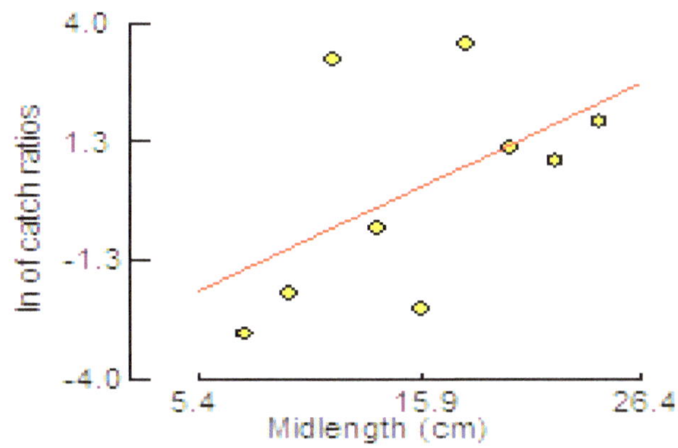

Figure 12. FISAT output of ln of catch ratios versus mid-length for *S. mystus* caught by gillnets of 3 and 5 cm mesh sizes in Stratum VII.

Table 3. Common selection factor (SF), common standard deviation (SD) and estimated optimum lengths (in cm), L_3 and L_5, of *S. membranaceus* and *S. mystus* for gillnet of 3 and 5 cm mesh sizes, respectively.

Species	SF	SD	L_3 (cm)	L_5 (cm)
S. membranaceus	3.9	5.8	11.8	19.7
S. mystus	3.6	5.8	10.9	18.2

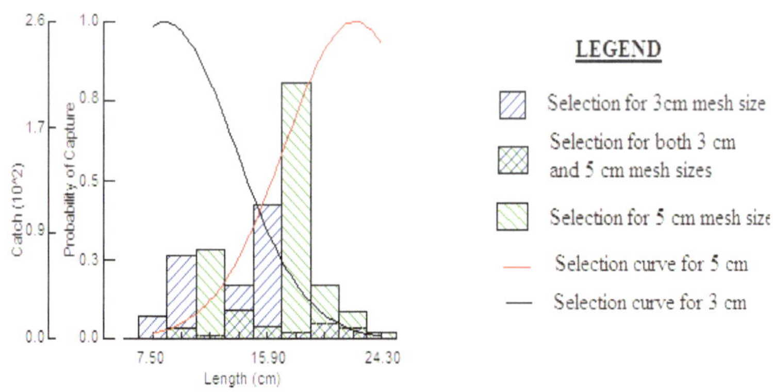

Figure 13. Log-normal gillnet selection curves obtained with FISAT of 3 cm and 5 cm mesh sizes for *S. mystus* in Stratum VII.

log-normal plot of *S. membranaceus* increased from 8.9 to 24.6 cm. The optimum length for capture of *S. mystus* also increased from 10.9 cm to 18.2 for the 3 and 5 cm mesh sizes (Figure 12). The selection ranges of *S. mystus* for the 3 and 5 cm mesh sizes were 5.4 to 24.3 cm and 7.5 to 26.4 cm. The plot of the regression of the natural logarithms of the catch ratios of mesh sizes 3 cm and 5 cm on the length interval midpoints for *S. mystus* is shown in Figure 13. Figure 14 shows new optimum length for capture of *S. mystus* in a log-normal plot of the same variables. The optimum length for capture of *S. mystus* in a log-normal plot increased from 8.3 to 22.3 cm for 3 and 5 cm mesh sizes. The selection factor (SF), standard deviation (SD) and optimum lengths (L_3 and L_5) of the two fish species are summarized in Table 3.

Relation between the catch per unit effort (CPUE) and precipitation during the period of study in Stratum VII of Lake Volta

The CPUE and precipitation showed a monthly inverse relation as depicted by Figure 15. In other words, where CPUE was increasing, precipitation (rainfall) was

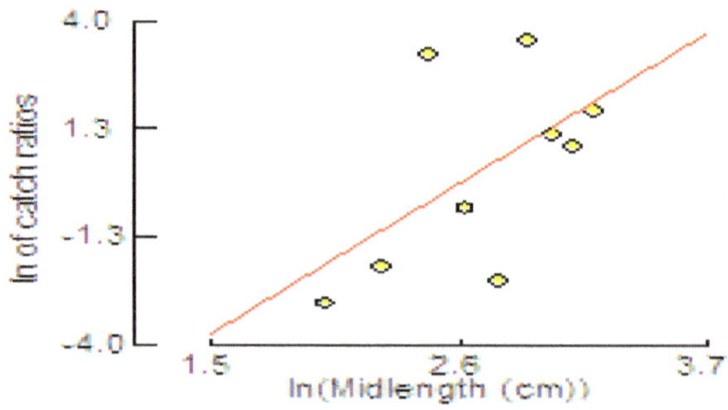

Figure 14. FISAT output of log-normal catch ratio versus midlength for *S. mystus* caught by gillnets of 3 and 5 cm mesh sizes in Stratum VII.

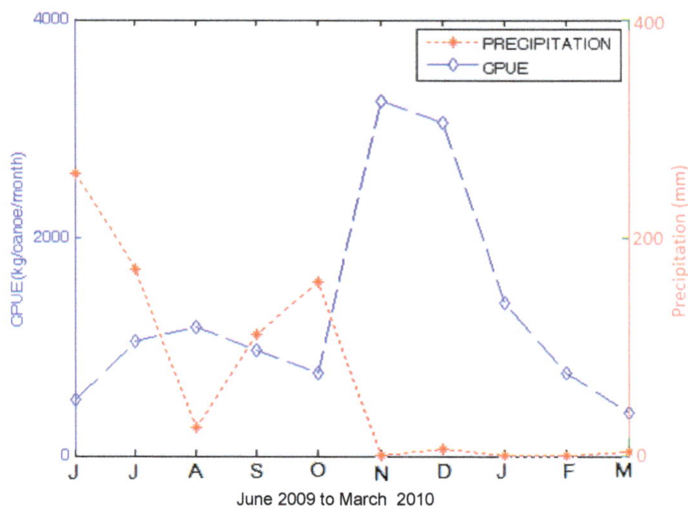

Figure 15. Relationship between CPUE and precipitation in Stratum VII of Lake Volta from June, 2009 to March, 2010.

decreasing and vice versa. While the peak of CPUE (3242.7 kg/canoe/month) occurred in November, 2009 that of precipitation (258.4 mm) was realized in June, 2009. The multi-linear regression analysis revealed a relationship (R^2 = 0.827) among the variables. Also, the test was significant (0.011) at 95% confidence level (α = 0.05) as shown in Table 4.

DISCUSSION

Physico-chemical condition of Stratum VII

The mean surface water temperature recorded for Stratum VII was 31.6°C. This dropped gradually to 29.0°C at the bottom representing a temperature difference of 2.6°C between the surface and the bottom waters. This observation indicates a weak thermal stratification. This phenomenon could be due to the riverine nature of Stratum VII of the lake. Clearly, the lake at the Yeji area is influenced principally by the Black Volta. Ofori-Danson and Antwi (1994) made a similar observation at the Akosombo gorge area. In general the temperature was optimal and characteristic of tropical lakes and did not exceed the critical value (40°C), lethal temperature to fish (Jeppe and Zwieten, 2005). The water of the Lake generally, had low pH value with slight variation in values with depth, with higher pH value on the surface. With a mean pH of 6.8 the Volta lake at Yeji exhibited a pH value within the World Health Organisation (WHO) range of 6.5 to 8.5 (WHO, 1993) a condition which may be described as near neutrality. This pH range is also suitable for fish growth and reproduction.

The vertical distribution of oxygen showed a decrease

Table 4. Multi-linear regression of pH, Temp (temperature) and DO (dissolved oxygen) on CPUE (catch per unit effort).

Model summary

Model	R	R square	Adjusted R square	Std. error of the estimate
1	0.909	0.827	0.740	2.76883

ANOVA

Model	Sum of Squares	df	Mean Square	F	Sig.
Regression	2219.930	3	73.243		
Residual	45.999	6	7.666	9.554	0.011
Total	265.727	9			

Coefficients

Model	Unstandardised Coefficients		Standardised Coefficients	t	Sig.
	B	Std. Error	Beta		
Constant	1648.368	498.047		3.310	0.016
Temp	-54.964	13.884	-0.746	-3.970	0.007
DO	-31.770	15.062	-0.431	-2.109	0.079
pH	40.786	13.884	0.554	2.946	0.026

with depth and was uniform. The surface waters of the lake were well oxygenated (9.2 mg/l) representing about 107.7% oxygen saturation while the bottom recorded 6.8 mg/L which could be translated into 79.6% oxygen saturation. Generally, the oxygen concentrations were satisfactory, being present in sufficient quantities (over 70% of saturation) to support fish life.

Mean Secchi disc transparency was 60.5 cm. The depth of light penetration was drastically reduced during the flood period (July to September) as also observed by Braimah (1995). The multi-linear regression analysis factors (pH, DO and Temp) on the CPUE indicated that all three partial coefficients are statistically significant ($P < 0.05$), thereby proving a strong relation of pH, temperature and dissolved oxygen on CPUE. Although, a decline in CPUE could be attributed to significant changes of any of the above variables as confirmed by the regression analysis (Table 4), the case in Stratum VII could be as a result of an increased number of canoes and a reduced water level. This is because pH, dissolved oxygen and temperature values have not reached lethal levels as long as the result of this study is concerned. DeGraaf and Ofori-Danson (1997) observed a similar trend of increased number of canoes coupled with a reduced water level affecting CPUE negatively.

Fish stock composition of commercial gillnet catches

Commercial gillnet catches comprised mainly *Oreochromis niloticus*, *Synodontis membranaceus*, *Alestes nurse*, *Bagrus bajad, Chrysichthys auratus*, *Labeo coubie*, *Citharinus citharus*, *Sierrathrissa leonensis* and *Schilbe mystus* during the study period, with no respect to the order in which they are listed. These fish were mostly caught by gillnet probably of their morphology, behaviour and trophic characteristics which rendered them prone to be caught by gillnet. Numerically, *S. membranaceus* was the highest (24.1%) as shown in Figure 3. Olaosebikan and Raji (1998) reported that *S. membranaceus* occur in both shallow and deep waters close to the shore where they feed on plankton, detritus and insects. *S. membranaceus* has replaced *Labeo coubie* and *Chrysichthys auratus* which were reported to be dominant (Ofori-Danson, 1999). *S. mystus* recorded 21% by number following *S. membranaceus* closely. Teugels et al. (1988) indicated that *S. mystus* are mostly caught by gillnet and are generally more active at night or in subdued light and feed from mid-water and surface waters on a wide variety of foods including fish, insects, shrimps, snails, plant seeds, and fruit. This may account partly for the high number of *S. mystus* in Stratum VII of Lake Volta. This is because gillnets were often set over night hence making the species vulnerable to capture. In terms of percentage composition by weight, *S. membranaceus* (7%) and *S. mystus* (4%) values were very low. Even though *S. membranaceus* and *S. mystus* contributed greatly in terms of numbers, this did not reflect in their unit weight. The reason may be that these species were dominated by small fish specimens.

The instability of the hydrological regime and increasing fishing efforts in the lake pose adverse consequences on composition of fish stock. Willis (1971) revealed that changes in the composition of fish stock continued to occur after the formation of the lake in 1964. The study also noted the disappearance of some species that were originally present (Goudswaards and Avoke, 1993b). They included *Gymnarchus niloticus, Lates niloticus, Steatocranus irvinei* and *Hepsetus odoe*. *Gymnarchus niloticus, Lates niloticus* and *Steatocranus irvinei* were not encountered throughout the study period.

Additionally, *L. niloticus* and *Hydrocynus* spp. were earlier observed to occur in small numbers in the lake (Vanderpuye, 1982). *L. niloticus and Hydrocynus* spp. are piscivores and therefore can change in their dominance and abundance should their prey such as those in the family Clupeidae (*Odaxothrissa mento, Pellonula leonensis* and *Sierrathrissa leonensis*) get over-exploited.

Gillnet catches in Stratum VII during the study period

The use of gillnets generally is affected by several factors including colour of net, mesh size, twine size and strength, and morphology and behaviour of fish among others. Fishermen in Stratum VII were much aware of these factors based on a preliminary interaction. However, they were more attached to twine strength, mesh size and net colour. It was observed that gillnet of mesh sizes 3.0 to 6.0 cm were mostly used; since small and big fish alike can be caught by the range of these nets. White colour was preferred by fishermen to other colours because it blended relatively well with the colour of the water to ensure maximum gilling, entangling, wedging and nagging of fish.

The highest monthly catch was recorded in November, 2009 (Figure 6). The possible reason could be the rise in water level of the lake during the rainy season. This agrees with the observation of Vanderpuye (1982). During rainy season, turbidity increases resulting in poor visibility of fish which probably make them vulnerable to capture. By March, 2010 the water level had retreated to about 1 to 1.5 m and this might have accounted for the significant reduction in monthly catch.

Assuming all other factors are constant, the current estimated mean CPUE of 8.0 kg/canoe/day can produce a total monthly catch of 396.1 to 3,242.7 kg (σ = 1003.9) in Stratum VII. However due to increasing number of fishing efforts, the quantity of catch may decrease in subsequent years if even uncontrollable factors such as climate, water dynamics and food remain unchanged.

Length distribution of fish stock in gillnet catches

The minimum and maximum total lengths of fish species caught with gillnets were 20 and 40 cm, respectively(Figure 5). Fish that recorded 20 cm total length contributed 50%, the highest percentage of fish caught with gillnets. Olaosebikan and Raji (1998) reported that *S. membranaceus* can reach a maximum total length of 50 cm. Additionally, the standard length of *S. membranaceus* was estimated to be 33 cm in the northern arm of Lake Volta (Ofori-Danson et al., 2001). *S. mystus* according to Teugels et al. (1988) grows to a maximum total length of 40 cm. Considering these reported values of lengths of *S. membranaceus* and *S. schilbe,* this study revealed rather low values of length. Therefore 50% of the fish caught with gillnets during the study period were probably juveniles. Although, fish remain smaller, grow slower and mature earlier as a fishery-induced evolution (Mollet, 2010), the small size of fish caught with gillnets in Stratum VII during the study was perhaps as a result of the deployment of undersize mesh gillnets.

Annual fish catches of 2008 and 2009 in Stratum VII

Catch data as a matter of policy are being taken in Stratum VII from five landing sites by the Yeji Fisheries Department. Data collected ranged from fish caught by all the 11 methods of fishing namely *atidza*, bamboo pipe, beach seine, cast net, gillnet, hook-and-line, *nifa nifa*, spear, trap, winch net, and *wangara* net (Braimah, 1998). All these methods yielded a total catch of 82.8521 × 10^{-3} metric tonnes (standard deviation, σ = 3259.5) and 106.6084 × 10^{-3} metric tonnes (σ = 4804.6) in 2008 and 2009, respectively.

The peak catches occurred in the rainy season (July and August), this could be related to availability of food since nutrients are flushed from inland and the catchment drainage, due to gravitational depression of the adjacent wet lands as reported by Nwankwo (2004) and Emmanuel et al. (2008). The low catches recorded in the past could be as a result of poor hydrological regime especially water level. Vanderpuye (1982) reported that annual catches of Lake Volta continued to decline year after year from 1969. On the contrary, in 1996, an estimated total annual catch of 22.422 metric tonnes was reported (DeGraaf and Ofori-Danson, 1997) for the Yeji sector and this increased to 74.5 metric tonnes in 2008 (FAO, 2008). Therefore it suggests that water level and fish food greatly affect annual catches in the lake.

Fish stock composition of 2008 and 2009 annual catches

The results revealed the following families from which the main commercial fish were caught with all gears operating in Stratum VII (Yeji Sector) of Lake Volta for 2008 and 2009. They included Mochokidae, Schilbeidae,

Cichlidae, Characidae, Bagridae, Cyprinidae, Citharinidae and Clupeidae (Table 2). The family Mochokidae which *S. membranaceus* belong contributed 13.61 and 12.59% by weight of fish catches recorded by the Fisheries Directorate, Yeji for 2008 and 2009, respectively (Table 2). Schilbeidae on the other hand, recorded as low as 5.05 and 7.43% by weight in 2008 and 2009 respectively. The reason for the low contribution by weight of fish in the families Mochokidae and Schilbeidae was not clear. However, Bagridae recorded the highest of 29.08 and 27.84% for 2008 and 2009, respectively. This could be attributed to the adoption of specialized gears like *nifa nifa* that target species belonging to this family (Goudswaard and Avoke, 1993a).

Gillnet selectivity of *S. membranaceus* and *S.mystus*

S. membranaceus and *S. mystus* were the most dominant species caught in commercial gillnet fishery in Stratum VII of lake Volta and target species for this study. This study provided initial selectivity estimates of gillnets used in Stratum VII of lake Volta to catch *S. membranaceus* and *S. mystus*. The optimum catch lengths increased with increasing mesh size which agreed with similar study conducted by Ahmed and Inpinjolu (2008) on *S. membranaceus, Citharinus citharus*, *Hydrocynus forskalii*, and *Distichodus rostratus* in lake Kainji, Nigeria.

The optimum catch lengths of 11.8 cm for *S. membranaceus* and 10.9 cm for *S. mystus* of mesh size 3 cm were less than the reported length of 26.7 cm (Ofori-Danson, 2002) and 13.3 cm (Teugels et al., 1988) at first sexual maturity for the species. Considering the recommendations of FAO (1995a) Code of Conduct for Responsible Fisheries and the Precautionary Approach to fisheries (FAO, 1995b), and the multi-species nature of the lake fishery the 3 cm mesh size gillnet was considered too low; it would be of immense help to discourage completely the use of 3 cm mesh size gillnets for reasons of sustainability and conservation of fisheries resources of the lake.

The optimum catch lengths of 19.7 cm for *S. membranaceus* with deployment of mesh size 5 cm gillnets must also be discouraged since it was less than the reported length of 26.7 cm at first sexual maturity. Although the optimum catch length of 18.2 cm for *S. schilbe* of mesh size 5 cm appeared greater than the reported length of 13.3 cm at first sexual maturity, due to the multi-species nature of the lake, gillnets of 5 cm mesh size may not be encouraged in order to ensure optimum protection and conservation of such species.

In the interest of conservation and judicious exploitation of the fishery, the use of 8 cm mesh size could be prescribed as the precautionary minimum mesh size so as to allow most fish species to reach sexual maturity and grow big enough before they are caught to prevent growth and recruitment overfishing. The adoption and enforcement of 8 cm mesh size gillnet as the minimum legal mesh size would not only be precautionary enough to enhance recruitment and increase fish production and conserve the lake fisheries through prevention of the catch of fish before they are sexually mature, but would also prevent loss of stock to the ecosystem with age.

Conclusions

The dominant commercial gillnet fish species observed in Stratum VII of Lake Volta during the study period were *S. membranaceus, S. mystus, Chrysichthys auratus, Sierrathrissa leonensis, Labeo coubie, Citharinus citharus, Bagrus bajad, Oreochromis niloticus* and *Alestes nurse*. The two most dominant species in numbers (*S. mystus* and *S. membranaceus)* was as a result of suitable physico-chemical conditions for their spawning in Stratum VII. The mean monthly gillnet catch during the study period was estimated to be in the range $0.3961 \times 10^{-3} - 3.2427 \times 10^{-3}$ metric tonnes ($\sigma = 1003.\ 9$), which was far lower than 2083.3×10^{-3} metric tonnes from earlier studies.

The optimum catch lengths for *S. membranaceus* and *S. mystus* were 11.8 and 10.9 cm respectively for the 3 cm mesh size gillnet. For the 5 cm mesh size gillnet the optimum catch lengths were 19.7 and 18.2 cm for *S. membranaceus* and *S. mystus* respectively. In view of the recommendations of code of conduct for responsible fisheries and the precautionary approach to fisheries, and the multi-species nature of the lake fishery the 3 cm mesh size gillnet was found too small. Also, for the sake of conservation and optimum protection of the fishery, the 5 cm mesh size gillnet for *S. mystus* may not be encouraged though the reported length of 13.3 cm appeared greater than the optimum length of catch.

ACKNOWLEGEMENTS

Authors thank the officers of the Fisheries Directorate of Ministry of Food and Agriculture, Yeji for providing monthly fish catch data and technical assistance in fish species identification; Ghana Meteorological Agency, Accra for providing us with mean monthly rainfall values; and Mr. Awudu Karimu, Headmaster of Atebubu Senior High School for allowing us to use their science resource centre laboratory. We also appreciate the efforts of the fishermen working at Jaklai and Site Area landing sites for helping us in the setting and removal of gillnets and allowing us to assess their fish catches.

REFERENCES

Agyenin-Boateng CE (1989). Report on the socio-economic conditions in the fishing communities in the Yeji area of the Volta Lake. FAO\ IDAF Project, GHA\88\004. Field Document, Rome, p. 89.

Ahmed YB, Inpinjolu JK (2008). The gillnet selectivity of *Citharinus citharus (*Pisces: Citharinidae*), Hydrocynus forskalii (*Pisces: Characidae*), Distichodus rostratus (*Pisces: Distichodontidae*)* and *Synodontis membranaceus (*Pisces: Mochokidae) in Kainji Lake, Nigeria. Nig. J. Fish. 5(1): 46- 62.

Atta-Kesson R, Atugba AR (2008). Legal framework of fish, fishing, fisherfolk, the law and fishing institutions. Available at www.revitalization.org

Braimah LI (1995). Recent developments in the fisheries of Volta Lake. CIFA Technical paper 30, Rome, pp. 115-134.

Braimah LI (1998). Full frame survey at Lake Volta. GHA\93\008. Technical Report. Inland Fisheries Division, Yeji. P 196.

Dankwa HR, Abban EK, Teugels GG (1999). Freshwater fishes of Ghana: Identification, distribution, ecological and economic importance. Water Research Institute – CSIR, Achimota, Ghana, 38(3): 1-53.

DeGraaf GJ, Ofori-Danson PK (1997). Catch and fish stock assessment in Stratum VII of Lake Volta. FAO Technical Paper, Rome, p. 107.

Emmanuel BE, Chukwu LO, Azeez LO (2008). Cast net design characteristics, catch composition and selectivity in tropical open lagoon. Afr. J. Biotechnol., 7(12): 2081-2089.

FAO (1995a). Code of conduct for responsible fisheries. Rome, FAO, p. 41.

FAO (1995b). Precautionary approach to fisheries. Part 1: Guidelines on the precautionary approach to capture fisheries and species introduction. FAO Fisheries Technical Paper. No. 350, Part 1. Rome, p. 52.

FAO (2008). Report on national fish production, imports, exports and consumption. Technical Paper. Rome, p. 8.

Goudswaard PC, Avoke SK (1993a). Catch and effort of commercial fishing methods in Lake Volta. FAO – UNDP Project (GHA / 88 / 004); Fisheries Department, Ghana.

Goudswaard PC, Avoke SK (1993b). Fish species in the Northern arm of Lake Volta. FAO – UNDP Project (GHA / 88 / 004); Fisheries Department, Ghana.

Holt SJ (1963). A method for determining gear selectivity and its application. INCAF spec. Publ., 5: 106-115.

Jeppe K, van Zwieten P (2005). Improving productivity in tropical lakes and reservoirs. University of Bergen, Norway and Wageningen University, the Netherlands.

Mollet F (2010). Intensive fishery activities in the North Sea In: *TheFishSite* News, Monday, 10 May, 2010.

Nwankwo DI (2004). Studies on the environmental preference of blue – green algae (Cyanophyta) in Nigerian coastal waters. In: Emmanuel BE, Chukwu LO, Azeez LO Cast net design characteristics, catch composition and selectivity in tropical open lagoon. Afr. J. Biotechnol., 7(12): 2081-2089.

Ofori-Danson PK (1999). Stock assessment of the five major commercial fish species in Yeji area (Stratum VII) of the Volta Lake. Ph.D Thesis, University of Ghana, Legon, p. 189.

Ofori-Danson PK (2002).Trophic relationships and spawning habits of post impoundment fish stocks of Lake Volta in Ghana. Ghana Sci. J,. 42: 61– 70.

Ofori-Danson PK, Antwi LAK (1994). Limnology of the Volta Lake (Akosombo gorge area) twenty-five years after impoundment. Tropical Ecol., 35(1): 69-82.

Ofori-Danson PK, DeGraaf GJ, Vanderpuye CJ (2001). Growth and mortality of the catfish, Hemisynodontis membranaceus (Geoffroy St. Hilaire), in the northern arm of Lake Volta,Ghana In: Fisheries Management and Ecology*,* Blackwell Sci. Ltd, 8: 37-45.

Olaosebikan BD, Raji A (1998). Field guide to Nigerian freshwater fishes. Federal College of Freshwater Fisheries Technology, New Bussa, Nigeria, p. 106.

Pauly D (1984).Fish population dynamics in tropical waters. A manual for use with programmable calculators. ICLARM stud. Rev., 8: 325.

Sparre P, Venema SC (1992). Introduction to tropical fish stock assessment. Part I Manual. FAO Fisheries Technical Paper No. 306.1, Rev. 1.FAO, Rome. P 376.

Teugels GG, Lévêque C, Paugy D, Traoré K (1988). State of knowledge of the fish fauna of coastal basins of Ivory Coast and western Ghana. Rev. Hydrobiol. Trop., 21(3): 221-237.

Vanderpuye CJ (1982). Further observations of the distribution and abundance of fish stocks in Volta Lake, Ghana Fish. Res., 1: 319-343.

WHO (1993). Guidelines for drinking-water quality, W. H. O. Geneva. 1.

Willis AE (1971). Stock assessment and fish biology. A report prepared for the Volta Lake research project. FAO\WS\C422. Rome, p. 33.

15

Sperm Quality and reproductive performance of male *Clarias Gariepinus* induced with synthetic hormones (Ovatide and Ovaprim)

Oguntuase Bukola Gbemisola[1] and Adebayo O. T.[2]

[1]Department of Ecotourism and Wildlife Management, Federal University of Technology, Akure, Ondo State, Nigeria.
[2]Department of Fisheries and Aquaculture Technology, Federal University of Technology, Akure, Ondo State, Nigeria.

Food, especially food protein source is highly essential for healthy living, therefore, this study was aimed at boosting fish seed production and it compared the effectiveness of two different synthetic hormones: Ovaprim and Ovatide on sperm quality and reproductive performance of male *Clarias gariepinus* and the resulting effects on percentage fertilization, hatchability, and survival. The study was made up of three experimental groups. The first group comprising four males was injected with 0.3 ml/kg of ovatide hormone, the second group with four males was injected with 0.3 ml/kg of ovaprim hormone; and the third group, the control group with four males was not given hormones. Sperm motility duration was higher in males induced with the synthetic hormones (29.5 and 25.5 s respectively) which were significantly different (P < 0.05) from the control (18.5 s). Motility percentage was observed to be higher too in the groups treated with the hormones and was different significantly (P < 0.05) from the control. Percentage egg fertilization was similar for the groups induced with the hormones (94.4 and 87.7% respectively) while the control (63.8%) differed significantly from them. Percentage hatchability was highest in eggs fertilized with milt from ovaprim induced males (66.3%) which was significantly higher than those of Ovatide induced males (59.7%) and the Control (14.4%). Milt volume was highest in ovatide induced males (1.1 ml) and was significantly different from ovaprim induced males (0.60) and the control (0.25). Survival rate was the same for all the treatment. The hormones gave similar results except for the milt volume. Treatment with the hormones yielded better result when compared to the control, it could be inferred therefore that, the use of synthetic hormones is better in reproduction of *C. gariepinus* in order to boost productivity thereby ensuring abundant food protein production.

Key words: Spermiation, reproductive performance, ovatide, ovaprim, male *Clarias gariepinus*

INTRODUCTION

Fish forms an important source of human diet as they provide proteins, fats and especially vitamins A and D. Special importance of fish is that they contain vitamin B, which is not present in the plant food. Fish is the good source of calcium; polyunsaturated fatty acids belonging to linolenic acid series (18:3) are present in fish (Crawford et al., 1989). Fish culture came in as a means of producing adequate fish food for human consumption on regular basis, in other words producing fish in and out of season. The practice of fish culture is very old. Huet (1972) traced the art of fish culture in ponds to the Egyptians, the Romans, the people of Indo-Pacific regions, and the Chinese. However the first written account of fish culture in ponds was by a Chinese fish farmer Fan Lai, in 475 BC (Chackroff, 1976).

Also in Nigeria, the first trace of fish farming was the practice by some missionaries in the early 1920's in Ilora, Oyo State, where fish was raised to supplement the protein intake of pregnant women (Bamidele, 2007). In Nigeria, fish alone contributes, on the average, 20 to 25% per caput animal protein intake and could be as high as 80% in coastal and riverrine communities (FAO, 2000). Aquaculture therefore remains the only viable alternative for increasing fish production in order to meet the protein needs of the people. No other group of vertebrates serves man in so many forms as fish (Bamidele, 2007). African catfish (*Clarias gariepinus*) is a hardy species for aquaculture purposes, it is widely accepted in the tropics and commands good commercial value. This species will not readily breed in captivity all year round and hence the need for artificial seed production using hormones.

In order to boost fish production on a level that will be able to serve people as reliable source of protein, there is a need to be able to reproduce fish on a large basis in and out of season to ensure regular supply. Artificial reproduction and selective breeding has become very popular nowadays in the age of science and technology as a means of ensuring large scale production of fish seeds throughout the year. Artificial propagation methods constitute the major practicable means of providing enough quality seed for rearing in confined fish enclosure such as fish ponds, reservoirs and lake (Charo and Oirere, 2000).

Reproduction in fish is controlled by several factors which include sex steroids in the regulation of reproductive processes (Kime, 1993). These reproductive processes are controlled through the brain-pituitary-gonadal axis. The brain is stimulated by environmental cues like water rise, temperature, feeding, rainfall, and photoperiod to release gonadotropin releasing hormones (Zohar et al., 2010). Ovulation and spermiation are effected as a result of the sex steroids that have been produced. Administration of gonadotropin releasing hormone analogue has been reported to increase levels of plasma sex steroids in fish (Zhuo et al., 2011).

The use of synthetic hormones in female fish is now popular as a means of artificially inducing the female fish in order to ovulate. However, Zhuo et al., 2011 has shown from his study that Gonadotropin releasing hormone analogue multiple injection potentially accelerated testicular maturation of male yellow catfish. The use of both synthetic and natural hormones brings about quick ovulation and higher percentage of hatched fish, though synthetic hormone gives higher yield than the natural hormones (Krol et al., 2006).

Gonadotropin Releasing Hormone analogue (GnRHa) is now the best available biotechnological tool for the induced breeding of fish. Ovaprim and Ovatide are both synthetic hormone preparation containing salmon gonadotropin releasing hormone analogue and domperidone (SGnRHa + Domperidone) which are usually used for spawning induction in catfishes to get quality seed (Sahoo et al., 2005). Synthetic hormones, and specifically, Ovaprim are known to significantly increase ovulation in African catfish (Sharaf, 2012).

As shown by researchers, synthetic hormones are best in inducing ovulation in female fish towards the yield of viable seeds on regular basis, there is a need to also cause the same effect of viable seed production by inducing spermiation in male fish. The use of hormones in female fish is gaining popularity each day while there is little work on inducing spermiation in male fish. This study was necessary to test and compare the effects of synthetic hormones on the milt quality of *C. gariepinus* and the resultant effects of the induced milt on seed yield and quality.

MATERIALS AND METHODS

The study was carried out in the Fish Farm of Federal University of Technology, Akure in Ondo State, Nigeria, between May and July, 2009. Ovaprim and Ovatide (SGnRHa + Domperidone) are synthetic hormones containing gonadotropin releasing hormone analogues and Domperidone that were used for the study; the synthetic hormones were from Aqualife Syndel International Inc., Vancouver, BC, Canada and Hemmo Pharma, India respectively. Matured males and female *C. gariepinus* between 8 and 9 months old and weighing between 1.0 to 1.2 kg were procured from the Federal University of Technology Akure Fish farm.

Brooder selection and maintenance

Hormonal treatments of the broodstocks were conducted in the month of May 2009 in order to coincide with the rainy season, which is the natural breeding period of the fish.

The broodstocks consisting of 12 males and 1 female were removed from the ponds and stocked in concrete tanks for a period of 4 weeks for acclimatization prior to induction. The broodstocks were selected on the basis of their morphological or external sexual characteristics (Ayinla et al., 1994).

The selected brood stocks were weighed individually, and the different weight recorded and they were stocked in plastic troughs of about 25 L capacity each.

Preparation of the spawning bowls

The spawning troughs of about 5 L capacity were collected, washed thoroughly and dried. The bowls were labeled as Control (1, 2, 3, and 4), Treatment A1, A2, A3, A4, and Treatment B1, B2, B3, and B4.

Preparation of hormones and injection of brooders

The quantity of Ovaprim and Ovatide used was commensurate with the weight of the male fish to be injected that is, 0.3 ml per kg of male fish.

Treatment A: Ovatide hormone was injected into the dorsal muscle of four male broodstocks serving as Replicate 1, 2, 3, and 4 accordingly.
Treatment B: Ovaprim was injected into the dorsal muscle of four male broodstocks serving as Replicate 1, 2, 3, and 4 accordingly.
Control: Four male broodstocks were not injected with hormones at all, while the female broodstock to be used for reproductive test was induced to ovulate with 0.5 ml/kg Ovaprim hormone. Administration of dosage and observation of latency period was according to Legendre (1986) and the specifications for the use of the synthetic hormones by the manufacturers. The injected area of each fish was massaged gently with a finger in order to distribute the hormone evenly into the muscle while slowly retracting the needle.

Dissection of male broodstock for gonad removal

The male broodstocks after observing the latency period of 12 h were removed from the troughs; they were placed dorsally on a wet towel. The fish was held firmly down to ensure careful removal of the testis. Using a sharp blade, the abdominal cavity of the fish was dissected and testes were carefully removed from where they were lying, at the ventral wall of the abdominal cavity. The testes were removed whole and cleaned with fluffy material.

Stripping and egg fertilization

After observing the 12 h ovulation period, the female broodstock was removed from the trough carefully and held firmly with a wet towel at both ends, the abdomen was then pressed carefully to release the eggs into a dry bowl, and the stripping was done towards the fish vent. After stripping, the spent female was carefully returned into the trough. 1 g each of the egg mass that is, about 700 eggs, (Viveen et al., 1985) was weighed out and the content of the testes (milt) was spilled on the eggs for fertilization, the eggs were then poured inside the labeled spawning troughs already containing 1 L of water for the sperm activation, and the fertilized egg was left in the spawning troughs for incubation with water temperature between 25 and 27°C.

Incubation and hatching

The fertilized egg masses were incubated in the spawning bowls for a period of 24 to 40 h. After this period, most of the larvae emerged and some eggs did not hatch at all. The percentage hatching rate was estimated 40 h after fertilization.

Survival

After hatching and determination of hatching rate, the unhatched eggs were siphoned out of the spawning bowls in order to ensure the survival of the hatched ones and the water was partially changed with utmost care. The larvae were daily observed after hatching for about 4 weeks to determine the rate of survival by estimating the dead and live ones.

Sperm evaluation

The sperm evaluation includes: gross (visual) and microscopic examination (as reviewed by Rurangwa et al., 2004; Cosson et al., 2008a, b; Cabrita et al., 2009)

The gross examination was based on visual and physical observation such as the milt volume by collecting the milt in graduated cylinder and the level was determined in milliliters. While the microscopic examination was carried out using Olympus model CX40, with magnification between X10 and X25 to determine other parameters such as: motility (duration and percentage). Motility percentage and duration were determined by observing water activated milt placed on a glass slide under microscope. The motile sperms were observed and expressed as a percent of non-moving sperm. Motility duration was determined as the period between movements of the sperm to cessation of any progressive movement expressed in seconds.

Analysis of data

The data obtained for each hormone treatment were compared by one-way analysis of variance (ANOVA) test to determine significant difference ($p = 0.05$), and treatment means were subjected to the Duncan's Multiple Range Test.

Calculations

The rates of fertilization (Adebayo, 2006), hatching, survival, and relative fecundity were calculated as shown below:

$$\text{Fertilization rate} = \frac{\text{No. of fertilized eggs}}{\text{Total no. of eggs counted}} \times 100$$

$$\text{Hatching rate} = \frac{\text{No. of eggs hatched}}{\text{Total no. of eggs in a batch}} \times 100$$

$$\text{Survival rate} = \frac{\text{No. of hatchlings alive up to larvae stage}}{\text{Total no. of hatchlings}} \times 100$$

RESULTS

Significant differences in spermiation success between control and hormonally induced groups were observed.

Sperm evaluation

The motility duration was expressed in seconds, and volume in ml. Milt was collected from all males originating from both experimental groups (Ovatide and Ovaprim treated males, as well as from the control).

Figure 1 shows the result of the motility duration of hormonally induced milt compared with control, motility duration was higher in the induced groups and was significantly different from the control ($p < 0.05$) as well as the motility percentage (Figure 2). Effect of the synthetic hormones was similar in terms of sperm motility percentage in the groups induced, which was different significantly from the control. Milt volume (Figure 3) was recorded to be highest in males treated with Ovatide, and was significantly different from the other treatments ($P < 0.05$).

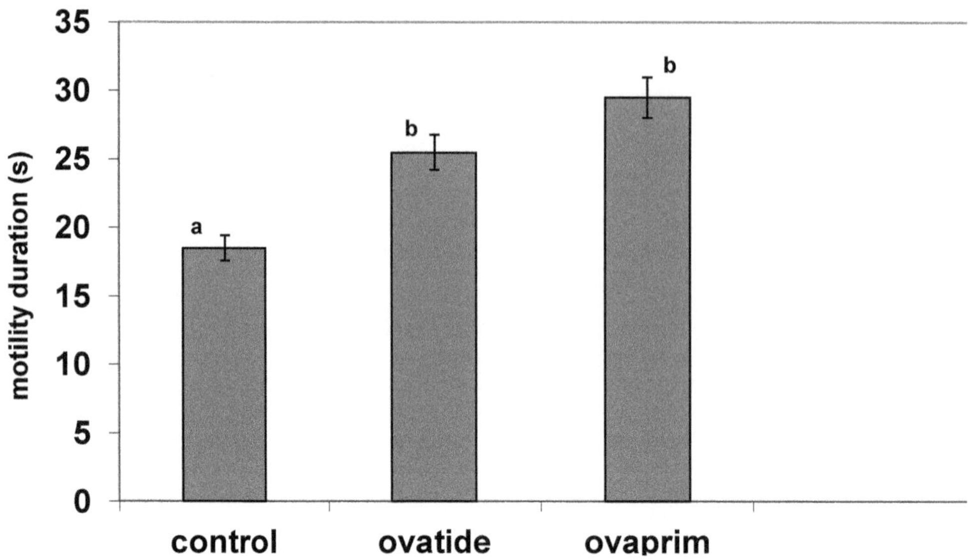

Figure 1. Motility duration of induced male *C. gariepinus* compared with the control.

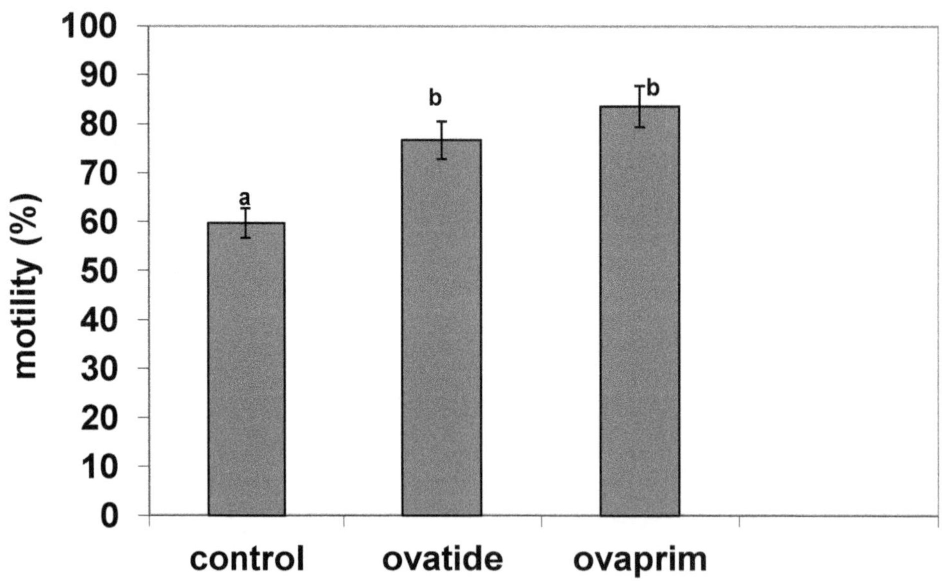

Figure 2. Motility percentage of induced male *C. gariepinus* compared with the control.

Effects of hormonally induced milt on fertilization, hatchability and survival.

Table 1 shows the effects of the milt collected from the hormonally induced groups and the control group on female *C. gariepinus* eggs.

Effects on female eggs showed that milt of male induced with the hormones had higher percentage fertilization, hatchability and survival which were significantly different from the control. The hatching period was the same for all the treatments, while the hatching time varied with the highest time observed from the control male.

DISCUSSION

The quality of fish sperm is as important as quality of female eggs for viable off-springs, various studies have been carried out on induced spawning in female fish with lesser attention on the male counterpart. Sperm morphology, density, volume, motility and fertilizing

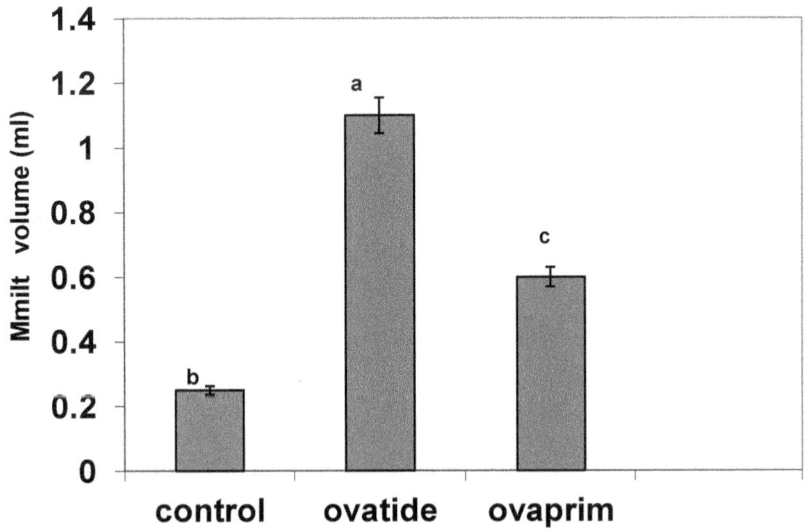

Figure 3. Milt volume of induced male *C. gariepinus* compared with the control.

Table 1. Reproductive performance of induced male sperm on fertility, hatchability and survival percentage of *C. gariepinus* egg.

Parameter	Control	Ovatide	Ovaprim
Fertilization (%)	63.81 ± 5.21^{a}	87.73 ± 3.38^{b}	94.40 ± 2.03^{b}
Hatchability (%)	14.14 ± 2.14^{a}	59.70 ± 0.88^{b}	66.37 ± 0.74^{c}
Survival (%)	61.91 ± 4.77^{a}	63.56 ± 13.56^{a}	77.38 ± 5.95^{a}
Hatching time	26.5 ± 0.50^{a}	23.5 ± 0.50^{a}	24 ± 00^{a}
Hatching period	72 ± 00	72 ± 00	72 ± 00

*Mean values in the same row having the same superscript are not significantly different.

capacity, as well as composition and osmolality of the seminal plasma are parameters commonly measured to assess sperm quality in fish (Alavi et al., 2004).

Fish often respond well to hormonal induction, in atlantic sturgeon, both sexually matured and immature males can be induced to spermiate by injection of gonadotropins. The gonadotropin-releasing hormone stimulates the synthesis and release of pituitary gonadotropin (Melamed et al., 1996), and GTH stimulates the production of steroid hormones in the gonads (Van der kraak et al., 1992). In some species such as *C. gariepinus,* hormonal treatments are the only way for controlling reproduction reliably. Treatment with GnRHa has been tested successfully in various marine and freshwater species and has been shown as an effective strategy for improving milt quantity and quality (Mylonas et al., 1997; Clearwater and Crim, 1998; Mylonas and Zohar, 2001; Zohar and Mylonas, 2001). GnRHa acts at a higher level of the hypothalamus pituitary gonad axis than gonadotropins. Consequently, GnRH can provide a more balanced stimulation of reproductive events and presumably a better integration of these events with other physiological functions, by directly or indirectly affecting the release of other hormones necessary for successful spermiation (Zohar and Mylonas, 2001).

This study has clearly shown that sperm of *C. gariepinus* gained better quality when induced with synthetic hormones as compared to the control, this agrees with the work of Adebiyi et al. (2013) that Ovaprim, a synthetic hormone produced best result in inducing spawning in a river Catfish (*Hemibagrus nemurus*).

Synthetic hormones gave a significant increase in milt volume and other sperm parameters as observed in this study which is in line with the findings of Crim et al. (1988) and Lin et al. (1996) for studies on Rainbow trout *Oncorrchnchus mykiss* and *Paramisgurnus dabryanus* respectively. It is also known, as opined by Mylonas et al. (1997), Zohar and Mylonas (2001) that, the treatment of fish with hormones by injection typically results in a short-term increase in milt volume, and changes in plasma steroids. It is possible that, in smelt, increase in milt volume were partially caused by milt hydration, indicating that seminal plasma volume was being increased at a faster rate than spermatozoa production. Sperm motility in teleost fish, is so far considered the best biomarker of milt quality (Lahnsteiner et al., 2004). One of the objectives of this study was to examine the quality of the sperm in terms of motility (duration and percentage); and volume. As it was in yellowtail flounder *Limanda ferruginea*, treatment with two different types of GnRHa-delivery systems increased the percentage of motile sperm from 20, 40 to 90% (Clearwater and Crim, 1998). There was an overall improvement in the sperm quality of the induced males in this study which is supported by the findings of Taghi et al. (2010) where male carps induced with synthetic hormones showed improved spermatological properties.

This study showed that there was higher percentage motility observed in induced male when compared to the control, this agrees with the work of Krol et al. (2006) which showed higher percentage of motile sperm stripped smelt spermatozoa of stimulated males than the control group.

The spermatozoa motility and its duration have great influence on successful fertilization.

In this study, percentage fertilization, hatching and survival rate were recorded to be higher in the treated groups, and were significantly different from the control, this agrees with the work of Krol et al. (2006) in which Progeny of smelt hatched with milt from Ovaprim induced male had higher survival than progeny hatched naturally and with the work of Ndimele and Owodeinde (2012) where the induced group of *C. gariepinus* showed significant increase in percentage fertilization, hatching rate, and surviva. Also the findings of Nwokoye et al. (2007) further substantiate the efficacy of synthetic hormones by recording highest mean number of fertilized eggs, hatchability, and survival in clariid catfish (*Heterobranchus bidorsalis*) injected with synthetic hormone (Ovaprim).

CONCLUSIONS AND RECOMMENDATION

The use of synthetic hormone is a better way of improving catfish production with respect to reproductive performance in aquaculture. Stimulation of males by Ovaprim and Ovatide significantly increased volume of milt, motility duration and percentage. The efficacy of these synthetic hormones was evident on the reproductive performance as tested on the female *C. gariepinus*.

The study has shown that, artificial stimulation of male *C. gariepinus* using synthetic hormone is more reliable in reproducing fish. The use of these hormones in male fish is a means of boosting reproductive performance and ensuring good and viable fish seeds. The results arising from this trial is important in encouraging fish seed production through the use of synthetic hormones.

REFERENCES

Adebayo OT (2006) Reproductive performance of African clariid catfish *Clarias gariepinus* broodtock on varying maternal stress. J. Fish. Int. 1(1-2):17-20.

Adebiyi FA, Siraj SS, Harmin SA, Christianus A (2013) Induced spawning of a river catfish *Hemibagrus nemurus.* Pertanika J. Trop. Agric. Sci. 36(1):71-78.

Alavi SMH, Cosson J, Karami M, Mojazi AB, Akhoundzadeh MA (2004). Spermatozoa motility in the persian surgeon, *Acipenser persicus*. Effects of pH, dillution rate, ions and osmolality. Reproduction 128:819-828.

Ayinla OA, Kayode OJ, Idoniboye-obu OJE, Oresegun A, Adidu VE (1994). Use of tadpole meal as substitute for fish meal in the diet of *Heterobranchus bidorsalis*. J. Aquac. Trop. 9(1):25-33.

Bamidele O (2007) Introduction to fish farming in Nigeria. Ibadan University Press, Publishing House. University of Ibadan, Nigeria, pp. 1-6.

Cabrita E, Robles V, Herraez P (2009) Sperm quality assessment. Methods in reproductive aquaculture. Marine and freshwater species. Cabrita E., Robles V., Herraez P. eds. CRC Press Taylor & Francis Group, Boca Raton, USA, ISBN 978-0- 8493-8053-2.

Charo H, Oirere W (2000). River-based artificial propagation of the African catfish *Clarias gariepinus*: An option for the small fish farmer. NAGA – The ICLARM Q. Jan-March 2(1):14-16.

Chackroff M (1976). Fresh water fish pond culture and management. VITA (Volunteers in Technical Assistance) Publications Manual 36E. Peace Corps Programme and Training. Journal manual IB.U.C Government printing office, Washington, D.C., pp. 191.

Clearwater SJ, Crim LW (1998). Gonadotropin releasing hormone-analogue treatment increases sperm motility, seminal plasma pH and sperm production in yellow tail flounder Pleuronectes ferrugineus. Fish physiol. Biochem. 19:349-357.

Cosson J, Groison AL, Suquet M, Fauvel C, Dreanno C, Billard R (2008a). Marine fish spermatozoa: racing ephemeral swimmers. Reproduction 136(3):277-294.

Cosson J, Groison A., Suquet M, Fauvel C, Dreanno C, Billard R (2008b) Studying sperm motility in marine fish: an overview on the state of the art. J. Appl. Ichthyol. 24(4):460-486.

Crawford MA, Doyle W, Williams G, Drury PJ (1989). The role of fats and EFAs for the structures in the growth of fetus sp. and neonate. In: The Role of Fat in Human Nutrition (Vergroesen, A.J and M. Crawford, eds.) pp: 81-115. Academic press, Canada.

Crim LW, Evans DM, Vickery BH (1988). Manipulation of the seasonal reproductive cycle of the Landlocked Atlantic Salmon *Salmon salar* by LHRH analogues administered at various stages of gonadal development. Can. J. Aquac. Fish. Sci. 40:61-67.

FAO (2000). FAO Fishery Department Country Profile. *Nigeria FID/CP/NIR Rev.* 4 March 2000. Projected population and fish demand/supply in Nigeria 1997-2025 *http://www.fao.org/fi/fcp/en/NGA/body.htm, March 2000.*

Huet M (1972). Textbook of fish culture, Fishing News (Books) Ltd., 23 Romont Avenue, west Byfleet Survey, England. Pp 436. (Translated by Henry Kahn from the Fourth French Edition of Traite de Pisciculture, Eds. Ch. De Wyngaert, Brussels, 1970. pp. 116-121.

Kime DE (1993). Classical and non-classical reproductive steroids in fish. Rev. Fish Biol. Fish. 3:160-180.

Krol J, Glogowski J, Demska-Zakes K, Hliwa P (2006). Quality of semen and histological analysis of testis in *Perca fluviatilis* L. during spawning period. Czech J. Anim. Sci. 51:220-226.

Lahnsteiner F, Berger B, Horvath A, Urbanyi B (2004). Studies on the semen biology and sperm cryopreservation in the Sterlet, Acipenser ruthenus L. Aquac. Res. 35:519-528.

Legendre M (1986). Seasonal changes in sexual maturity and fecundity and human chorionic gonadotropin (HCG) induced breeding of the catfish, *Heterobranchus longifilis* Val (Vlariidae) reared in Ebic Lagoon, Ivory Coast. Aquaculture 55:201-213.

Lin F, Ciereszko A, Dabrowski K (1996). Sperm production and cryopreservation in Muskellunge after carp pituitary extract and human chorionic gonadotropin injection. Prog. Fish-cult. 58:32-37.

Melamed P, Gur G, Elizur A, Rosenfeld H, Sivan B, Rentier-Delrue F, Yaron Z (1996). Differential effects of gonadotropin- releasing hormone, dopamine and somatostatin and their second messengers on the mRNA levels of gonadotropin IIb subunit and growth hormone in teleost fish, tilapia. Neuroendocrinol. 64:320-325.

Mylonas CC, Zohar Y (2001) Endocrine regulation and artificial induction of oocyte maturation and spermiation in basses of the genus morone. Aquaculture 202:205-220.

Mylonas CC, Gissis A, Magnus Y, Zohar Y (1997). Hormonal changing male white bass (*Morone chrysops*) and evaluation of milt quality after treatment with a sustained-release GnRHa-delivery system. *Aquaculture* 153:301-313.

Ndimele PE, Owodeinde FG (2012). Comparative Reproductive and Growth performance of *Clarias gariepinus* and its hybrid induced with synthetic hormone and pituitary gland of *Clarias gariepinus.* Turk. J. Fish. Aquat. Sci. 12: 619-626.

Nwokoye CO, Nwuba LA, Eyo JE (2007). Induced propagation of African clariid catfish, *Heterobranchus bidorsalis* using synthetic and homoplastic hormones. Afr. J. Biotechnol. 6(23):2687-2693.

Rurangwa E, Kime DE, Ollevier F, Nash JP (2004). The measurement of sperm motility and factors affecting sperm quality in cultured fish. Aquaculture 234:1-4, 1-28.

Sahoo SK, Giri SS, Sahu AK (2005). Induced Spawning of Asian Catfish, *Clarias batrachus* (Linn): effects of various latency periods and SGnRHa and domperidone doses on spawning performance and egg quality. Aquacult. Res. 36: 1273.

Sharaf SM (2012). Effect of GnRHa, Pimozide and Ovaprim on ovulation and plasma sex steroid hormones in African catfish *Clarias gariepinus. Theriogenology* 77(8):1709-1716.

Taghi S, Mohammad RI, Amin G (2010). The effect of Different Hormonal Treatments on Semen Quality Parameters in cultured and Wild carp. Turk. J. Fish. Aquat. Sci. 11:595-602.

Van der kraak G, Suzuki K, Peter RE, Itoh H, Kawauchi H (1992). Properties of common carp gonadotropin I and gonadotropin II. Gen. Comp. Endocrinol. 85:217-229.

Viveen WJ, Richterm CJ, Van Oodot PG, Janssen JA, Huisman EA (1985). Practical manual for the culture of the African catfish *Clarias gariepinus.* Section for Research and Technology, Ministry for Development Cooperation, The Hague, Netherlands, 128 pp..

Zhuo Q, Zhang Y, Huang W, Liu X, Li Y, Zhu P, Lu D, Lin H (2011). Gonadotropin releasing hormone analogue multiple injection potentially accelearted testicular maturation of male yellow catfish (*Pelteobagrus fluridruco* Richardson) in captivity. Aquacul. Res. 42:1-14.

Zohar Y, Mylonas CC (2001). Endocrine manipulation of spawning in cultured fish: from hormones to genes. Aquaculture 197:99-136.

Zohar Y, Munoz-cueso JA, Elizor A, Kah O (2010). Neuroendocrinology of reproduction in teleost fish. General and comparative Endocrinology 165:438-455.

Microplanktic assemblages (sarcodines and alveolates) in the central and southeastern Aegean Sea (NE Mediterranean)

H. Miliou[1]*, K. Kapiris[2] and I. Castritsi-Catharios[3]

[1]Department of Applied Hydrobiology, Faculty of Animal Science and Aquaculture, Agricultural University of Athens, Iera Odos 75, 11855, Athens, Hellas.
[2]Hellenic Centre for Marine Research, Institute of Marine Biological Recourses, Agios Kosmas Hellinikon, 16604, Athens, Hellas.
[3]Department of Zoology - Marine Biology, Faculty of Biology, National and Kapodistrian University of Athens, Panepistimiopolis, 15784, Athens, Hellas.

Spatial changes in the structure of microplanktic assemblages (sarcodines and alveolates) in the surface waters of coastal regions in the central and southeastern Aegean Sea were estimated during late summer to early autumn. Tintinnidae (Ciliophora), as well, Acanthometridae (Radiolaria) and Globiderinidae (Foraminifera) exhibited a higher abundance in the central Aegean Sea, compared to that in the more oligotrophic southeastern area of the same sea. Multivariate analysis revealed that the samples from the eastern Cretan Sea (South Aegean Sea) were highly distinguished from all the others and showed the highest densities of polycystines (Radiolaria), specifically of the families Thalassicollidae, Thalassosphaeridae, Sphaerozoidae (Collodaria), as well as of Actinommidae (Spumellaria) and Theoperidae (Nassellaria). In addition, the families Ceratiaceae and Prorocentraceae (Dinoflagellata) were more abundant in the southeastern than in the central Aegean Sea. It is indicated that the hydrographic conditions prevailing in the eastern Cretan Sea affected the structure of the microplanktic community in the surface layer. It is proposed that the assemblage of the identified radiolarian families belonging to Collodaria, Spumellaria (S) and Nassellaria (N), with a high S/N ratio, could be used as biological proxy of weak upwelling systems in the warm stratified waters of the oligotrophic eastern Mediterranean.

Key words: Eastern Mediterranean, microplankton, sarcodines, alveolates, abundance, oligotrophic ecosystems.

INTRODUCTION

The Aegean Sea has been characterized as oligotrophic, like the whole eastern Mediterranean Sea (Krom et al., 1991). It is generally accepted that the degree of oligotrophy increases along an easterly gradient in the Mediterranean, which is clearly reflected in a general quantitative reduction of vertical fluxes (Heussner and Monaco, 1995). In the eastern regions of the Mediterranean, waters remained stratified throughout most of the year and presumably pico- and small nanoplankton play a most important role in the primary production (Li et al., 1993; Ignatiades et al., 1995). In such oligotrophic ecosystems, microzooplanktic rather than mesozooplanktic organisms are expected to be the main herbivores (Calbet and Landry, 2004). In an oligotrophic ecosystem, the microplankton community is expected to be characterized by a high diversity among the protest primary and secondary producers as well as

*Corresponding author. E-mail: elenmi@aua.gr.

predators (Margalef, 1994). Furthermore, protozoan communities in the eastern Mediterranean includes picophagous species, which feed mostly on bacteria, as well as nano- and microphagous species, which generally feed on protists (Hausmann et al., 2002).

In addition, acquisition of phototrophy through maintenance of algal endosymbionts or algal organelles is common among aquatic phagotrophic protists (Stoecker et al., 2009). Most protists with acquired phototrophy are mixotrophic; they combine heterotrophy and phototrophy, and so we need to go beyond traditional concepts of 'plant' and 'animal' functions to progress in our understanding of aquatic microbial ecology. Acquired phototrophy occurs primarily among radiolarians, foraminiferans, ciliates and dinoflagellates, and is crucial in the sarcodines (radiolarians and foraminiferans) to their contributions to carbonate, silicate, strontium, and carbon flux in subtropical and tropical oceans, particularly in oligotrophic regions (Caron et al., 1995; Michaels et al., 1995; Dennett et al., 1999; Stoecker et al., 2009). On the other hand, a high percentage of dinoflagellates that bear chloroplasts also contain food vacuoles and therefore, dinoflagellates, in general, are included in the microzooplankton along with the animal-like protists by several authors (Jyothibabu et al., 2003). Microzooplankton is an important component of the diet of mesozooplankton, which is in turn consumed by larval fish (Hunter, 1981). Furthermore, there is now growing evidence that fish larvae directly eat protozoa of microplankton, particularly the alveolates (ciliates and dinoflagellates) (Figueiredo et al., 2005). Thus, the abundance and composition of sarcodines and alveolates in microplankton may have important implications to fisheries of both pelagic and benthic organisms.

Although Aegean Sea is considered as oligotrophic, some relatively eutrophic areas do exist. For example, Stergiou et al. (1997) estimated that nutrient concentrations as well as plankton and benthos abundance and/or productivity are all by one order of magnitude higher in the N-NW than in the S-SE Aegean Sea. The Cretan Sea in the southernmost basin of the Aegean Sea is thought to be among the few marine ecosystems that could be characterized as strongly oligotrophic through long periods of the year (Ignatiades, 1998; Psarra et al., 2000). Apart from the low nutrient levels, impoverished phytoplankton populations and low mesozooplankton standing stocks were recorded (Gotsis-Skretas et al., 1999). A significant retention of vertical biogenic particle flux has been recorded in the Cretan Sea, since microzooplanktic communities (phaeodarians and foraminiferans) rapidly decreased net sedimentation from the photic layer, by ingesting and repackaging liable particles several times during descent, and thus restricting food supply to benthic organisms (Wassmann et al., 2000). The high water column temperatures in the Cretan Sea further enhance the decomposition rate of organic matter.

However, Duineveld et al. (2000) argued that benthic-pelagic coupling does exist in the Cretan Sea, although the amplitude of the seasonal signal is small. Circulation pattern is rather complex in the Aegean Sea, where water masses with different characteristics enter and may get uplifted in some areas, enriching surface with nutrients and enhancing primary productivity (Psarra et al., 2000; Tselepides et al., 2000). As a result, significant spatial differences in the abundance and structure of pelagic and benthic communities are generated (Psarra et al., 2000; Duineveld et al., 2000).

Distribution of microzooplanktic species is partly controlled by temperature and hydrographic structures and their interrelations on water mass fertility; some species are characteristic of upwelling systems and/or of fertile areas (Venec-Peyre and Caulet, 2000). Despite that microzooplanktic organisms play an important role in the biogenic matter fluxes in the oligotrophic systems, and may also serve as biological signatures (Boltovskoy, 1999), little is known about their distribution and abundance in the Aegean Sea, but emphasis has been given mainly on phytoplankton and mesozooplankton (Stergiou et al., 1997). Furthermore, studies on microzooplankton have been performed in relatively eutrophic gulfs (Kormas et al., 1998; Mihalatou and Moustaka-Gouni, 2002), but not in the islands and particularly in the more oligotrophic southeastern part of the Aegean Sea.

The aim of the present study was to estimate the abundance of radiolarians, foraminiferans, tintinnids and dinoflagellates in the surface waters of coastal areas in the central and southeastern Aegean Sea during the stratification period. We tried to evaluate spatial changes in the structure of their communities (on family level) based on perturbations in the productivity levels and hydrographic features aforementioned. Furthermore, the possible role of certain microplanktic taxa in the microbial loop and bentho-pelagic coupling is discussed. The results of this study are the first ones carried out on the study area and could be used in the future as a useful tool for comparison purposes, among other regions of the Mediterranean basin.

MATERIALS AND METHODS

Microplankton (20 to 200 μm) was sampled from sixteen stations at coastal areas in the central and southeastern Aegean Sea in 1994, during late summer to early autumn (Figure 1, Table 1). The bottom depth of these stations ranged from 12 to 100 m and the samples were collected by NISKIN bottle at depth of 10 to 30 m (Table 1). Specimens were preserved by adding borax buffered formalin to a final concentration of 2% immediately after sampling. A 2 L sample was used for protistan enumeration and identification down to family level, as in other microplankton studies in Eastern Mediterranean (Dolan et al., 2002; Kormas et al., 1998). The sample was concentrated to a volume of 100 ml by filtering through membrane polycarbonate filters (NUCLEPORE) of 5 μm diameter pore size. Filtration was performed at low pressure (250 mmHg). The organisms remaining on the filter were then very gently washed

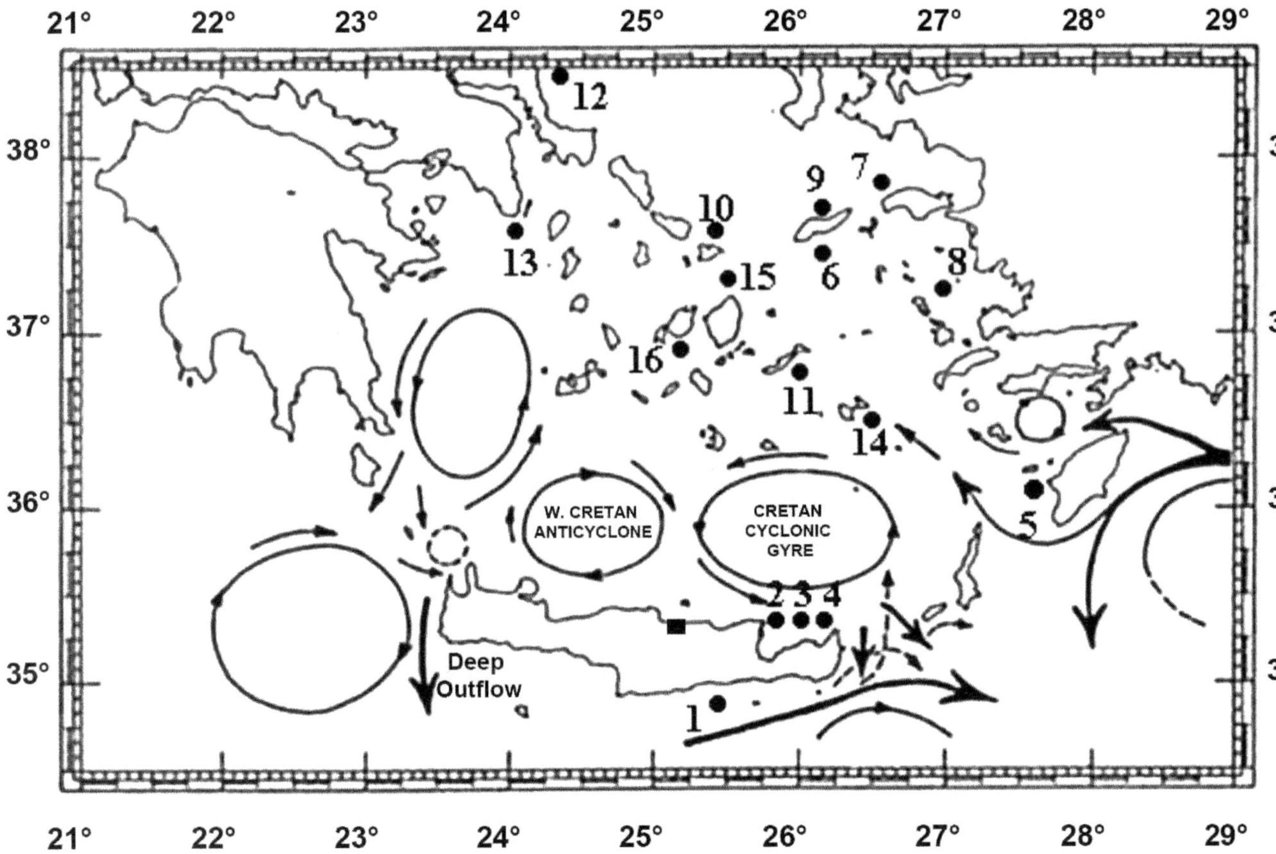

Figure 1. Sampling stations (•) in the Aegean Sea (basic map from Souvermezoglou et al., 1999) and city of Heraklion (■).

Table 1. Location, date, coordinates and sampling depth of the stations in the islands of the Aegean Sea.

Location	Date	Longitude (E)	Latitude (N)	Sampling depth (m)	Bottom depth (m)	T (°C)	S (psu)
Gaidouronisi	10/1994	25°35'46"	34°59' 56"	30	70	22.5	40
West Sitia, Crete	08/1994	25°46' 59"	35°20' 14"	20	33.5	25	39
Central Sitia, Crete	08/1994	25°59' 22"	35°12' 44"	20	37.5	25	37
East Sitia, Crete	08/1994	26°15' 26"	35°17' 58"	10	12	25.6	39
Rhodos	10/1994	27°33' 17"	36°12' 14"	10	20	23.9	42
Fourni	10/1994	26°70' 90"	37°32' 43"	30	80	23	40
Samos	10/1994	26°37' 58"	37°46' 26"	30	87.5	22	40
Lipsi	10/1994	26°45' 58"	37°24' 11"	30	67.5	22.4	40
Ikaria	10/1994	26°60' 53"	37°38' 17"	30	100	23	39
Mykonos	09/1994	25°22' 15"	37°29' 49"	10	20	22	40
Amorgos	09/1994	25°56' 12"	36°50' 48"	30	95	22.5	41
Kimi	08/1994	24°11' 55"	38°29' 39"	10	20	25.5	40
Cav°d'Oro	08/1994	24°20' 57"	37°39' 24"	10	20	25.5	37
Kalymnos	10/1994	26°26' 34"	36°33' 48"	30	55	22	40
Naxos	09/1994	25°27' 70"	37°10' 23"	30	55	23.5	41
Paros	09/1994	25°11' 40"	36°57' 39"	20	32.5	22	40

Temperature (T) and salinity (S) recorded at the surface of the water column.

into a 100 ml settling chamber using a fine tipped glass pipette and some water which run through the filter. For settling (2 days), the Utermöhl method (Utermöhl, 1958) was used. The counting was performed by using an inverted-phase microscope. The

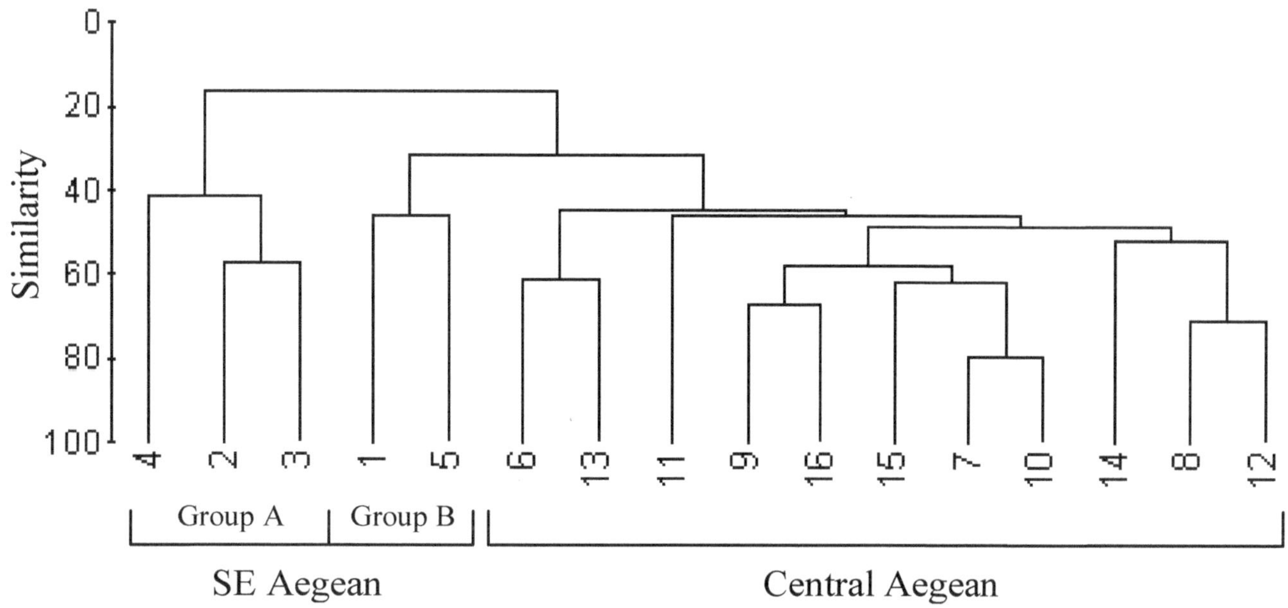

Figure 2. Bray-Curtis similarity indices of the abundance of microplanktic families at 16 stations in southeastern and central Aegean Sea. Dendogram using group averaged clustering. Group A: stations in eastern Cretan Sea; Group B: the other stations in SE Aegean Sea.

Table 2. Average abundance of microplanktic families (cells l^{-1}) collected in Aegean Sea.

Families	SE Aegean	Central Aegean				
	Av. Abund	Av. Abund	Av. Diss	Diss/SD	Contrib%	Cum%
Tintinnidae	1.00	11.55	12.18	2.35	15.50	15.50
Thalassicollidae	18.80	0.09	12.09	1.65	15.38	30.88
Thalassosphaeridae	15.60	0.45	10.42	1.38	13.26	44.14
Prorocentraceae (10^3)	3.40	0.55	7.45	1.16	9.48	53.62
Sphaerozoidae	4.80	0.09	5.76	0.84	7.32	60.95
Ceratiaceae (10^3)	4.40	0.18	5.46	1.02	6.94	67.89
Actinommidae	6.00	0.00	4.67	0.72	5.94	73.83
Acanthometridae	0.40	1.45	4.57	0.83	5.81	79.64
Globigerinidae	0.20	0.91	3.78	1.01	4.81	84.45
Dinophysiaceae (10^3)	0.60	0.18	3.24	0.77	4.12	88.57
Theoperidae	1.00	0.18	2.60	0.66	3.31	91.88

SIMPER analysis showing discrimination of the samples in central and SE Aegean Sea. Av. Abund: average abundance, Av. Diss: average dissimilarity between all pairs of inter-group samples, Diss/SD: dissimilarity/standard deviation of contribution, Contrib%: percentage contribution, Cum%: cumulative percentage.

classification of microplankton is presented according to Systema Naturae 2000 (Brands, 2000), except that of Radiolaria Polycystinae that follows the scheme proposed by Boltovskoy (1999).

The Primer statistical software package was used to perform multivariate analysis on the abundance data (Clarke and Warwick, 1994). A hierarchical cluster analysis based on the Bray-Curtis similarity index was calculated after log transformation of data. The dendrograms were formed with the grouping average method, and analysis of similaritie (ANOSIM) was used for testing the significance of grouping. In order to determine discriminating species, the similarity percentage (SIMPER) analysis was performed.

RESULTS

Multivariate analysis on the abundance data of microplanktic organisms (sarcodines and alveolates) revealed a clear separation (average dissimilarity = 78.58) of the samples collected in the central and southeastern Aegean Sea (Figure 2). The ANOSIM test proved that the samples of these two regions differed significantly ($R = 0.811$, $p = 0.1\%$). The SIMPER routine (Table 2) showed that Tintinnidae (phylum: Ciliophora, class: Spirotrichea) contributed (15.5%) the most to the

Table 3. Average abundance of microplanktic families (cells l^{-1}) collected in SE Aegean Sea.

Families	Group A	Group B				
	Av. Abund	Av. Abund	Av. Diss	Diss/SD	Contrib%	Cum.%
Thalassosphaeridae	26.00	0.00	17.64	5.69	23.30	23.30
Thalassicollidae	30.67	1.00	14.23	2.29	18.79	42.09
Sphaerozoidae	8.00	0.00	9.69	1.23	12.79	54.88
Ceratiaceae (10^3)	6.67	1.00	6.68	1.80	8.82	63.69
Actinommidae	9.67	0.50	6.40	0.93	8.45	72.14
Prorocentraceae (10^3)	4.00	2.50	5.90	1.41	7.79	79.93
Theoperidae	1.67	0.00	3.42	0.65	4.51	84.44
Dinophysiaceae (10^3)	0.33	1.00	3.15	1.16	4.15	88.60
Tintinnidae	0.67	1.50	2.28	1.11	3.01	91.61

SIMPER analysis showing discrimination of the samples in two groups. Group A: stations in eastern Cretan Sea; Group B: the other stations in SE Aegean Sea. Av. Abund: average abundance, Av. Diss: average dissimilarity between all pairs of inter-group samples, Diss/SD: dissimilarity/standard deviation of contribution, Contrib%: percentage contribution, Cum%: cumulative percentage.

average dissimilarity, being less abundant in the SE Aegean Sea. Acanthometridae (phylum: Radiozoa, subphylum: Radiolaria, class: Acantharea) and Globigerinidae (phylum: Foraminifera, class: Foraminifera) also displayed a reduced abundance in the SE Aegean Sea.

Furthermore, the stations (2, 3 and 4, Sitia bay) in the eastern Cretan Sea (Group A) were distinguished (average dissimilarity = 75.74) from the other in the SE Aegean Sea (Group B), and even more from those in the central Aegean Sea (average dissimilarity = 85.29). The discriminating families that contributed cumulatively 88.6% to the average dissimilarity between Groups A and B (Table 3) belonged to the classes Polycystinea (phylum: Radiozoa, subphylum: Radiolaria) and Dinophyceae (phylum: Myzaozoa, infraphylum: Dinoflagellata). The assemblage of polycystines (Order Collodaria: Thalassicollidae, Thalassosphaeridae and Sphaerozoidae, Order Spumellaria: Actinommidae, Order Nassellaria: Theoperidae) showed the highest densities in the stations of Group A. Dinoflagellates were more abundant in the stations of SE Aegean Sea and specifically Ceratiaceae and Prorocentraceae in the Group A, while Dinophysiaceae in the Group B.

DISCUSSION

Tintinnidae may constitute one major component of marine planktic ciliates, as they can respond quickly to phytoplankton pulses composed mainly of nanoplankton (Psarra et al., 2000). The abundance of tintinnids in the Aegean Sea, estimated in this study, was much lower than that reported for eutrophic regions in the Atlantic (Dolan and Gaggegos, 2001). Dolan et al. (2002) referred that the concentrations of tintinnids declined from west-to-east, between the Moroccan upwelling system and the oligotrophic eastern Mediterranean, paralleling declines in primary production and increases in the depth of the chlorophyll maximum layer. The average abundance of tintinnids in the central Aegean Sea (late summer to early autumn) can be compared to that recorded (September) in the central basin (35 to 38°N, 8 to 18°E) of the Mediterranean (10 to 20 cells l^{-1}) (Dolan et al., 2002). The densities of tintinnids in the SE Aegean Sea are within the levels (frequently <5 cells l^{-1}) recorded during summer to early autumn in the coast (north Libanon) of the eastern-most (35°E) basin of the Mediterranean (Abboud-Abi, 2002). The decline in the abundance of tintinnids from the central to the south Aegean Sea seems to be influenced by the north-south gradient of increasing oligotrophy, deepening on the chlorophyll maximum layer in this sea (Ignatiades et al., 2002; Stergiou et al., 1997).

Foraminifera occupy circum-global latitudinal climatic provinces (Stoecker et al., 2009) with densities varying from <1 to >100 m^3 cells (Caron and Swanberg, 1990). The family Globigerinidae contains spinose species, which live mainly in the photic zone; most of them host symbiots and prefer a carnivorous diet (tintinnids, copepods, etc.) (Hemleben et al., 1989; Boltovskoy, 1999). Globigerinidae showed a higher density in the central than southeastern Aegean Sea, probably due to a relatively increased availability of their prey, such as tintinnids. In the eastern Mediterranean (north Libanon), at latitudes similar to those of the SE Aegean Sea, the foraminiferal densities in the surface layer were lower than 0.5 cell l^{-1} during summer to early autumn (Abboud-Abi, 2002).

Acantharea and Polycystinea represented Radiolaria in the surface waters of the Aegean Sea. Phaeodarians (Radiolaria) were not found in this study probably because they are typically deep-water animals, usually peaking in both abundance and diversity below 200 m (Boltovskoy, 1999; Wassmann et al., 2000).

Acantharians are common in tropical/subtropical oceans and are generally restricted to the illuminated upper layer (Caron and Swanberg, 1990; Boltovskoy, 1999). Algal symbionts are found nearly in all acantharian species, including all members of the family Acanthometridae (Febvre et al., 2000). Acantharian densities ranged from 4 to 7 cells l^{-1} in the equatorial Atlantic and North Pacific central gyre and up to 30 cells l^{-1} in the more eutrophic southeastern Atlantic, although lower (<1 cell l^{-1}) or higher (50 cells l^{-1}) densities have been reported (Febvre et al., 2000; Boltovskoy, 1999; Stoecker et al., 2009). The values obtained in this study for the abundance of Acanthometridae are found at the lower limits of this range, probably due to the oligotrophy of the Aegean Sea, particularly in its southeastern part. In the latter, acantharians showed densities as those recorded (0.3 cell l^{-1}) in the surface waters of the highly oligotrophic Levantine Sea (Abboud-Abi, 2002). It has been shown that acantharians consumed mostly tintinnids in the surface waters of the oligotrophic Sargasso Sea (Swanberg, 1979), and this seems also to be verified in the Aegean Sea.

The polycystine radiolarians Thalassicollidae, Thalassosphaeridae, and Sphaerozoidae belong to Collodaria, an Order lacking a skeleton, or provided only with scattered spicules (Boltovskoy, 1999, Anderson et al., 2000). The family Sphaerozoidae includes colonial species that form colonial gelatinous masses (Swanberg, 1979). Due to their fragility, members of this Order preserve poorly or not at all in net plankton samples, and information on their distribution is limited (Ignatiades et al., 1995; Swanberg, 1979; Dennett et al., 2002). It seems that they have tropical/subtropical distribution ranges in the three major oceans (Dennett et al., 2002; Swanberg, 1979; Yuasa et al., 2005; Rogers and De Deckker, 2007). Collodaria (for example, *Physematium, Sphaerozoum, Collozoum* and *Thalassicolla*) have commonly been found in oligotrophic regions of the Atlantic, particularly in the Sargasso and Caribbean Seas (Swanberg, 1983; Swanberg and Anderson, 1985). In the Mediterranean, Collodaria have been studied many years ago (Swanberg, 1979; Haeckel, 1887), but have received little attention until recently (Abboud-Abi, 2002).

The densities of the shelled polycystines are typically around 0.3 to 1 cells l^{-1}, but values exceeding 50 cells l^{-1} have been recorded in some productive areas (Boltovskoy, 1999; Caron and Swanberg, 1990). In the eastern Cretan Sea, polycystines exhibited high densities (76 cells l^{-1}), whereas their majority (86%) belonged to Collodaria including colonial gelatinous polycystines (8 cells l^{-1}). For the latter, underwater video recordings revealed densities up to 13 cells l^{-1} in the central North Pacific (Dennett et al., 2002). In the oligotrophic Levantine Sea (north Libanos), the densities of polycystines in the plankton net tow samples (55 μm mesh size) were usually lower than 1 cell l^{-1} during spring to early autumn, and maximum values (up to 6 cells l^{-1}) were due to proliferation of one or two species belonging to colonial Collodaria, *Sphaerozoum* spp. and *Collozoum* sp. (Abboud-Abi, 2002). Similarly, the densities of the polycystines encountered in this study for the samples collected by Niskin bottles were around 1 cell l^{-1} in most of the stations.

The enhanced abundance of polycystines in the stations (Group A) of the eastern Cretan Sea (northeastern of the city Helaklion), may be attributed to the multiscaled circulation pattern and the complex hydrographic structure prevailing in the Cretan Sea, which links with the NW Levantine Sea and the SE Ionian Sea. Within this basin, circulation is dominated by a mesoscale dipole; an anticyclone in the West and a cyclone in the East (Georgopoulos et al., 2000). The intrusion (at intermediate depths of 200 to 700 m) of the Trans Mediterranean Water (TMW) from the east has important implications to this system, because it is a water mass that is older and comparatively richer in nutrients (Souvermetzoglou et al., 1999). The persistent mesoscale dipole circulation can act as a pump, upwelling TMW into the euphotic zone and transporting it southward up to the shelf. This creates a persistent water flow (jet) in a SE and ESE direction that inhibits offshore transport of material (Tselepides et al., 2000). As a result, an offshore increase in oligotrophy has been observed (along a transect that extended north of the city Heraklion) from a coastal ecosystem with consistently higher standing stock and productivity rates, to a more stressed oligotrophic pelagic environment (Psarra et al., 2000).

Plankton studies on shelled polycystines have shown that in the surface oceanic waters, nassellarians usually exceed spumellarians in abundance (Boltovskoy et al., 1993), although slightly in the oligotrophic southern Sargasso Sea (Spaw, 1979). This was true in this study for the samples of the central Aegean Sea. Actinommidae dominated the shelled spumellarians in the Aegean Sea, as in the Sargasso Sea (Spaw, 1979). This family has an important biostratigraphic role with species that are included in radiolarian upwelling assemblages (Benson, 1966; Nigrini, 1991). In the shelf of the eastern Cretan Sea, Actinommidae exhibited an important abundance of around 10 cells l^{-1}. It seems that in the warm stratified waters of the Cretan Sea, spumellarians (S) respond more than nassellarians (N) to the weak upwelling systems, resulting in high S/N ratios. Benson (1966) found in the sediments of the Gulf of California that the S/N ratios increased under coastal upwelling conditions, while it decreased (<1) under open oceanic upwelling conditions resulting in high diatom production.

All colonial polycystines, and most of the solitary ones, possess photosynthetic symbionts that are commonly dinoflagellates (Swanberg, 1983; Stoecker et al., 2009; Caron et al., 1995). The exploitation of a symbiotic relationship may account in part for the high abundance of the polycystines observed in the eastern Cretan Sea. However, their predator activity seems to also play a considerable role, particularly that of the families Thalassosphaeridae and Sphaerozoidae (Swanberg, 1983; Swanberg et al., 1986). In *Physematium muelleri* (Thalassosphaeridae), a highly opportunistic predator collected in the Caribbean and Sargasso Seas, the most common large prey identified were silicoflagellates (42%) or dinoflagellates (63%), respectively, as well as tintinnids and copepods (Swanberg et al., 1986).

Dinoflagellates showed increased densities in the southeastern than central Aegean Sea, as it is characteristic in the oligotrophic waters during summer (Krom et al., 1991; Jyothibabu et al., 2003; Stoecker et al., 2009). The enhanced abundance of polycystines in the eastern Cretan Sea, and particularly that of collodarians, coincided with the highest densities of dinoflagellates observed in this study. Unexpectedly high levels of phytoplankton, dominated by dinoflagellates, have been recorded during the summer of the same year (July to August 1994) in stratified waters at the shelf of the Cretan Sea (north of Heraklion), attributed to the TMW intrusion (Psarra et al., 2000). It seems that an enrichment in nutrients of the surface waters under vertical stability resulted in higher densities of the dinoflagellates that belonged to the families Ceratiaceae and Prorocentraceae, in the stations located in the Cretan Sea (Group A) compared to the other in the southeastern Aegean Sea (Group B). Furthermore, Silva et al. (2009) inferred that dinoflagellates dominated phytoplankton under persistent and weak upwelling events in stratified waters.

Dinophysiaceae includes species that carry out required phototrophy, by sequestering plastids or multiple organelles from their algal prey, mainly cryptophytes, or indirectly from their ciliate prey *Myrionecta rubra,* and form blooms usually associated with upwelling mixing events with nutrient delivery (Stoecker et al., 2009). This family did not follow the increase of Ceratiaceae and Prorocentraceae in the shelf of the eastern Cretan Sea, probably due to the stability of water column during summer. Ciliates have shown close relationships to pico- or nanoplankton abundance in the Aegean Sea (Kormas et al., 1998; Mihalatou and Moustaka-Gouni, 2002). Tintinnidae and Dinophysiaceae showed a lower abundance in the stations of Group A than B in the southeastern Aegean Sea. This might be attributed to a relatively reduced availability of their favorable prey, but competitor interactions or predator activity should not be disregarded.

Polycystines, particularly Collodaria, dominated sarcodines and overwhelmed tintinnids in the eastern Cretan Sea. The density of Globigerinidae and Acanthometridae was similar between the two groups of stations in the southeastern Aegean Sea, although they consume a wide range of prey almost similar to that of polycystines. Thus, competitive interactions may also play an important role in the enhanced abundance of polycystines in the eastern Cretan Sea. Furthermore, in polycystines, there is evidence that some dinoflagellate symbionts produce sterols, which could render the host less vulnerable to predators or diseases (Stoecker et al., 2009). It seems that mixotrophy gives polycystines a competitive advantage over strict heterotrophic or other mixotrophic animal-like protists.

Polycystines appear to play a diverse and important role in the microbial loop of the eastern Cretan Sea. It has been shown that polycystines constitute a significant fraction of the total primary production in the surface waters of certain oligotrophic regions, even the colonial Collodaria that are usually ignored (Caron et al., 1995; Dennett et al., 2002). Present results indicate that the mixotrophic polycystines may be associated with the unexpectedly high levels of primary production recorded in the coast of the Cretan Sea during July to August 1994 (north of Heraklion) (Psarra et al., 2000). Furthermore, a significant contribution of polycystines in the total microbial biomass and the vertical flux of biogenic matter from the surface waters have also been suggested, particularly as far as the colonial Collodaria is concerned (Stoecker et al., 2009; Caron et al., 1995; Michaels et al., 1995). Thus, a participation of polycystines in the bentho-pelagic coupling detected (Duineveld et al., 2000) in the Cretan Sea (along a transect north of Heraklion) could be expected. In addition, during the surveys of this study, sponges (Porifera) were collected concomitantly from the seabed, and the demosponge *Spongia officinalis* was numerous at the eastern Cretan Sea and scarce in all the other stations (Castritsi-Catharios et al., 2011).

Present results showed that polycystine radiolarians constitute an important component of the microplankton in the surface waters of the Aegean Sea, particularly in the Cretan Sea. It is concluded that the hydrographic features and the episodic pulses of nutrients in the surface waters at the coast of the eastern Cretan Sea, due to the intrusion of the TMW, its upwelling, the SE direction of the mesoscale jet and the eastern eddy, could affect the structure of the microplanktic community in the highly oligotrophic southeastern Aegean Sea. It is proposed that the assemblage of the radiolarian families belonging to Collodaria, Spumellaria (S) and Nassellaria (N), with a high S/N ratio in the surface layer could be used as biological proxy of weak upwelling systems in the warm stratified waters of the oligotrophic eastern Mediterranean. Venec-Peyre and Caulet (2000) inferred that interpretation of planktic foraminiferal assemblages in terms of changes in the intensity of productivity and/or upwelling conditions is not easy as in the case for radiolarian assemblages.

The potential contribution of Collodaria in the oligotrophic ecosystems of the eastern Mediterranean should not be disregarded in biogeographical, biogeochemical and bioproductivity studies. Further research on the taxonomic composition, abundance and biomass of microplankton, especially in marine regions where they have been hitherto ignored, such as the

Aegean Sea, could assess the larval fish recruitment. In addition, it could improve the knowledge concerning the abundance and distribution of certain benthic organisms like sponges, since microplankton may be included among their main food resources (Ribes et al., 1999), apart its potential role in the vertical fluxes.

REFERENCES

Abboud-Abi SM (2002). Annual cycle of the microzooplankton communities in the waters surrounding the Palm Island Nature Reserve (north Lebanon), with special attention to tintinnids. Medit. Mar. Sci., 3/2: 55-76.

Anderson OR, Nigrini C, Boltovskoy D, Takahashi K, Swanberg NR (2000). Class Polycystinea. In: An Illustrated Guide to the Protozoa, JJ. Lee et al (Eds.), Second Edition. Society of Protozoologists, Lawrence, Kansas, 2: 994-1022.

Benson RN (1966). Recent *Radiolaria* from the Gulf of California. PhD dissertation, University of Minnesota, Minneapolis.

Boltovskoy D, Alders VA, Abelmann A, (1993). Annual flux of Radiolaria and other shelled plankters in the eastern equatorial Atlantic at 853 m: seasonal variations and polycystine species specific responses. Deep-Sea Res. I, 40: 1863-1895.

Boltovskoy D (1999). Zooplankton *of the* South Atlantic Ocean. Backhuys Publishers, Leiden, The Netherlands, pp. 1706.

Brands SJ (2000). (comp.) (1989-2010). Systema Naturae. The Taxonomicon. Universal Taxonomic Services, Zwaag, The Netherlands. [http://sn2000.taxonomy.nl/]. Access date: April 10, 2010.

Calbet A, Landry MR (2004). Phytoplankton growth, microzooplankton grazing, and carbon cycling in marine systems. Limnol. Oceanogr., 49: 51-57.

Caron DA, Michaels AF, Swanberg NR, Howse A (1995). Primary productivity by symbiont-bearing planktonic sarcodines (Acantharia, Radiolaria, Foraminifera) in surface waters near Bermuda. J. Plankton Res., 17: 103-129.

Caron DA, Swanberg NR (1990). The ecology of planktonic sarcodines. Rev. Aquat. Sci., 3: 147-180.

Castritsi-Catharios J, Miliou H, Kapiris K, Kefalas E (2011). Recovery of the commercial sponges in the central and southeastern Aegean Sea (NE Mediterranean) after an outbreak of sponge disease. Med. Mar. Sci., 12(1): 5-20.

Clarke KR, Warwick RM (1994). Change in Marine Communities: An Approach to Statistical Analysis and Interpretation. Plymouth Marine Lab. U.K., pp. 172.

Dennett MR, Caron DA, Michaels AF, Gallager SM, Davis CS (2002). Video plankton recorder reveals high abundances of colonial Radiolaria in surface waters of the central North Pacific. J. Plankton Res., 24: 797-805.

Dennett MR, Caron DA, Murzov SA, Polikarpov IG, Gavrilova NA, Georgieva LV, Kuzmenko LV (1999). Abundance and biomass of nano- and microplankton during the 1995 Northeast Monsoon and Spring Intermonsoon in the Arabian Sea. Deep-Sea Res. II, 46: 1691-1717.

Dolan JR, Claustre H, Carlotti F, Plounevez S, Moutin T (2002). Microzooplankton diversity: relationships of tintinnid ciliates with resources, competitors and predators from the Atlantic Coast of Morocco to the Eastern Mediterranean. Deep-Sea Res. I, 49: 1217-1232.

Dolan JR, Gaggegos CL (2001). Estuarine diversity of tintinnids (planktonic ciliates). J. Plankton Res., 23: 1009-1027.

Duineveld GCA, Tselepides A, Witbaard R, Bak RPM, Berghuis EM, Nieuwland G, Van Der Weele J, Kok A (2000). Benthic-pelagic coupling in the oligotrophic Cretan Sea. Prog. Ocean., 46: 457-480.

Febvre C, Febvre J, Michaels A (2000). Acantharia Haeckel, 1881. In: An Illustrated Guide to the Protozoa, JJ. Lee et al (Eds.), Second Edition. Society of Protozoologists, Lawrence, Kansas, 2: 783-803.

Figueiredo de GM, Nash RDM, Montagnes DJS (2005). The role of the generally unrecognised microprey source as food for larval fish in the Irish Sea. Mar. Biol., 148: 395-404.

Georgopoulos D, Chronis G, Zervakis V, Lykousis V, Poulos S, Iona A (2000). Hydrology and circulation in the Southern Cretan Sea during the CINCS experiment (May 1994-September 1995). Prog Ocean., 46: 89-112.

Gotsis-Skretas O, Pagou K, Moraitou-Apostolopoulou M, Ignatiades L (1999). Seasonal horizontal and vertical variability in primary production and standing stocks of phytoplankton and zooplankton in the Cretan Sea and the Straits of the Cretan Arc (March 1994-Junuary 1995). Prog. Ocean., 44: 625-649.

Haeckel E (1887). Report on the Radiolaria collected by H.M.S. Challenger during the Years 1873-76. Report on the Scientific Results on the Voyage of H.M.S. Challenger during Zool. Edinburg, 18: 1873-1876.

Hausmann K, Hülsmann N, Polianski I, Schade S, Weitere M (2002). Composition of benthic protozoan communities along a depth transect in the eastern Mediterranean Sea. Deep-Sea Res. *I*, 49: 1959-1970.

Hemleben C, Spindler M, Anderson OR (1989). Modern planktonic foraminifera. Springer-Verlag, Berlin, pp. 363.

Heussner S, Monaco A (1995). The MTP sediment trap experiments: highlights on particle fluxes through the eutrophic and oligotrophic systems of the Mediterranean Sea. In: Second workshop of the Mediterranean targeted project, P. Wassmann, A. Tselepides (Eds.), extended abstracts. Greece, Heraklion, pp. 233–239.

Hunter JR (1981). Feeding ecology and predation of marine fish larvae. In: Marine fish larvae. Morphology, ecology and relation to fisheries, R. Lasker (Ed.), Seattle, University of Washington Press, pp. 33-77.

Ignatiades L (1998). The productive and optical status of the oligotrophic waters of Southern Aegean Sea (Cretan Sea). Eastern Mediterranean. J. Plankton Res., 20: 985-995.

Ignatiades L, Geogopoulos D, Karydis M (1995). Description of the phytoplanktonic community of the oligotrophic waters of the SE Aegean Sea (Mediterranean). Mar. Ecol., 16: 13–26.

Ignatiades L, Psarra S, Zervakis V, Pagou K, Souvermezoglou E, Assimakopoulou G, Gotsis-Skretas O (2002). Phytoplankton size-based dynamics in the Aegean Sea (Eastern Mediterranean). J. Mar. Syst., 36: 11-2823.

Jyothibabu R, Madhu NV, Maheswaran PA, Nair KKC, Venugopal P, Balasubramanian T (2003). Dominance of dinoflagellates in microzooplankton community in the oceanic regions of the Bay of Bengal and the Andaman Sea. Curr. Sci., 84: 1247-1253.

Kormas KA, Kapiris K, Thessalou–Legaki M, Nicolaidou A (1998). Quantitative relationships between phytoplankton, bacteria and protists in an Aegean semi – enclosed embayment (Maliakos Gulf, Greece). Aquat. Microb. Ecol., 15: 255–264.

Krom MD, Kress N, Brennen S, Gordon LI (1991). Phosphorus limitation of primary productivity in the eastern Mediterranean Sea. Limnol. Ocean., 36: 424-432.

Li WKW, Zonar T, Yacobi YZ, Wood AM (1993). Ultraphytoplankton in the eastern Mediterranean Sea: towards deriving phytoplankton biomass from flow cytometric measurements of abundance, fluorescence and light scatter. Mar. Ecol. Prog. Ser., 102: 79–87.

Margalef R (1994). Through the looking glass: how marine phytoplankton appears through the microscope when graded by size and taxonomically sorted. Sci. Mar., 58: 87–101.

Michaels AF, Caron DA, Swanberg NR, Howse FA, Michaels CM (1995). Planktonic sarcodines (Acantharia, Radiolaria, Foraminifera) in surface waters near Bermuda: abundance, biomass and vertical flux. J. Plankton Res., 17: 131-163.

Mihalatou HM, Moustaka-Gouni M (2002). Pico-, nano-, microplankton abundance and primary productivity in a eutrophic coastal area of the Aegean Sea, Mediterranean. Int. Rev. Hydrobiol., 87: 439–456.

Nigrini C (1991). Composition and biostratigraphy of radiolarian assembladges from an area of upwelling (northwestern Arabian Sea, LEG 117). Proceedings of the Ocean Drilling Program, Scientific Results, 117: 89-126.

Psarra S, Tselepides A, Ignatiades L (2000). Primary productivity in the oligotrophic Cretan Sea (NE Mediterranean): seasonal and interannual variability. Prog. Ocean., 45: 187-204.

Ribes M, Coma P, Gili JM (1999). Natural diet and grazing rate of the temperate sponge *Dysidea avara* (Demospongiae, Dendroceratida)

throughout an annual cycle. Mar. Ecol. Prog. Ser., 176: 179-190.
Rogers J, De Deckker P (2007). Radiolaria as a reflection of environmental conditions in the eastern and southern sectors of the Indian Ocean: A new statistical approach. Mar. Micropal., 65: 137–162.
Silva A, Palma S, Oliveira PB, Moita MT (2009). Composition and interannual variability of phytoplankton in a coastal upwelling region (Lisbon Bay, Portugal). J. Sea Res., 62: 238-249.
Souvermetzoglou C, Krassakopoulou E, Pavlidou A (1999). Temporal variability in oxygen and nutrient concentrations in the southern Aegean Sea and the Straits of the Cretan Arc. Prog. Ocean., 44: 573-600.
Spaw JM (1979). Vertical distribution, ecology and preservation of recent polycystine radiolaria on the north Atlantic Ocean (southern Sargasso sea region). PhD dissertation, Rice University.
Stergiou KI, Christou ED, Georgopoulos D, Zenetos A, Souvermetzoglou C (1997). The Hellenic Seas: physics, chemistry, biology and fisheries. Oceanogr. Mar. Biol.- Annu. Rev., 35: 415-538.
Stoecker DK, Johnson MD, de Vargas C, Not F (2009). Acquired phototrophy in aquatic protists. Aquat. Microb. Ecol., 57: 279–310.
Swanberg NR (1979). The ecology of colonial radiolarians: their colony morphology, trophic interactions and associations, behavior, distribution, and the photosynthesis of their symbiots. PhD dissertation, Massachusetts Institute of Technology, Woods Hole Oceanogr. Institution.
Swanberg NR (1983). The trophic role of colonial radiolarian in oligotrophic oceanic environments. Limnol. Ocean., 28: 655–666.
Swanberg NR, Anderson OR (1985). The nutrition of radiolarians: Trophic activity of some solitary Spumellaria. Limnol. Ocean., 30: 646-652.
Swanberg NR, Anderson OR, Lindsey JL, Bennett P (1986). The biology of *Physematium muelleri:* trophic activity. Deep-Sea Res., 33: 913-922.
Tselepides A, Zervakis V, Polychronaki T, Danovaro R, Chronis G (2000). Distribution of nutrients and particulate organic matter in relation to the prevailing hydrographic features of the Cretan Sea (NE Mediterranean). Prog. Ocean., 46: 113-142.
Utermöhl H (1958). Zur Vervollkommung der quantitätiven Phytoplankton Methodik. Mitteilungen Internationale Vereinigung der Theoretische angewanden. Limnol., 9:1-38.
Venec-Peyre MT, Caulet JP (2000). Paleoproductivity changes in the upwelling system of Socotra (Somali Basin, NW Indian Ocean) during the last 72,000 years: evidence from biological signatures. Mar. Micropal., 40: 321-344.
Wassmann P, Ypma JE, Tselepides A (2000). Vertical flux of faecal pellets and microplankton on the shelf of the oligotrophic Cretan Sea (NE Mediterranean Sea). Prog. Ocean., 46: 241-258.
Yuasa T, Takahashi O, Honda D, Mayama S (2005). Phylogenetic analyses of the polycystine Radiolaria based on the 18s rDNA sequences of the Spumellarida and the Nassellarida. Europ. J. Protist., 41: 287–298.

Diversity of the edible fishes of the Lagos Lagoon, Nigeria and the public health concerns based on their Lead (Pb) content

F. E. Ajagbe[1], A. O. Osibona[2] and A. A. Otitoloju[1]*

[1]Department of Zoology, Faculty of Science, University of Lagos, Lagos, Nigeria.
[2]Department of Marine Sciences, Faculty of Science, University of Lagos, Lagos, Nigeria.

In this study, a total of eighteen fish species were recorded in the Lagos Lagoon, Nigeria during a one year survey. Among the animals species recorded, the following, *Galeoides decadactylus, Dentex gibbosus, Elops lacerta, Scomber japonicus, Ethmalosa fimbriata, Lutjanus agennes, Caranx senegallus, Callinectes amnicola, Paeneus notialis* and *Mytilus edulis* were considered to be abundant in the Lagos lagoon based on the frequency of occurrence in fishermen catches. The analysis of the lead content in muscles of the edible fisheries revealed that the animals accumulated measurable quantity of lead in the edible parts. The level of lead detected in most (12 out of 18 species) of the fisheries species were found to be lower than the daily allowable concentration of 2.0 µg/g standard recommended by the World Health Organisation (WHO). Fishes species with high level of lead in their edible parts include Senegal jack, *C. senegallus* (2.188 µg/g), Bobo croaker, *Pseudotolithus elongatus* (2.024 µg/g), Cassava croaker, *Pseudotolithus senegalensis* (3.157 µg/g), crayfish, *Penaeus notialis* (25.46µg/g), edible mussel, *Mytilus edulis* (17.69 µg/g) and crab, *Callinectes amnicola* (10.19 µg/g). The mean levels of lead in *Mytilus edulis, Penaeus notialis*, and *Callinectes amnicola* collected from the Lagos lagoon were about 8 to 24 times higher than the WHO daily allowable standards for lead in sea foods and are therefore unsafe for human consumption. The need for regular monitoring of the levels of metals and other contaminants in edible fisheries resources and collaboration between environmental, health and food agencies to avert human tragedies due to lead accumulation are discussed.

Key words: Bioaccumulation, lead, industrial pollution, aquatic ecosystem, biomonitoring.

INTRODUCTION

Lagos lagoon is a major water body in the Lagos metropolis. It cuts across the southern part of the metropolis, linking the Atlantic Ocean (in the west and south) and Lekki Lagoon (in the east). It is about 6354.708 sq km in area and 285 km in perimeter. The Lagos Lagoon consists of three main segments such as Lagos harbor, the metropolitan end and Epe division segments. The lagoon impacts significantly on the lives of the Lagosians enormously. It is used primarily for fishing, aquaculture, sand mining and recreation activities. It also provides a good platform for inland waterways transportation which has potentialities of reducing transportation problems in Lagos metropolis and a place of abode for the some of the indigenous fishing communities such as the Ilajes and Ijaws (Stilts housing). The Lagos lagoon is one of the four major lagoons in the Nigerian coastal system, and a significant tourist attraction centre in Lagos State.

Due to the level of primary production in the lagoon, it is usually inhabited periodically by fish species of fresh water and marine origins, in search of food and good nursery ground. It has been reported that about 34% of fishes observed in this ecosystem are of freshwater origin and the remaining being marine (Amadi, 1991). Mullets (Mugilidae), sardine and bonga (Clupeidae) were classified as permanent fishes in this habitat (Fagade and

*Corresponding author. E-mail: bayotitoloju@yahoo.com.

Olaniyan, 1974; Amadi, 1990). The fisheries productivity of the Lagos Lagoon system is put above 75 kg/ha/year and the fishermen from the Nigerian coastal lagoon and estuaries contribute more than 50% of the current domestic fish production of about 800,000 metric tons (Onefeghare, 1990). It was also reported that about 28 metric tonnes/ha/year of oysters are produced annually from this ecosystem (Afinowi, 1985). As a result of lagoon's rich fauna, especially in fishes, fishing activities are usually witnessed on a daily basis. Recently, the state government is proposing using the Lagos Lagoon for cage culture as a form of contribution to food security and research. Fish and fisheries products are generally regarded as an important part of a healthy diet. In the developing countries, fish and fisheries products apart from being a source of cheap animal protein, are widely consumed as they have high quality protein and other essential nutrients, and are low in saturated fat while containing Omega 3 fatty acids (NSPFS, 2005).

In Nigeria, most investigations of the Lagos Lagoon have focused on the sediments and benthic communities (Ajao and Fagade, 1991), the diversity and density of macrobenthic fauna in the western part of the lagoon (Oyenekan, 1975), trend of heavy metals concentrations in Lagos Lagoon (Don-Pedro et al., 2004), the food and feeding interrelationships of the fishes (Fagade and Olaniyan, 1973) and the biology of some fishes and the fisheries (Fagade and Olaniyan, 1974). There is however, a need to update most of these studies and provide current information on the diversity of the edible fishery and the effects of pollutants on the health and abundance of these organisms inhabiting the Lagos Lagoon.

By virtue of its position, the Lagos Lagoon is surrounded by the densely populated (about fifteen million people) and highly industrialized Lagos metropolis, making it a convenient dumping site for numerous industrial and domestic wastes. According to Singh et al. (1995), an estimated 10,000 m^3 of industrial effluents are discharged into the Lagos Lagoon per day. Oyewo (1998) estimated levels of heavy metals discharged into drains/canals/streams and subsequently into the Lagos Lagoon as follows: Fe – 161,718 Kg, Mn – 205,989 Kg, CO – 15,683 Kg, Zn – 7026 Kg, Cr – 5285 Kg, Pb – 2259 Kg, Ni – 6124 Kg, Cd – 538 Kg and Hg – 278 Kg per annum. These estimates also confirmed that the industries are the major source of metal contaminants in the drains, streams and lagoon, since the graded prominence of metal types was similar in the sampled effluents and aquatic systems. The continued discharge of all sorts of untreated waste materials into the lagoon threatens the state of ecological equilibrium and diversity of fisheries resources in the lagoon (Don-Pedro et al., 2004; Otitoloju et al., 2007). According to Singh et al. (1995) the use of the lagoon as a dump for waste materials has reduced annual fish production in the Lagos Lagoon by over five folds between 1970 and 1990. Therefore, there is the need for more studies to establish the present state of pollution in the Lagos Lagoon, the diversity of biological resources and the level of pollutants such as heavy metals which can be detrimental to the health of the consumers (Chindah et al., 2000). In recognition of these needs and in pursuance of the sustainable use of coastal water and the adjacent land, the Federal Government of Nigeria put in place an Action Plan on water pollution control and biological diversity conservation in the country. Internationally, collaborative efforts were also made with the West African sub-region under the Gulf of Guinea Large Marine Ecosystem (GOGLME) Project aimed at monitoring coastal water in terms of pollution and biological diversity conservation. On the basis of the above-mentioned, this present study was carried out to investigate:

(i) The diversity of edible fisheries resources of the Lagos Lagoon.
(ii) The level of pollution-indicator heavy metal (Lead) in the edible parts (muscles) of the fisheries resources collected from different parts Lagos Lagoon, as well as, in sediment and water samples.
(iii) The potential public health risk based on the level of metals in the edible fisheries resources, derived bioaccumulation factors and internationally acceptable maximum allowable limits/standards of metals in food.

MATERIALS AND METHODS

Description of the study area and sampling design

The Lagos Lagoon complex stretches from Cotonou in the Republic of Benin and extends to the fringes of the Niger Delta in Nigeria along its 257 km course (Hill and Webb, 1958). The main body of the lagoon complex which serve as the study site is located between longitude 3° 22' and 3° 40'E and 6°17' and 6°28' N (Figure 1).

Fifteen different stations were sampled on the Lagos Lagoon which covered Oworoshokoki – Ikorodu, Osumare, Ibese, Oreta, Aabo, Tin Can, Apapa, Iganmu, Iddo, Lagos mainland, Okobaba, Okobaba trans Lagos mainland, UNILAG Lagoon front. The fifteen stations were marked with the aid of a Global Geopositioning System (GPS) and recorded as coordinates. Sampling operations were carried out on open fibre-glass boat motorized with a 75 hp Yahama outboard engine.

Collection of edible fishes resources in the Lagos Lagoon

Edible fishes of the Lagos Lagoon were procured from the fishermen that were fishing on the lagoon on a monthly basis from October 2009 to September 2010 at fifteen different stations. The sampling gears used for fishing by the fishermen include basket traps, baited long lines and set gill net. A total of at least 12 samples of each fish species were collected for the analysis. The samples were transported in an ice chest to University of Lagos, Ecotoxicology Laboratory for proper identification and examination.

Identification of fish species

In the laboratory each specimen was identified (using Field Guide to the Commercial Marine Resources of the Gulf of Guinea, Food

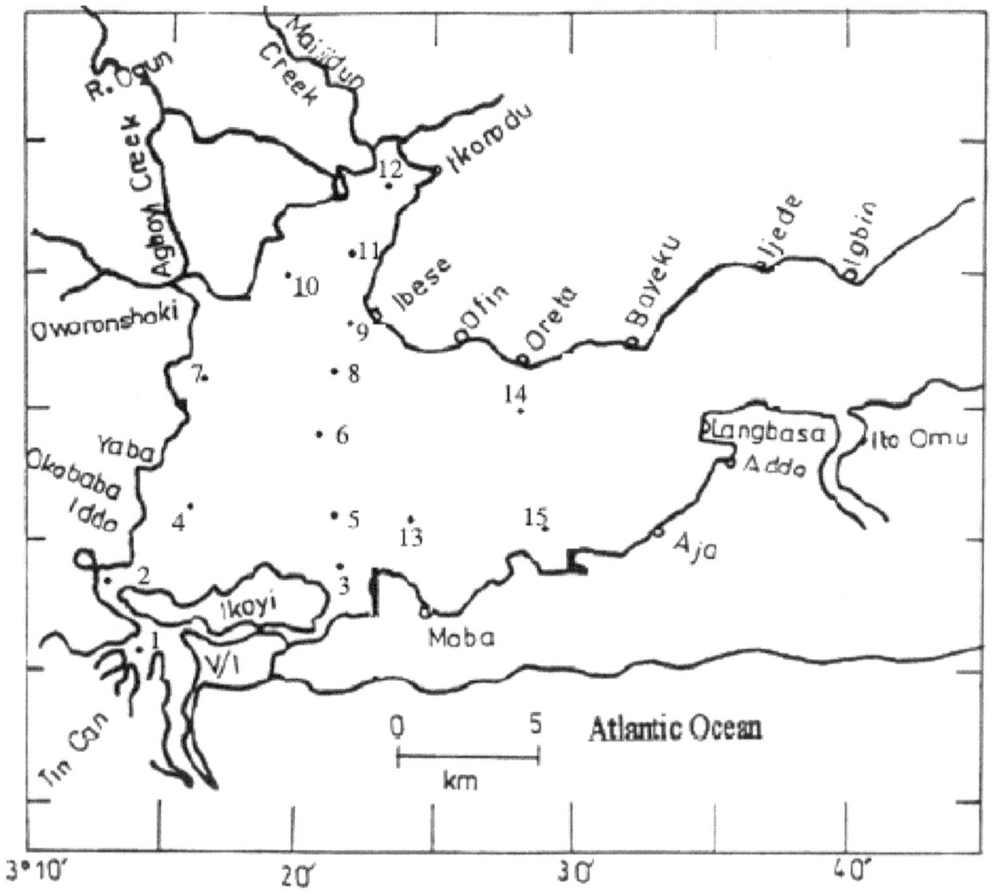

Figure 1. Lagos lagoon showing the sampling stations.

and Agriculture Organisation Species Identification Sheets for Fishery Purposes (FAO, 1990).

Chemical analyses of sample (water, sediment and animals) for lead (Pb) content

Digestion of samples

Water samples were filtered and digested using standard digestion method procedure (APHA/AWWA/WPCF, 1995). Sediment samples were normalised for particle size and digested following method provided by Ageiman and Chau (1976). The animal muscles were filleted and homogenized. A portion (10 g wet weight basis) from the homogenate was digested using the method described by FAO/SIDA (1986). All digestates and extracts obtained were filtered through Whatman No.1 filter paper and made up to the 25cc mark in volumetric flasks.

Atomic absorption spectrophotometry (AAS)

The Pb concentration in each digested sample was determined by comparing their absorbance's with those of standards (solutions of known Pb concentration) using an Alpha-cathodeon AAS. For data quality assurance, 10 samples of each blanks and muscle of fish spiked with known amount of analytical standard Pb solution were analyzed and recoveries calculated. The recoveries were 101.2 to 102.3% for blanks and 100 to 101.8% for fish muscle. In addition, diluted factory prepared AAS standard solutions were run as samples for accuracy check after every five measurements.

All digested samples were analyzed three times to obtain a mean value. The Pb concentration in each digested sample was determined by comparing their absorbance's with those of standards (solutions of known Pb concentration) using an Alpha-cathodeon AAS. Analytical blanks were run in the same way as the samples and concentrations prepared in the same acid matrix. For data quality assurance, the results were checked by analyzing standard reference material (SRM, TIL1 – Tilapia muscle) obtained from the National Reference Lab and factory prepared AAS standard Pb solutions. The results indicated good agreement between certified and the analytical values (Table 1). The recoveries of the certified standard reference material and AAS solution was 103 and 98% respectively.

Statistical analysis

Analysis of variance (ANOVA) was used to compare means of Pb concentrations in the fishes species of the lagoon using SPSS 17. Further analysis was carried out only where there was a significant difference at the 5% ($P<0.05$) level of significance (taken as minimum requirement). Further analyses were conducted where necessary by comparing pairs of mean (that is, individual treatment

Table 1. Concentration of metals in standard reference materials (TIL1 – Tilapia muscle) and standard AAS Pb solution.

Parameter	TIL1 (mg/kg)	AAS solution (µg/L)
Certified	0.066	10.3
SE	0.006	0.12
Observed*	0.068	10.1
SE	0.007	0.1
Recovery	103	98

*The value is the average of three determinations.

Table 2. Concentration of Pb in sediment and water samples collected from the different sampling stations.

Stations	Concentration in sediment (µg/g)	Concentration in water (µg/L)
1	0.005	0.012
2	0.001	0.008
3	0.007	0.01
4	0.01	0.008
5	0.001	0.025
6	0.001	0.05
7	0.001	0.025
8	0.001	0.02
9	0.008	0.03
10	0.002	0.015
11	0.05	0.035
12	0.003	0.026
13	0.002	0.02
14	0.001	0.018
15	0.005	0.015
Mean	0.0065	0.0211
*SD	0.0124	0.0113

*SD – Standard Deviation.

means with controls or themselves based on least significant differences (LSD) at 0.05 and 0.001 levels of significance.

Bioconcentration factor (BCF)

Bioconcentration factor (BCF) was estimated as the ratio of the concentration of the metal in animal tissue to the concentration of metal in the lagoon water.

$$\text{Bioconcentration factor (BCF)} = \frac{\text{Concentration in animal tissue}}{\text{Concentration in water}}$$

Bioaccumulation factor (BAF)

Bioaccumulation factor (BAF) was also estimated as the ratio of the concentration of the metal in animal tissue to the concentration of metal in the lagoon sediment.

$$\text{Bioaccumulation factor (BAF)} = \frac{\text{Concentration in animal tissue}}{\text{Concentration in sediment}}$$

RESULTS

Lead concentrations in lagoon sediment and water samples

The results of the Pb concentrations in sediment and water samples collected from the sampling stations are provided in Table 2. The results show that the level of Pb in the environmental media range from 0.001 to 0.05 µg/g for sediment and 0.008 to 0.035 µg/L for water samples. The mean concentration of Pb in the sediment samples was 0.0065 µg/g while the mean concentration in the water was 0.0211 µg/L.

Fisheries resources of the Lagos Lagoon

The results of the survey of the edible fisheries of the Lagos Lagoon are presented Table 3. The survey

Table 3. Fisheries resources of the Lagos lagoon during the survey.

S/N	Scientific name of fish sample	Frequency of occurrence in fishermen catch	Common name of fish sample
1	*Galeoides decadactylus*	+++	Threadfins
2	*Pomadasys jubelini*	++	Sompat Grunter
3	*Dentex gibbosus*	++	Pink dentex
4	*Elops lacerta*	+++	Ten ponders/West African Ladyfish
5	*Lisa falcipinnis*	+++	Sicklefin mullet
6	*Scomber japonicas*	++	Chub mackerel
7	*Ethmalosa fimbriata*	+++	Bonga shad
8	*Lutjanus agennes*	+++	African red snapper
9	*Caranx senegallus*	++	Senegal jack
10	*Sphyraena barracuda*	+	Great barracuda
11	*Pseudotolithus elongates*	++	Bobo Croaker
12	*Pseudolithus senegalensis*	++	Cassava Croaker
13	*Mormyrus rume*	+	Mormyrids
14	*Synodontus clarias*	+	Mandi/Upside-down catfishes
15	*Tilapia guineensis*	+++	Guinean Tilapia
16	*Mytilus edulis*	++	Edible Mussels
17	*Penaeus notialis*	++	Crayfish
18	*Callinectes amnicola*	+++	Crab

+++: Very high, ++: High, +: Seldom.

revealed fifteen fish species belonging to the class Actinopterygii, Crab (*Callinectes* species), class Malacostraca, Crayfish (*Penaeus notialis)*, Class Malascotraca and Mussel (*Edulis mytilus*), Class Palecypoda (Bivalva).

Bioaccumulation studies

The results of the level of Pb accumulated in the edible part (muscle) of the fish from the Lagos Lagoon are presented in Table 4. The results showed that the edible fish contained measurable quantities of Pb in their edible parts ranging from 0.046 to 25.46 μg/g. The highest concentration (25.46 μg/g) of Pb was found in the crayfish, *P. notialis* while the lowest concentration (0.046 μg/g) was detected in threadfins, *Galeoides decadactylus*.

On the basis of the computed bioaccumulation factor with reference to the concentration of lead in the sediment, the crab, *Callinectes amnicola* was found to have the highest ability to accumulate Pb in the muscle followed by *Lutjanus agennes* and then *Pseudotolithus senegalensis*. Comparisons of the derived concentration factors showed that the bioaccumulation factors are about three to four folds higher than the bioconcentration factors, which were derived based on Pb concentration in the water samples (Table 4). Statistical comparisons by ANOVA of the metal level in the fish samples and sediment showed that the levels of Pb detected in the fish species were significantly ($P < 0.05$) higher than the concentrations detected in the sediment or water samples.

DISCUSSION

In this study, fifteen edible fish species, one species each of crab, crayfish and mussel were recorded during the period of study. Among the species recorded the following *G. decadactylus, Dentex gibbosus, Elops lacerta, Scomber japonicus, Ethmalosa fimbriata, Lutjanus agennes, Caranx senegallus, Callinectes amnicola, P. notialis and Mytilus edulis* were considered to be of commercial important in Nigeria. Nine of these edibles species are of brackish water origin, while four of them are found in both freshwater and brackish water environment, and only two of the species are basically freshwater species. The results of the fisheries diversity of the Lagos lagoon is in agreement with the study by Fagade and Olaniyan (1973) who reported on the food and feeding habits of commonly found fish species of the Lagos Lagoon. Out of the twenty six recorded fish species in the study, twelve of the species such as *G. decadactylus, Pomadasys jubelini, Elops lacerta, Lisa falcipinnis, Scomber japonicas, Ethmalosa fimbriata, Lutjanus agennes, Caranx senegallus, Sphyraena barracuda, Pseudotolithus elongates, P. senegalensis,* and *Tilapia guineensis* were encountered during the study. The other species that were not recorded during this study may have been missed during the survey or their abundance may have been reduced drastically due

Table 4. The mean concentration of lead (Pb) accumulated by edible fisheries resources of the Lagos Lagoon.

Fish sample	Common names of fish sample	Mean concentration of lead in fish sample (µg/g)	Mean concentration of lead in water sample (µg/L)	Mean concentration of lead in sediment sample (µg/g)	BCF	BAF
Galeoides decadactylus	Threadfins	0.046±0.02	0.02±0.011	0.0065±0.012	2.3	7.08
Pomadasys jubelini	Sompat Grunter	0.054±0.03	0.02±0.011	0.0065±0.012	2.7	8.31
Dentex gibbosus	Pink dentex	0.068±0.01	0.02±0.011	0.0065±0.012	3.4	10.46
Elops lacerta	Ten ponders/West African Ladyfish	0.074±0.02	0.02±0.011	0.0065±0.012	3.7	11.39
Lisa falcipinnis	Sicklefin mullet	0.079±0.01	0.02±0.011	0.0065±0.012	3.95	12.15
Scomber japonicas	Chub mackerel	0.852±0.15	0.02±0.011	0.0065±0.012	42.6	131.08
Ethmalosa fimbriata	Bonga shad	0.894±0.12	0.02±0.011	0.0065±0.012	44.7	137.54
Lutjanus agennes	African red snapper	1.616±0.43	0.02±0.011	0.0065±0.012	80.8	248.62
**Caranx senegallus*	Senegal jack	2.188±0.65	0.02±0.011	0.0065±0.012	109.4	366.62
Sphyraena barracuda	Great barracuda	1.824±0.54	0.02±0.011	0.0065±0.012	91.2	280.62
**Pseudotolithus elongates*	Bobo Croaker	2.024±0.64	0.02±0.011	0.0065±0.012	101.2	311.39
**Pseudotolithus senegalensis*	Cassava Croaker	3.157±0.74	0.02±0.011	0.0065±0.012	157.9	485.69
Mormyrus rume	Mormyrids	0.938±0.15	0.02±0.011	0.0065±0.012	46.9	144.31
Synodontus clarias	Mandi/Upside-down catfishes	1.973±0.66	0.02±0.011	0.0065±0.012	98.65	303.54
Tilapia guineensis	Guinean Tilapia	1.412±0.55	0.02±0.011	0.0065±0.012	70.6	217.23
**Mytilus edulis*	Edible Mussels	17.69±2.54	0.02±0.011	0.0065±0.012	884.5	2721.54
**Penaeus notialis*	Crayfish	25.46±3.46	0.02±0.011	0.0065±0.012	1273	3916.92
**Callinectes amnicola*	Crab	10.19±2.25	0.02±0.011	0.0065±0.012	509.5	1567.69

*Lead concentration detected in fishes species higher than the World Health Organisation (WHO) daily allowable concentration of 2.0 µg/g in sea foods.

to the problem of pollution as a result of the discharge of untreated waste materials into the lagoon.

Out of the various fish species obtained during the survey, *Callinectes amnicola* was found to have the highest ability to accumulate Pb in its body tissues followed by *L. agennes* and *P. senegalensis* in that order. This observation is in agreement with the findings of Otitoloju and Don-Pedro (2002) who reported the high level of accumulation of Pb by some benthic animals of the Lagos Lagoon. Baron (1995) reported that bioaccumulation occurs as a result of competing rates of chemical uptake and elimination, the latter comprising of biotransformation and excretory process. The biotransformation of metals in animal tissues involve the synthesis of low-molecular weight proteins e.g. metallothionien in the animal tissues which then form complexes with the metals, where the resultant metal-protein complexes formed are not water soluble, they become sequestered in lipids or fatty tissues resulting in metal bioaccumulation (Langston and Zhou, 1987; George, 1989). The practical implications of the results of the findings of this study is that in the wild, animals which occupy higher trophic level including man, feeding on organisms which have accumulated the environmental metal, Pb in their body tissues, may in turn acquire a greater concentration of the metal in their bodies. The concentration of toxin along the food chain is known as biomagnification. The transfer of toxins along the food chain will be particularly hazardous to man when they consume edible fishes from the Lagos Lagoon which have high concentration of the environmental lead (Pb) in their body tissues.

The analysis of the lead content in muscles of the edible fisheries revealed that the animals accumulated measurable quantity of Pb in the edible parts. The level of Pb detected in most (12 out of 18 species) of the fisheries species were found to be lower than the daily allowable concentration of 2.0 µg/g standard recommended by the FAO/WHO (1989) and WHO (1997). On this basis, most of the fisheries species analyzed in this study can be considered safe for consumption with regards to lead content since the detected levels of Pb were below the recommended standard. It is however important to note that the 2.0 µg/g standard is based on the FAO recommended annual per capita fish consumption rate of 13 kg. However, bearing in mind that consumption of fish among members of coastal / urban areas like Lagos is usually several folds higher than that of water "deprived" / rural populations (Amao et al., 2006). According to FAO (1989), the annual per capita fish consumption in African countries including Nigeria is usually above 20 kg, especially among members of coastal communities. Therefore, exposure of people from the Lagos area to lead (Pb) arising from consumption of fishes that were captured from the lagoon with high lead content is likely to be higher than the daily allowable levels. Therefore, these fish species that are currently contain lower level of Pb in the muscle may still pose a public health risk due to consumption of higher quantity of fish than the recommended annual per capita fish consumption rate of 13 kg in the area.

Other fish species encountered during the survey which contained higher level of Pb than the allowable 2.0 µg/g are *C. senegallus* (Senegal jack) which had a mean Pb level of 2.188 µg/g, *P. elongatus* (Bobo croaker) 2.024 µg/g; *P. senegalensis* (Cassava croaker) 3.157 µg/g; *P. notialis* (crayfish) 25.46 µg/g; *M. edulis* (edible mussel) 17.69 µg/g and *C. amnicola* (crab) 10.19 µg/g. The mean levels of Pb in *M. edilis, P. notialis* and *C. amnicola* were about 8 – 24 times higher than the recommended standard for Pb in sea foods. Therefore, consumption of these species collected from the Lagos lagoon might pose real public health risks especially among members of the fishing communities who have a large number of menu in which crabs, mussels, oysters and periwinkles constitute the dominant portion of the dishes.

Edible fishes from the Lagos Lagoon serve as a cheap source of protein forming part of the diet of the coastal populations (especially residents of riverine communities) in Nigeria. The resources of the Lagos Lagoon and coastal region of the country are being exposed to high degree of contaminants and toxicants, which when consumed may lead to conditions such as anaemia, renal toxicity, hearing loss, low Intelligent Quotient (IQ), hypertension and behavioural disorders. (Bellinger, 2004). The public health implication of consuming metal contaminated fishes from the Lagos lagoon is that organisms higher in the food chain such as man that feed on them may in turn be exposed to very high concentration of these pollutants accumulated in the animal tissues. The consumption of fisheries species with high concentration of heavy metals has been reported to have caused serious cases of human tragedies such as the Minamata disease and Itai-Itai disease tragedies (Kurdland, 1960; Varma et al., 1976). The lesson from these human tragedies is the obvious need for continuous monitoring of edible organisms for their heavy metal contents to avert such occurrences. It is however, regrettable to note that most regulators in many developing countries such as Nigeria have learnt little from these lessons and do not carry out any form of monitoring of fisheries species for their metal contents before they are allowed for human consumption. It is therefore important for environmental regulatory agencies such as Ministry of Environment and Environmental Protection Agencies (EPAs) to regularly monitor the levels of metals and other contaminants that are being discharged into the aquatic ecosystem which can in turn be accumulated by fisheries resources. The need for the Food and Drug Administration Agencies e.g. NAFDAC in Nigeria to also collaborate with the environmental agencies by monitoring edible fisheries species for their pollutant contents is also required. The results of such measurements when compared against set standards and maximum allowable limits can form the basis for allowing or disallowing the sales and consumption of sea foods based on the level of pollutant accumulated in the body tissues and therefore help in averting human tragedies such as the Minamata disease in these developing countries.

ACKNOWLEDGEMENTS

The authors are grateful to staff of the Analytical Laboratory, University of Lagos for the analysis of samples and Prof. Olusola Ogunwenmo of Babcock University for assisting in reviewing the manuscript.

REFERENCES

Afinowi MA (1985). The mangrove oyster, *Crassostrea gasar* (Adason): Cultivation and potential in the Niger Delta. In: Wuleox B. H. R. and Power C. B. (eds.) Workshop Proceedings of the Mangrove Ecosystem of the Niger Delta, University of Port Harcourt, p. 87.

Ageiman H, Chau ASY (1976). Evaluation of extraction techniques for the determination of metals in aquatic sediments. Analyst, 101: 761-767.

Ajao EA, Fagade SO (1991). A study of sediments and communities in Lagos Lagoon, Nigeria. Oils Chem. Poll., 7: 85-117.

Amadi AA (1990). A comparative ecology of estuaries in Nigeria. Hydrobiologia., 208: 27-38.

Amadi AA (1991). The coastal and marine environment of Nigeira. Aspects of ecology and management. NIOMR Techn. Pap., 76: 1-34.

Amao JO, Oluwatayo IB, Osuntope FK (2006). Economics of fish demands in Lagos state, Nigeria. J. Hum. Ecol., 19(1): 25-30.

American Public Health Association-American Water Works Association-Water Environment Federation (APHA-AWWA-WPCF) (1995). Standard methods for the examination of water and waste water (16th Edition) American Public Health Association, p. 10400.

Baron MG (1995). Bioaccumulation and Bioconcentration in aquatic Organnisms: In: Handbook Of Ecotoxicology (Hoffman DH, Rattner BA, Burton BA, Jr and Cairns GA Jr.(eds) CRC Press Inc. Lewis Publishers, London, pp. 652-662.

Bellinger DC (2004). Lead. Pediatrics, 113: 1016-1022.

Don-Pedro KN, Oyewo OE, Otitoloju AA (2004). Trend measurement of heavy metals in Lagos lagoon ecosystems, Nigeria. W. Afr. J. Appl. Ecol., 5: 103-114.

Fagade SO, Olaniyan CIO (1973). The food and feeding interrelationship of the fishes in the Lagos lagoon. J. Fish Biol., 5: 205-225.

Fagade SO, Olaniyan CIO (1974). Seasonal distribution of the fish fauna of the Lagos lagoon. Bull. De IFAN Ser. A, 36(1): 244-254.

FAO (1989). A Regional Survey of the Aquaculture Sector in the Sahara. ID 19077. Project Report, ADCP/REP/89/37.

FAO (1990). Field guide to the commercial marine resources of the gulf of Guinea. FAO species identification sheets for fishery purposes. RAFR, p. 268.

FAO/SIDA (1986). Manual of methods in aquatic environmental research, part 9. Analyses of metals and organochlorines in fish. FAO Fish/Technical Paper, p. 212.

FAO/WHO (1989). Evaluation of certain food additives and the contaminants mercury, lead and cadmium. WHO Techn. Report Ser., 505: 123.

George SG (1989). Cadmium effects on place liver xenobiotics and detoxification system dose-response. Aquat. Toxicol., 15: 303-310.

Hill MB, Webb JE (1958). The ecology of Lagos lagoon II. The topography and physical features of Lagos harbour and Lagos lagoon. Phil. Trans. Bull., 241: 319-333.

Langston WJ, Zhou M. (1987). Cadmium accumulation, distribution and metabolism and in the gastropod *Littorina littoria:* The role of metal binding proteins. J. Mar. Biol. Assoc. UK., 67: 585-601.

National Special Prorgamme for Food Security (NSPFS) (2005). Farming Nigeria's water. A compilation of the newsletter of aquaculture and inland fisheries project, 1(21): 34.

Onefeghare FA (1990). Nigerian Wetlands; An overview. In: Akpata, T.V.I. and Okali, D. U. U. (eds.). Nigerian Wetlands. UNESCO/MAB Port-Harcourt, pp. 14-26.

Otitoloju AA, Don-Pedro KN (2002). Bioaccumulation of Heavy metals (Zn, Pb, Cu and Cd) by *Tympanotonus fuscatus var radula* (L.) exposed to sublethal concentrations of the test metal compounds in laboratory bioassays. W. Afr. J. Appl. Ecol., 3: 17-29.

Otitoloju AA, Don-Pedro KN, Oyewo EO (2007). Assessment of potential ecological disruption based on heavy metal toxicity, accumulation and distribution in media of the Lagos Lagoon. Afr. J. Ecol., 45: 454-463.

Oyenekan JA (1975). A survey of Lagos lagoon benthos (with particular reference to the mollusca). M.Sc Thesis. University of Lagos, Nigeria, p. 137.

Oyewo EO (1998). Industrial sources and distribution of heavy metals in Lagos lagoon and their biological effects on estuarine animals. Ph.D Thesis. University of Lagos, p. 274.

Singh J, Hewawassam H, Moffat D (1995). Nigeria: Strategic options for redressing industrial pollution. Industry and Energy Division, West Central Africa Department, 1: 45.

WHO (1997). Recommended limit for metals in finfish. Environment health; criteria No. 70, Principles of safety and assignment of food additives and contamination in food. Technical Report Series 505, Geneva, p. 309.

Rheological and structural properties of sea cucumber *Stichopus japonicus* during different heating temperature

Lianfeng Liu, Zhaohui Zhang*, Qian Liu, Bofeng Yang, Jinfa Huang and Xin Gao

College of Food Science and Engineering, Ocean University of China, Qingdao 266003, P. R. China.

Changes in tissue structure and rheological properties of sea cucumber meat during different heating temperature were studied. Tissue structure was observed using an optical microscope and scanning electron micrograph, and its characteristic values (Wm, Dm, Rm, C*) were enumerated using image processing and analysis technique. Rheological properties (E_0, τ_1, η_1, rupture strength) and Texture profile analysis (TPA) parameters were obtained by stress-relaxation experiment and texture profile analysis experiment. The results indicated that the structure in heated samples had greatly changes comparing with the raw sample. Muscle fibers in heated samples shrank and subsequently reassembled, and the voids in muscular tissue tended to expand with heating temperature. The characteristic values (Dm and C*) gradually increased with heating temperature, however the Wm and Rm showed irregular change. Rheological properties (E_0 and rupture strength) in heated samples were smaller than those in the raw samples and expressed the largest value at 70°C; however τ_1 and η_1 gradually increased with heating temperature. Rheological properties (E, τ, η and rupture strength) are mainly correlated with the characteristic values (Dm and C*). These results confirmed that the changes in texture and rheological properties of sea cucumber meat were caused by thermal denaturation of proteins.

Key words: Sea cucumber, structure, rheological properties and thermal denaturation.

INTRODUCTION

Sea cucumber (*Stichopus japonicas)* has become important cultivated aquatic species in China and other Asian countries in recent years (Conand, 2004). In general, the body wall of raw sea cucumber is hard to masticate. However, it becomes tender after cooking. In food science, studies of sea cucumber mainly focused on the food feed, nutrition values, biological activity and structure of acid polysaccharide (Slater et al., 2009; Zamora and Jeffs, 2011; Stonik et al., 1998; Chen et al., 2011). However, few studies have considerations on the structural and rheological properties in sea cucumber meat. Previous studies on rheological and texture properties of catfish, abalone and salmon stated that factors contributing to the textural properties of meat are muscle fibers and connective tissue (Jiang et al., 2008; Gao et al., 2001; Kong et al., 2008), and there are quantitative correlations between rheological properties and structural characteristic values (Gao et al., 2002; 2003). In the case of heating treatment, tissue structure and textural properties from meat have significantly difference and the denaturation of proteins was assumed to contribute to these differences. Saito et al (2002) reported that there is a larger proportion of collagen against the total amount of protein in sea cucumber meat. There are few studies related on rheological properties of sea cucumbers after heating at different temperature. Therefore, the structural and rheological properties of heated sea cucumber meat were studied, and correlations between structural characteristic values and

*Corresponding author. E-mail: zhangzhh@ouc.edu.cn.

rheological properties were analyzed. In this paper, tissue structure was observed by an optical microscopy and scanning electron micrograph. Structural characteristic values were enumerated using image processing and analysis technique. Rheological properties, including stress-relaxation measurement, texture profile analysis and rupture strength experiment, were taken by using texture meter measurements.

MATERIAL AND METHODS

Raw sea cucumber (*S. japonicas*), harvested in Qingdao, Shandong Province, China (average weight was 101.91 ± 18.42 g) were purchased at a retail store. Cleared away their viscera, the body meat were heated at 50, 70 and 100℃ for 15 min, respectively. The upper part of the meat was used as sample for these experiments after being cooled to room temperature.

Cooking loss

Cooking loss was calculated as the percent weight reduction of the heated samples comparing with the raw samples.

Tissue structure

Preparation for optical microscopic

Raw and heated samples were cut into 5 × 5 × 5 mm^3 rectangular prism blocks, embedded in Tissue-Tek (O.C.T. compound) at -20℃, and then formed into 10 mm cubes (Niitu and Hiramoto, 1982). These blocks were trimmed and sliced into sections of 16 µm thickness using a cryostat (model CM1900; Leica Co. Ltd, USA). Then the sections were mounted on a glass slide and stained with Van Gieson staining (Kageyama and Watamabe, 1988). Light micrographs were taken with an Olympus BX51 optical microscopic.

Preparation for scanning electron microscopy (SEM)

Raw and heated samples were separated into 3 × 3 × 3 mm^3 rectangular prism blocks and fixed with glutaaldehyde phosphate buffer. The fixed samples were dehydrated with ethanol and substituted by isoamyl acetate, and then substituted samples were dried by critical point dry method and coated with gold palladium in a vacuum (Sukumar, Kohzoh, Maysuno, 2008). Observation was conducted by scanning electron microscopy (JSM-840; JEOL Co. Ltd, Japan).

Image processing and analysis technique

The morphometric analysis of muscle structure was carried out by a software package (Image Pro Plus 5.1, Media Cybernetics, Inc.) which could converse the optical micrographs of raw and heated samples to binary images. A binary image works only on two gray scales levels, black (0 to 127) and white (128 to 255). From the binary morphology, the following structural characteristic values were calculated:

Wm: width of myofibrils (µm),
Dm: distance between myofibrils (µm),
Rm: ratio of total myofibrils in full image domain (%),
Rvm: ratio of total voids in full image domain (%),
Cm: circularity of void area between myofibres.

The values of these morphometry parameters were calculated by the method using in chilled abalone meat (Gao et al., 2003), and defined $C^* = 4\pi/Cm$. The maximum value of real circle boundaries becomes one, 0.79 for square boundaries and 0.60 for equilateral triangular boundaries. The larger is the value of C^*, the more similar to a circle.

Rheological properties

Stress-relaxation measurement and analysis

Stress-relaxation experiment was carried out using a TMS-PRO Texture Analyzer (Food Technology Corporation, USA) at room temperature. Cylindrical probe, 4 mm in diameter, was used to compress the samples at a constant strain of 0.02. Figure 1 shows the typical stress-relaxation curve of sea cucumber meat.

Stress-relaxation curve was analyzed with the following progressive approximate method (Saito et al., 2002). Instantaneous modulus E_0 was calculated from the load when compression reached the maximum. The approximate equation of the stress-relaxation could be expressed as follows:

$$\sigma(t) = e_0 \sum_{i=1}^{n} E_i \exp(-t/\tau_i) \quad (1)$$

Where $\sigma(t)$ is the stress at relaxation time, e_0 is the constant strain, τ is the time, E_i is the elastic modulus of the i-th element, and τ_i is the stress-relaxation time of the i-th element. η_i is related to the viscosity of the i-th element, η_i and E_i as shown in Equation 2:

$$\tau_i = \eta_i / E_i \quad (2)$$

E_0 is defined as follows

$$E_0 = E_1 + E_2 + \ldots + E_n \quad (3)$$

Texture profile analysis (TPA)

Texture profile analysis of sea cucumber meat was done by the same apparatus used for stress-relaxation measurement. The double compression cycle test was performed according to the following testing procedure: cylindrical probe (4mm i. d.), final strain 0.50, time interval between first and second compression 5 s. The pre-speed, test-speed, and post-speed, were 0.5 mm/s, 1.0 mm/s, and 5.0 mm/s, respectively.

The following were quantified TPA parameters (Caine et al., 2003; Duan et al., 2010): Hardness (N) is peak force in the first compression cycle; springiness (*dimensionless*) is ratio of the length duration of force input during the second compression to that during the first compression; cohesiveness is the ratio of the positive force area during the second compression to that during the first compression; chewiness (N), is the product of hardness, springiness and cohesiveness.

Rupture strength experiment

Rupture strength was measured by the same texture analyzer at room temperature. The cylindrical probe (2 mm i. d.) compressed the samples at a rate of 2 mm/s. The peak of the force-time curve was regarded as the rupture strength value.

Statistical analysis

Statistical calculations were performed using the Excel Statistics

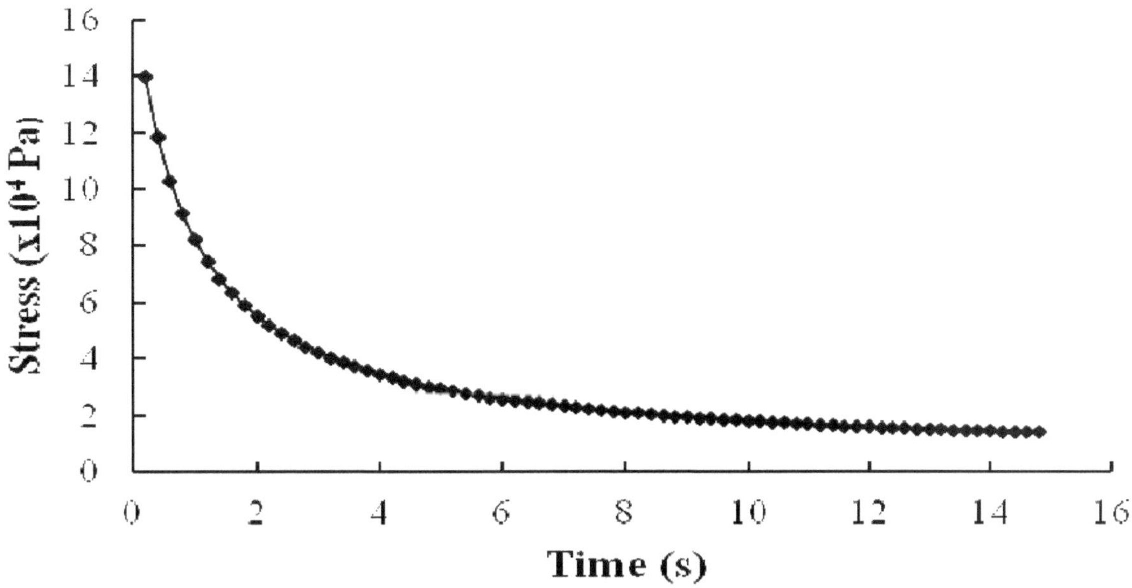

Figure 1. Typical stress-relaxation curve of sea cucumber meat.

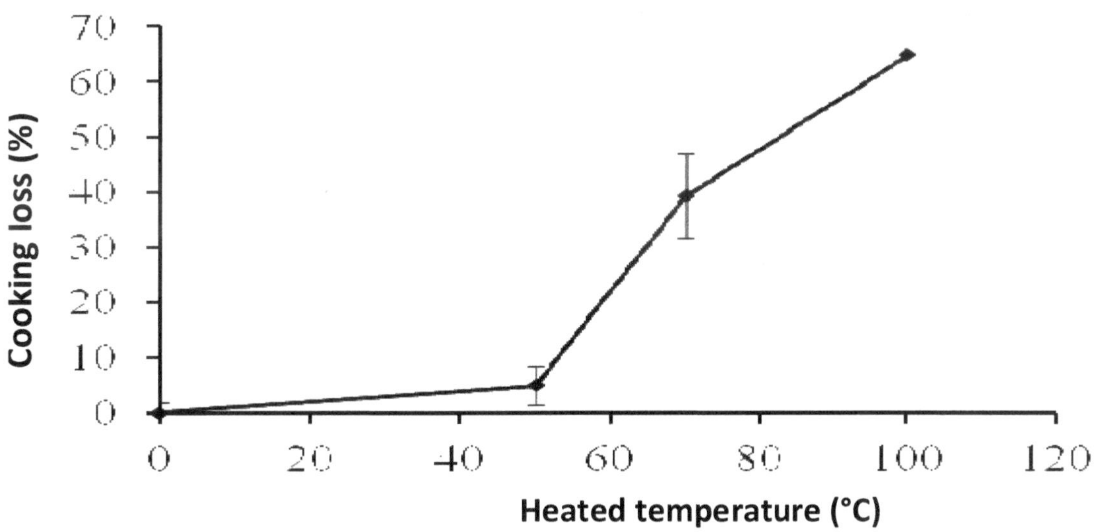

Figure 2. Cooking loss of sea cucumber meat at difference heating temperature.

software (Micro. Co.USA). Significantly ($P< 0.05$) difference property and between variable correlation were investigated with *SPSS* 17.0 statistical package. All data were expressed as means ± standard deviations.

RESULTS

Cooking loss

Figure 2 shows the cooking loss of sea cucumber at different heating temperature. The cooking loss only had lower value (4.88%) when raw sea cucumber meat was heated at 50°C. The rapidly increasing of cooking loss occurred in the range 50 to 70°C. After heating at 100°C, the cooking loss had increased to approximately 64%, and moisture content decreased from 92.87 to 79.83%. Water-soluble components, like gelatin changed from collagen, also flowed out with water during thermal treatment.

Muscle structure

Muscle structures of raw and heated samples observed

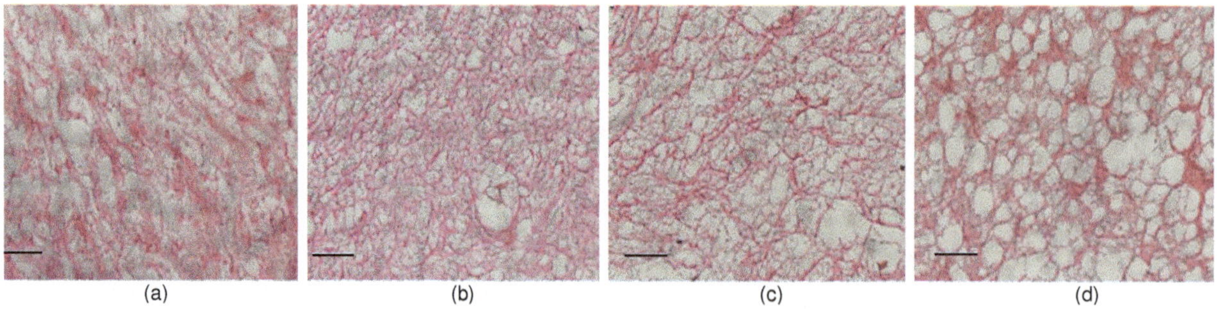

Figure 3. Light microscopy of raw and heated sea cucumber stained with Van Gieson staining. (a) raw sea cucumber, (b) heating at 50°C (c) heating at 70°C, (d) heating at 100°C. Bars: 30 μm.

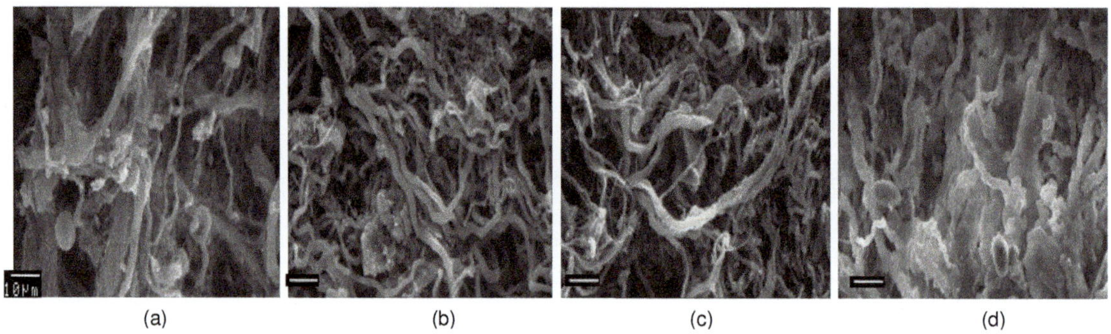

Figure 4. Views of raw and heated sea cucumber by a scanning electron micrographic. (a) Raw sample, (b) heating at 50°C, (c) heating at 70°C, (d) heating at 100°C.

by an optical microscopy are shown in Figure 3. The muscle fibers were stained with Van Gieson staining which dyed myofibrils yellow and collagen fibers red. Neither raw nor heated sample was dyed yellow, and raw sample was dyed red mostly. These results confirmed that collagen fibers were the mainly fibers in sea cucumber meat. For raw sample, muscle fibers in tissue were dense, and these fibers nearly parallel ordered with each other. In contrast, muscle fibers in heated samples shrank and subsequently reassembled, and the voids in muscular tissue tended to expand with heating temperature. After heating at 100°C, muscular tissue was porous and the voids in tissue were larger than those in the raw samples.

To illustrate the detail changes that occurred in the muscle fibers of heated samples, we studied with a SEM (Figure 4). Compared with Figure 3, the SEM photographs revealed the microstructural changes of fibers in sea cucumber meat after heating. Fibers bundles in raw sample were rich and loosely, and the gaps between fibers were large. However, the gaps in heated samples were smaller than those in raw sample. For heated samples, the SEM photographs were significantly different. Fibers bundles in tissue aggregated and the gaps between fibers become smaller after heating at 50°C, while coagulated blocks of muscular fibers were observed at 70°C. With continuous increase of the heating temperature, these fibers completely denatured and disintegrated, and then formed close structure at 100°C.

Enumerating muscle structure

For a marine mollusk, textural properties were thought to depend on tissue structure. Therefore, in this work, the images in Figure 3 are converted into binary morphology (Blatta et al., 2004) and are shown in Figure 5. Binary morphology works on two gray scale levels, white and black. In Figure 5, white pixels represented muscle fibers, while black pixels represented the voids in tissue. Using the image processing and analysis technique, the average values of Wm, Dm, Rm, Rvm and C^* were enumerated, and given them in Table 1. The Rm from heated samples was smaller than that from the raw samples. However the C^* from heated samples was bigger than that from the raw sample, which indicated that the extent of voids in heated meat approached to a real circle. The Dm and C^* gradually increased accompanied with increasing of heating temperature, however the Wm and Rm showed irregular changes. When fish were heated, cooking loss was significantly correlated with shrinkage ratio and the collagen solubility enhanced with heating (Kong et al., 2008). The correlations between cooking loss and structural characteristic

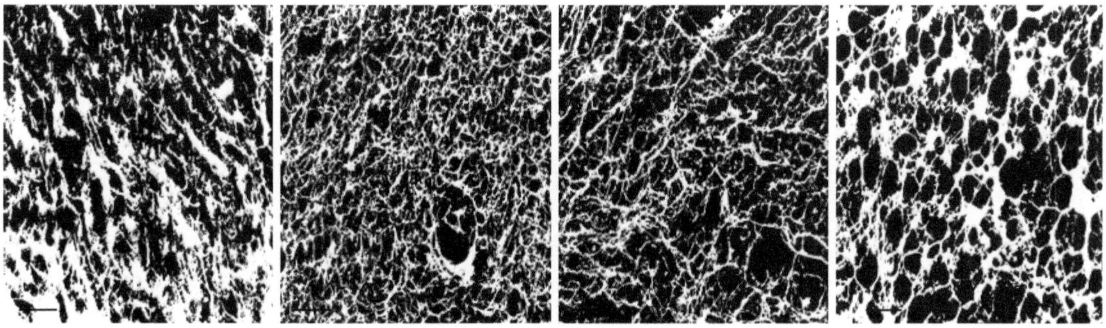

(a) (b) (c) (d)

Figure 5. The binary morphology images of raw and heated sea cucumber samples. (a) Raw sample, (b) heating at 50℃, (c) heating at 70℃, (d) heating at 100℃. Bars: 30 μm.

Table 1. Structural characteristic values of sea cucumber heated at different temperature.

Parameter	Wm (μm)	Dm (μm)	Rm (%)	Rvm (%)	C*
Raw sample	6.71 ± 2.01^{a}	7.12 ± 2.65^{b}	41.99 ± 2.84^{a}	58.01 ± 2.84^{b}	0.39 ± 0.13^{c}
Heated sample					
50℃	2.38 ± 0.59^{b}	4.45 ± 1.42^{c}	33.22 ± 8.73^{ab}	66.78 ± 8.73^{ab}	0.53 ± 0.12^{b}
70℃	2.17 ± 0.63^{b}	6.64 ± 2.24^{bc}	30.70 ± 3.06^{b}	69.30 ± 3.06^{a}	0.52 ± 0.14^{b}
100℃	7.97 ± 4.12^{a}	15.9 ± 5.09^{a}	39.16 ± 2.64^{ab}	60.84 ± 2.64^{ab}	0.76 ± 0.07^{a}

The different superscripts of the same column values are significantly different ($p < 0.05$).

Table 2. Correlations between cooking loss and structural characteristic values.

	Cooking loss	Wm	Dm	Rm	C*
Cooking loss	1				
Wm	0.329	1			
Dm	0.829	0.789	1		
Rm	-0.097	0.907	0.459	1	
C*	0.875	0.356	-0.820	-0.033	1

values in sea cucumber meat are shown in Table 2. For sea cucumber meat, cooking loss was positively correlated with Dm ($r = 0.829$) and C^* ($r = 0.875$), but not significantly correlated to Wm or Rm. It is confirmed that positively correlations between characteristic values (Dm and C^*) and cooking loss existed in sea cucumber meat.

Rheological properties

Rheological properties, such as elasticity modulus $\boldsymbol{E}$, relaxation time $\boldsymbol{\tau}$, viscosity $\boldsymbol{\eta}$ and rupture strength, obtained from raw and heated samples are showed in Table 3. Rheological properties greatly changed during heating treatment. Elasticity modulus ($\boldsymbol{E_0}$, $\boldsymbol{E_1}$, $\boldsymbol{E_2}$) from raw sample were higher than those from heated samples. For heated samples, this elasticity expressed a highest value at 70℃. Relaxation times ($\boldsymbol{\tau_1}$) for heated samples tended to increase with heating temperature, while viscous modulus ($\boldsymbol{\eta_1}$) had the same reaction with heating temperature as relaxation time. Rupture strength for sea cucumber samples was the similar to elasticity modulus, and the highest value existed at 70℃.

On the other hand, the textural properties of meat also could be described by TPA parameters (Palka and Daun, 1999). TPA parameters obtained from raw and heated sea cucumber samples are given in Table 4. Table 4 also gives a comparison of significant values obtained from different heating temperature. Hardness values in heated samples were lower than that in raw sample, while springiness was much larger than raw sample. Hardness values in heated samples were significantly different, same significant difference in springiness. The hardness from heated samples expressed a highest value at 70℃, however springiness expressed the lowest value at the same temperature.

Table 3. Rheological properties of sea cucumber meat heated at different temperature.

Parameter	E_0 (×10^5Pa)	E_1 (×10^5Pa)	E_2 (×10^5Pa)	τ_1 (s)	τ_2 (s)	η_1 (×10^6Pa·s)	η_2 (×10^6Pa·s)	Rupture strength (N)
Raw sample	3.53 ± 1.95^a	1.85 ± 0.55^a	2.31 ± 1.41^a	27.93 ± 12.52^b	1.44 ± 0.07^b	2.95 ± 0.29^{ab}	0.33 ± 0.21^a	31.95 ± 6.91^a
Heated sample								
50℃	1.26 ± 0.34^a	0.30 ± 0.08^b	0.96 ± 0.28^a	14.43 ± 5.26^b	1.25 ± 0.08^b	0.46 ± 0.29^b	0.12 ± 0.04^a	17.04 ± 4.79^b
70℃	2.75 ± 1.63^a	1.64 ± 0.73^a	1.12 ± 0.90^a	37.83 ± 15.78^b	1.78 ± 0.34^{ab}	6.01 ± 3.70^{ab}	0.18 ± 0.14^a	17.65 ± 4.40^b
100℃	0.98 ± 0.51^a	0.83 ± 0.43^b	0.16 ± 0.08^a	97.22 ± 29.27^a	2.11 ± 0.28^a	7.42 ± 3.59^a	0.03 ± 0.02^a	2.17 ± 0.07^c

The different superscripts of the same column values are significantly different ($p < 0.05$).

Those results kept consistent with rheological properties (***E*~0~**, ***E*~1~** and rupture strength). Cohesiveness tended to increase during heating; however chewiness expressed an irregular change.

DISCUSSION

In a previous paper, we published a large amount of experimental data demonstrating that thermal denaturation of muscle proteins is related to tissue structural changes(Gao et al., 2002; 2003). It is well known that raw sea cucumber meat contains a higher content of collagen than protein and the degree of aggregation for protein molecules increased with heat. Cui et al. (2007) has confirmed that the thermal stability temperature of pepsin-solubilized collagen which was extracted from sea cucumber meat was 57℃. In this study, the optical photographs revealed that muscular tissue in sea cucumber meat had greatly changed during different heating temperature (Figure 3). While SEM microphotographs also showed the denatured degrees for muscle fibers increased with heating temperature (Figure 4). After heating from 50 to 100℃, muscle fibers in sea cucumber meat respectively aggregated closely, formed blocks and denatured disintegrated. Therefore, it was confirmed that the presence of collagen and its denatured degree have greatly influence on muscular tissue for rich-collagen meat.

Structural characteristic values (Table 1) obtained from raw and heated sea cucumber meat has significantly changed with heating temperature. The gradually increasing of Dm and C^* suggested that the profile of voids in muscular tissue expanded after heating from 50 to 100℃, which was consistent with those voids tended to larger in optical photographs. As early mentioned, Dm and C^* were greatly influenced by cooking loss. It is well known that collagen is changed into gelatin which is a water-soluble fluid substance and muscle fibers is assembled and aggregated each other by heat denaturation. Therefore, it is supposed that the larger voids in heated sea cucumber meat are caused by the outflow of denatured protein for collagen-rich meat. In this paper, the texture properties were described in two fields: Rheological properties (Table 2) and TPA parameters (Table 3). Rheological properties ($\boldsymbol{E_0}$, $\boldsymbol{\tau_1}$, $\boldsymbol{\eta_1}$) and TPA parameter from sea cucumber meat were greatly changed after heating from 50 to 100℃. Matumoto and Yamano (1987) has reported that elastic modulus is greater when elastic force is smaller, whereas elastic modulus is smaller when elastic force is greater. As with the elastic, viscosity, $\boldsymbol{\eta}$, is greater when viscous force is smaller. Therefore, we concluded that the heated sea cucumber meat was more elastic than the raw meat and, similarly, more viscous than the raw meat. TPA parameter springiness also confirmed that the heated meat was more elastic than the raw meat.

Previous studies have reported that structural diversity of muscle connective tissue existed specifically and structural differences in connective tissue among fish species were clarified in relation with muscle firmness (Farahnaky et al., 2010; Johnston et al., 2000). In this present study, rheological properties, such as rupture strength, and TPA parameter hardness from heated sea cucumber meat has a largest value after heating at 70℃. Shrinkage of muscle fibers were observed at the same temperature by optical and SEM microphotographs. It is well known that rupture strength and hardness is related chiefly to firmness and energy as the meat was being ruptured. In previous studies, the extent of muscle fiber deformation during cooking depends on the compression stress applied by collagen fibers and the resistance of the muscle fibers to compression, while the compression force applied by collagen network on muscle fibers bundles depends upon the amount of collagen present and its thermal solubility (Lepetit et al., 2000). It was confirmed that the largest value of rupture strength or hardness in heated samples was caused by denaturation and aggregation of muscle proteins and the subsequent shrinkage and dehydration (Kong et al., 2007; 2008; Chen et al., 2011). Meanwhile, solubilization and gelatin of collagen contribute to the decrease of rupture

Table 4. TPA parameters of sea cucumber meat heated at different temperature.

Parameter	TPA parameters			
	Hardness (N)	Springiness (-)	Cohesiveness (-)	Chewiness (N)
Raw sample	32.85 ± 3.87^{a}	0.27 ± 0.05^{c}	0.18 ± 0.05^{b}	1.67 ± 0.61^{b}
Heated sample				
50°C	8.86 ± 1.86^{b}	0.54 ± 0.14^{b}	0.07 ± 0.01^{c}	0.38 ± 0.24^{b}
70°C	26.34 ± 7.87^{a}	0.40 ± 0.09^{bc}	0.36 ± 0.07^{a}	3.97 ± 1.69^{a}
100°C	3.79 ± 0.74^{b}	1.03 ± 0.03^{a}	0.43 ± 0.06^{a}	1.73 ± 0.45^{b}

The different superscripts of the same column values are significantly different ($p < 0.05$).

Table 5. Correlations between structural characteristic values and rheological properties.

	Wm	Dm	Rm	C^*	E_0	τ_1	η_1	Rupture strength
Wm	1							
Dm	0.789	1						
Rm	0.907	0.459	1					
C^*	0.356	0.820	-0.033	1				
E_0	-0.046	-0.439	0.205	-0.833	1			
τ_1	0.694	0.989*	0.327	0.862	-0.464	1		
η_1	0.304	0.765	-0.098	0.726	-0.258	0.847		
Rupture strength	-0.187	-0.715	0.205	-0.985*	0.866	-0.777	-0.707	1

*.Correlation is significant at the 0.05 level.

strength and hardness after heating at 100°C.

The heated sea cucumber meat, which was immersed in hot water and experienced an outflow of denatured proteins and moisture due to heating, formed unique muscular structure. Simultaneously, rheological properties reflected by $\boldsymbol{E_0}$, $\boldsymbol{\tau_1}$, $\boldsymbol{\eta_1}$ and rupture strength also greatly changed. Gao et al. (2002) reported that rheological properties were negatively correlated with characteristic values of myofibrils in abalone meat. Correlations analyses were conducted to evaluate the relationships between rheological properties and structural characteristic values. In Table 5, Dm was highly positively correlated with $\boldsymbol{\tau_1}$ (r = 0.989) and $\boldsymbol{\eta_1}$ (r = 0.989), was negatively correlations with rupture strength (r = -0.715). C^* was negatively correlated with instantaneous modulus $\boldsymbol{E_0}$ (r = -0.833), rupture strength (r = -0.985). However, Wm and Rm were not significantly correlations with rheological properties. In other words, $\boldsymbol{E}$ and rupture strength were negatively correlated with characteristic values (Dm, C^*), which are consistent with Gao report. However, $\boldsymbol{\tau}$ and $\boldsymbol{\eta}$ were positively correlated with characteristic values (Dm). Therefore, rheological properties reflect by $\boldsymbol{E}$, $\boldsymbol{\tau}$, $\boldsymbol{\eta}$ and rupture strength are mainly correlated with the characteristic values (Dm and C^*) in heated sea cucumber meat.

Conclusion

Based on these results, we confirmed that the muscular structure in heated sea cucumber meat due to cooking loss and denaturation of collagen. Rheological properties ($\boldsymbol{E}$, $\boldsymbol{\tau}$, $\boldsymbol{\eta}$ and rupture strength) and TPA parameters were greatly changed after heating. Rheological properties ($\boldsymbol{E_0}$ and rupture strength) in heated meat expressed the largest value at 70°C. The shrinkage and dehydration of muscle fibers were assumed to result in the largest value. Cooking loss positively correlated to structural characteristic values (Dm, C^*). Meanwhile, rheological properties reflect by $\boldsymbol{E}$, $\boldsymbol{\tau}$, $\boldsymbol{\eta}$ and rupture strength are mainly correlated with the characteristic values (Dm and C^*). These results suggested that the change in texture and rheological properties of sea cucumber during heating was caused by denaturation of proteins. Compared with raw sea cucumber meat, heated meat had higher springiness, viscosity and lower hardness.

ACKNOWLEDGEMENTS

This work was supported by the National Natural Science Foundation of China (No. 31071631) and Qingdao Technology Development Project (No. 2011-5-028-QT).

REFERENCES

Blatta RJ, Clarka AN, Courtney J, Tullye C, Tuckera AL (2004). Automated quantitative analysis of angiogenesis in the rat aorta model using Image-Pro Plus 4.1. Comput. Methods Progr. Biomed. 75:75-79.

Caine WR, Aalhus JL, Best DR, Dugan MER, Jeremiah LE (2003). Relationship of texture profile analysis and Warner-Bratzler shear force with sensory characteristics of beef rib steaks. Meat Sci. 64:339.

Chen SG, Xue CH, Yin LA, Tang QJ, Yu GL, Chai WG (2011). Comparison of structures and anticoagulant activities of fucosylated chondroitin sulfates from different sea cucumbers. Carbohydr. Polym. 83:688-696.

Conand C (2004). Advances in Sea Cucumber Aquaculture and Management, Food and Agriculture Organization of the United Nations. pp 25-69.

Cui FX, Xue CH, Li ZJ, Zhang YQ, Dong P, Fu XY, Gao X (2007). Characterization and subunit composition of collagen from the body wall of sea cucumber Stichopus japonicus. Food Chem. 100:1120-1125.

Duan X Zhang, M, Mujumdar AS, Wang SJ (2010). Microwave freeze drying of sea cucumber (Stichopus japonicus). J. Food Eng. 96:1-497.

Farahnaky A, Askari H, Majzoobi M, Mesbahi GH (2010). The impact of concentration, temperature and Ph on dynamic rheology of psyllium gels. J. Food Eng. 100:294-301.

Gao X, Ogawa H, Tashiro Y, Iso N (2001). Rheological properties and structural changes in raw and cooked abalone meat. Fisheries Sci. 67:314-320.

Gao X, Tashiro Y, Ogawa H (2002). Rheological properties and structural changes in steamed and boiled abalone meat. Fisheries Sci. 68:499-500.

Gao X, Zhang ZH, Tang ZX, Tahiro Y, Ogawa H (2003). The relationships between rheological properties and structural changes of chilled abalone meat. J. Ocean Univ. China 2:171-176.

Jiang MK, Wang YF, Santen EV, Chappell JA (2008). Evaluation of textural properties of channel catfish (Ictalurus punctatus Rafinesque) fillet with the natural contour method. Food Sci. Technol. 41:1548-1554.

Johnston IA, Alderson R, Sandham C, Dingwall A, Mitchell D, Selkirk C, Nickell D, Baker R, Robertson B, Whyte D, Springate J (2000). Muscle fibre density in relation to the colour and texture of smoked Atlantic salmon (Salmo salar L.). Aquaculture 189:335-349.

Kageyama K, Watanabe Y (1988). Manual of Histologic Techniques. IgakuSyoin Ltd, Tokyo. pp. 92-95.

Kong FB, Tang JM, B Rasco, Crapo C (2007). Kinetics of salmon quality changes during thermal processing. J. Food Eng. 83:510-520.

Kong FB, Tang JM, Lin MS, Rasco B (2008). Thermal effects on chicken and salmon muscles: tenderness, cook loss, area shrinkage, collagen solubility and microstructure. Swiss Soc. Food Sci. Technol. 41:1210-1222.

Lepetit J, Grajales A, Favier R (2000). Modeling the effect of sarcomere length on collagen thermal shortening in cooked meat: consequence on meat toughness. Meat Sci. 54(3):239-250.

Matumoto S, Yamano S (1987). The rheology of foods. In: Watanabe S (ed). The Tenderness of Chicken Meat. Food Material Research Institute Press, Tokyo. pp. 13-15.

Niitu T, Hiramoto U (1982). Laboratory Lecture of Biology. Maruzen Press, Tokyo. pp. 215-261.

Palka K, Daun H (1999). Changes in texture, cooking losses, and myofibrillar structure of bovine M. semitendinosus during heating. Meat Sci. 51:237-243.

Saito M, Unisaki NK, Urano N, Kimura S (2002). Collagen as the Major Edible Component of Sea Cucumber (Stichopus japonicus). J. Food Sci. 67:1319-1322.

Slater MJ, Jeffs AG, Carton AG (2009). The use of the waste from green-lipped mussels as a food source for juvenile sea cucumber, Australostichopus mollis. Aquaculture 292:219-224.

Stonik VA, Ponomarenko LP, Makarieva TN, Boguslavsky VM, Dmitrenok AS, Fedorov SN, Strobikin SA (1998). Free sterol compositions from the sea cucumbers Pseudostichopustrachus, Holothuria (Microtele) nobilis, Holothuriascabra, Trochostomaorientale and Bathyplotesnatans. Comparative. Biochem. Physiol. 120:337-347.

Zamora LN, Jeffs AG (2011). Feeding, selection, digestion and absorption of the organic matter from mussel waste by juveniles of the deposit-feeding sea cucumber, Australostichopus mollis. Aquaculture 317:223-228.

Gill arch occupation models of parasite communities of *Barbus martorelli* (Teleostean: Cyprinidae)

TOMBI Jeannette and BILONG BILONG Charles Félix

Laboratory of Parasitology and Ecology, Faculty of Science, University of Yaounde I, P. O. BOX 812, Yaounde, Cameroon.

Changes in *Barbus martorelli* gill arch occupation models by its parasite component community were investigated for nine months. The work revealed three different categories of models. The first category follows the combined action of the respiratory water current and the colonized surface of each gill arch. The second category based on the mode of infestation of Monopisthocotylean larvae while the third category is established as a result of perturbations. Thus, the occupation models of *B. martorelli* gill arches by its microfauna are constantly changing.

Key words: *Barbus martorelli*, gill parasites, arch occupation models, perturbations, month.

INTRODUCTION

For many decades now, fish parasites have been the subject of several ecological community studies (Smith, 1969; Rohde, 1979; Silan and Maillard, 1989; Dzika, 1999; Howard et al., 2002; Poulin, 2002). The specific richness of fish gill parasites is generally high and as a result, several species share a limited surface area (Koskivaara et al., 1991). The regular ecological problem concerned here is to understand how these communities exploit their biotope. It has been observed that gill parasites are attached mostly to specific sites within the hosts. The study of the occupation models of gill arch preferences is usually done globally without considering variations with time. For instance, various works carried out globally within a given time, have resulted in different occupation models: Bagge and Valtonen (1999) found that *Dactylogyrus nanus* and *Dactylogyrus micracanthus* occurred mostly on the second gill arch of *Rutilus rutilus*; Dzika (1999) observed the presence of *Pseudodactylogyrus anguillae* on the median arches and *Pseudodactylogyrus bini* on the three anterior gill arches of *Anguilla Anguilla.* Koskivaara et al. (1991) found that while *Dactylogyrus suecicus* and *Dactylogyrus similis* were more abundant on both second and third gill arches, *Dactylogyrus fallax* was more abundant on the first and second gill arches of *R. rutilus.* Although Euzet and Combes (1998) and Kearn (1998) showed that monogeneans commonly exhibit very precise site specificity, fish parasites equally respond strongly to physico-chemical characteristics of the aquatic environment and also to modifications in the physiological and biological conditions of the host (Ferrari-Hoeinghaus et al., 2006). According to Vidal-Martinez and Poulin (2003); Tavares and Luque (2004), these factors might influence parasite community structure. For these reasons, Combes (1995) recommended the analysis of ecological variations within a very short period because, the overall stability can dissimulate many local perturbations. The aim of this paper is to determine if the

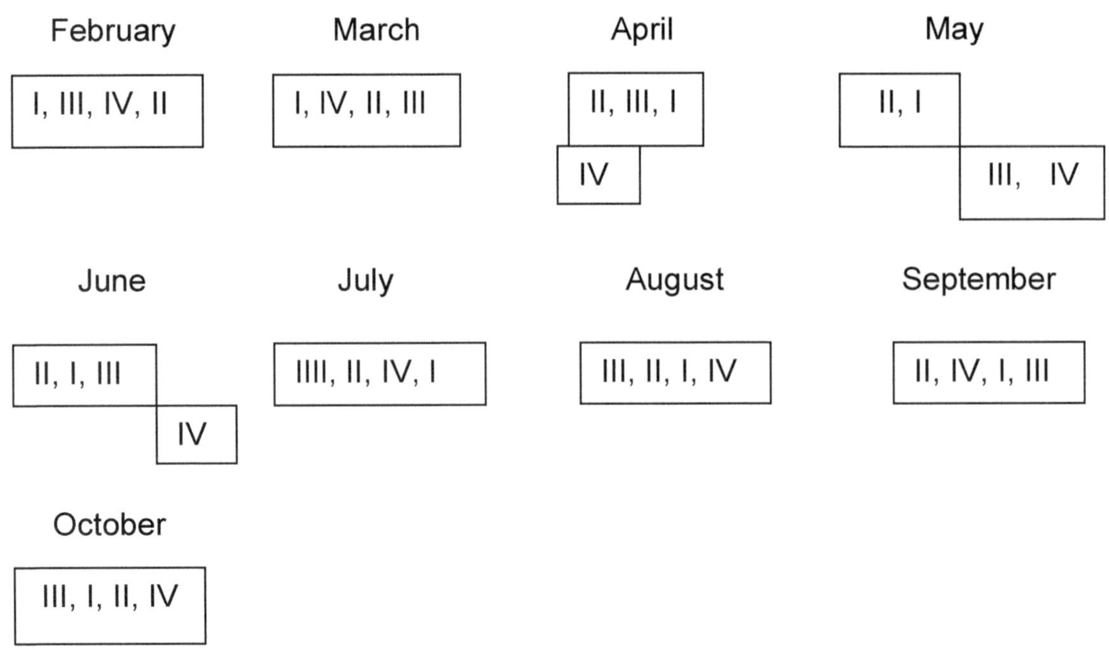

Figure 1. *Dactylogyrus bopeleti* monthly gill arch occupation models (gill arches from anterior to posterior are numbered I to IV).

gill arch occupation models of *B. martorelli* parasite communities are affected by various perturbations over a period of nine months.

MATERIALS AND METHODS

From the months of February to October 1998, 270 host specimens were collected using mesh gillnet (1 cm by 1 cm) in the Foulou watercourse at Nkolfoulou locality (3°53'N, 11°34'E), at the outskirts of Yaounde, the capital of the Republic of Cameroon. Two samples were collected monthly and an effort was made to catch at least 30 fishes per month. After capturing a fish, it was immediately fixed and kept in 10% formalin before transferring from the sampling site to the laboratory. In the laboratory, the gills from each side were dissected and the gill arches from the anterior to the posterior were placed in various Petri dishes numbered from I to IV respectively, prior to examining under a stereoscopic microscope. Myxosporidian cysts present on the filaments or on the bony part of each gill arch were counted. The monogeneans collected were mounted on slides on which a drop of hematoxylin (eosin) was added to stain the parasites. Parasite species were identified and counted.

Component community and infra-population were defined according to Combes (1995). The Student t-test was used to compare two monthly mean parasitic charges of various gill arches. For each parasite species, the different models were then established, on one hand, based on the number of helminthes or protozoan cysts harbored monthly by each arch, and on the other hand, on the t-test results. The model is termed anteroposterior if the parasitic charge reduces progressively from the first arch (anterior position) towards the fourth (posterior position). The term equipartition is used when the number of individual species does not vary statistically from one gill arch to another, while unequipartition is used for cases where the number of gill arch parasites varies statistically.

RESULTS

A total of six parasite species were collected on the gill of *Barbus martorelli*. Four of them (*Dactylogyrus bopeleti*, *D. insolitus*, *D. simplex* and *D. maillardi*) were monogeneans and the other two (*Myxobolus barbi* and *M. njinei*) were myxosporidia. The analysis based on the infrapopulation size and the monthly evolution of the occupation models of gill arches provided various results depending on each parasite species.

It was observed that for six months (February, March, July, August, September and October),

Dactylogyrus bopeleti displayed an equipartition occupation pattern ($P>0.5$). During the remaining three months, this parasite indicated an unequipartition pattern comprising three models namely: The gill arch II was statistically more colonized than the three others but, a significant difference was observed only between arches II and IV in April; the gill arches I and II have statistically similar parasitic charge and were more parasitized than the gill arches III and IV in May; the difference of infestation between the gill arches I, II, III compared with IV were observed in June (Figure 1).

In the case of *D. insolitus*, an equipartition pattern was observed during the months of March, April, June, July and September ($P > 0.5$). This equipartition was perturbated, as the only model observed more than once, that is, "III, II, IV, I" was in June and July (Figure 2). Unequipartition pattern was observed during the remaining four months. In February, arch I was statistically more parasitized than arch IV. Although the

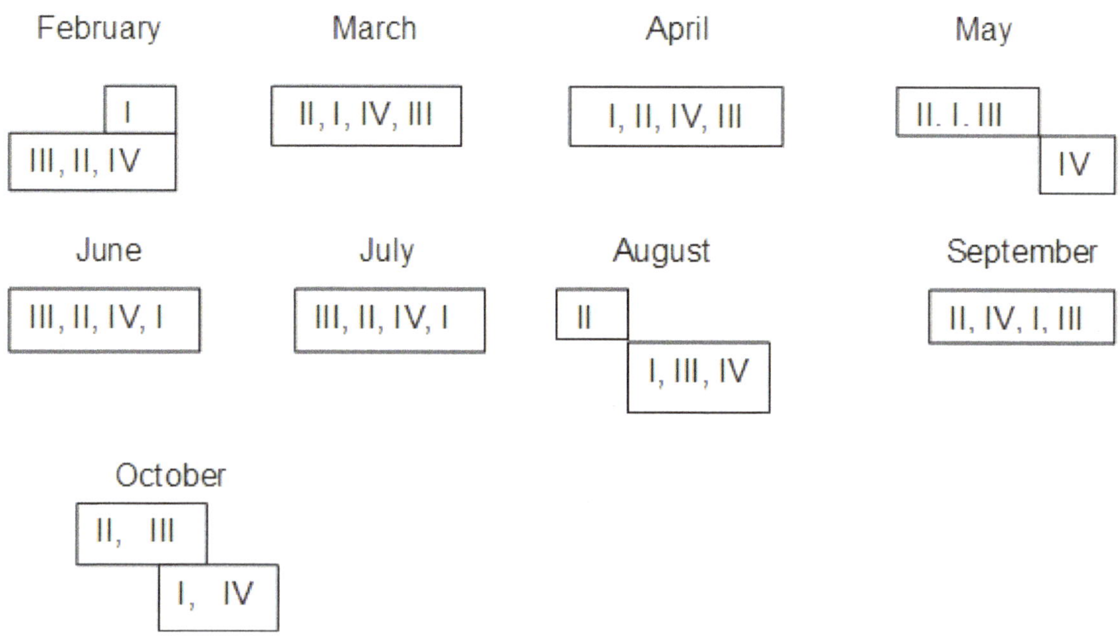

Figure 2. *Dactylogyrus insolitus* monthly gill arch occupation models (gill arches from anterior to posterior are numbered I to IV).

parasite loads of the arches II and III were superior to that of arch IV, the difference remained statistically insignificant (P > 0.5). Arch IV was statistically less infected than the three others whose parasitic charge was statistically insignificant in May (P>0.5). In August, arch II was statistically more infested than the three others whose parasitic loads remained statistically equal. In October, the median gill arches II and III harbored statistically more parasites than arches I and IV however, the parasitic loads of arch IV remained less important.

D. simplex exhibited an equipartition spatial occupation pattern presenting different models for six months (February, May, June, July, August and September) in addition, an unequipartition pattern was observed for the rest of the months (March, April, October). In March, arch I was statistically less colonized than the three others whose parasitic charges were statistically insignificant. Though a high parasitic charge of arch IV was observed compared to arches III and II respectively and a high parasitic load of arch III was observed compared to arch II, the difference was insignificant. In April, arch IV was statistically less infested than arches II and III, their respective parasite loads being statistically equal to that of arch I. Moreover, all the various models observed for *D. simplex* occurred once during the study period. In October, arch II was more colonized than the three others but, a significant difference was noted only between arches I and II (Figure 3).

Considering *D. maillardi,* the gill arch colonization showed several perturbations. An equipartition pattern was observed during the months of February, June, August and September (P > 0.5). This species was completely disappeared from all the arches in July. An unequipartition was noted for the remaining months. *D. maillardi* was absent on arch III in the month of March. In April, arch I was less colonized than the three others but a significant difference was observed only between arches I and III. In the months of May and October, arch IV statistically harbored more individuals (Figure 4).

The colonization of the four gill arches of *B. martorelli* by *Myxobolus barbi* was less perturbated. The models observed were «I, IV, II, III» in February; « I, II, III, IV » in April, July, August and October; «I, II, IV, III » in May, June and September. The only case of unequipartition was observed in March. Within this month, arch IV was statistically less colonized than arches I and II and no significant difference was observed for the three anterior arches (Figure 5).

An unequipartition of *M. njinei* cysts on the different gill arches was highly perturbated. This myxosporidia disappeared during the months of March for arch IV, April for arch IV, May for arches II, III, IV, June for arches III, IV and July for arches II and IV while an equipartition was observed during February, September and October (Figure 6).

DISCUSSION

This work has revealed that, the gill parasites of *Barbus martorelli* colonized the four arches in two patterns: Equipartition which consisted of equal statistical distribution of individuals of the same species on the four pairs of gill arches; and unequipartition, when one or

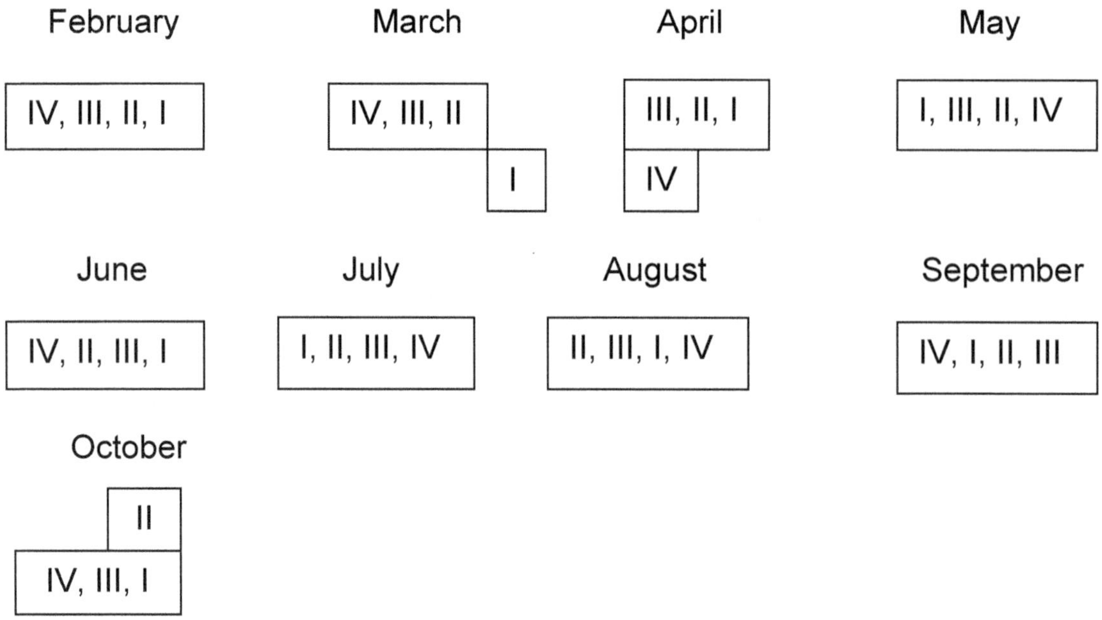

Figure 3. *Dactylogyrus simplex* monthly gill arch occupation models (gill arches from anterior to posterior are numbered I to IV).

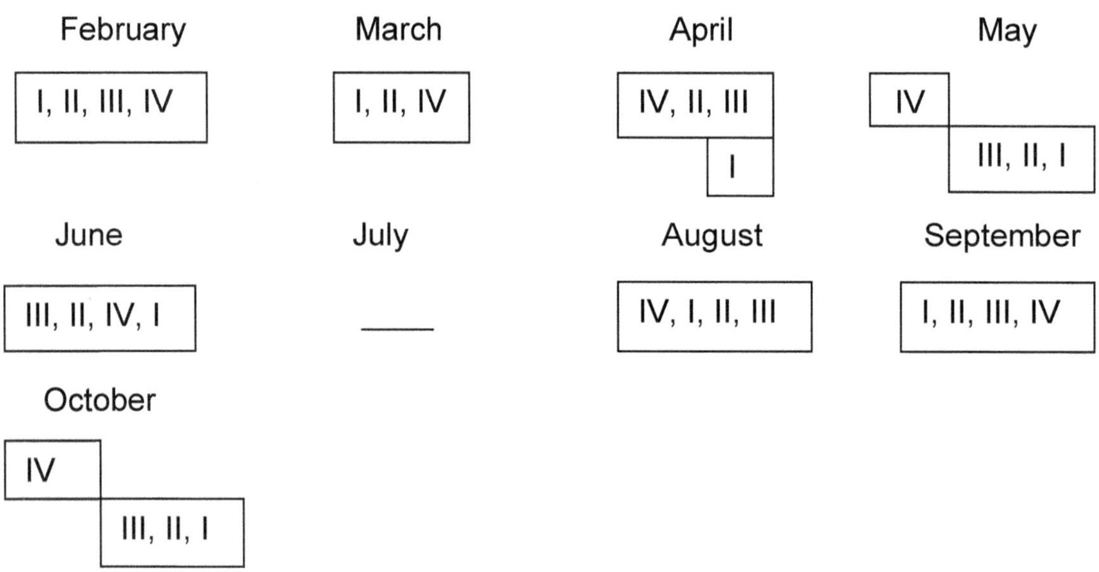

Figure 4. *Dactylogyrus maillardi* monthly gill arch occupation models (gill arches from anterior to posterior are numbered I to IV).

more gill arches were statistically more colonized than the others by the same parasite species. During the study period, each parasite species adopted both patterns of colonization. However, each pattern had several models. Although very little studies are found in literature, some authors nevertheless had obtained similar results. For instance, Bilong Bilong et al. (2004) showed that in *Hemichromis fasciatus*, the monogenean *Onchobdella voltensis* adopted equipartition during the months of June and November despite some gill arches being statistically more or less infested throughout the year. These authors also revealed that in the same host, *Cichlidogyrus longicirrus* occupied the four gill arches with equal statistical values for seven months and an unequipartition with different models was obtained during the months of November and December 1989 and then in March, July and October 1990. Koskivaara et al. (1992) studied the gills of *R. rutilus* between February 1988 and April 1989

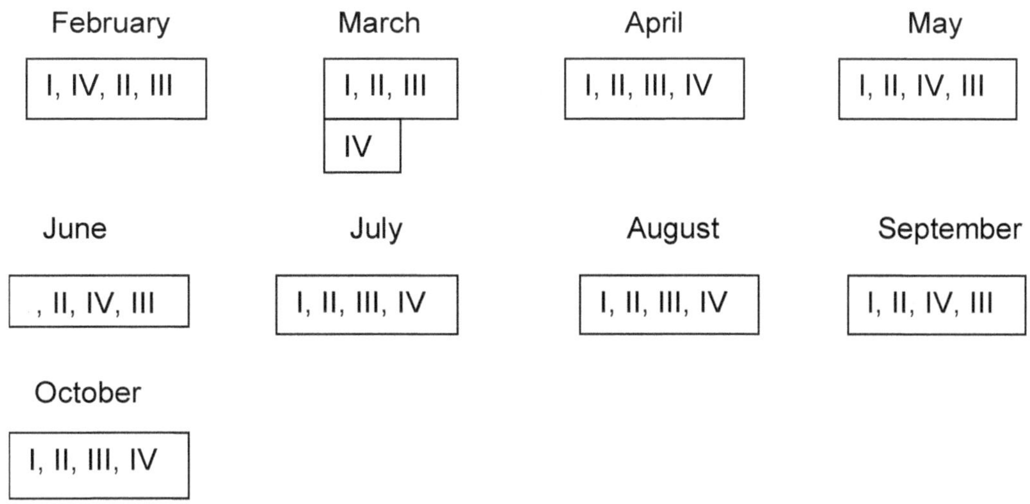

Figure 5. *Myxobolus barbi* monthly gill arch occupation models (gill arches from anterior to posterior are numbered I to IV).

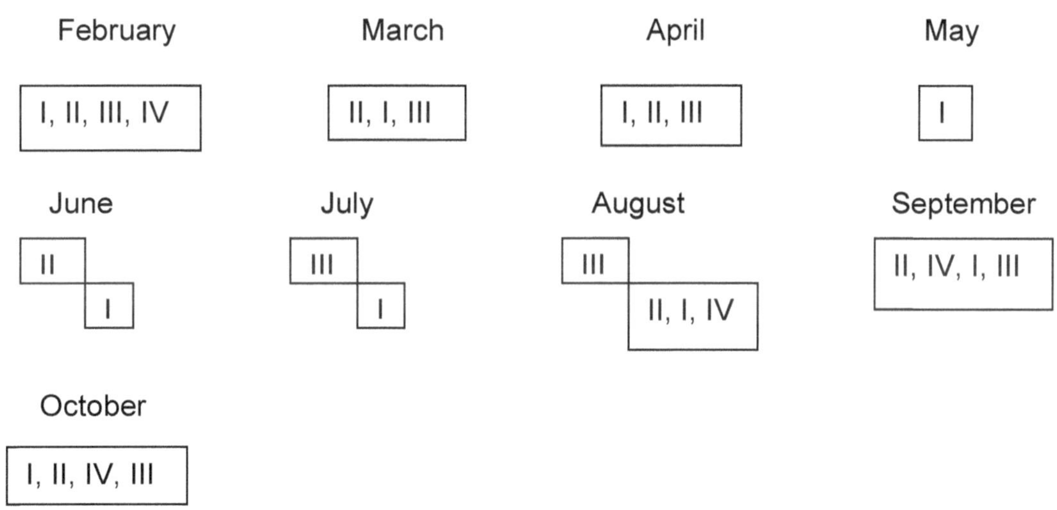

Figure 6. *Myxobolus njinei* monthly gill arch occupation models (gill arches from anterior to posterior are numbered I to IV).

and noted some differences in the occurrence of *Dactylogyrus nanus* on the four gill arches when comparing preferences between the months of May and June. In May, this monogenean had a statistically lower abundance on the first than on the other arches but, in June there were no significant differences between the various arches.

This study has shown that the spatial occupation patterns of arches (equipartition and unequipartition) by the gill parasites of *B. martorelli* presented three different categories of models. The first involved, on one hand, those in which the concentration of parasites reduced in the anteroposterior direction and, on the other hand, those in which the parasites concentrated more on the median arches. This type of model can be explained based on either the respiratory water current or the colonized gill surface of each arch. According to Guitiérrez and Martorelli (1994) and Lo and Morand (2001), the median arches II and III were more infested due to the large volume of ventilated water current flowing across. Additionally, Koskivaara et al. (1991) estimated that, the preference of median arches was linked to the large colonizing gill surface available to the parasites. For Bilong Bilong and Tombi (2004), the surfaces of the first three arches of *B. martorelli* seemed to be equivalent, while arch IV often presented the smallest colonizing surface. The combined action of the water current and the colonizing arch surface might explain the establishment of the first category of models.

The second category of models involved those in which

the fourth arch, which is the smallest and the least ventilated was more infested. This implies that in the case of monogenean concerned, this type of models could reflect the colonization pattern of oncomiracidium of Monopisthocotylea. Actually, Combes and Jourdane (2003) had shown that these infesting larvae generally were first attached on the body of their host before migrating towards the gill. This second category of models might imply that, after attaching to the fish body, the migration of oncomiracidium towards the gills was done in the posteroanterior direction.

The last category of models concerned those in which some arches were either not colonized at all or those in which the colonization was done in disorder, that is, «I, III, V, II »; «I, IV, II, III»; «III, II, IV, I»; «II, I, IV, III». In this last category, arch colonization was neither done in the direction of respiratory water current, nor in the direction of colonization arch surfaces and was not depend on the colonization pattern of gill by oncomiracidium of Monopisthocotylea. Such models might reveal that spatial structure of *B. martorelli* gill parasites had undergone several perturbations. These perturbations were also identified through the constant changing observed in the specific occupation models of gill arches; consequently no species adopted a statistically dominant model. The term perturbation in this case, refers to ruptures that could be either lightly or violently, scarcely or frequently occurring in any environment. For tropical species, it is known that the presence or abundance of parasites is influenced by both host and environmental factors (Esch et al., 1990). However, to our knowledge, these perturbations cannot be clearly associated with specific causal factors or environmental variations.

Therefore, it can be concluded that *B. martorelli* component community adopts three main categories of arch occupation models. Each category presents several types of models and no species indicated a statistically dominant model during the study period.

REFERENCES

Bagge AM, Valtonen ET (1999). Development of monogenean communities on the gills of roach flies (*Rutilus rutilus*). Parasitology 118:479-487.

Bilong Bilong CF, Tombi J (2004). Hétérogénéité du système branchial de *Barbus martorelli* Roman, 1971 (Poisson Cyprinidae) et modèle de croissance. J.C.A.S. 4(3):191-286.

Bilong Bilong CF, Atyame Ntem CM, Njine T (2004). Structure de la guilde des monogènes parasites branchiaux du poisson *Hemichromis fasciatus au* lac municipal de Yaoundé. J.C.A.S. 4(1):33-40.

Combes C (1995). Interactions durables. Ecologie et évolution du parasitisme. Collection d'écologie, n° 26. Paris. Ed. Masson: P. 524.

Combes C, Jourdanes J (2003). Taxonomie écologie et évolution des métazoaires parasites - taxonomy, ecology and evolution of metazoan parasites. (Livre hommage à Louis Euzet) Tome I. PUP, Perpignan France. pp. 161-201.

Dzika E (1999). Microhabitat *of Pseudodactylogyrus anguillae* and *P. bini* (Monogenea: Dactylogyridae) on the gills of large-size European eel *Anguilla anguilla* from Lake Gaj, Poland. Folia Parasit. 46:33-36.

Esch GW, Shostak AW, Marcogliese DJ, Goater TM (1990). Patterns and processes in helminthes parasite communities: An overview. In: Parasites communities: Patters and processes, Esh GW, Bush AO and Aho JM (Eds). Chapman and Hall, London, UK. pp. 1-20

Euzet L, Combes C (1998). The selection of habitats among the monogenea. Int. J. Parasitol. 28:1645-1652.

Ferrari-Hoeinghaus AP, Takemoto RM, Oliveira LC, Makrakis MC, Baumgartner G (2006). Host-parasite relationships of monogeneans in gills of *Astyanax altiparanae* and *Rhamdia quelen* of the São Francisco Verdadeiro river, Brazil. Parasite 13:315-320.

Guitiérrez PA, Martorelli SR (1994). Seasonality distribution and preference sites of *Demidospermus Valenciennesi* (Monogenea: Ancryrocephalidae). Res. Rev. Parasitol. 54:259-261.

Howard SC, Donnelly CA, Kabatereine NB, Ratard RC, Brooker S (2002). Spatial and intensity - dependant variations in associations between multiple species helminth infestions. Acta Trop. 83:141-149.

Kearn GC (1998). Parasitism and the Platyhelminths. Chapman and Hall, London, P. 544.

Koskivaara M, Valtonen ET, Prost M (1991). Dactylogyrids on the gills of roach in Central Finland: features of infection and species composition. Int. J. Parasitol. 21:565-572.

Koskivaara M, Valtonen ET, Vuori KM (1992). Microhabitat distribution and coexistence of *Dactylogyrus* species (Monogenea) on the gills of roach. Parasitology 104:273-281.

Lo CM, Morand S (2001). Gill parasites of *Cephalophobis argus* (Teleostei: Serranidae) from Moorea (french Polynesia): site selection and coexistence. Folia Parasit. 48:30-36.

Poulin R (2002). The evolution of monogenea diversity. Int. J. Parasitol. 32:245-254.

Rohde K (1979). A critical evaluation of intrinsic and extrinsic factors responsible for niche restriction in parasites. Am. Nat. 114(5):648-671.

Silan P, Maillard C (1989). Biology of Serranicotyle labracis, ectoparasite of Dicentrarchus (Teleostei): Contribution to the study of its populations. Mar. Biol. 103:481-487.

Smith JW (1969). The distribution of one Monogenean and two copepod parasites of Whiting, Merlangius merlangus (L.), caught in British waters. Nytt. Mag. Zool.17:57- 63.

Tavares LER, Luque JL (2004). Community ecology of the metazoan parasites of White Sea catfish, Netuma barba (Osteichthyes: Ariidae), from the coastal zone of the state of Rio de Janeiro, Brazil. Braz. J. Biol. 64(1):169-176.

Vidal-Martinez VM, Poulin R (2003). Spatial and temporal repeatability in parasite community structure of tropical fish hosts. Parasitology. 127:387–398.

Distribution, diversity and taxonomy of marine angelfishes (Pomacanthidae) of Tamilnadu, Southeast coast of India

Mayavan Veeramuthu Rajeswari and Thangavel Balasubramanian

Centre of Advanced Study in Marine Biology, Faculty of Marine Sciences, Annamalai University, Parangipettai 608502, India.

In the present investigation, a total of 5 species belonging to three genera such as *Pomacanthus*, *Centropyge* and *Apolemichthys* were reported from the 10 selected stations of Tamilnadu, south east coast of India. The diversity studies revealed that the species diversity, richness and evenness were comparatively higher at Rameshwaram and Kanniyakumari due to the rocky shore and coral reef ecosystem. Species composition study revealed that the smoke angel (*Apolemichthys xanthurus*) was found to be dominant group which constituting 57% of total fishes. The results of the present study evidenced that the occurrence and distribution of marine angel fishes were higher in Gulf of Mannar than other region of Tamilnadu. The present findings clearly explained about distribution pattern of marine angel fishes which could be useful for better understanding of the status of its diversity along southeast coast of India and also highlights the need for effective conservation measures of these commercially important marine ornamental fish group.

Key words: South east coast of India, distribution, diversity, taxonomy, marine angelfishes, pomacanthidae.

INTRODUCTION

The marine ornamental fishes are one of the most popular attractions in worldwide due to their adaptability to live in confinement. The trade of marine ornamentals has been expanding in recent years and has grown into a multimillion dollar enterprise. The worldwide market of marine ornamental fishes has shown a steady increase over the past few years and the annual trade varies between 2 and 7 billion US$ (Jayalal and Ramachandran, 2012). A total of 1471 species of ornamental fish are traded globally (Wabnitz et al., 2003). Of which, angelfishes (Pomocanthidae) are the next most important group than damsel and anemone fish (33%), consisting of about 25% of the total trade (Sadovy, 1992). Pomacanthus angel fishes are one among the most highly prized of the coral reef fishes which contains 8 genera and 82 species worldwide (Debelius et al., 2003; Nelson, 2006). Marine angelfishes (Pomacanthidae) are among the most spectacularly coloured and widely recognized groups of coral reef-associated fishes. They are found on tropical reef bottoms, widespread in the Atlantic, Indian and pacific oceans (Murugan and Durgekar, 2008). Marine angelfishes have deep compressed bodies, ctenoid scales, a single un-notched dorsal fin and a small mouth with brush-like teeth. The best way to identify them is to look for the spine in front of the gill plates (Alwany, 2009). Marine angel fishes contain an elongate extension on the hind margin of the soft dorsal and anal fins and rounded or crescent-shaped caudal fin. Most angelfishes are pygoplites which feed mainly on sponges as adults, but *Genicanthus* sp. feed mainly on zooplankton and *Centropyge* sp. are grazers on algae and detritus (Khalaf and Kochzius, 2002). As pelagic spawners, marine angelfish releases tiny buoyant eggs into water, which then become part of the plankton. The eggs float freely with currents until hatching, a high number falling victim to planktonic feeders (Thresher, 1982).

Tamil Nadu is one among the important maritime states on the east coast of India, with a coastline of 1,076 km, 13 coastal districts and 591 fishing villages. The Coromandel coast extends 357 km, from Pulicat to Point Calimere, and Palk Bay extends 294 km from Point Calimere to Dhanushkodi. The Gulf of Mannar, rich in biodiversity and hosting a variety of marine species, is a marine biosphere reserve, and extends 335 km from Dhanushkodi to Kanniyakumari (Murugan and Durgekar, 2008). So far, Venkataramani and Jawahar (2004) reported about 113 marine ornamental fish and their biodiversity and standing stock biomass from Gulf of Mannar region. Although marine angelfish plays a very significant role in ornamental fish trade, there is no attempt made to study its occurrence, distribution and diversity along Tamil Nadu coast, Southeast coast of India. Despite its prominent status among the coral reef fishes, the biology and ecology of the Pomacanthidae has not been yet studied in details, especially in the Southeast coast of India. In view of that, present investigation was aimed to be the first detailed investigation to understand the distribution pattern and diversity of marine angel fishes along Tamil Nadu coast, Southeast coast of India.

MATERIALS AND METHODS

Collection of data

To understand the occurrence, distribution and diversity of marine angel fish, survey was made between January 2008 and December 2009 in the 10 selective stations along southeast coast of India. Collection of marine ornamental fish has been done in two ways such as incidental fishing and target fishing. The incidental fishing includes use of traps, shore seine operations and country trawl nets, and target fishing includes skin diving with scoop nets and gill nets. Target fishing has been carried out for the collection of fishes during low turbidity and calm periods. Gill netting with skin diving was practiced when the conditions are calm in the reef regions. In addition, data were also collected by regular enquiry from the local fish traders and fish collectors in the respective landing centres.

Description of the study area

The survey was conducted from 10 stations along Tamil Nadu, Southeast coast of India (Figure 1). Station 1 is located in Chennai, (Lat: 12°51'58.58"N, Lon: 80°14'40.70"E) which is one among the major landing centre in Southeast or Coramental coast of India. Station 2 is located in Puducherry (11°55'40.21"N, 79°47'26.13"E) is rocky shore area, rich in algae and echinoderms. Station 3 is located in Cuddalore (11°45'15.95"N, 79°47'27.63"E) which is polluted due to the chemical, petrochemical, pharmaceutical biocide, fertilizer and metal processing industrial discharges. Station 4 situated in Nagapattinam (10°45'40.19"N, 79°50'56.80"E) is also one of the major fish landing centres in Tamil Nadu. Station 5 is located in Thondi (09°44'1.32"N, 79° 0'59.06"E) in the Palk Strait which is being influenced mainly by the northeast monsoon and has strong potential of both living and non-living resources. The coastal area has a very minimal wave action and turbidity of the seawater is moderately low and also they are rich in nutrients hence, it is a potential zone for ornamental fishes and many fish collectors were involved in ornamental fish collection and trade. Station 6 is situated in Kilakkarai (09°13'43.43"N, 78°47'33.02"E) is rich in seaweeds, seagrass and coral reefs which is also a potential zone for ornamental fishery resources. Station 7, Rameshwaram (9°16'50.96"N, 79°19'3.66"E) is an island connected to the main land by the Pamban Bridge which is rich in seaweed *Sargassum* sp. and the fish collectors and traders were extensively involving in ornamental fish trading. Station 8, Tuticorin (8°47'38.94"N, 78°9'36.58"E) is one of the well-known and age old fishery ports of India especially by virtue of production of valuable and good quality pearls and chanks in addition to marine ornamental fish production. Station 9 is located in Thiruchendur (8°29'19.16"N, 78° 7' 26.57"E) contains rocky bottom with shallow waters which is also famous for its live ornamental fish trade. Station 10, Kanniyakumari (8°5'7.28"N, 77°33'11.75"E) is situated at the southernmost end of the India where the meeting place of Arabian Sea, Bay of Bengal and Indian Ocean. The fish collectors and traders were also extensively involving in ornamental fish trading in this area.

Data analysis

The data were analysed with approached to various statistical methods such as univariate, graphical/distributional and multivariate methods. The computer programme PRIMER (Ver.6.1) was used for univariate and multivariate data analysis. Species richness was expressed by considering the number of species (D), and species diversity and homogeneity were determined using the Shannon-Wiener diversity index (H') and the evenness index (J') (Pielou, 1966). One-way ANOVA was carried out with SPSS program.

RESULTS

Distribution, diversity and taxonomy of marine angel fishes

The survey was made on the quantitative assessment of marine angel fish distribution and diversity in southeast coast of India. In the present survey, a total of 5 species belong to three genera were recorded viz., Pomacanthus (3 species), Centropyge (1 species) and Apolemichthys (1 species). The taxonomical position and distinctive characters of the marine angel fishes reported were presented below.

Taxonomical position of marine angel fishes (Pomacanthidae)

Kingdom: Animalia
Phylum: Chordata
Class: Actinopterygii

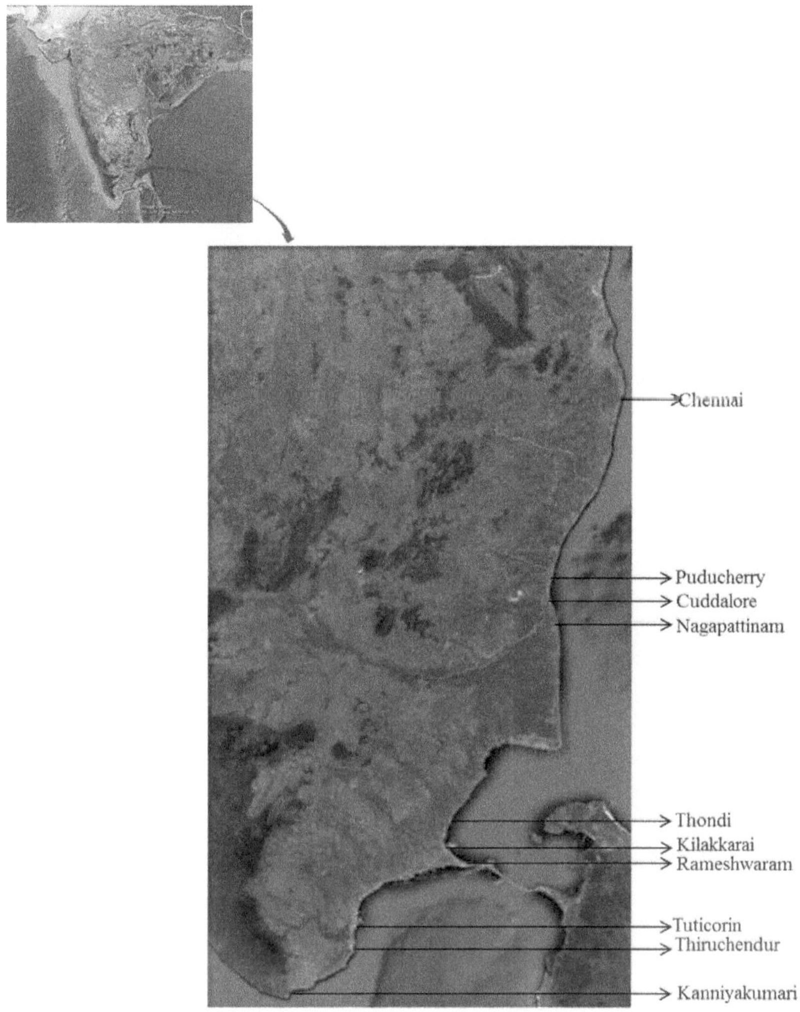

Figure 1. Study area.

Order: Perciformes
Family: Pomacanthidae.

Apolemichthys xanthurus (Bennett, 1832) (Smoke angel)

Distinctive characters

Dorsal spines: 14 - 15; dorsal soft rays: 16 - 18; anal spines: 3; anal soft rays: 17 - 19. Body oval to square and latterly compressed; eyes are distinct and small mouth. Pre-opercle bears strong spine. Lateral line is crossing from the upper part of the body.

Colour

Head is dark brown or black in colour. Snout lighter; a yellow spot present very close to eye. Caudal fin is yellow in colour; dorsal and anal fins are dark brown to black with white margin; adjacent caudal peduncle region also black. Body pale grey or light brown; black spots on scales larger dorsally, resulting in dark colour on upper sides grading to pearly white ventrally.

Habitat and biology

Inhabits in coral-rich areas, singly or in pairs. Frequently observed outer reef edge and reef slopes at depths of 15 m. Feeds mainly on sponges and sea squits. Maximum size is about 25 cm (Diagram 1).

Centropyge nox (Bleeker, 1853) (Midnight angel)

Distinctive characters

Dorsal spines (total): 14 - 15; dorsal soft rays (total): 16 - 17; anal spines: 3; anal soft rays: 16 - 17. Body is oval and latterly compressed; distinct eye and small mouth. Pre -opercle bears strong spine. Lateral line is crossing

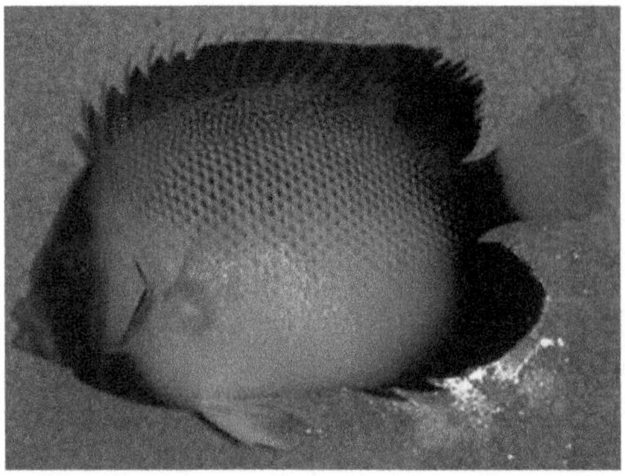

Diagram 1. *Apolemichthys xanthurus* (Bennett, 1832) (Smoke angel). Local Name: Yellow tail Angel, Smoke angel.

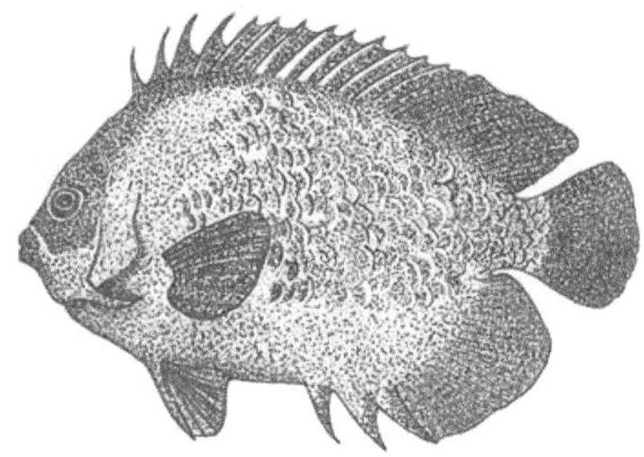

Diagram 2. *Centropyge nox* (Bleeker, 1853) (Midnight angel). Local Name: Midnight Angel.

from the upper part of the body.

Colour

Dark brown to black in colour; caudal fin with a narrow white margin. Uniformly black at all stages, sometimes a yellowish on the pectoral fin base but has a rare xanthic form in which most of the body is dark yellow.

Habitat and biology

Common in coral rich areas of outer reef slopes, occasionally sighted in lagoons and passages. Prefers habitats with richly mixed invertebrate growth. Usually occur singly or in pairs. Feeds on algae, forms harems of 3 to 7 individuals, maximum length is about 10.0 cm.

Characters

Body oval to square and laterally compressed; distinct eye and small mouth. Pre-opercle bears strong spine. Creamy white colour with anterior and posterior thick black vertical patch; lateral line crossing from the upper part of the body (Diagram 2).

Pomacanthus semicirculatus **(Cuvier, 1831) (Koran Angel)**

Distinctive characters

Dorsal fin with 13 spines and 20 to 23 rays. Anal fin with 3 spines and 18 to 21 rays. Pectoral *fin* rays 19 to 21. A filament posteriorly on dorsal and anal fins.

Colour

Greenish body with dark flecks. Bright blue streaks on cheek, and posterior part of the body with a blue outline and lips are yellow. Juveniles are looks dark blue with white and pale blue crescents on their sides.

Habitat and biology

Adults occur in coastal reefs with heavy coral growth to

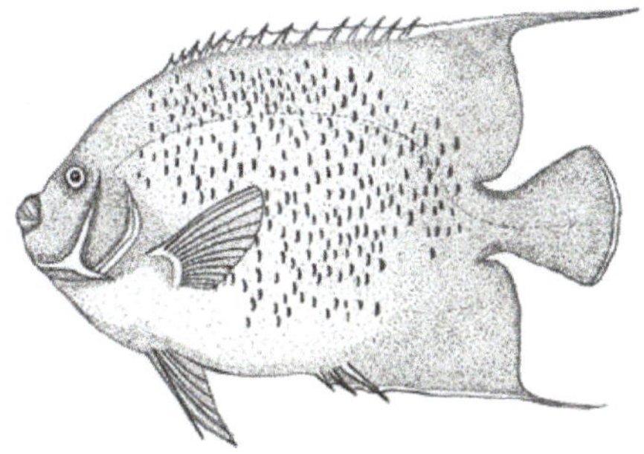

Diagram 3. *Pomacanthus semicirculatus* (Cuvier, 1831) (Koran Angel). Local Name: Alla Fish.

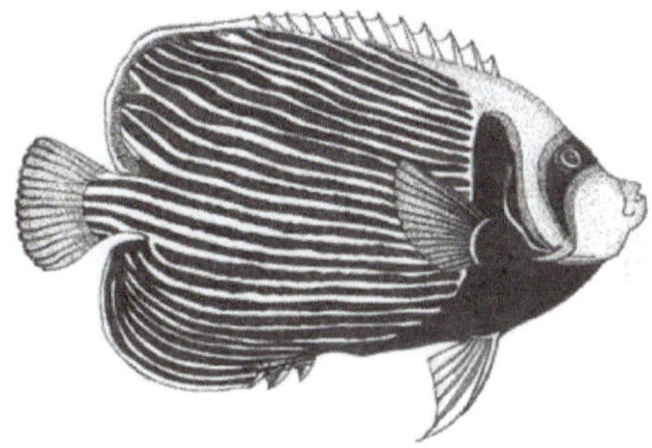

Diagram 4. *Pomacanthus imperator* (Bloch, 1787) (Emperor Angel). Local Name: Emperor angel fish.

depths of 30 m. Solitary, but sometimes in pairs. Maximum size is about 38 cm.

Characters

Dorsal profile straight to slope towards front, body compressed. Mouth is small and prominent upwards; distinct eye, body with excellent seven blue lines bending towards down to tip of the soft dorsal fin (Diagram 3).

Pomacanthus imperator **(Bloch, 1787) (Emperor Angel)**

Distinctive characters

Dorsal fin with 13 to 14 spines and 18 to 21 rays. Anal fin with 3 spines and 18 to 21 rays. Pectoral fin rays are 19 to 20. Body depth 1.6 to 1 .8 in standard length. Soft dorsal fin develops a point with growth. Posterior margin of preopercle finely serrate. Caudal fin rounded.

Colour

Sides with alternate blue and yellow stripes. Dorsal and caudal fin are mainly yellow and anal blue or black. Black face mask with blue edge. Juveniles are deep blue with curved circular white bands.

Habitat and biology

Solitary, usually seen on exposed outer reef areas near edges and caves in rich coral areas up to 25 m. Feeds primarily on sponges. Maximum size is about 40 cm.

Characters

Laterally compressed and rhomboid. Pre-opercle spine longer and prominent with serration on margin. Dorsal fin acute and soft at the end, body purplish blue with yellow stripes. Snout and chick bluish in colour. Anal fin semicircled or blended at the end (Diagram 4).

Pomacanthus annularis **(Bloch, 1787) (Bluering Angel)**

Distinctive characters

Dorsal spines (total): 13; dorsal soft rays (total): 20 to 21;

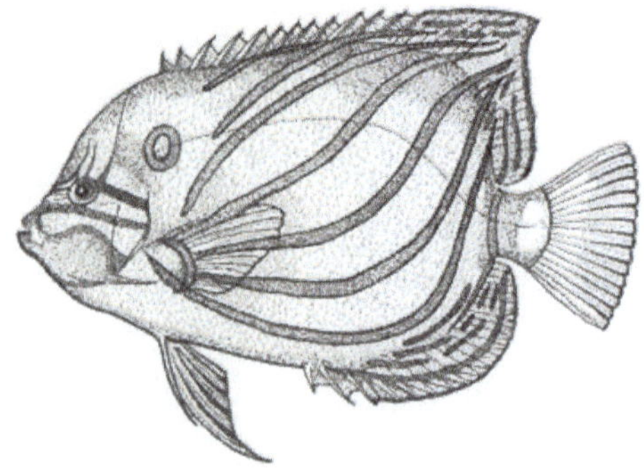

Diagram 5. *Pomacanthus annularis* (Bloch, 1787) (Bluering Angel). Local Name: Ring angel, Vannathi meen.

anal spines: 3; anal soft rays: 20. Juveniles black with alternating white and blue curved well-spaced stripes on the sides. Adults golden brown or orange with well-spaced curved horizontal stripes radiating from the pectoral-fin base area, running along the sides towards the posterior portion of the dorsal fin.

Colour

Two similar blue stripes run horizontally across the face, one running through the eye, from above the snout to the edge of the operculum. A blue ring is behind and slightly above the edge of the operculum. Caudal fin is white with bright yellow margin.

Habitat and biology

Inhabits coastal reefs to at least 30 m. Adults often found in pairs inside caves, Juveniles settle in very shallow inshore habitats with short filamentous algae growth on rock or dead coral substrates. Feeds on sponges and tunicates. Undergoes a complete colour transformation from the juvenile to adult stage. Maximum size is about 45.0 cm.

Characters

Dorsal profile straight to slope towards front, body compressed. Mouth is small and prominent upwards; body with excellent seven blue lines bending towards down to tip of the soft dorsal fin (Diagram 5).

Diversity of marine angel fishes

In the present study, abundance of marine angel fishes between January 2008 and December 2009 in 10 stations along Tamilnadu coast, Southeast coast of India was investigated. During the study period, a total of 5 species belong to three genera namely *Apolemichthys xanthurus (*Smoke angel*)*, *Centropyge nox* (Midnight angel), *Pomacanthus semicirculatus* (Koran angel)*, Pomacanthus annularis* (Bluering angel) and *Pomacanthus imperator* (Emperor angel) were recorded. Of these, smoke angel was found to be the dominant followed by midnight angel, koran angel, bluering angel and emperor angel was found to be lower in the order with 5 species in all seasons at 10 stations. The population density of marine angel fishes ranged between 1638 and 10,656 number of fishes with maximum during summer'08 and minimum during postmonsoon'08 (Figure 2).

The studies on station wise abundance of marine angel fishes revealed that maximum was recorded at Tuticorin and minimum at Puducherry during the study period (Figure 3). The higher abundance of marine angel fishes reported in Tuticorin, Kilakkarai, Rameshwaram and Thiruchendur was due to the occurrence of shallow water, coral reefs and rocky bottom. The koran angel, bluering angel and emperor angel were recorded in higher numbers only in Gulf of Mannar region especially in Rameshwaram, Kilakkarai and Tuticorin whereas emperor angel was only in Thiruchendur due to the rocky shore. The species wise abundance studies revealed that the smoke angel was found to be reported at all the 10 stations throughout the year followed midnight angel was predominant and almost recorded in all the 10 stations in throughout the year. Compared to other marine angel fishes, smoke angel (*Apolemichthys xanthurus*) possessing high market demand, easy to acclimatize to any environmental condition and ready to accept any kind of feed.

The percentage compositions of marine angel fish in all the 10 stations were presented in Figure 4. The results revealed that the smoke angel was found to be the dominant group by constituting 57% of the total fishes recorded followed by midnight angel (14%), koran angel (14%), bluering angel (10%) and emperor angel (9%). To study the diversity of marine angel fishes in south east coast of India, three biodiversity indices, namely, Simpson diversity, Shannon diversity and Brillouin

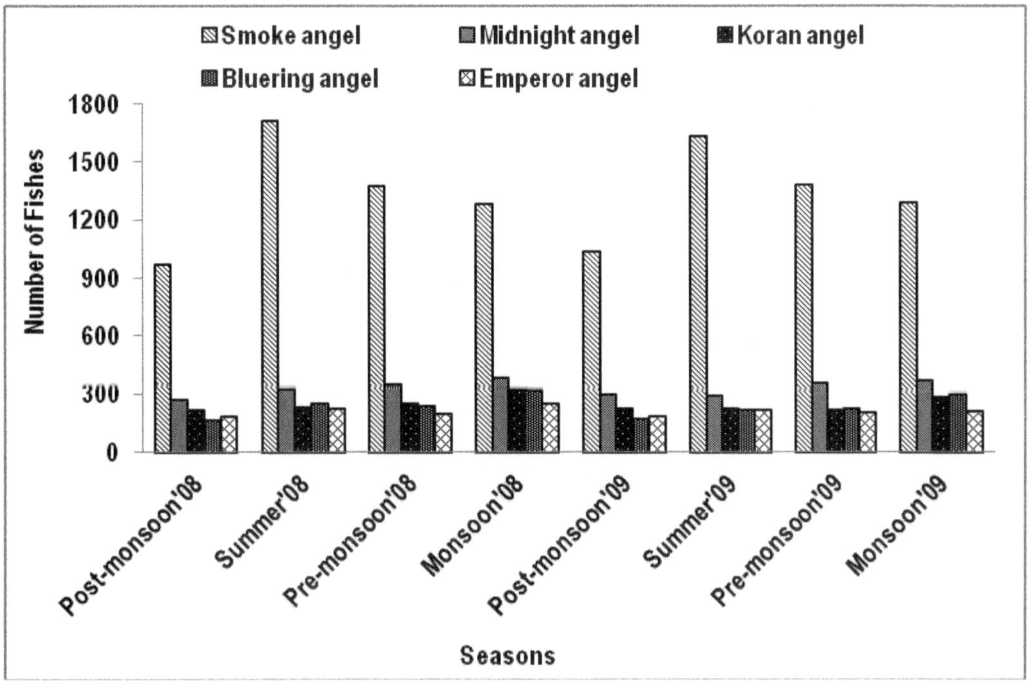

Figure 2. Abundance of marine angelfishes on various seasons between January 2008 and December 2009 along Tamilnadu, southeast coast of India.

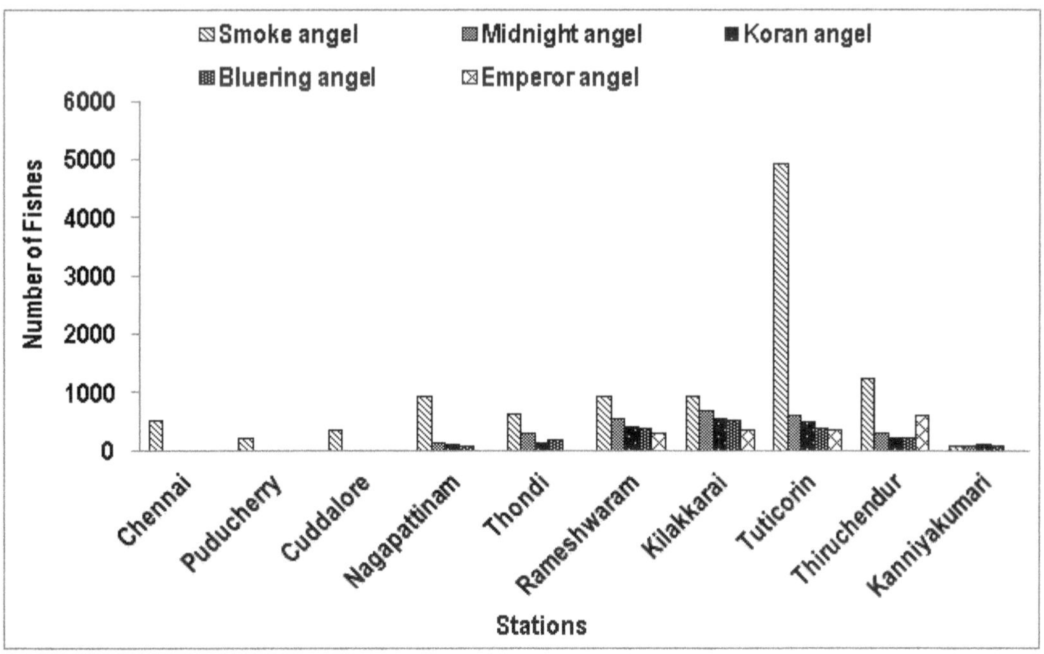

Figure 3. Abundance of marine angelfishes at various stations between January 2008 and December 2009 along Tamilnadu, southeast coast of India.

diversity, Simpson evenness, Evenness and Brillouin evenness were employed. The marine angel fish diversity (H) was ranged between 0.187 and 2.298 with maximum at Rameshwaram and minimum at Chennai. The richness (d) ranged between 0.281 and 1.169 with maximum at Kanniyakumari and minimum at Chennai and the evenness ranged between 0.170 and 1 with maximum at Kanniyakumari and minimum at Nagapattinam (Table 1).

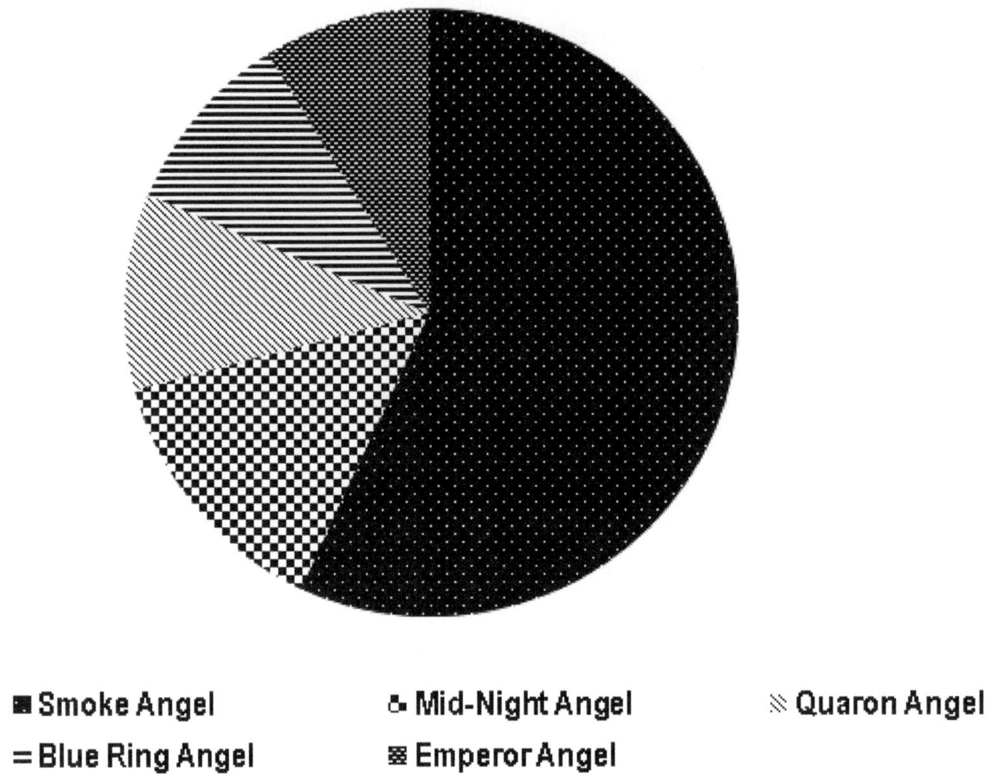

Figure 4. Percentage composition of marine angel fish species recorded during study period along Tamilnadu, southeast coast of India.

Table 1. Diversity indices of marine angel fish along Tamilnadu, Southeast coast of India.

Stations	Diversity		Richness		Evenness	
	Max	Min	Max	Min	Max	Min
Chennai	1.378783	0.187176	1.027797	0.281266	0.970951	0.187176
Pondy	1.378783	0.266765	1.027797	0.314658	0.970951	0.266765
Cuddalore	1.459148	0.242292	1.116221	0.303413	0.970951	0.242292
Nagapattinam	1.408311	0.270512	0.834065	0.253085	0.823812	0.170674
Thondi	2.119451	0.803072	1.027797	0.522379	0.991964	0.506682
Kilakkarai	2.297499	2.11075	1.022489	0.729842	0.989479	0.90181
Rameshwaram	2.298543	2.032264	1.022489	0.746089	0.989929	0.875248
Tuticorin	1.683164	1.026483	0.805989	0.648559	0.724899	0.442082
Thiruchenthur	2.105318	1.468818	1.038921	0.785276	0.906711	0.632585
Kanyakumari	2.092625	0.591673	1.169614	0.434294	1	0.591673

In order to ascertain the diversity pattern graphically, K-dominance plot drawn combine for all the 10 stations did not show any pronounced variation between the stations (Figure 5). The trend observed in diversity index was quite evident here as well. The plots drawn for 10 stations rose slowly due the presence of more number of species. As the percentage contribution of each species was added, the curve extended horizontally giving 'S' shape, before reaching the cumulative 100%.

The MDS plot drawn clearly revealed that the postmonsoon'08 and postmonsoon'09 were grouped at the highest level of similarity (98%) and followed by monsoon'08 (Figure 6). Among the 5 species of recorded marine angel fishes, smoke angel was grouped separately with higher level of similarity followed by midnight angel (Figure 7). Similarly, Tuticorin was grouped separately with highest level followed by Rameshwaram and Kilakkarai (Figure 8). The above

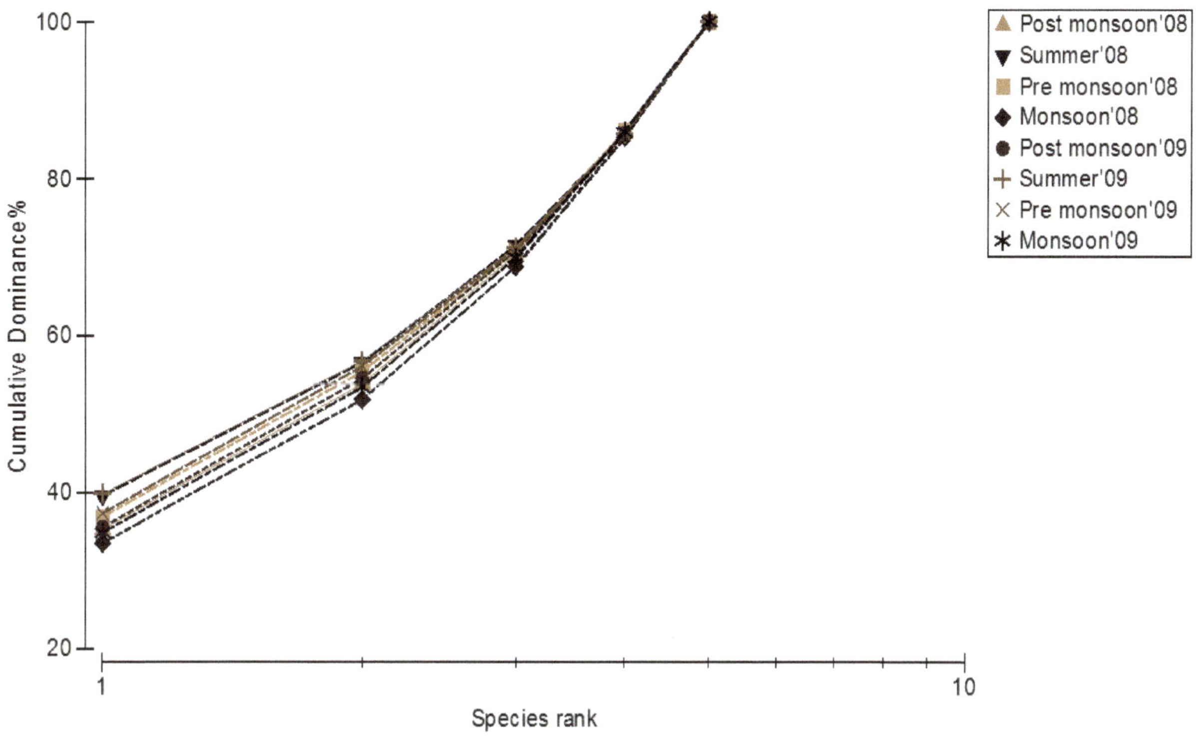

Figure 5. K - Dominance curve drawn for abundance of marine angel fishes at various seasons between January 2008 and December 2009.

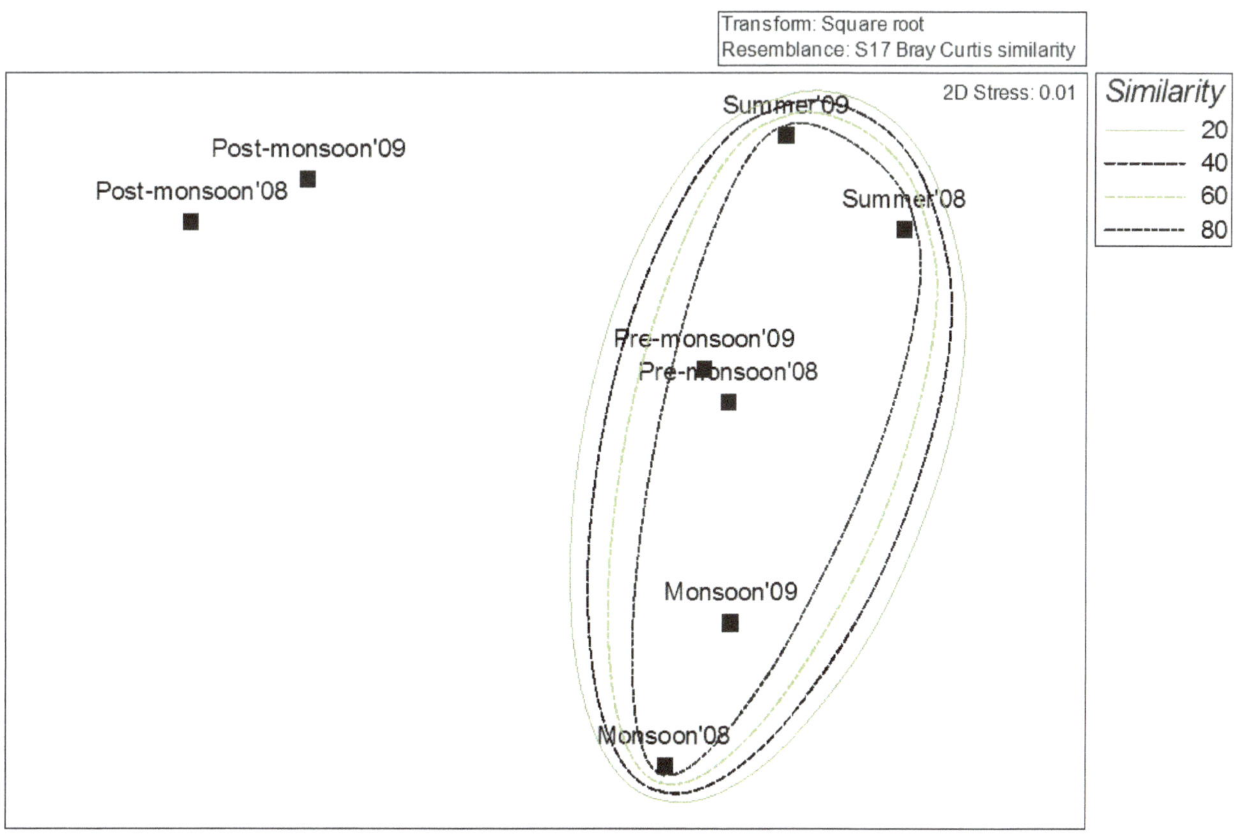

Figure 6. MDS ordination plot for abundance of marine angel fishes at various seasons.

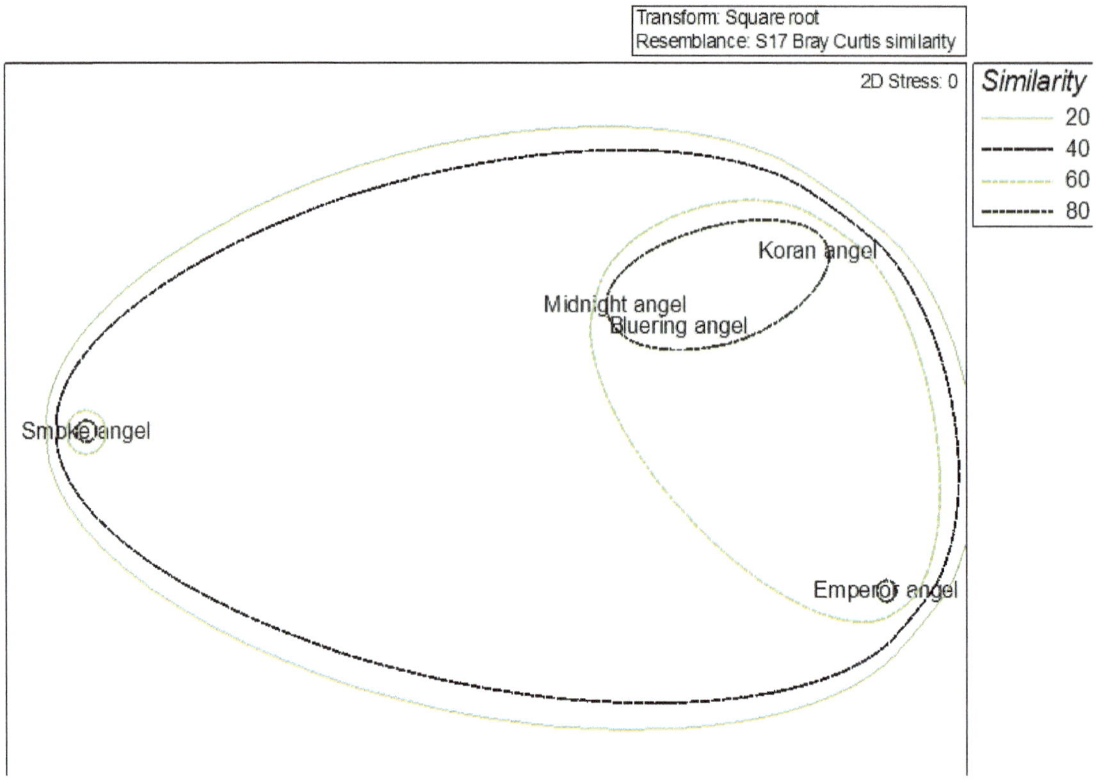

Figure 7. MDS ordination plot for species wise abundance of marine angel fishes recorded along Tamilnadu, southeast coast of India.

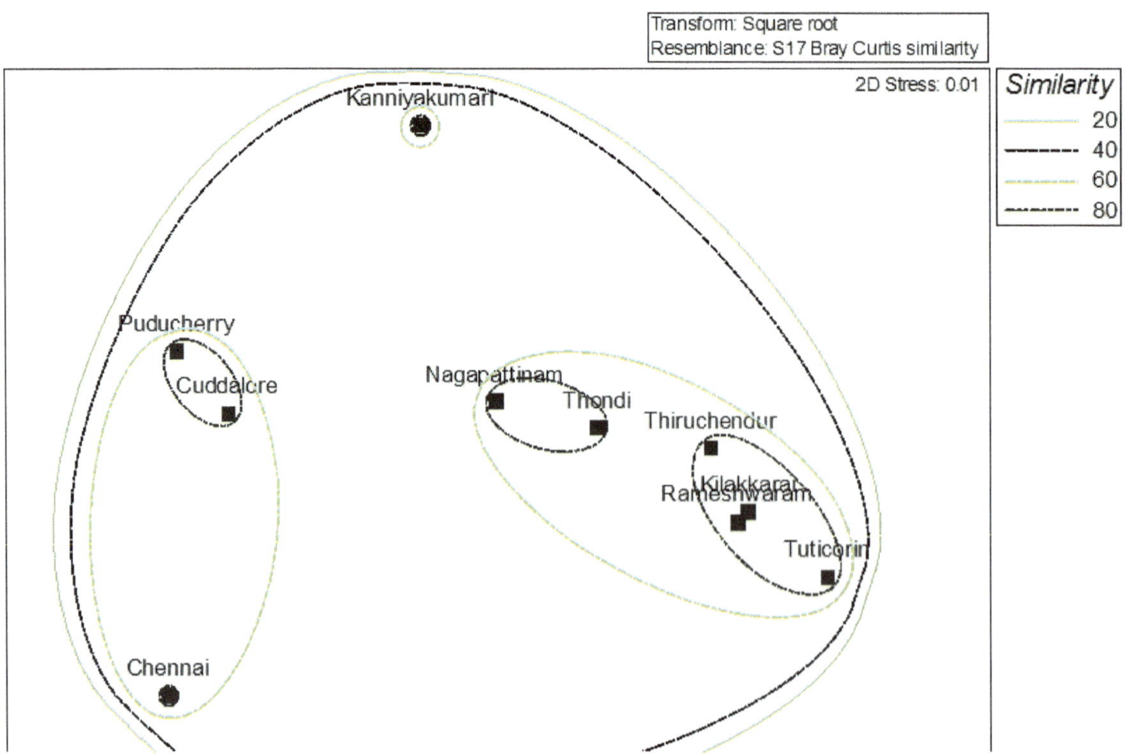

Figure 8. MDS ordination plot for station wise abundance of marine angel fishes recorded along Tamilnadu, southeast coast of India.

patterns of grouping of stations were quite evident in MDS due to the fact that the Rameshwaram and Kilakkarai were more lying on the one side of the map and other stations on the other side. The stress value which was overlying on the MDS was also vary (0.01) indicating the excellent ordination pattern of the stations

DISCUSSION

Marine angelfishes of the family Pomacanthidae are typical coral reef inhabitants and are economically important as high priced ornamental fishes (Moenich, 1991). The family Pomacanthidae has circumtropical distribution with 88 species in eight genera (Chung and Woo, 1998; Debelius et al., 2003). In the present study, a total of 5 species of marine angelfishes belongs to 3 genera such as Pomacanthus (3), Centropyge (1) and Apolemichthys (1) were recorded between 1 and 20 m depth along Tamilnadu coast, South east coast of India. Bellwood et al. (2004) reported about 24 Pomacanthid species with two individuals of each species and at least two species from each of the eight genera in Australian waters. However, Alwany (2009) recorded about 6 species of angelfishes which occurred at the depth ranged between reef flat (0.5 to 1 m) in Shalateen region in the northern Red Sea. Khalaf and Disi (1997) reported 5 species of angelfishes in the Gulf of Aqaba. This gives evidence that the angelfish community is relatively the same inside the fringing reefs of Egyptian coast of the Red Sea. It seems that the healthy coral communities of the northern Red Sea, a low diversity of species of angelfishes and number of individual was higher than other areas, such as the French Polynesia and the Great Barrier Reef. The closed - system of Red Sea may be the important factor, which leads to scarcity of fish species indicates that the number of individuals was high because of the healthy coral communities of the northern Red Sea (Alwany, 2009).

The results of the present study indicated that, there are five species of marine angelfishes were reported to occur at the depth ranged between 0.5 to 1 and 20 m. Similarly, Alwany et al. (2007) also stated that the abundance and number of species of the angelfishes were increased with increasing depth in Shalateen region in the northern Red Sea. The occurrence of marine angel fish studies concealed that the smoke angel was distributed uniformly in all the 10 stations throughout year whereas midnight, blue ring, Koran and emperor angel were distributed at some specific locations. The diversity of marine angel fishes were analysed using various diversity indices such as Simpson diversity, Shannon diversity, Brillouin diversity and evenness test revealed that the diversity and evenness was comparatively higher in Rameshwaram and Kanyakumari respectively. Venkataramani and Jawahar (2004) recorded about 113 marine ornamental finfish species in Gulf of Mannar and also claimed that this region could be exploited more for ornamental fish trade in India.

The abundance of marine angel fishes in this study was comparatively higher especially in coral reef rich regions such as Gulf of Mannar comprising of Thondi, Rameshwaram, Kilakkarai, Tuticorin and Thiruchendur compared to other regions. In addition to this, MDS ordination plots were also revealed that the diversity was higher in the postmonsoon season and Rameswaram and Kilakarai were grouped separately with highest level of similarity (98%). The abundance and distribution of fishes is dependent on several distinct factors such as recruitment, habitat structure, food availability, and environmental factors (Jones, 1991; Williams, 1991). Roberts and Ormond (1987) stated that most of the coral reef fishes tend to increase in both abundance and number of species with increasing depth on fringing reefs. In contrast, Olivotto et al. (2006) stated that the vertical distribution of angelfishes was increasing with depths, where the highest average abundance of angelfishes was recorded on 20 m (93.0 ± 6.1 fish/600 m^3). However, due to unavailability of the comparable published literature, it is safe to say this is the first as well as pioneer attempt on this commercially important ornamental fish group which could be useful for better understanding of the status of diversity and distribution pattern of marine angelfishes along Tamil Nadu coast, southeast coast of India.

ACKNOWLEDGEMENTS

The authors are very much grateful to the Ministry of Earth Sciences - COMAPS for providing fund and fellowship. The authors are also grateful to the authorities of Annamalai University for providing facility to carry out the work.

REFERENCES

Alwany MA (2009). Distribution and feeding ecology of the angelfishes (Pomacanthidae) in Shalateen region, Red Sea, Egypt. Egypt J. Auat. Biol. Fish 13:79-91.

Alwany MA, El-Etreby SG, Hanafy MH (2007). Distribution and abundance of butterfly fishes along the Gulf of Aqaba, Egypt J. Fish. Aqua. Sci. 2(6):395-402.

Bellwood DR, Herwerden L, Konow N (2004). Evolution and biogeography of marine angelfishes (Pisces: Pomacanthidae). Mol. Phyl. Evol. 33:140-155.

Chung KC, Woo NYS (1998). Phylogenetic relationships of the Pomacanthidae (Pisces: Teleostei) inferred from from allozyme variation. J. Zool. Lond. 246:215-231.

Debelius H, Tanaka H, Kuiter RH (2003). Angelfishes: A comprehensive guide to Pomacanthidae. TMC Publishing, Chorleywood, UK. p 208.

Jones GP (1991). Post recruitment processes in the ecology of coral reef fishes populations: a multifactorial perspective. In: P.F. Sale (ed.) the ecology of coral reef fishes. New York Academic Press, San Diego, pp. 294 – 328. Khalaf MA, Disi AM (1997). Fishes of the Gulf of Aqaba. Mar. Sci. Station, Aqaba, Jordan. 252 pp

Khalaf MA, Kochzius M (2002). Community structure and biogeography of shore fishes in the Gulf of Aqaba, Red Sea. Helgol. Mar. Res. 55:252-284.

Moenich DR (1991). Marine angels - *Pomacanthus*. TFH Mag. 39:68-81.

Murugan A, Durgekar R (2008). Beyond the Tsunami: Status of fisheries in Tamil Nadu, India: A snapshot of present and long-term trends. UNDP / UNTRS, Chennai and ATREE, Bangalore, India. P. 75.

Nelson JS (2006). Fishes of the World, 4th edn, John Wiley and Sons, NewYork, USA, P. 601.

Olivotto I, Holt SA, Carnevali O, Holt GJ (2006). Spawning, early development, and first feeding in the lemon peel angelfish *Centropyge flavissimus*. Aquaculture 253:270-278

Jayalal L, Ramachandran A (2012). Export trend of Indian ornamental fish industry. Agric. Biol. J. N. Am. 3(11):439-451

Roberts CM, Ormond RF (1987). Habitat complexity and coral reef fish diversity and abundance on Red Sea fringing reefs. Mar. Ecol. Prog. Ser. 41:1-8

Sadovy S (1992). A preliminary assessment of the marine aquarium export trade in Puerto Rico. Proc. 7th Int. Coral Reef Symp. Guam 2:1014-1022

Thresher RE (1982). Courtship and spawning in the emperor angelfish *Pomacanthus imperator*, with comments on reproduction by other pomacanthid fishes. Mar. Biol. 70:149-156

Venkataramani VK, Jawahar P (2004). Resource assessment of ornamental reef fisheries of Gulf of Mannar, Southeast coast of India. Final report - ICAR / NATP / CGP / Project. P. 66.

Wabnitz C, Taylor M, Green E, Razak T (2003). From ocean to aquarium. UNEP-WCMC, Cambribge, U.K. P. 64.

Length-weight relationships and condition factors of *Alestes baremoze*, *Brycinus nurse* and *Schilbe intermedius* from the lower reaches of White Volta River (Yapei), Ghana

Seth Mensah Abobi[1] **and Werner Ekau**[2]

[1]Department of Fisheries and Aquatic Resources Management, University for Development Studies, P. O. Box TL 1882 Tamale, Ghana.
[2]Leibniz Center for Tropical Marine Ecology (ZMT), Fahrenheitstr. 6 D-28359 Bremen, Germany.

A study to assess the length-weight relationships and the condition factors of 3 abundant and commercial fish species namely *Alestes baremoze*, *Brycinus nurse* and *Schilbe intermedius* in the lower reaches of the White Volta River (Yapei) was conducted from October 2011 to March 2012. Using $W = aL^b$, the length-weight relationships of the species were calculated. Condition factors of the species were obtained using the formula: $W*100/L^b$. A total of 284 *A. baremoze* were measured with total mid length size range of 10.5 to 44.5 cm. A total of 197 *B. nurse* were measured and ranged from 6.5 to 27.5 cm total mid-length whereas 175 of *S. intermedius* were measured and had total mid length size range of 8.5 to 26.5 cm. The slope *b* of the length-weight relationships of the 3 species were within the acceptable range of 2.5 to 3.5 and can be used to approximate the weight for the 3 species in the lower reaches of the White Volta River. The condition factors variation of *A. baremoze* and *B. nurse* showed a pronounced change in December 2011, when the surface water temperature was lowest. *S. intermedius* had no significant changes in the condition factors over the study period.

Key words: White Volta River, total mid-length, size range, species, surface water temperature.

INTRODUCTION

The Lake Volta is the most important inland fishery area in Ghana. According to FAO statistics, inland capture fisheries contributed 27 of total Ghanaian fish production in 2009 (FAO FishStat Plus). It is estimated that the reservoir provides 90% of the national freshwater fish production (Abban, 1999). The Lake Volta in Ghana is an immense reservoir created after a completion of the Akosombo Dam on the Volta River in 1964. It is the largest man-made lake by surface area in the world. It has a length of 520 km and covers about 8,500 km^2, or 3.2% of Ghana's total land area. The reservoir stores about 149 km^3 of water (van Zwieten et al., 2011). The lake Volta extends from the Akosombo Dam in southeastern Ghana to the town of Yapei in the Northern region of Ghana. Yapei is thus the terminal point for the lake Volta and the lower reaches of the White Volta River (WVR) starts from there. According to Braimah (2003) some 300 000 fisherfolk depend on the Volta reservoir for their livelihood. The Volta reservoir is endowed with fisheries resources. Dankwa et al. (1999) identified 121 species in the Volta reservoir.

The Lake Volta and its tributary Volta Rivers in the last

4 decades have undergone great changes in its ecology, limno-chemistry and socio-economy. Increased pressure on land along the banks has led to high rates of deforestation. This has resulted in increased soil erosion leading to the transportation of high loads of silt and nutrients through rivers into the lake, thereby contributing to its eutrophication. Furthermore, wetlands bordering the lake are being converted into agricultural land or land for grazing cattle, and therefore may not be able to act as natural filters for nutrients and silt, and now do not provide breeding grounds for many fish species (Ofori-Danson et al., 2001). As a result of these problems, the Lake Volta Research and Development Project (VLR and DP) was undertaken under the Food and Agricultural Organization and United Nations Development Programme during the first decade of the lake's existence (FAO and UNDP 1971, 1979). These studies came to an end in 1978. Since then, systematic data collection from the Lake Volta and Volta Rivers natural resources has been lacking. There have been calls for renewed studies to facilitate their management due to declining catches.

A. baremoze, B. nurse and *S. intermedius* are very abundant from October to March in the Yapei stretch of the White Volta River. During this period, these fishes constituent a significant part of the fish diet of the local inhabitants. In this way, they provide a source of animal protein.

The importance of the Yapei fishery cannot be underestimated. Yapei is 52 km from the Northern regional capital Tamale. Tamale is the third largest city in Ghana with a population of 537,986 (Wikipedia, 2012). Yapei is easily accessible as it is on the Accra-Kumasi-Tamale trunk road. Yapei is one of the 9 main sources of smoked fish to the Tamale Central Market (Obodai et al., 2009). The Tamale Central Market is the largest market center in the metropolis and majority of the inhabitants buy their food stuffs from this market. It thus makes Yapei an important supplier of fish especially smoked fish. Fishermen at the Yapei stretch of the White Volta River practice unregulated and unselective harvesting of fish with fishermen harvesting all sizes of fish without regard to the sustenance of the fishery. These practices targeted at more catches for more income. Fish can only be harvested at the maximum sustainable yield (MSY) when all the biological parameters are known.

Length-weight relationship and condition factor are of great importance to the fishery industry as they help to predict the best length and time suited to harvest a particular species of fish. The study therefore focused on providing information on the length-weight relationships and condition factors of *A. Baremoze*, *B. Nurse* and *S. Intermedius* from White Volta River, Ghana.

MATERIALS AND METHODS

Study area

The samples were conducted at the Yapei stretch of the White Volta River, which is a major landing site in the Northern Region of Ghana. 3 non-overlapping landing sites namely Pataplapei, Porturto and Aglassipei were selected (Figure 1) to provide a representative overview of the fisheries in the area.

Fish sampling and measurement

The samples of the selected fish species were taken at random from landings and their total length recorded to the nearest 0.1 cm using a fish measuring board while the body weight was recorded to the nearest 0.01 g using a balance after using tissue to mop off water from the surface of the specimens.

The measured lengths were later grouped into class intervals for analysis. The length-weight equation; $W= aL^b$ as described by Ricker (1975) was used to establish the length-weight relationship of the species measured, where W is the weight (g), L is the total length (cm), *a* is a constant determined empirically and *b* is the slope of the equation. The slope *b* has a numerical value which is mostly between 2.5 and 3.5 and is often close to 3 (Bagenal and Tesch, 1978; Carlander, 1969; Gayanilo and Pauly, 1997; Froese, 2006; Weatherley and Gill, 1987). Under isometric growth *b* equals to 3 but isometric growth in fish is rare (Bolger and Connolly, 1989; McGurk, 1985).

The equation relating weight-length gives some indication of the state of wellbeing of a fish in a population which is known as the condition factor. Fish in good condition have higher condition factor value than those in poor condition.

The condition factor (K) is defined as:

$$K=\frac{\text{weight (g)}}{\text{length (cm)^b}}\times 100$$

Modified from Fulton (1902) formula of K: W/L^3 (W: weight and L: Total length)

The total body weight (in grams) of the ungutted fish was used; the length was total length (TL) in centimeters. The exponential b value is the slope of the length-weight equation. Having weighed and measured over hundreds of the selected fish species from the lower reaches of the White Volta River, the condition factor (K) was multiplied by 100 in order to bring the K value close to unity.

Physicochemical parameters

Using Garmin GPS 60, the coordinates of the sampling points in the water were stored and at each time of sampling, the sampling points were traced back using the GPS tracking system. All the parameters were taken monthly at all the landing site. The readings and the water samples were taken between 10.00 am and 12.00 pm GMT.

The surface water temperature was measured using Hanna Checktemp1 (a digital pocket thermometer). 3 readings were made at each landing site (20 m from the mouth of the landing site, and then in the middle of water and 20 m from the mouth of the opposite end of the landing site) and the average was taken as the mean surface water temperature for the site.

Dissolved oxygen concentration was measured using the Azide modification of Winkler method. The bubble free water samples were taken from the center of the water at each landing site and treated on the field by removing the stopper from the 250 ml BOD bottle and adding 2 ml of Manganese (II) Suphate ($MnSO_4$) (Winkler 1) to the water samples, followed by 2 ml alkali-iodate-azide (Winkler 2). The stopper was restored at once and the contents were mixed by inverting the bottle for about 3 min. The precipitates were allowed to settle. The samples were stored in ice chest and

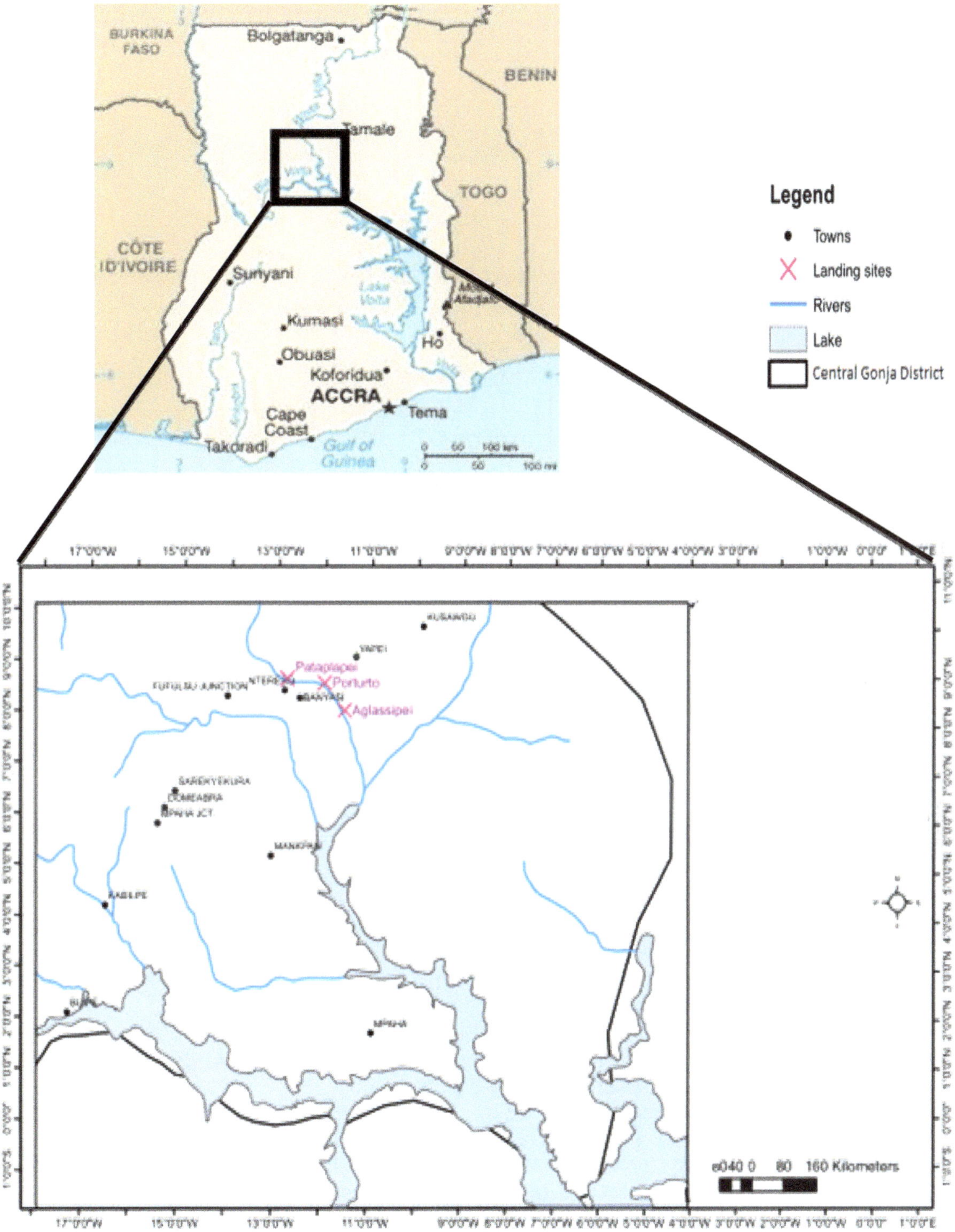

Figure 1. Map of Ghana with an insert of the detailed study area showing Yapei, and the 3 landing sites; Pataplapei, Porturto, and Aglassipei.

transported to the laboratory. In the Water Research Institute (WRI) laboratory Tamale, Ghana, 2 ml concentration of sulphuric acid was added along the neck and shaken till dissolution was complete. Hundred (100) ml of the solution was taken and titrated with M/80 sodium thiosulphate solution to a straw yellow colour. 1 to 2 ml of starch solution was added and the titration continued to the point where the blue colour changed to colourless. The calculation was as follows: mg/l O_2 = Vol. of thiosulphate used × 101.6 / Vol. of sample used. pH meter and combination of electrode (pH Meter-Jenway Model 3510) was used to measure the pH at middle of the

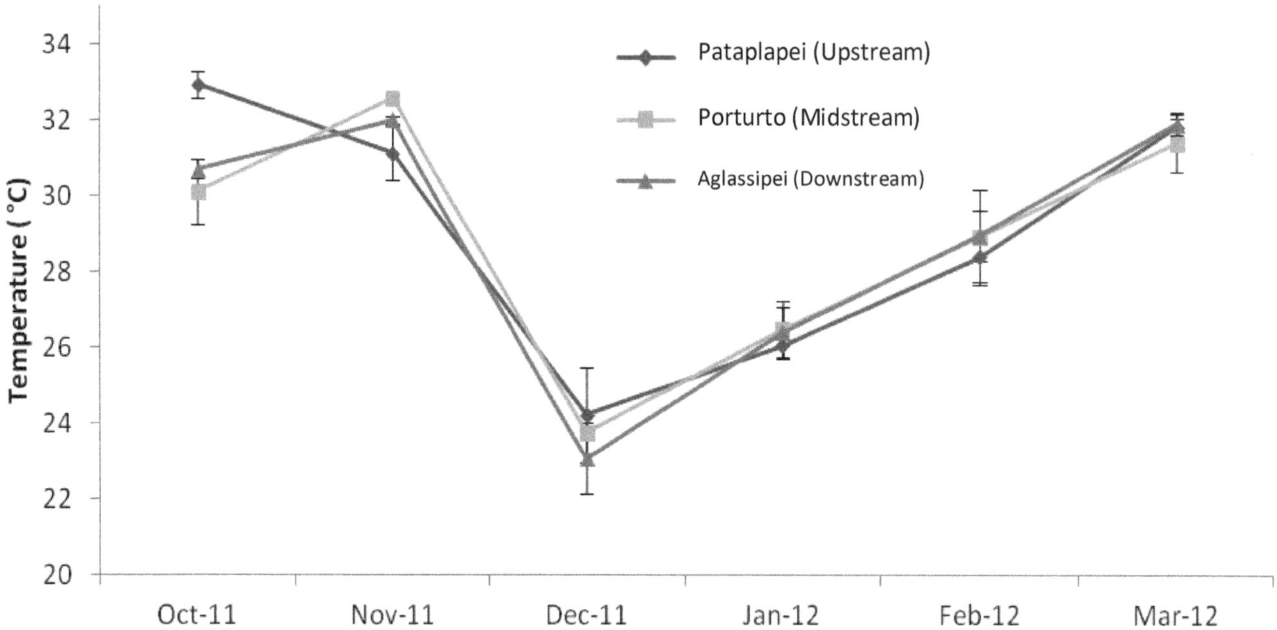

Figure 2. Mean and SD (±) of the surface water temperature of 3 landing sites in the lower reaches of the White Volta River from October 2011 to March, 2012.

water. The readings were also compared to the observations made with pH strips and colour charts.

The water depths of 3 points at the center (the coordinates stored on the GPS) of the water and perpendicular to the landing site were measured monthly with Speedtech Hand Eco sounder.

Data analysis

Descriptive statistics of the Software STATISICA 8.1 and Microsoft Excel 2010 were used to analyze catches data for means, frequencies, percentages, length-weight relationships and condition factors.

RESULTS

Physico-chemical condition of lower reaches of the white volta river during the period of study

Monthly mean temperature values recorded were between 23.1and 32.9°C. The mean temperature of the whole 8.84 km stretch (Pataplapei, Porturto and Aglassipei) during the 6 months period of study was 28.8°C. In the landing site Pataplapei, the highest temperature was recorded in October 2011 whereas the landing sites, Porturto and Aglassipei had the highest temperature readings in November 2011. All the 3 landing sites had their lowest and striking changes in temperature in December 2011 (Figure 2).

The dissolved oxygen ranged from 2.7 to 4.7 mg/l whereas pH values ranged between 6.8 and 7.86.

Water levels

The highest water depth was 6.5 m and it was recorded in October 2011 at the downstream Aglassipei. The lowest was 3.9 m and it was recorded in March 2012 at upstream Pataplapei. All the landing sites had the highest water level in October 2011 and the water level consistently decreased month by month in all the sites and subsequently the lowest water level at all the sites was recorded in March 2012 (Figure 3). Pataplapei which is upstream, is the shallowest. It is followed by Porturto (midstream). Aglassipei has the deepest water levels.

Length frequency distributions of the three fishes

The size frequency histograms for *A. baremoze*, *B. nurse* and *S. intermedius* is shown in Figures 4 to 6 respectively. Over the 6 months period of study, *A. baremoze* had the largest sample size with a total of 284 fish measured. The total length ranged from 10.5 to 44.5 cm. 2 cohorts of the species were observed throughout the period of the study. The first cohort had total lengths around 10.5 to 15.5 cm and the second cohort 27.5 to 33.5 cm. The first cohort remained within the same size range from month to month whereas the second cohort had a slight shift in size range. This shift was realized from January to March when the peaks of the total lengths moved to the left towards a smaller size range around 28.5 cm. Length distributions of *A. baremoze* as

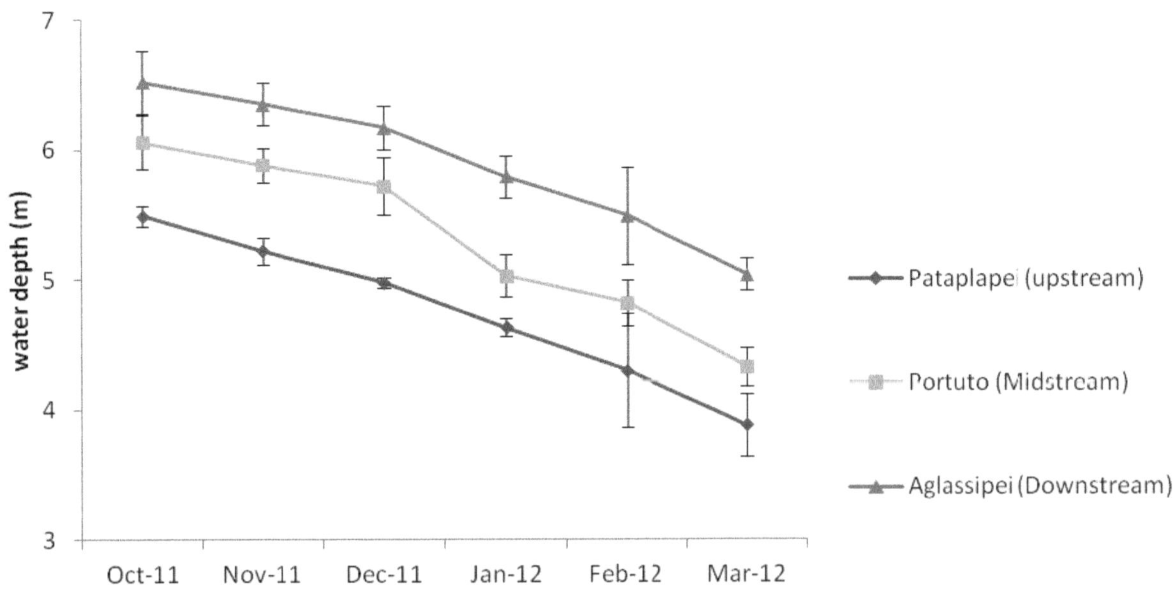

Figure 3. Mean and SD (±) of the water depth of 3 landing sites in the lower reaches of the White Volta River from October 2011 to March 2012.

shown in Figure 4 depicts that different sizes of the species prefer different areas of the lower reaches of the white volta river. Aglassipei which is downstream had more of smaller sizes of the species with total length around 12.5 cm. Porturto (midstream) had the species evenly distributed over the entire size range. Pataplapei (upstream) had more of the middle sizes of the species with most of their total lengths around mean of the entire size range of the *A. baremoze* in the lower reaches of the white volta river.

A total of 197 of *B. nurse* were measured over the 6 months period of the study. *B. nurse* total length ranged from 6.5 to 27.5 cm. One cohort was observed over the entire period of the study. The cohort had a mean total length around 15.5 cm. The following observations are apparent from the *B. nurse* distribution. Firstly, a higher percentage of the species is concentrated at the middle of the size range. Secondly, larger sizes with total lengths between 25.5 and 27.5 cm were only observed in October to November, 2011, suggesting that possibly a cohort was facing or dying out. The next observation is that, the smaller sizes with total lengths around 6.5 to 8.5cm were also observed only in the last 2 months (February to March, 2012) of the study, a possible indication of an emerging cohort.

One hundred and seventy five (175) of *S. intermedius* from the lower reaches of the white volta river were measured over the 6 months period. The *S. intermedius* total lengths ranged from 8.5 to 26.5 cm. One cohort with a wide size range (total length traverses 17.5 and 21.5 cm) was observed from October 2011 to March 2012. Though, there were little differences in size ranges among the sites, smaller sizes of the species were more concentrated at Porturto which had seemingly an emerging cohort appearing in the last 2 months (February and March, 2012) of the study.

Length-weight relationships

The length to weight relationships of *A. baremoze*, *B. nurse* and *S. intermedius* are presented in Figure 7. A total of 283 *A. baremoze* were measured. *B. nurse* and *S. intermedius* were 197 and 175 respectively. Exponential value of the length-weight relationship 'b' in *A. baremoze* was 2.9502. In *B. nurse*, the 'b' value was 3.0737 whilst *S. intermedius* had a highest 'b' value of 3.4592.

Condition factor

The condition factor is an indicator of the wellbeing of a fish. The monthly condition factors of *A. baremoze*, *B. nurse* and *S. intermedius* from the lower reaches of the white volta river over the 6 months study period are shown in Figures 8 to 10, respectively.

The highest condition factors of *A. baremoze* were in the month December 2011 in all the 3 landing sites. *A. baremoze* condition factors variation over the 6 months period were similar in the landing sites Pataplapei and Aglassipei which had relatively higher condition factors than the landing site Porturto. January to March 2012 mean condition factor of Pataplapei was higher than the

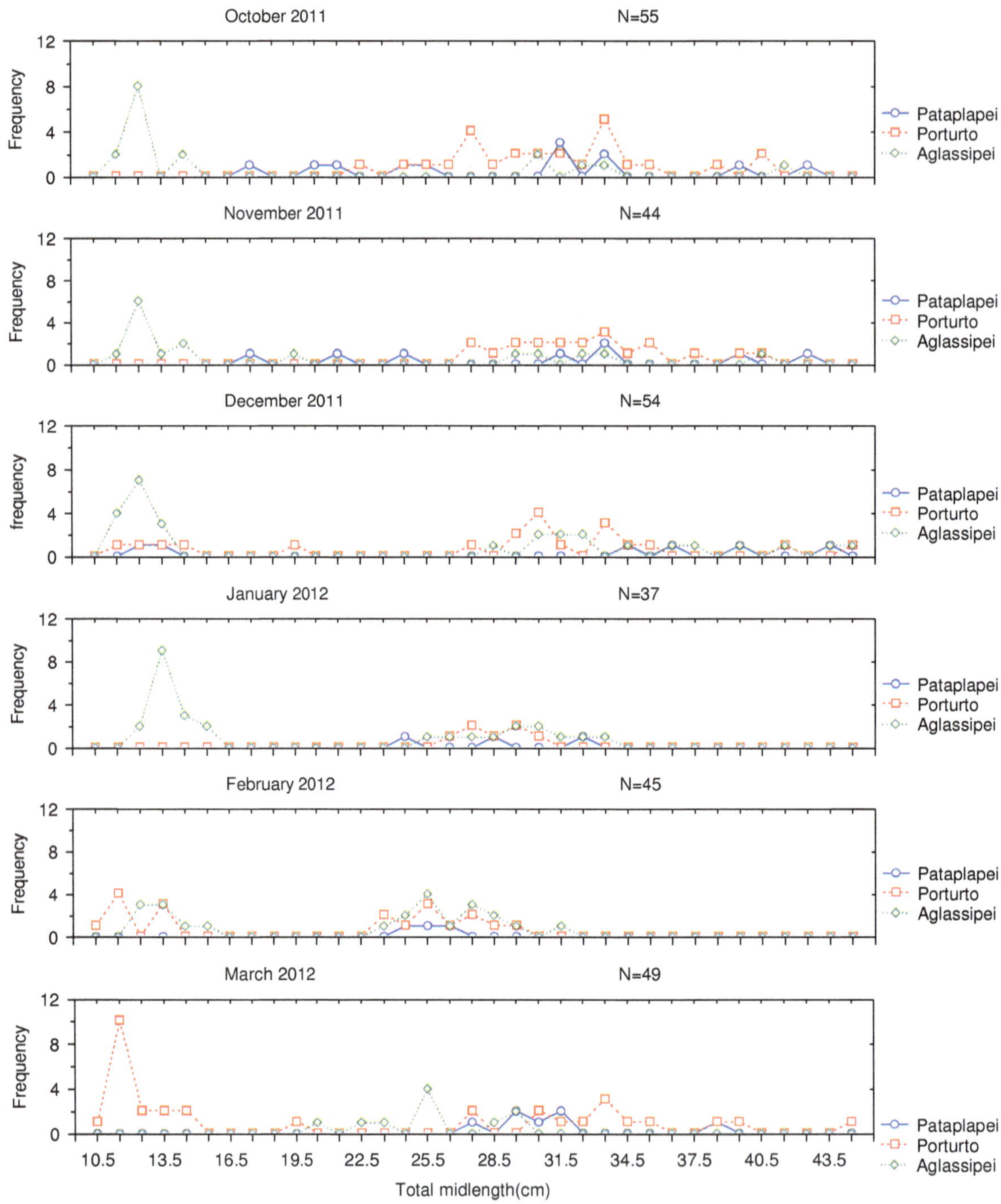

Figure 4. Length frequency distribution for *A. baremoze* from 3 landing sites in the lower reaches of White Volta River during the period of the study (October 2011 to March, 2012).

mean condition factor in October to November, 2011. Aglassipei also had slightly higher mean condition factor in January to March, 2012 than the mean condition factor of October to November, 2011. Porturto on the other hand was distinct in its monthly variation when compared to the other 2 sites. It had relatively little variability over the 6 months period and mean condition factor in October to November, 2011 was rather higher than that of January to March, 2012.

In B. *nurse*, condition factors of Pataplapei did not vary in October and November, 2011. In December 2011 the condition factor declined and then in the month January 2012, the condition factor rose to its highest in that site during the study period. The site had very little variation

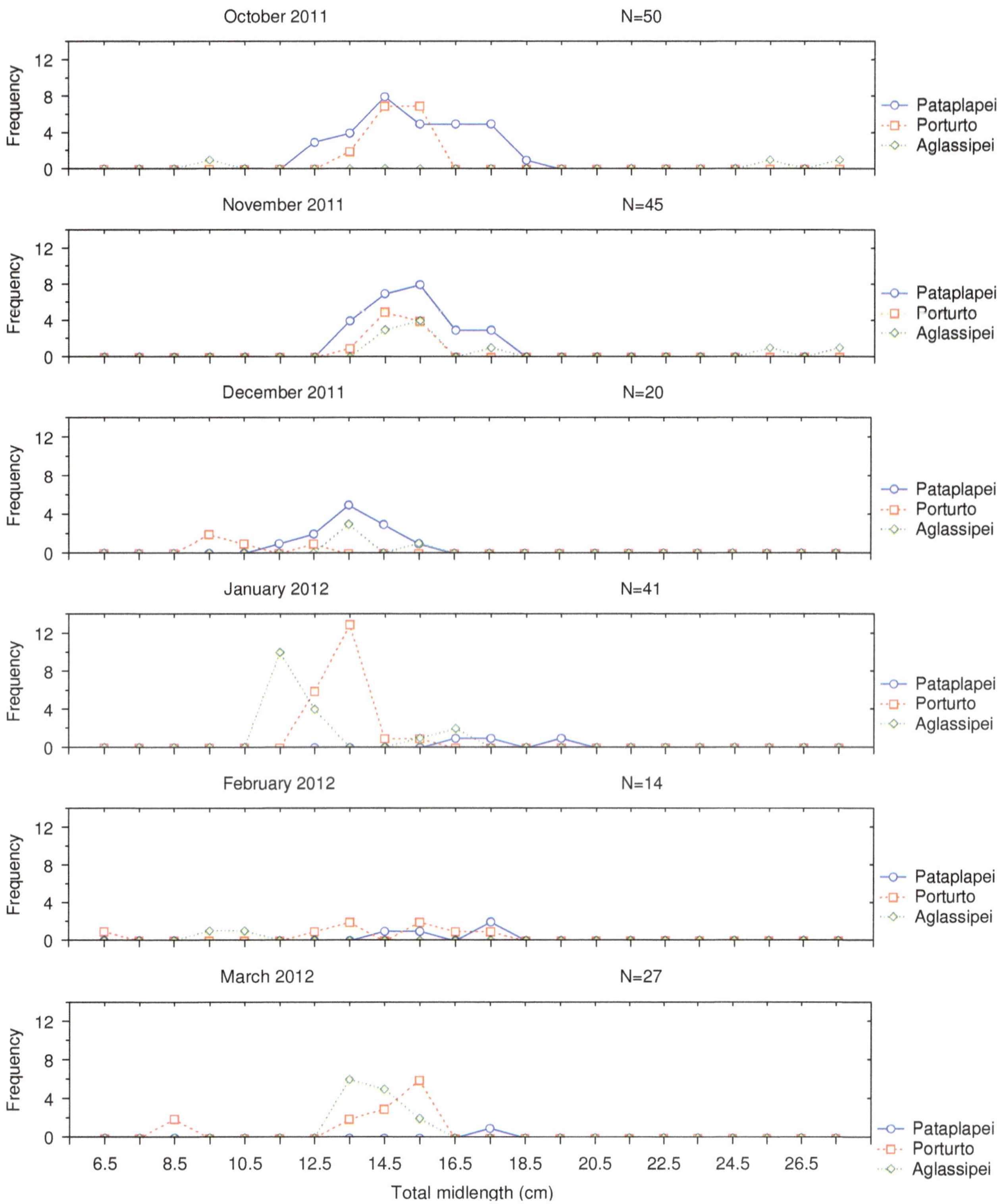

Figure 5. Length frequency distribution for *B. nurse* from 3 landing sites in the lower reaches of White Volta River during the period of the study (October 2011 to March 2012).

of the condition factors over February and March, 2012. Also, in the landing site Porturto the condition factors remained fairly the same in the October to November, 2011. There was a sharp decline in the condition factor in December 2011 and then in January 2012 increased and then decreased in February 2012 and sharply elevated at

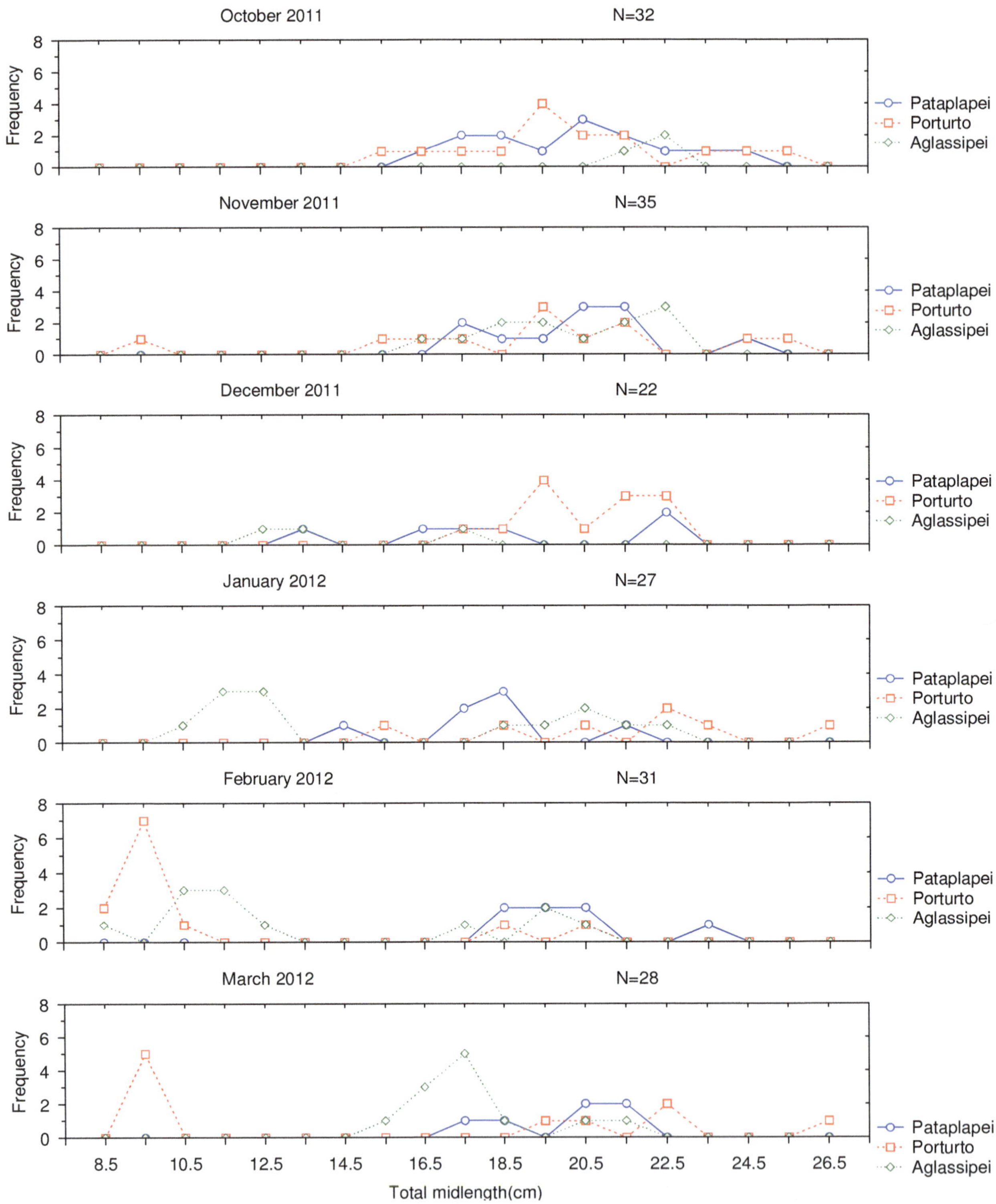

Figure 6. Length frequency distribution for *S. intermedius* from 3 landing sites in the lower reaches of White Volta River during the period of study (October 2011 to March 2012).

end of the study in March 2012. In Aglassipei, the condition factors were lower than those of Pataplapei and Porturto throughout the period of study. The monthly pattern was that of a normal variability or increase-decrease, starting with a mean condition factor of 0.81 in October 2011 and then increased in November 2011,

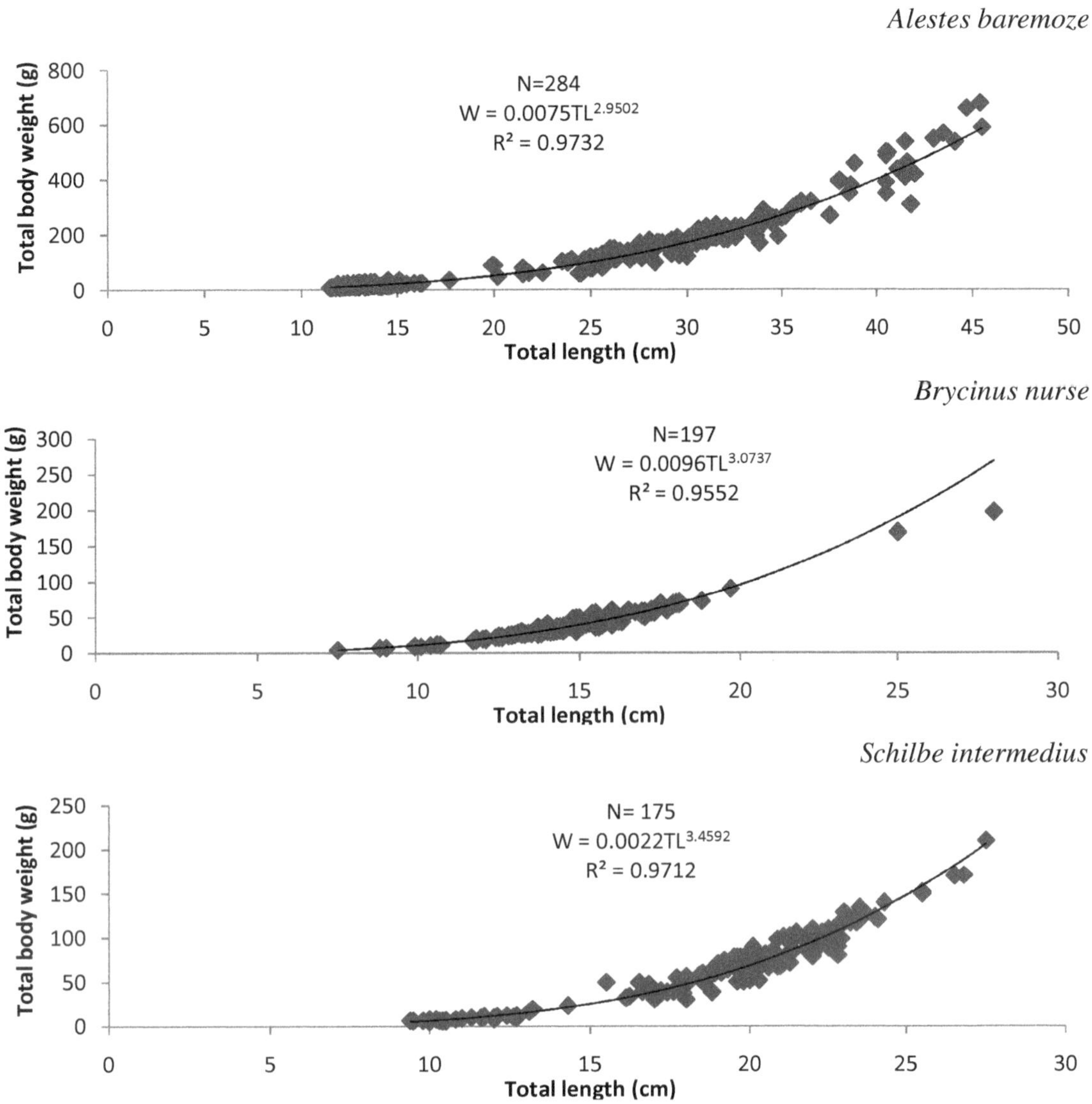

Figure 7. Length-weight relationship of 3 species in the lower reaches of the White Volta River during the study period (October 2011 to March 2012). N, Number of fish; W, Weight (grams) and TL, Total Length (cm).

declined in December 2011. This pattern of increase-decrease continued at similar rate till the end of the study, March 2012.

The condition factors of *S. intermedius* among the 3 landing sites were not very distinct. Pataplapei had the highest condition factors of the species. It started with 0.21 in October 2011 and then continuously went up over the months to 0.27 in January 2012 and slightly declined to 0.26 in February 2012 and also to 0.24 in March 2012. There was very little monthly variability of the condition factors between the landing sites Porturto and Aglassipei with the condition factor values varying between 0.20 and 0.23. Porturto for instance had the same condition factor of 0.22 in November 2011, December 2011 and January 2012. Pataplapei had fairly higher condition factors during the dry season (January 2012 to March 2012) than the other post flood season and also, than the other landing sites.

DISCUSSION

A. baremoze

The numerical value of *b* of the length-weight relationship of *A. baremoze* obtained from this study is within the acceptable range of 2.5 and 3.5 which is typical for tropical fish stocks (Carlander, 1969; Froese, 2006) and was very close to 3. It also conforms to what www.fishbase.org has published on the species. The value

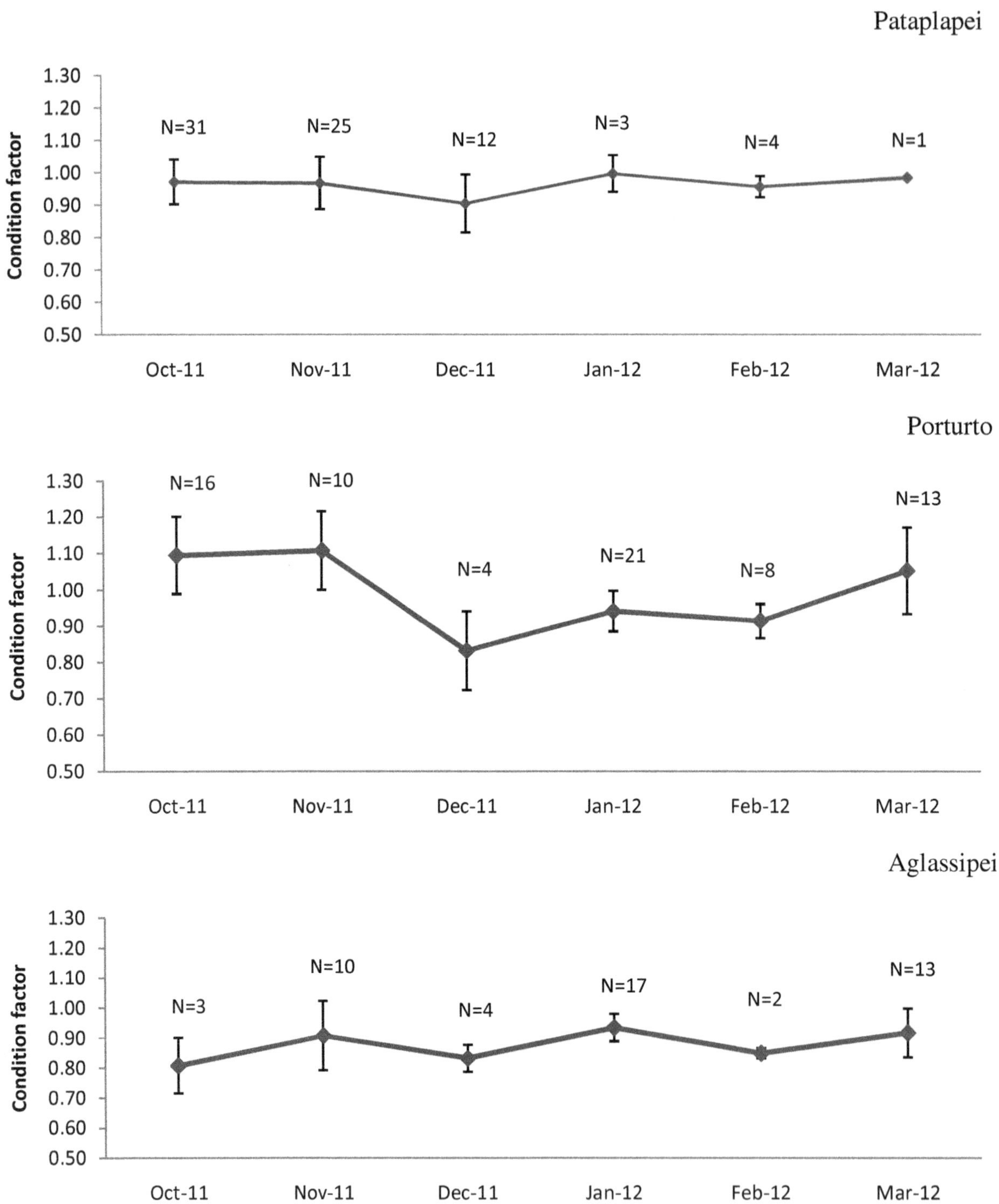

Figure 8. Condition factors ($K=W*100/L^b$) of *A. baremoze* from 3 landing sites in the lower reaches of White Volta River over the 6 months study period (with their standard deviations as bars in the graph).

can thus be used for approximation of length-weight relationship of *A. baremoze* populations in the lower reaches of the White Volta River. Condition factor defined as the state of wellbeing of the fish reflects through variations, some information on the physiology of the fish. It is assumed that the higher the value, the better the state of wellbeing. Nutritionally, it is assumed that increases or higher condition factor reflects improvement in nutritional status of the fish whereas a decline or lower condition factor could mean malnutrition (Koskela et al., 1997). In using the *b* for the condition factor (W/L^b) calculation, there were some differences among the

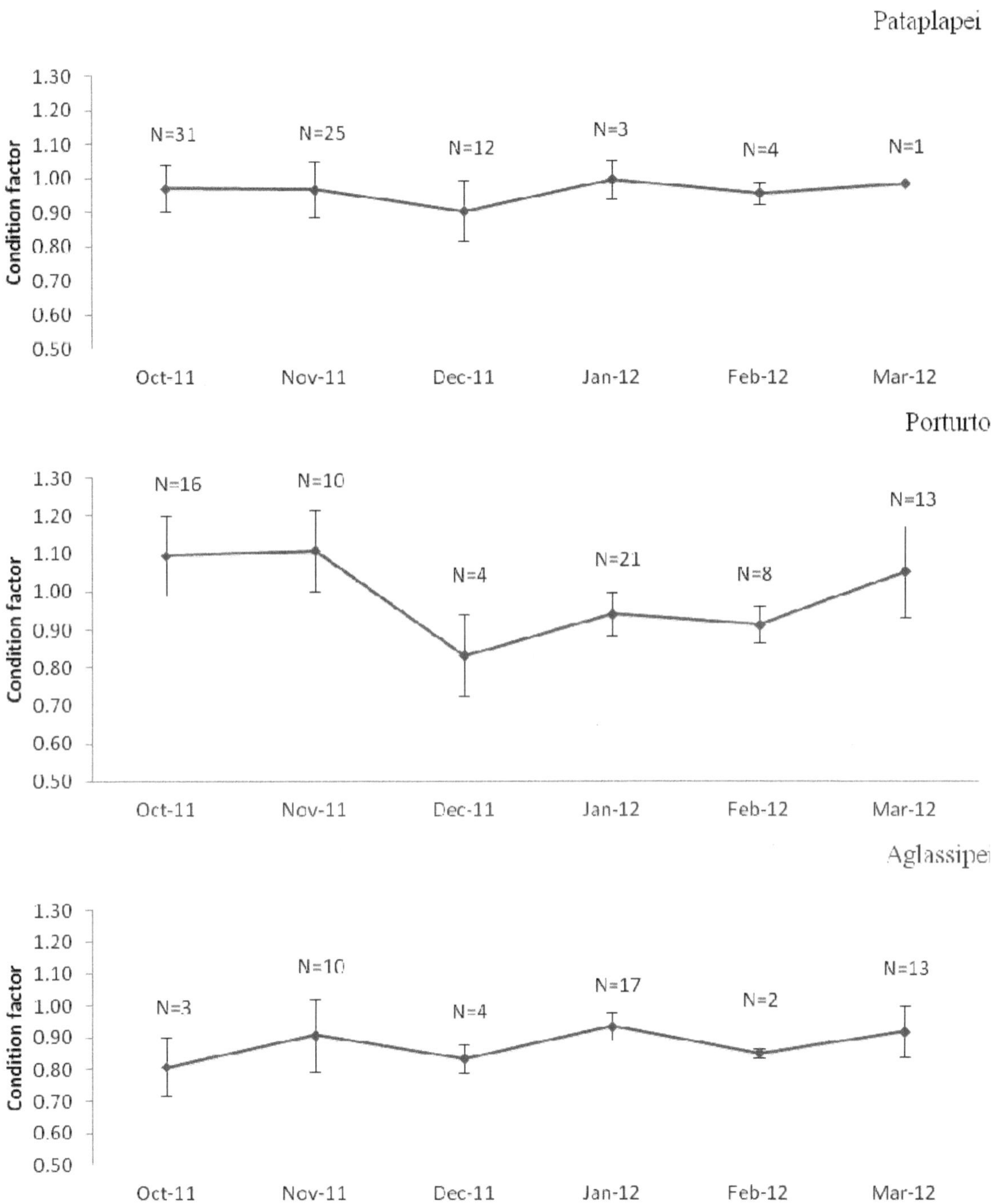

Figure 9. Condition factors ($K=W*100/L^b$) of *B. nurse* from 3 landing sites in the lower reaches of White Volta River over the 6 months study period (with their standard deviations as bars in the graph).

landing sites. On average, the sites Pataplapei which is upstream and Porturto which is midstream had higher condition factors than the downstream site Aglassipei. The highest condition factors which were recorded in December 2011 could be related to the low temperature observed in the month as it was this same month December 2011 that the lowest temperatures of the study were recorded. Moreover, it is reported by Chinci World Atlas (2012) that December is the coolest month in Yapei due to the influence of the North - easterly winds otherwise known as harmattan. The low temperatures possibly influence and enhance some food items and their sources for *A. baremoze* and such a providing more food for a better wellbeing of the species populations in the lower reaches of the white volta river at Yapei during the month December. It is also possible that the species

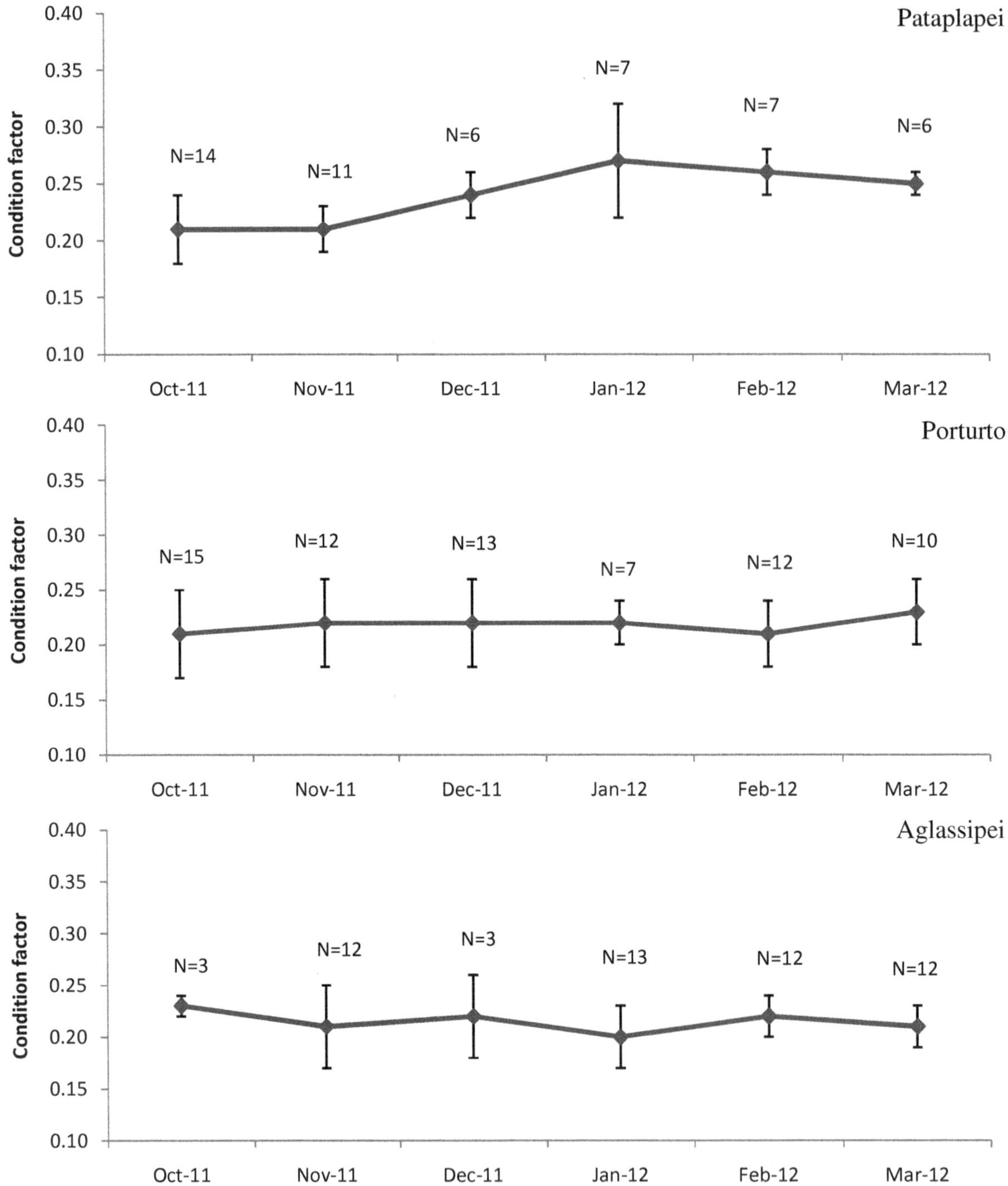

Figure 10. Condition factors ($K = W*100/L^b$) of *S. intermedius* from 3 landing sites in the lower reaches of White Volta River over the 6 months study period (with their standard deviations as bars in the graph).

due to low sample size and as the highest standard deviations of the species condition factors in all the 3 landing sites were obtained in the same month December, 2011.

It was evident in at least 2 landing sites namely Pataplapei (upstream) and Aglassipei (downstream) that the months January to March 2012 (dry season) on average had higher condition factors than the months October to November, 2011, suggesting *A. baremoze* in those 2 sites had a better state of wellbeing in dry season than major part of the post flood (October to December). The post flood season is the time that water level draws down and sediments and materials transported by runoff to the water would be settling. It is thus presumable that starting from December when much suspended materials would have settled; a better state of wellbeing of the species should be anticipated.

B. nurse

From the length-weight relationship of *B. nurse*, the slope *b* of 3.0737 is within the acceptable range of 2.5 to 3.5 and thus offers an approximate account of the length-weight relationship and as such the results of the length-weight relationship of *B. nurse* in the lower reaches of the White Volta River can be useful for assessing the species growth conditions. It was similar to *b* value of *A. baremoze* observed in this study. In comparing the state of wellbeing using condition factors over the 6 months period of study and also among the landing sites, it was realised that unlike the condition factors in the previous species *A. baremoze*, which was highest in December 2011, the lowest condition factors of *B. nurse* in upstream Pataplapei and midstream Porturto landing sites were rather in the month December, 2011. This shows that the 2 species although belong to the same family Alestidae, may have quite a distinct pattern of variation in the condition factors when they occur in the same habitat. The relatively lower condition factors of the species in the downstream Aglassipei as compared those of Pataplapei and Porturto may not as result of any peculiar differences in physicochemical characteristic as no significant differences in temperature, dissolved oxygen concentration or pH were observed among the landing sites during the period of the study. It is thus possible to have the same species occurring within similar locations and within common physicochemical characteristic and yet may have some differences in the state of wellbeing otherwise the condition factor. It may be necessary to study different stages of the species among the sites; since Aglassipei had relatively small sizes and large sizes of the species occurring at different times whereas the upstream Pataplapei and the midstream Porturto had all the species sampled at the middle of the size range.

S. intermedius

Length-weight relationship of the species showed that the *b* of 3.4592 though was within the acceptable range of 2.5 and 3.5, it depicted a strong allometric type of growth among the *S. intermedius* populations in the lower reaches of the White Volta River. It is also comparable to the *b* value of 3.1398 Ofori-Danson (2005) reported on the species at Yeji, Lake Volta of Ghana. The significance of this higher *b* value of the *S. intermedius* populations in the lower reaches of the white volta river is that it gives an indication that the species may be growing more allometrically than *A. baremoze* and *B. nurse*. This is agreeable as *A. baremoze* and *B. nurse* belong to the same family and have very similar features and as such similar growth types but *S. intermedius*, a catfish belonging to the family S. dae expectedly grew quite differently. Regarding monthly variations in mean condition factors of *S. intermedius* among the landing sites; very little variability was observed, meaning the species during the period of the study were not impacted by factors that could results in substantial changes of wellbeing. The month of December which was a notable period of changes in the condition factors of *A. baremoze* and *B. nurse* was not period of any significant change in *S. intermedius* condition factors. Pataplapei which is upstream and the shallowest part of the whole stretch had higher condition factors during the dry season (January 2012 to March 2012) than other the post flood season and also than the other landing sites. This can be as a result of its greatly reduced water levels during dry seasons, which might have exposed some food items for the species to feed on, and thus species in this site being in a better state of wellbeing than those from the other 2 sites.

Although, all the 3 species are semi-pelagic omnivorous, the differences in growth among the species can be related to the species physiology, some physicochemical parameters, as well as nutrition. There may be more linkages or complexities among these factors

Conclusion

The length frequencies had shown that *B. nurse* of the family Alestidae and *S. intermedius* of the family S.dae have similar size ranges with their maximum total length around 27 cm. Their occurrence was also identical as 1 cohort was realized in both species during the period of study. *A. baremoze* of the family Alestidae, on the other hand, was quite distinct in size range and occurrence. It had maximum total length of 45 cm and 2 cohorts were observed during the period of the study. The length-weight relationship is a useful tool in fishery assessment that helps in predicting weight from length required in yield assessments (Garcia et al., 1998), and in the calculation of the standing crop biomass (Martin-Smith, 1996). According to Carlander (1969); Gayanilo and Pauly (1997) and Froese (2006), the slope *b* values may range between 2.5 to 3.5. The *b* of length-weight relationships of all the 3 species were within the 2.5 to 3.5 range which is typical for tropical fish stocks. Therefore, for the 3 species, the *b* values can be used in length-weight relationships to approximate the weight of the species. Weight gain of the species will be at the same rate as growth in length. In terms of rate of growth, all the 3 had high growth rates and thus the findings on the selected species in this study support the widely-held assumption that tropical fishes are generally fast growing. The monthly variation in condition factors was very pronounced in December 2011 when the surface water temperatures were lowest. *A. baremoze* had sharp increase in condition factors in December and these increases were the highest in all the landing sites. Therefore, in December, *A. baremoze* species of a given length in the lower reaches of the white volta river are heavier than the other remaining 5 months of the study. *B. nurse* on the contrary had its lowest condition factors in 2 of the 3 landing sites in December 2011. It is probable that the low temperatures might not have directly influenced condition factors or the state of wellbeing of the fishes but rather it may have triggered other unknown factors that brought the dissimilar changes in the condition factors between the 2 species. *S. intermedius* had little variability in condition factors and even in December 2011 there was no change in its variation. Ultimately, significant drop in temperature has corresponding effects on the state of wellbeing of the species in the lower reaches of the white volta river, but it should be noted that, the temperature effects on the condition factors of the different species are not direct and are also not comparable/identical.

ACKNOWLEDGEMENT

Authors express their sincere thanks to the Deutscher Akademischer Austausch Dienst (DAAD) for the funds that made the study possible. Special thanks also go to the Leibniz Center for Tropical Marine Ecology (ZMT), Bremen, Germany for the basic equipment provided for the field data collection.

REFERENCES

Abban EK (1999). Integrated development of artisanal fishes. Integrated Development of Artisanal Fisheries Project. GHA/93/008. P. 43.

Bolger T, Connolly PL (1989). The selection of suitable indices for the measurement and analysis of fish condition. J. Fish Biol. 34:171-182.

Braimah LI (2003). Fisheries Management Plan for the Volta Lake, Accra, Ghana. Ministry of Food and Agriculture, Directorate of Fisheries. P. 77.

Carlander KD (1969). Handbook of freshwater fishery biology. The Iowa State University Press, Ames, IA. 1:752.

Chinci World Atlas (2012). www.chinci.com/travel/pax/q/2293844/Yapei/GH/Ghana/0/#(Last accessed: 30.07.2012).

Dankwa HR, Abban EK, Teugels GG (1999). Freshwater fishes of Ghana: identification, distribution, ecological and economic importance. Annls.Sci. zool. P. 283.

FAO: UNDP (1971). Physico-chemical conditions of Lake Volta, Ghana. Report prepared for the Government of Ghana by FAO of the UN acting as executing agency for the UNDP. Based on the work of C.W. Czernin-Chudenitz. FI: SF:GHA:10:Technical Report. Rome: FAO. P. 77.

FAO: UNDP (1979). Volta Lake research and development project, Ghana. Project findings and recommendations. Terminal Report. Report prepared for the Government of Ghana by FAO of the UN Acting as Executing Agency for UNDP. FAO:FI:DP:GHA:67:510:71:535. Rome: FAO. P. 75.

Froese R (2006). Cube law, condition factor and weight-length relationships: history, meta-analysis and recommendations. J. Appl. Ichthyol. 22:241-253.

Fulton T (1902). Rate of growth of seas fishes. Sci. Invest. Fish. Div. Scot. Rept. P. 20.

Gayanilo FC, Pauly D (1997). The FAO-ICLARM Stock Assessment Tools (FISAT). Reference manual. Rome, FAO. P. 262.

Koskela J, Pirhonen J, Jobling M (1997). Growth and feeding responses of a hatchery population of brown trout (Sulrno trutta L.) at low temperatures. Ecol. Freshw. Fish. 6:116-121.

Martin-Smith KH (1996). Length/weight relationship of fishes in a diverse tropical freshwater community, Sabah, Malaysia. J. Fish Biol. 49:731-734.

McGurk MD (1985). Effects of net capture on the post preservation morphometry, dry weight, and condition factor of Pacific herring larvae. Transactions of the Am. Fish. Soc. 114:348-355.

Obodai EA, Muhammad BA, Obodai GA, Opoku E (2009). Effect of Fuel wood on the Quality of Smoked Freshwater Fish Species Sold in Tamale Central Market, Northern Region, Ghana. Ethiopian J. Environ. Stud. Manage. 2(2):27-35.

Ofori-Danson PK (2005). An assessment of the purse-seine (winch-net) fishery in Lake Volta, Ghana. Lakes & Reservoirs: Res. Manage. 10:191-197.

Ofori-Danson PK, Vanderpuye CJ, de Graaf GJ (2001). Growth and mortality of the catfish, *Hemisynodontis membranaceus* (GeoffroySt.Hilaire), in the northern arm of Lake Volta, Ghana in: Fisheries Management and Ecology, Blackwell Sci. Ltd. 8:37-45.

Ricker WE (1975). Computation and interpretation of biological statistics of fish populations. Bulletin of the Fisheries Research Board of Canada .191:382.

van Zwieten PAM, Béné C, Kolding J, Brummett R, Valbo-Jørgensen J (2011). Review of tropical reservoirs and their fisheries – The cases of Lake Nasser, Lake Volta and Indo-Gangetic Basin reservoirs. FAO Fisheries and Aquaculture Technical Paper. Rome, FAO. 557:148.

Weatherley AH, Gill HS (1987). The biology of fish growth. Academic Press, London (UK). P. 443.

Wikipedia (2012). http://en.wikipedia.org/wiki/Tamale,_Ghana (last accessed: 16.07.2012).

www.fishbase.org (Last accessed: 11.08.2012).

Nitrite intoxication of *Clarias gariepinus* at different water temperatures

Ajani Funmilola[1]* and Adeyemo Olanike Kudirat[2]

[1]Department of Animal Science and Fisheries Management, Bowen University, Iwo, Osun state, Nigeria
[2]Department of Veterinary Public Health and Preventive Medicine, University of Ibadan, Ibadan, Nigeria.

***Clarias gariepinus* (Burchell, 1822) of mean weight 320±11.2 g and total length 60.0±1.0 cm were exposed to nitrite (0.2 mg/l NO_2) for 48 h at 27 and 35°C in order to investigate the effects of nitrite poisoning at these water temperatures. The effects of nitrite exposure on fish was assessed on selected haematological and biochemical indicators of the blood. The test was performed in a semi-static assay for 48 h. Fish were kept in thermostart –controlled water baths each containing 15 L of test solution. Four groups each containing 8 specimens of 3 month old *C. gariepinus* were exposed to nitrite at the different water temperatures. Nitrite exposed fish showed lower haematocrit value (PCV) at both experimental temperatures compared with controls. Significantly higher PCV values were recorded in fish with nitrite at 27°C (36.67±0.57%) when compared with fish at 35°C (33.33±0.57%). A significant difference was also recorded for haemoglobin level and the erythrocyte counts. Leucocyte count in fish with nitrite at 27^0 C was significantly higher ($P<0.05$) compared to the count at 35°C. For Plasma Biochemical parameters, significant differences were observed between the 2 groups for K^+ concentration (27°C, 4.9±0.10 mmol/l; 35°C, 4.1±0.12 mmol/l). Statistically significant differences were observed in plasma Cl^- and Na^+ concentrations in fish at both temperatures. It has been established from this study that nitrite is more toxic at higher temperature.**

Key words: Haematology, plasma biochemistry, *Clarias gariepinus*, temperature.

INTRODUCTION

Nitrite is an intermediate product in bacterial nitrification processes. An inbalance in either of these processes can lead to elevated ambient nitrite concentrations (Eddy and William, 1987). Nitrite problems are typically more likely in closed, intensive culture systems due to insufficient, inefficient or malfunctioning of filtration systems (Durborow et al., 1997). High nitrite concentrations in ponds occur more frequently when temperature is fluctuating, which result in the breakdown of the nitrogen cycle due to plankton and bacterial activity. If nitrite levels exceed that which resident bacteria can rapidly convert to nitrate, build-up of nitrite occurs and brown blood disease is a risk. This may result in mass fish mortality (Svobodova et al., 2005a). Nitrite is actively taken up through the gills and enters the blood stream (Margiocco et al., 1983; Jensen et al., 1987). From the blood plasma, nitrite diffuses into red blood cells where it oxidizes the Iron in haemoglobin to produce methaemoglobin which is incapable of oxygen transport. Methaemoglobin turns the blood to a chocolate-brown colour, so nitrite poisoning is often called a brown blood disease.

Nitrite toxicity in fish is influenced by a large number of external and internal factors, among which the chloride concentration in water plays a major role (Lewis and Morris, 1986; Svobodova et al., 2005b; Pistekova et al., 2005). Adeyemo et al. (2003) documented the haemato-

*Corresponding author. E-mail: funmilolajani@yahoo.com.

Abbreviations: **LC_{50},** Median lethal concentration; **RBC,** erythrocyte count; **PCV,** haematocrit value; **Hb,** haemoglobin; **WBC,** leucocyte count; **MCV,** mean corpuscular volume; **MCHC,** mean corpuscular haemoglobin; **ANOVA,** analysis of variance.

logical response of *Clarias gariepinus* to changes in acclimation temperature. Kroupova et al. (2006) worked on nitrite intoxication of common carp (*Cyprinus carpio*) at different water temperatures but no information is known about the influence of temperature on nitrite toxicity of the most cultured fish species in Nigeria, *C. gariepinus*.

Generally the water temperature is an important factor that determines chemical toxicity (Cairns et al, 1975), and it has multiple effects on the physiology of fish. Therefore, the aim of this study is to assess the influence of water temperature on the nitrite toxicity in *C. gariepinus*

MATERIALS AND METHODS

Experimental fish

C. gariepinus of mean weight 320±11.2 g and mean total length of 60.0±1.0 cm were used for the study. The fish were collected in the morning between 8 – 9 a.m. They were obtained from Bowen fish farm and maintained for 2 weeks in 2 circular plastic tanks (300 L) with dechlorinated tap water. They were fed with 3 mm CHI pelleted feed. Feeding was stopped 24 h prior to the commencement of the bioassay. Four days before the start of the experiment, the fish were divided into four groups and acclimated to 27 and 35°C. Temperatures around 27 and 35°C prevail during the growing season in Nigeria. During the experimental period, the fish were not fed.

Experimental procedure and fish sampling

The test was performed in a semi-static assay for 48 h. Fish were kept in thermostat-controlled water baths, each containing 15 L of test solution. The nitrite concentration was obtained by adding $NaNO_3$ to dechlorinated tap water (Odiete, 1999). The dose of nitrite represented the median lethal concentration (LC_{50}) for *C. gariepinus* at a similar chloride water concentration and similar relative weight of fish according to Ajani (2006).

During the acclimation and experimental period, the range of the basic chemical indices of water taken were alkalinity 45mg/l, dissolved Oxygen 8.3mg/l, phosphate 30mg/l, hardness 72.4mg/l, chloride 19.2mg/l, ammonia 55.8mg/l and the pH 7.0 using HACH freshwater aquaculture test kit (FF-1A).

Four groups each containing 8 specimens of three-month old *C. gariepinus* was exposed to nitrite at different water temperatures (27 and 35°C). Each treatment were replicated.

Group A_1, (0. 2 mg/l) NO_2, t =27±0.4°C
Group B_1, (control 1) traces NO_2, t =27±0.4°C
Group A_2, (0.2 mg/l) NO_2, t =35±0.3°C
Group B_2, (control 2) traces NO_2; t =35±0.3°C

Nitrite and chloride content were checked twice during the test and measured values did not differ from the nominal value by more than 7%. Also, traces NO_2 in the control groups were below detection limit.

Forty *C. gariepinus* (5 from each group) with their replicates were examined to determine the haematological and biochemical profiles of blood. The blood samples were taken after 48 h exposure in heparinized bottles by inserting a syringe needle (2 ml) at the ventral midline just posterior to the anal fin at angle 45° until it penetrate the caudal vessels lying between adjacent hermal arches (Morgan and Iwama, 1997). They were centrifuged at 3000G for 5 min and the plasma were separated and analyzed.

The Erythrocyte count (RBC), haematocrit (PCV) haemoglobin (Hb), leucocyte count (WBC), platelets, Mean corpuscular Volume (MCV) and the Mean Corpuscular Haemoglobin (MCHC) were determined in the blood samples according to Jain (1986). Also, white blood cell differential; lymphocytes, neutrophils, monocytes and eosinophils. For plasma biochemistry, plasma Na^+, K^+ and Cl^- were determined as described by Jain (1986).

Statistical analysis

Statistical software SPSS package (Version 11) was used to determine differences between the test groups. Two - way analysis of variance (ANOVA) was applied to determine whether there were any significant differences in measured variables between nitrite-exposed fish at different temperatures and control fish.

RESULTS

Fish mortality

No mortality was observed in the control groups. High fish mortality was noticed during nitrite poisoning in the first 24 h in groups A_1 (35%) and A_2 (50%) when compared with the control groups. However, the difference between groups. A_1 and A_2 were not statistically significant ($P > 0.05$).

Vital signs

Nitrite – treated *C. gariepinus* at 27 and 35°C showed typical symptoms of methaemoglobinemia which was manifested by erratic and disoriented behaviour and brown colouring of the blood and gills. The intensity of the colouration was temperature dependent. .

Haematological variables

The result of the haematological parameters is presented in Table 1.

Haematological variables

The PCV values were significantly lower ($P < 0.05$) in the control of group B_2 *C. gariepinus* compared with control group B_1. Significantly lower ($P < 0.05$) PCV values were found in fish of A_1 compared with B_1. No significant difference in PCV values were recorded in fish of A_1 compared with A_2 .No significant difference was observed in the haemoglobin level ($P < 0.05$) in fish of group A_2 compared to the control group B_2. A significant decrease was observed ($P > 0.05$) in fish of group B_1 compared to A_1. Significantly lower Haemoglobin values were obtained in fish of group B_2 compared with group B_1.

No significant decrease was observed in the erythrocyte counts in group B_1 compared to group A_1. It increased significantly ($P < 0.05$) in group B_2 compared to

Table 1. Haematological indices of *C. gariepinus* after 48 h nitrite exposure at 27 and 35°C.

Indices	27°C		35°C	
	A_1 (Nitrite)	B_1 (Nitrite – free)	A_2 (Nitrite)	B_2 (Nitrite – free)
PCV (%)	36.67 ± 0.58^a	47.67 ± 0.11^c	33.33 ± 0.57^a	38.33 ± 0.32^b
Hb (g/dl)	12.37 ± 0.82^b	15.67 ± 0.06^a	11.10 ± 0.10^b	11.56 ± 0.05^b
RBCC (10^6/l)	3.41 ± 0.02^c	3.48 ± 0.11^c	3.03 ± 0.01^a	3.11 ± 0.04^b
WBC (10^9 /l)	1.70 ± 1.52^c	1.58 ± 3.20^b	1.6 ± 1.43^{ab}	1.49 ± 0.91^a
Platelets (10^9/l)	1.07 ± 1.00^a	1.16 ± 0.61^c	1.24 ± 0.21^d	1.01 ± 0.57^a
Lymphocyte (10^3 /mm^3)	35.00 ± 1.00^a	66.33 ± 1.52^b	38.00 ± 1.02^a	65.00 ± 1.40^b
Neutrophil (10^3 /mm^3)	60.00 ± 0.08^b	29.67 ± 0.50^a	57.01 ± 0.12^b	32.01 ± 2.10^a
Monocyte (10^3/mm^3)	2.33 ± 0.57^a	3.33 ± 1.52^b	2.33 ± 0.10^a	1.33 ± 0.57^a
Eoslnophil (10^3/mm^3)	3.00 ± 2.00^a	3.00 ± 2.00^a	3.00 ± 1.00^a	2.00 ± 1.00^a
MCV (fl)	107.5 ± 10.00^a	136.9 ± 8.00^b	110.0 ± 1.00^a	123.0 ± 12.00^b
MCH (pg)	36.27 ± 3.61^a	45.02 ± 0.81^b	36.63 ± 2.07^a	37.17 ± 6.34^a
MCHC(g/l)	33.73 ± 0.96^a	32.87 ± 1.08^a	33.30 ± 2.11^a	30.16 ± 4.21^a

Mean followed by the same superscript in the same row are not significantly different (P > 0.05)

Table 2. Blood plasma indices of *C. gariepinus* after 48 h nitrite exposure at different water temperature.

	27°C		35°C	
	B_1 (Nitrite – free)	A_1 (Nitrite)	B_2 (Nitrite – free)	A_2 (Nitrite)
K^+ (mmol/l)	3.0 ± 0.2^a	4.9 ± 0.10^b	3.2 ± 0.24^a	4.1 ± 0.12^b
$C1^-$ (mmol/l)	93.0 ± 1.21^a	92.0 ± 0.72^a	89.0 ± 1.42^a	52.0 ± 1.11^b
Na^+ (mmol/l)	129.0 ± 1.00^b	135.0 ± 0.86^b	131.0 ± 0.72^b	90.0 ± 0.64^a

Mean followed by the same superscript in the same row are not significantly different (P > 0.05)

group A_2. A significant decrease was observed in group B_1 compared to group B_2.

Leucocyte count in group A_1 is significantly higher ($P < 0.05$) compared to the count in group A_2. Significant differences were observed among groups A_2 and B_2 and B_1 and B_2.

Highly significantly lower ($P < 0.05$) platelet counts were found in fish of group A_1 compared with groups A_2. Also, significant difference ($P < 0.05$) was observed in platelet count among groups A_2 and B_2; B_1 and B_2. For the white blood cell differential count, no significant difference were observed in lymphocyte in group A_1 compared to the count in group A_2 ; and B_1 and B_2 . However, significant differences were observed among groups A_1 and B_1; A_2 and B_2. Also, no significant difference was observed in neutrophils B1 and B2 while significant differences ($P < 0.05$) were observed in groups A_1 and B_1; A_2 and B_2.

No significant difference was recorded among the various groups for monocyte and eosinophil. Also, no significant difference was observed among the test groups for MCHC and only B_1 was significantly higher for MCH.

Plasma biochemical parameters

The results of the plasma biochemistry of *C. gariepinus* after 48 h nitrite exposure at different water temperatures is presented in Table 2.

Significantly lower values of plasma K^+ concentration were obtained for groups B_1 and B_2 compare with groups A_1 and A_2 Significantly higher differences were observed in groups A, B_1 and B_2 compared with A_2 for $C1^-$ and Na^+. No statistically significant differences were observed in plasma $C1^-$ and Na^+ concentrations at the lower temperature

DISCUSSION

Mortality was observed during nitrite poisoning at both temperatures; 27°C group A_1 - mortality 23% and at 35°C group A_2 - mortality 30%. However, no significant influence of temperature on mortality in *C. gariepinus* was found. Nitrite - exposed fish showed lower PCV values at both experimental temperatures compared with controls.

The decrease in PCV recorded in fish exposed to nitrite at higher temperature compared to the one at lower temperature may be caused by red cell shrinkage as reported by Jensen et al. (1987). The RBC shrinkage is connected with the efflux of K^+ from the red blood cells. The K+ efflux seems to result from activation of a K^+/Cl^- cotransporter that is normally involved in cell volume regulation (Jensen, 1990). The activation of K^+/Cl^- efflux draws osmotically obligated water out of the cells and hence induces erythrocyte shrinkage (Jensen, 1990). RBC shrinkage is usually followed by the loss of haemoglobin solubility, resulting in haemoglobin crystals and structural damage to erythrocytes (Jensen et al., 1987). Nitrite poisoning caused a marked increase in plasma K concentrations in both test groups compared with controls. The rise in plasma K^+ is due to the Nitrite stimulated release of K^+ from skeletal muscle and red blood cells (Jensen 1990; Knudsen and Jensen, 1997).

The efflux of K^+ from RBC seems to result from the activation of a K^+/Cl^- cotransporter that is normally involved in cell volume regulation (Jensen, 1990, 1992).

A significant decrease in plasma Na^+ concentration was observed in group A_2 (35°C) compared with group A_1 (27°C). This result deviated from the observation of Jensen et al. (1987) and Kroupova et al. (2006), who observed slight decrease between the two groups.

Plasma Cl^- levels in this study was affected by nitrite poisoning. A significant decrease was observed at higher temperature (35°C) and this is in consonance with the findings of Jensen et al. (1987) and Knudsen and Jensen (1997). Nitrite is a competitive inhibitor of chloride uptake and vice versa. Thus, chloride influx is reduced due to the presence of nitrite in ambient water. In freshwater teleosts nitrite exposure is followed by several osmo-regulation responses as hyponatremia, hypochloremia (Jensen et al., 1987), branchial chloride cell failure (Gaino et al., 1984) and inhibition of chloride uptake (Williams and Eddy, 1986).

No significant difference was observed in the K^+ and Cl^- between the two control groups at varying temperatures. Adeyemo et al. (2003) corroborated this when she worked on *C. gariepinus* at varying temperatures. A significant difference recorded in the PCV between the control groups (27 and 35°C) was in line with the results of Adeyemo et al. (2003). However, no significant difference was observed between the controls of Na^+, K^+ and Cl-. It can be suggested that *C.gariepinus* has a high adaptive ability. Nitrite-intoxicated fish at both temperatures recorded significantly lower values for lymphocytes. This may be attributed to its inhibition in producing antibodies. Also, MCV, MCH and MCHC were temperature independent.

Conclusion

It has been established from this study that mortality during nitrite poisoning are not related to water temperature in *C. gariepinus*. This may be due to its hardy nature. However, other haematological and biochemical variables were altered during nitrite exposure at higher temperature.

REFERENCES

Adeyemo OK, Agbede SA, Olaniyan AO, Shoaga OA (2003). The haematological response of *C. gariepinus* to changes in acclimation temperature. Afr. J. Biochem. Res., 6: 105 – 108.

Cairns JR, Heath AG, Parker BC (1975). The effect of temperature upon the toxicity of chemicals to aquatic organisms. Hydrobiologica 47: 135-171.

Durborow RM, Crosby DM, Brunson MW (1997). Nitrite in Fish Ponds. SRAC Publication No. 462.

Eddy FB, Williams EM (1986). Nitrite and Freshwater Fish. Chem. Ecol., 3: 1 – 38.

Kroupova H, Machova J, Plackova V, Flajshans M, Svobodova Z, Poleszczuk G (2006). Nitrite Intoxication of Common Carp (*Cyprinus Carpio* L.) at different water temperature. Acta Vet. Brno., 75: 561 – 569.

Lewis WM, Morris DP (1986). Toxicity of nitrite in fish: A review. Trans. Amer. Fish. Soc. 115: 183 – 195.

Margiocco C, Arillo A, Mensi F, Shenone G (1983). Nitrite bioaccumulation in *Salmo gairdneri* Rich and Haematological consequences. Aquat. Toxicol., 3: 261 – 270.

Pistekova V, Voslarova E, Svobodova Z (2005). Nitrite toxicity to *Danio rerio:* effect of chloride concentration during acclimation and in toxic tests. Acta. Vet. Brno., 74: 435 – 440.

Svobodova AZ, Machova J, Drastichova J, Groch L, Luskova V, Poleszczuk G, Velisek J, Kroupova H (2005b). Haematological and biochemical profile of Carp blood following nitrite exposure at difference concentration of chloride. Aquac. Res., 36(12): 1177 – 1184.

Svobodova Z, Machova J, Poleszczuk G, Huda., Hamackova J, Kroupova H (2005a). Nitrite Poisoning of fish in aquaculture facilities with water-recirculating systems: three case studies. Acta Vet. Brno. 74: 129-137.

Gaino E, Arillo A, Mensi P (1984). Involvement of the Gill Chloride Cells of Trout under Acute Nitrite Intoxication. Comp. Biochem. Physiol. A., 77(4): 611 – 617.

Jensen FB, Andersen NA, Heisler N (1987). Effects of Nitrite Exposure on Blood Respiratory Properties, Acid – Base and Electrolyte Regulation in Carp (*Cyprinus Carpio*). J. Comp. Physiol., 157b: 533 – 542.

Jensen FB (1990). Nitrite and Red Cell Function. In Carp: Control Factors for Nitrite Entry, Membrane Potassium Ion Permeation, Oxygen Affinity and Methaemoglobin Formation. J. Exp. Biol., 152: 149 – 166.

Knudsen PK, Jensen FB (1997). Recovery from Nitrite-induced Methaemoglobinaemia and Potassium Balance Disturbance in Carp. Fish Physiol. Biochem., 16(1): 1 – 10.

Williams EM, Eddy FB (1986). Chloride Uptake in Freshwater Teleosts and its Relationship to Nitrite Uptake and Toxicity. J. Comp. Physiol. B., 156: 867 – 872.

Ichthyofauna characteristics of a tropical low brackish open Lagoon in South-western Nigeria

Emmanuel, B. E. and Osibona, A. O.

Marine Sciences Department, University of Lagos, Akoka, Lagos, Nigeria.

The fish fauna of Lekki lagoon in south-western part of Nigeria was studied from March 2006 to February 2008. Sampling was carried out once a month using different fishing gears and the market was visited to confirm the fish species diversity from the lagoon. A total of 16,960 specimens made up of juveniles and adults caught were identified and classified. They comprised of eighty-one species belonging to 40 families, 56 genera and 14 orders. Variation in the mesh of fishing gears used greatly influenced fish catch in the lagoon. The fish species encountered were of three ecological origin and seasonal occurrence in the Lekki Lagoon. There were eleven fishes of euryhaline origin which occurred mostly throughout the year, fifty-six fishes of freshwater origin and ten fishes of marine origin. The dominant fish species in the catches of the fisher folks between March 2006 and February 2008 were *Chrysichthys nigrodigitatus, Chrysichthys filamentosus, Tilapia guineensis, Tilapia zillii, Gymnarchus niloticus, Mormynus rume, Elops lacerta, Liza falapinnis, Clarias agboyiensis, Polydactylus quadrifilis, Synodontis clarias* and *Cynolossus senegalensis*. The high number of species (81) recorded for Lekki lagoon in this study confirms the fact that the lagoon is a transition zone between brackish and freshwater systems.

Key words: Ichytofauna, low brackish, fishing gears, Lekki lagoon.

INTRODUCTION

Lekki lagoon is in the eastern part of Lagos State, Nigeria. This lagoon impacts significantly on the lives of Lagosians. It is one of the four major lagoons in the Nigerian coastal system. It is used primarily for fishing, aquaculture, sand mining and recreation activities. The lagoon is also important in conservation terms because of the great diversity of endemic species. Fish and fishing related activities are among the main occupation of the majority of the people in the surrounding communities. The fishery productivity of the Lekki lagoon is put above 42.1 kg/ha/yr (Emmanuel, 2009). The fishery is diverse with about 35 landing sites and more than 6000 fishers. According to Olaosebikan and Raji (1998) the freshwater food fishes found in Nigeria are about 268 different species. They inhabit over 34 well-known freshwater bodies (rivers, lakes, reservoirs and lagoons), which constitute about 12% of Nigeria's total surface area which is put at 94,185,000 ha (Ita, 1993). Oguzie (1997) produced a key to some of the freshwater fishes of Nigeria as adopted from Boulenger (1916) and Olaosebikan and Raji (1998). The key agreed with the identification method published by Leveque et al. (1991) on the freshwater fishes of the NILO-Sudan river basin in Africa. Welman (1948) produced a list of 181 species of fishes that could be found in Nigeria inland waters. It was further reported that there are about 145 species of fish in the areas of the Kainji Lake basin. The report also revealed that Anambra, Kaduna and Sokoto/Rima Rivers have 23, 28 and 22 species respectively. Cross River, Ogun and Osun Rivers have 39, 23 and 23 fish species respectively. Olaosebikan and Raji (1998) published a list of African freshwater fishes to include 976 species, referable to 185 genera and 43 families. Ita (1993) reported that an estimated 230 species of fish have been recorded from the rivers of Nigeria, but no record is available on the species present in Rivers like Benin and Calabar which all empty directly into the ocean.

Knowledge of fish biology and species composition of different water bodies is necessary to enhance the management of water resources. Until now, published information on the fishes and fisheries of the lagoons of the south–western Nigeria has been mainly on Lagos lagoon (FAO, 1969; Fagade, 1969; Fagade and Olaniyan, 1974; Solarin, 1998; Emmanuel et al., 2008a, b). Recently, fishing pressure on the Lekki lagoon is high and there have been clearance on the landward side for agriculture. The swamp forests and rain forests of the area have been largely destroyed. All these have impacted greatly on the lagoon ecosystem in terms of the physical structure and physico-chemical parameters, which invariably result in changes in the fishery resources and making the conservation status of the area to be unprotected. Apart from the work of Kusemiju (1973) on catfish's biology and fishery over three decades ago, no major work has been done on the fishes and fisheries of the Lekki lagoon. Most recent investigations on the lagoon have focused on condition factors, fat and protein content of five fish species (Saliu et al., 2007) and population parameter of *Macrobrachium vollenhoevenii* (Abohweyere and Falaye, 2008). There is no literature record of the ichthyofauna composition of the Lekki lagoon. There is however, a need for studies on the abundance and diversity of the ichthyofauna of the Lekki lagoon, to serve as database information of the composition of the various fish species in the lagoon. Therefore, the present study is a preliminary investigation of the fish fauna composition and abundance of this major lagoon, with the view to appraising the economic importance of this lagoon in the area of fish protein supply in the area.

MATERIALS AND METHODS

Description of the study site

The Lekki lagoon is one of the largest lagoons in West Africa and it supports a major fishery. The lagoon is located between Lagos and Ogun States of Nigeria and lies between longitude 4° 00' and 4° 15' E and between latitude 60° 25' and 6° 37'N (Figure 1). The lagoon has a surface area of about 247 km^2 and it is mostly shallow (less than 3.0 m deep), the maximum depth being 12.25 m. Lekki lagoon is a freshwater environment fed by River Oni in the north eastern part and by Rivers Osun and Saga in the north western parts of the lagoon. It opens into the sea via the Lagos lagoon and harbour. The lagoon is transitional in that it connects three south western states (Ondo, Ogun and Lagos). The lagoon is part of an intricate system of waterways made up of lagoons and creeks that are found along the coast of South-western Nigeria from the Dahomey border to the Niger Delta.

The two distinct seasons (dry and rainy) are observable in the lagoon which is typical of the southern part of Nigeria.

The vegetation around the Lekki lagoon consists mainly of stilt rooted trees, a dense undergrowth of shrub and raphia palms (*Raphia sudanica*) and oil palms (*Elaeis guineensis*). The floating grass (*Saccarum* sp) occurred on the periphery of the lagoon while coconut palms (*Cocos nucifera*) are widely distributed in the surrounding villages. Some parts of the lagoon are covered by floating plant like the water lettuce, *Pistia stratiotes*, duck weed, *Lemna* sp and the water hyacinth, *Eichhornia crassipes* are always found in the periphery and are distributed all over the lagoon during the dry season especially December, January, February and early March of the year.

Fish sampling procedure

Fish collections were made through the assistance of local fishermen using gillnets, castnets, longlines, basket and bamboo traps. The fishing gears were set overnight (that is, they were set a day prior to the sampling) and were retrieved on the sampling day for data collection. Sampling was done once a month for two years (March, 2006 to February, 2008). Fish were identified to species level using the available literature on the coastal waters of Nigeria (Schneider, 1990; Holden and Reed, 1991; Olaosebikan and Raji, 1998).

Fish species analysis

All fish species collected were counted to determine species abundance. Numerical abundance of the fish species was observed and noted. The abundance of each species was estimated according to the following criteria as described by Benech et al. (1983):

≥ 10% = dominant
1 to 9% = subdominant
< 1% (but caught more than once) = occasional
< 1% (and caught only once) = rare.

In the market, the catch abundance was recorded using these keys: 1 (very abundant), 2 (abundant), 3 (few) and 4 (rare) as described by Emmanuel (2009).

RESULTS

General survey of the fish fauna in Lekki lagoon

A total of 16,960 specimens made up of juveniles and adults caught with different fishing gears types in the Lekki lagoon, were identified and classified. They comprised eighty-one species belonging to 40 families, 56 genera and 14 orders. Decapod crustaceans comprised the freshwater prawns (*M. vollenhovenii* (Herklots) and *Macrobrachium macrobrachion* (Herklots) and the swimming crab *Callinectes amnicola* (De Rocheburne). Table 1 shows a list of fish order, family and species, the size and weight ranges of the fish specimens. Anabantidae was represented by species

Figure 1. Map of Lekki Lagoon and its environs.

Table 1. Fish species composition in Lekki Lagoon, their weight and length.

Order/Family/Species	Total length range (cm)	Standard length (cm)	Weight (g)
Perciformes			
Anabantidae			
Ctenopoma petherici (Gunther, 1864)	6.0 - 13.5	4.5 - 11.0	20.0 - 95
Centropomidae			
Lates niloticus (Linne, 1762)	10.5 - 60.0	8.2 - 55.0	40.0 - 4000
Carangidae			
Caranx hippos (Linnaeus, 1766)	5.5 - 59.5	4.0 - 53.5	9.0 - 3900
Trachinotus teraia (Cuvier, 1832)	9.5 - 45.0	7.0 - 40.5	20.0 - 3600
Cichlidae			
Tilapia guineensis (Bleeker, 1862)	4.0 - 21.0	3.0 - 19.5	2.81 - 700
Tilapia zilli (Gervais, 1848)	5.5 - 24.0	3.5 - 18.40	5.94 - 530
Tilapia mariae (Boulenger, 1899)	5.7 - 16.0	3.5 - 14.5	3.0 - 410
Chromidotilapia guntheriguntheri (Sauvage, 1882)	5.0 - 12.0	3.0 - 9.5	4.0 - 200
Sarotherodon melanotheron (Rupell, 1852)	4.4 - 16.0	3.0 - 14.0	3.70 - 375
Oreochromis niloticus (Linne, 1758)	6.8 - 26.0	4.5 - 23.0	13.0 - 857
Hemichromis fasciatus (Peters, 1852)	4.4 - 12.0	3.2 - 10.0	4.16 - 120
Hemichromis bimaculatus (Gill, 1862)	3.9 - 10.0	2.5 - 8.5	3.39 - 30.0
Eleotridae			
Eleotris vittata (Dumeril, 1858)	7.8 - 14.5	5.5 - 10.5	7.95 - 150.0
Kribia nana (Boulenger, 1961)	3.4 - 4.5	2.5 - 3.5	3.40 - 10.0
Channidae			
Parachanna obscura (Gunther, 1861)	14.1 - 30.8	11.5 - 28.5	89.5 - 400
Parachanna africana (Steindachner, 1879)	13.0 - 29.5	10.2 - 27.0	79.8 - 390
Pomadasysidae			
Pomadasys jubelini (Cuvier, 1830)	9.3 - 22.0	7.0 – 19.5	19.75 - 309
Lutjanidae			
Lutjanus dentatus (Dumeril, 1860)	9.3 - 22.0	7.0 - 18.0	20.20 - 311
Polynemidae			
Polydactylus quadrifilis (Cuvier, 1829)	18.0 - 105.0	16.0 - 85.0	20.70 - 6000
Gobiidae			
Bathygobius soporator (Valenciennes, 1873)	12.90 - 14.60	10.90 - 12.5	24.8 - 40.0
Goboides ansorgii (Boulenger, 1909)	15.40 - 32.50	13.00 - 30.10	25.0 - 65.0
Spyraenidae			
Sphyraena barracuda (Walbaum, 1792)	30.9 - 103.0	27.5 - 87.0	98.5 - 4000
Monodactylidae			
Psettias sebae (Cuvier, 1931)	5.5 - 10.5	3.5 - 8.2	10.5 - 65.0
Distichodontidae			
Ichthyborus monodi (Pellegrin, 1929)	6.0 - 16.5	4.0 - 14.0	16.9 - 66.5
Rajiformes			
Dasyatidae			
Dasyatis garouaensis (Stauch and Blanc, 1962)	35.0		600
Polypteriformes			
Polypteridae			
Polypterus senegalus senegalus (Cuvier, 1829)	9.0 - 30.0	7.0 - 28.0	20.5 - 150.0
Erpetoichthys calabaricus (Smith, 1866)	20.2 - 35.5	18.0 - 33.5	19.5 - 50.2
Elopiformes			
Elopidae			
Elops lacerta (Valenciennes, 1846)	11.0 - 27.0	9.2 - 25.0	9.5 - 241

Table 1. Contd.

Osteoglossiformes			
Pantodontidae			
Pantodon buchholzi (Peters, 1876)	5.6 - 10.0	3.0 - 7.0	10.0 - 20.0
Notopteridae			
Papyrocranus afer (Gunther, 1868)	12.2 - 55.5	10.5 - 52.0	12.5 - 1069
Xenomystus nigri (Gunther, 1868)	12.0 - 45.0	10.0 - 42.0	13.0 - 1050
Osteoglossidae			
Heterotis niloticus (Cuvier, 1829)	14.5 - 54.5	12.0 - 50.5	20.0 - 2000
Mormyriformes			
Mormyridae			
Mormyrus rume (Valennciennes, 1846)	12.5 - 48.0	9.5 - 46.5	15.6 - 868
Mormyrus macrophthalmus (Gunther, 1866)	12.2 - 30.1	9.2 - 46.5	20.0 - 600
Hippopotamyrus pictus (Marcusen, 1864)	5.5 - 15.5	4.0 - 12.5	15.0 - 50.5
Hippopotamyrus psittasus	6.5 - 25.0	5.0 - 23.0	18.0 - 75.9
Hyperopisus bebe (Lacepede, 1803)	15.6 - 50.0	12.5 - 48.0	20.5 - 850
Mormyrops anguilloides (Linnaeus, 1758)	9.1 - 63.3	7.0 - 60.0	5.8 - 2453
Marcusenius senegalensis (Steindachner, 1870)	9.6 - 27.3	7.0 - 25.3	10.0 - 248
Pollimyrus adspersus (Gunther, 1866)	5.2 - 9.6	3.5 - 7.2	17.0 - 50.0
Marcusenius brucii (Boulenger, 1910)	6.3 - 30.8	5.0 - 28.5	12.1 - 515
Brienomyrus longianalis (Boulenger, 1901)	16.0 - 30.8	14.0 - 28.5	50 - 610
Gnathonemus petersii (Gunther, 1862)	15.0 - 35.0	13.5 - 33.0	48.5 - 590
Mormyrops caballus (Pelledrin, 1927)	9.1 - 46.0	7.1 - 44.2	15.8 - 850
Gymnarchidae			
Gymmarchus niloticus (Cuvier, 1829)	35.0 - 120.0	32.5 - 117	89.0 - 3000
Clupeiformes			
Clupeidae			
Pellonula afzeliusi (Johnels, 1954)	4.0 - 10.1	3.0 - 8.0	5.0 - 26.0
Ethmalosa fimbriata (Bowdich, 1825)	8.70-14.70	6.80-11.40	5.35-32.26
Characiformes			
Citharinidae			
Citharinus latus (Muller and Troschal, 1845)	7.0 - 46.0	5.0 - 43.5	25.5 - 1065
Cithranus citharus (Goeffrey Saint Hilane, 1809)	10.0 - 50.0	8.0 - 47.5	45.0 - 2010
Hepsetidae			
Hepsetus odoe (Bloch, 1794)	7.5 - 30.5	5.6 - 28.8	9.26 - 856
Characidae			
Alestes macropthalmus (Gunther, 1867)	20.5 - 30.6	18.2 - 29.0	45.0 - 150.0
Alestes baremose (de Joannis, 1835)	10.5 - 40.5	8.5 - 38.2	20.2 - 300
Brycinus nurse (Ruppell, 1832)	5.3 - 20.5	3.8 - 18.2	5.2 - 212
Brycinus longipinnus (Gunther, 1864)	4.8 - 10.7	3.0 - 8.8	3.6 - 41.6
Siluriformes			
Bagridae			
Chrysichthys Walkeri	5.5 - 36.5	3.5 - 33.6	4.36 - 724
Chrysichthys nigrodigitatus (Lacepede, 1803)	5.8 - 42.5	4.0 - 40.5	5.0 - 1500
Chrysichthys filamentosus (Boulenger, 1912)	5.6 - 38.5	3.8 - 36.8	4.0 - 798
Parauchenoglanis akiri (Risch, 1987)	10.0 - 12.5	8.0 - 10.0	15.9 - 45.6
Auchenoglanis occidentalis (Valenciennes, 1840)	15.0 - 20.0	13.0 - 18.0	24.0 - 50.1

Table 1. Contd.

Schilbeidae			
Schilbe mystus (Linne, 1758)	7.0 - 21.0	5.8 - 19.0	4.15 - 119.5
Schilbe uranoscopus (Ruppell, 1832)	6.2 - 28.5	5.0 - 26.5	7.61 - 360
Clariidae			
Clarias gariepinus (Burchell, 1822)	20.0 - 50.5	17.0 - 46.8	78.00 - 1920
Clarias jaensis (Boulenger, 1909)	10.2 - 20.0	7.5 - 17.9	22.8 - 64.25
Clarias agboyiensis (Sydenham, 1980)	11.2 - 21.0	8.5 - 18.7	21.9 - 72.96
Clarias anguillaries (Line, 1758)	9.0 - 34.5	7.2 - 31.5	17.0 - 65.0
Heterobranchus longifilis (Valenciennes, 1840)	40.5 - 50.0	37.8 - 48.5	1002 - 2100
Malapteruridae			
Malapterurus electricus (Gmelin, 1789)	13.0 - 16.5	11.5 - 14.0	60.6 - 89.8
Malapterurus minjiraya (Sagua, 1987)	14.0 - 17.5	12.2 - 15.0	64.7 - 92.0
Mochokidae			
Synodontis eupterus	4.5 - 22.0	3.0 - 20.0	9.2 - 218
Synodontis clarias (Linne, 1758)	5.5 - 22.5	3.8 - 21.0	10.2 - 316
Synodontis couterti (Pellergrin, 1906)	6.5 - 20.6	4.2 - 18.0	10.5 - 212
Synodontis filamentosus (Boulenger, 1901)	5.5 - 18.6	3.0 - 16.2	8.9 - 200
Mugiliformes			
Mugilidae			
Liza falcipinnis (Valenciennes, 1836)	13.0 - 26.5	10.5 - 19.2	41.6 - 200
Mugil cephalus (Linnaeus, 1758)	12.5 - 20.5	10.0 - 18.0	41.2 - 360
Synbranchiformes			
Mastacembelidae			
Caecomastacembelus decorsei (Pellegrin, 1919)	14.2 - 36.5	12.5 - 35.0	20.0 - 96.0
Pleuronectiformes			
Citharidae			
Citharus linguatula (Linnaeus, 1758)	10.30 - 15.0	8.0 - 13.2	9.0 - 15.9
Cynoglossidae			
Cynoglossus senegalensis (Kaup, 1858)	15.6 - 54.0	13.8 - 49.2	20.0 - 460.0
Gonorychiformes			
Phractolaemidae			
Phractolaemus ansorgii (Boulenger, 1901)	10.5 - 19.0	8.0 - 17.0	17.0 - 56.0
Decapoda			
Palaemonidae			
Macrobrachium vollenhoveni	6.4 - 13.0	3.00 - 600*	6.06 - 32.4
Macrobrachium macrobrachion	6.4 - 12.00	3.00 - 5.50*	6.04 - 28.29
Portunidae			
Callinectes amnicola	3.4 - 15.5**		19.5 - 115.5

*Carapace length, **Carapace width.

(*Ctenopoma petherici*)). Currently the species was caught mainly with double funnel traps in and around the River Saga's mouth. Dasyatidae was represented by only one species *Dasyatis garouaensis* (Stanch & Blanc) and being listed in the Lekki Lagoon for the first time. The species was caught with boat seine in the sandy area of the Agan in the centre of the lagoon. The length of 65.0 cm and a weight of 400 g were recorded for the species.

The family mormyridae was represented with twelve species. This family had the highest species diversity in the lagoon. Only two specimens of *H. longifilis* were recorded throughout this study. The fish species and the target fishing gear are shown in Table 2.

Fish species and their ecological origin

The fish species were of three ecological origin and seasonal occurrence in the Lekki Lagoon. There were eleven fishes of euryhaline origin which occurred mostly throughout the year, fifty six fishes of freshwater origin and ten fishes of marine origin occurred in Lekki Lagoon during this study (Table 3).

Fish species dominance in Lekki lagoon

The dominant fish species in the catches of the fisher folks during the year between March 2006 and February 2008 were *Chrysichthys nigrodigitatus, Chrysichthys filamentosus, Tilapia guineensis, T. zilli, Gymnarchus niloticus, Mormynus rume, Elops lacerta, Liza falapinnis, Clarias agboyiensis, Polydactylus quadrifilis, Synodontis clarias* and *Cynolossus senegalensis.* Other fish species observed in the fisher-folks catches during the rainy season period were *Ethmalosa fimbriata, Lates niloticus, Tilapia marie, Sarotherodon melanotheron, Chromidotilapia guntheri guntheri, Oreochromis niloticus, Pomadasys jubelini, Lutjanus agenes, Eleotris vitata, Caranx hippos, Hemichromis bimaculatus, Bathgobius soporator, Pantodon buchholzi, Papyrocranus afer, Sphyreana barracuda, Xenomystus nigri, Mormyrops anguilloides, Marcusenius bruci, Pellonula afzeluisi, Citharinus latus, Trachinotus teraia, Cithanus cithanus, Alestes macrophthalmus, Brycinus nurse, Schilbe uranoscopus, Cithanus linguatula, Citharus linguatula, M. vollenhoevenii* and *M. macrobrachion.* Few of the lagoon species were recorded for dry season; there were *Parachanna africana, Xenomystus nigri, Hippopotamyrus pictus, Hyperopisus bebe, Marcusenius senegalensis, Polymus adspersus, Brienomyrus longuanalis, Alestes baremose, Parauchenoglanis akiri, Hepsetus odoe, Clarias anguillaries, Malapterurus minijiraya, Synodontis filamentosus* and *Phractolaenus ansorgii.* Fish species dominance and seasonal variation is shown in Table 4.

DISCUSSION

This study provides general information on the fish species composition and assemblages of Lekki lagoon. However, the wet season is typically the period of the highest fish abundance and species richness in the area as well as the period of highest population densities of juvenile species and a single attempt to examine year-to-year variation in this area did not find significant differences in fish assemblage structure (Emmanuel, 2009). Due to the relationship between the sea and inland waters, the lagoon is characterized by environmental unpredictability. The high number of species recorded in wet season shows that Lekki lagoon is an important spawning and nursery area. The nursery function of lagoons has been documented throughout the world (Blaber and Blaber, 1980; Lenanton, 1982; Robertson and Duke, 1987; Blaber and Milton, 1990; Emmanuel and Kusemiju, 2005; Emmanuel et al., 2008a, b). The lagoon supported a high diversity of fish species of fresh, brackish and marine water origins. They must have had ability to adapt and cope with the variable conditions (salinity and temperature) of the lagoon. Many fishes have evolved specialized physiological and behavioral adaptation (Zander et al., 1999). Increasing the specialization of these adaptations to cope with localized conditions may be a strategy to exist as a permanent lagoon resident. More fish species were recorded in this study compared to Kusemiju (1973) report over 3 decades ago in the lagoon where 28 species were recorded.

The interaction between fish and its habitat (physical, as well as biological characteristics) may be factors in predicting changes in overall abundance, breeding population size and other aspects of population structure (Koutrakis et al., 2000; Caldeira et al., 2001; Emmanuel and Onyema, 2007; Emmanuel, 2009). In addition to this, Beneech et al. (1983) reported that fish communities studied are not generally equivalent to ichthyocoenoses because the description of any fish community is a biased image arising from the sampling of a group of fishes in a particular environment at a given period. This may be so of the report of Kusemiju (1973) where a trawl was used as against diverse gears used in this study. The ichthyofauna of Lekki lagoon with 81 species belonging to 40 families, 56 genera and 15 orders appear richer than 60 species from 34 families of Lagos Lagoon (Solarin, 1998). The occurrence of fishes of both marine and euryhaline origins gave a sign of environmental alteration (salt water intrusion). Several workers have recognized the phenomena of an optimum salinity level would cause variation in the number of species available (Hesse et al., 1951; Hedgepth, 1957).

Macan (1963) cited by Kusemiju (1973) observed that where masses of fresh water and seawater adjoined, species from each invaded the brackish region between them, but few penetrated far and the number of species at a midway between the two was small. Motwani and Kanwai (1970) worked in the completely freshwater environment and recorded 82 species of fish belonging to eighteen families at Kainji Lake, Nigeria. The high number of 81 species recorded for Lekki Lagoon in this study confirms the fact that the lagoon is a transitional zone between brackish and freshwater systems (Emmanuel, 2009).

Table 2. Fish species and the fishing gear used to catch them.

Species	Fishing gear types						
	GN	LL	CN	BS	BT	BAT	LN
Ctenopoma petherici (Gunther, 1864)	x				x		
Lates niloticus (Linne, 1762)	x	x		x			
Caranx hippos (Linnaeus, 1766)	x	x		x			
Trachinotus teraia (Cuvier, 1832)	x	x		x			
Tilapia guineensis (Bleeker, 1862)	x	x	x	x	x		
Tilapia zilli (Gervais, 1848)	x		x	x	x		
Tilapia mariae (Boulenger, 1899)	x		x	x	x		
Chromidotilapia gunthериguntheri (Sauvage, 1882)	x		x	x	x		
Sarotherodon melanotheron (Rupell, 1852)	x		x	x			
Oreochromis niloticus (Linne, 1758)	x		x	x			
Hemichromis fasciatus (Peters, 1852)	x		x	x	x		
Hemichromis bimaculatus (Gill, 1862)	x		x	x	x		
Eleotris vittata (Dumeril, 1858)				x	x		
Kribia nana (Boulenger, 1961)					x		
Parachanna obscura (Gunther, 1861)	x			x	x		
Parachanna africana (Steindachner, 1879)	x			x	x		
Pomadasys jubelini (Cuvier, 1830)	x			x			
Lutjanus dentatus (Dumeril, 1860)	x		x	x			
Polydactylus quadrifilis (Cuvier, 1829)	x	x		x			
Bathygobius soporator (Valenciennes, 1873)	X			x			
Goboides ansorgii (Boulenger, 1909)	x			x			
Sphyraena barracuda (Walbaum, 1792)	x	x	x	x			
Psettias sebae (Cuvier, 1931)	x	x	x	x			
Ichthyborus monodi (Pellegrin, 1929)	x		x	x			
Dasyatis garouaensis (Stauch & Blanc, 1962)	x	x		x			
Polypterus senegalus senegalus (Cuvier, 1829)					x		
Erpetoichthys calabaricus (Smith, 1866)					x		
Elops lacerta (Valenciennes, 1846)	x		x	x			
Pantodon buchholzi (Peters, 1876)					x		
Papyrocranus afer (Gunther, 1868)	x		x	x			
Xenomystus nigri (Gunther, 1868)	X		x	X			
Heterotis niloticus (Cuvier, 1829)	x		x	x			
Mormyrus rume (Valennciennes, 1846)	x		x	x	x		
Mormyrus macrophthalmus (Gunther, 1866)	x		x	x	x		
Hippopotamyrus pictus (Marcusen, 1864)	x		x	x	x		
Hippopotamyrus psittasus (Boulenger, 1897)	x		x	x	x		
Hyperopisus bebe (Lacepede, 1803)	x		x	x	x		
Mormyrops anguilloides (Linnaeus, 1758)	x		x	x	x		
Marcusenius senegalensis (Steindachner, 1870)	x		x	x	x		
Pollimyrus adspersus (Gunther, 1866)	x		x	x	x		
Marcusenius brucii (Boulenger, 1910)	x		x	x	x		
Brienomyrus longianalis (Boulenger, 1901)	x		x	x	x		
Gnathonemus petersii (Gunther, 1862)	x		x	x	x		
Mormyrops caballus (Pelledrin, 1927)	x		x	x	x		
Gymmarchus niloticus (Cuvier, 1829)	x	x	x	x			
Pellonula afzeliusi (Johnels, 1954)					X		x
Ethmalosa fimbriata (Bowdich, 1825)	x		x	x	x		
Citharinus latus (Muller & Troschal, 1845)	x			x			
Citharinus citharus (Goeffrey Saint Hilane, 1809)	x			x			
Hepsetus odoe (Bloch, 1794)	x	x	x	x	x		

Table 2. Contd.

Alestes macropthalmus (Gunther, 1867)	x		x	x	x	
Alestes baremose (de Joannis, 1835)	x		x	x	x	
Brycinus nurse (Ruppell, 1832)	x		x	x	x	
Brycinus longipinnus (Gunther, 1864)	x		x	x	x	
Chrysichthys walkeri	x	x	x	x	x	X
Chrysichthys nigrodigitatus (Lacepede, 1803)	x	x	x	x	x	X
Chrysichthys filamentosus (Boulenger, 1912)	x	x	x	x	x	X
Parauchenoglanis akiri (Risch, 1987)	X	x	x	X	x	
Auchenoglanis occidentalis (Valenciennes, 1840)	x			x	x	
Schilbe mystus (Linne, 1758)	x		x	x		
Schilbe uranoscopus (Ruppell, 1832)	x		x	x		
Clarias gariepinus (Burchell, 1822)	x	x	x	x	x	
Clarias jaensis (Boulenger, 1909)	x	x	x	x	x	
Clarias agboyiensis (Sydenham, 1980)	x	x	x	x	x	
Clarias anguillaries (Line, 1758)	x	x	x	x	x	
Heterobranchus longifilis (Valenciennes, 1840)	x	x	x	x	x	
Malapterurus electricus (Gmelin, 1789)	x			x		
Malapterurus minjiraya (Sagua, 1987)	x			x		
Synodontis eupterus	x	x	x	x		
Synodontis clarias (Linne, 1758)	x	x	x	x		
Synodontis couterti (Pellergrin, 1906)	x	x	x	x	x	
Synodontis filamentosus (Boulenger, 1901)	x	x	x	x	x	
Liza falcipinnis (Valenciennes, 1836)	x		x	x	x	
Mugil cephalus (Linnaeus, 1758)	x		x	x	x	
Caecomastacembelus decorsei (Pellegrin, 1919)						
Citharus linguatula (Linnaeus, 1758)	x		x	x	x	
Cynoglossus senegalensis (Kaup, 1858)	x		x	x		
Phractolaemus ansorgii (Boulenger, 1901)	x		x	x	x	
Macrobrachium vollenhoveni	x		x	x	x	
Macrobrachium macrobrachion	x		x	x	x	
Callinectes amnicola	x		x	x	x	

GN= gillnet; CN = cast net; LL = long line; BT = basket trap; BAT = Bamboo trap; LN = lift net.

Table 3. Fishes and their ecological origin in Lekki lagoon.

Freshwater origin	Euryhaline origin	Marine origin
Ctenopoma petherici	*Chrysichthys nigrodigitatus*	*Caranx hippos*
Lates niloticus	*Ethmalosa fimbriata*	*Trachinotus teraia*
Tilapia zillii	*Tilapia guineensis*	*Pomadasys jubelini*
Tilapia marie	*Pellonula afzeluizi*	*Lutjanus dentatus*
Chromidotilapia guntheri guntheri	*Elops lacerta*	*Sphyraena barracuda*
Oreochromis niloticus	*Liza falcipinnis*	*Ethmalosa fimbriata,*
Hemichromis bimaculatus	*Sarotherodon melanotheron*	*Liza falcipinnis*
Eleotris vittata	*Polydactylus quadrifilis*	*Mugil cephalus*
Kribia nana	*Bathygobius soporator*	*Citharus linguatula*
Parachanna obscura	*Cynoglossus senegalensis*	*Cynoglossus senegalensis*
Parachanna africana	*Psettias sebae*	
Ichthyborus monodi		
Dasyatis garouaensis		
Polypterus senegalus senegalus		

Table 3. Contd.

Erpetoichthys calabaricus
Pantodon buchholzi
Papyrocranus afer
Hippopotamyrus pictus
Hyperopisus bebe
Mormyrops anguilloides
Marcusenius senegalensis
Pollimyrus adspersus
Marcusenius brucii
Brienomyrus longianalis
Gnathonemus petersii
Mormyrops caballus
Gymnarchus niloticus
Citharinus latu
Auchenoglanis occidentalis
Hepsetus odoe
Citharinus citharus
Alestes macropthalmus
Alestes baremose
Brycinus nurse
Brycinus longipinnus
Chrysichthys walker
Chrysichthys filamentosus
Parauchenoglanis akiri
Schilbe mystus
Schilbe uranoscopus
Clarias gariepinus
Clarias jaensis
Clarias agboyiensis
Clarias anguillaries
Heterobranchus longifilis
Malapterurus electricus
Malapterurus minjiraya
Synodontis eupterus
Synodontis clarias
Synodontis courteti
Synodontis filametosus
Caecomastacembelus decorsei
Phractolaemous ansorgii

Table 4. Fish dominance and abundance in Lekki Lagoon.

Order/Family/Species	Mar 2006	Apr	May	Jun	Jul	Aug	Sept	Oct	Nov	Dec	Jan	Feb 2007
Perciformes												
Anabantidae												
Ctenopoma petherici (Gunther, 1864)	2	2	4	4	4	3	4	3	2	1	3	3
Centropomidae												
Lates niloticus (Linne, 1762)	4	4	4	4	4		3	3	3	4	4	3
Carangidae												
Caranx hippos (Linnaeus, 1766)	4	4	4	3	3	2	4	3	4	3	4	3

Table 4. Contd.

Trachinotus teraia (Cuvier, 1832)	3	3	3	3	3	2	2	3	2	2	2	3
Cichlidae												
Tilapia guineensis (Bleeker, 1862)	1	3	1	1	1		1	1	1	1	1	1
Tilapia zillii (Gervais, 1848)	2	3	2	3	1		1	1	4	1	4	2
Tilapia mariae (Boulenger, 1899)	4	2	3	4	4		4	3	2	3	1	3
Chromidotilapia guntheriguntheri (Sauvage, 1882)	3	3	4	4	4		2	4	1	1	2	4
Sarotherodon melanotheron (Rupell, 1852)	3	3	4	4	4		2	4	4	2	4	4
Oreochromis niloticus (Linne, 1758)	4	4	4	4	4		3	2	4	2	3	4
Hemichromis fasciatus (Peters, 1852)	3	3	3	4	4		4	4	3	3	2	2
Hemichromis bimaculatus (Gill,1862)	3	3	3	4	4		3	4	4	4	4	4
Eleotridae												
Eleotris vittata (Dumeril, 1858)	4	3	3	4	4		2	3	2	-	4	4
Kribia nana (Boulenger, 1961)	4	4	4	4	4		4	4	4	-	4	4
Channidae												
Para*channa obscura* (Gunther, 1861)	2	3	2	4	4	2	3	3	2	1	2	4
Parachanna africana (Steindachner, 1879)	2	3	4	4	4	4	4	4	4	4	4	4
Pomadasysidae												
Pomadasys jubelini (Cuvier,1830)	3	3	4	3	3	3	3	3	3	3	3	3
Lutjanidae												
Lutjanus dentatus (Dumeril, 1860)	4	4	4	3	4	4	4	4	3	4	4	4
Polynemidae												
Polydactylus quadrifilis (Cuvier, 1829)	3	3	3	3	4	3	3	3	3	3	3	3
Gobiidae												
Bathygobius soporator (Valenciennes, 1873)	4	3	3	4	4	3	4	4	4	4	4	4
Goboides ansorgii (Boulenger, 1909)	4	4	3	4	4	3	4	4	3	3	4	4
Spyraenidae												
Sphyraena barracuda (Walbaum, 1792)	2	4	3	3	4	3	3	3	3	3	3	2
Monodactylidae												
Psettias sebae (Cuvier, 1931)	3	3	4	3	4	3	3	4	3	3	3	4
Distichodontidae												
Ichthyborus monodi (Pellegrin,1929)	4	4	3	3	4	4	4	4	3	3	4	4
Rajiformes												
Dasyatidae												
Dasyatis garouaensis (Stauch and Blanc, 1962)	4	4	4	4	4	4	4	4	3	4	4	4
Polypteriformes												
Polypteridae												
Polypterus senegalus senegalus (Cuvier, 1829)	3	3	3	4	4	2	3	4	2	2	3	3
Erpetoichthys calabaricus (Smith, 1866)	3	3	4	4	4	4	4	4	1	1	1	4
Elopiformes												
Elopidae												
Elops lacerta (Valenciennes, 1846)	3	2	2	3	3	2	3	3	1	4	2	1
Osteoglossiformes												
Osteoglossidae												
Heterotis niloticus	4	4	3	2	2	2	3	3	3	4	4	4
Pantodontidae												
Pantodon buchholzi (Peters, 1876)	4	4	3	3	3	3	3	3	3	4	4	4
Notopteridae												
Papyrocranus afer (Gunther, 1868)	3	2	3	2	4	3	3	3	3	1	3	3

Table 4. Contd.

Xenomystus nigri (Gunther, 1868)	2	3	3	3	4	4	4	4	2	3	4	3
Mormyriformes												
Mormyridae												
Mormyrus rume (Valennciennes, 1846)	2	2	2	2	2	2	4	4	2	2	1	3
Mormyrus macrophthalmus (Gunther,1866)	2	3	4	2	4	2	4	4	4	4	2	4
Hippopotamyrus pictus (Marcusen, 1864)	3	4	4	4	4	4	4	4	4	4	1	2
Hyperopisus bebe (Lacepede, 1803)	3	4	4	4	4	4	4	4	4	2	4	4
Mormyrops anguilloides (Linnaeus, 1758)	3	4	4	4	4	4	2	2	2	4	4	4
Marcusenius senegalensis (Steindachner, 1870)	3	4	4	4	4	4	4	4	4	4	4	4
Pollymus adspersus (Gunther, 1866)	3	4	4	4	4	4	4	4	4	4	4	4
Marcusenius brucii (Boulenger, 1910)	3	4	4	2	4	4	2	4	4	4	4	4
Brienomyrus longianalis (Boulenger, 1901)	3	4	4	4	4	4	4	4	4	3	3	3
Gnathonemus petersii (Gunther, 1862)	3	4	3	2	2	4	1	4	1	2	4	4
Mormyrops caballus (Pelledrin, 1927)	3	3	4	3	3	3	2	2	2	3	3	4
Hippopotamyrus psittasus (Boulenger, 1897)	3	4	3	3	3	3	3	3	3	3	3	4
Gymnarchidae												
Gymnarchus niloticus (Cuvier, 1829)	3	3	3	2	2	3	2	2	1	1	1	4
Clupeiformes												
Clupeidae												
Pellonula afzeliusi (Johnels, 1954)	3	3	4	3	2	1	3	4	4	4	1	4
Ethmalosa fimbriata (Bowdich, 1825)	3	3	3	3	1	1	1	3	1	4	3	3
Characiformes												
Citharinidae												
Citharinus latus (Muller and Troschal, 1845)	4	4	4	4	3	3	3	3	3	4	4	4
Cithranus citharus (Goeffrey Saint Hilane, 1809)	4	4	4	4	4	4	4	4	3	3	4	4
Hepsetidae												
Hepsetus odoe (Bloch, 1794)	3	3	3	4	4	4	4	4	3	3	2	3
Characidae												
Alestes macrophthalmus (Gunther, 1867)	3	3	3	3	3	3	3	3	4	3	4	1
Alestes baremose (de Joannis, 1835)	3	3	3	4	4	2	4	4	4	3	3	1
Brycinus nurse (Ruppell, 1832)	3	3	4	4	4	3	3	3	2	3	4	2
Brycinus longipinnus (Gunther, 1864)	3	4	4	4	4	4	4	4	1	4	3	3
Siluriformes												
Bagridae												
Chrysichthys Walkeri (Gunther, 1899)	2	3	4	4	4	1	4	2	4	1	1	1
Chrysichthys nigrodigitatus (Lacepede, 1803)	1	1	1	1	1	1	1	1	1	1	1	1
Chrysichthys filamentosus (Boulenger, 1912)	1	1	1	1	1	2	1	4	2	2	2	2
Parauchenoglanis akiri (Risch, 1987)	4	4	4	4	4	4	4	4	3	2	4	3
Auchenoglanis occidentalis (Valenciennes, 1840)	4	4	4	4	4	3	4	4	3	2	3	3
Schilbeidae												
Schilbe mystus (Linne, 1758)	3	2	2	2	2	2	1	1	3	2	3	3
Schilbe uranoscopus (Ruppell, 1832)	3	3	4	4	4	4	4	4	4	-	4	3
Clariidae												
Clarias gariepinus (Burchell, 1822)	2	2	2	4	2	1	4	3	3	1	1	1
Clarias jaensis (Boulenger, 1909)	3	3	4	4	2	4	4	1	1	1	3	2
Clarias agboyiensis (Sydenham, 1980)	3	3	4	4	4	2	1	4	4	1	4	2
Clarias anguillaries (Line, 1758)	3	3	1	1	1	4	4	4	4	4	4	3

Table 4. Contd.

Heterobranchus longifilis (Valenciennes, 1840	4	4	2	4	4	4	4	4	3	4	4	4
Malapteruridae												
Malapterurus electricus (Gmelin, 1789)	3	3	3	2	2	2	3	4	2	2	2	3
Malapterurus minjiraya (Sagua, 1987)	3	3	4	4	4	4	4	4	4	4	4	3
Mochokidae												
Synodontis eupterus	3	3	4	4	4	4	4	4	1	4	4	3
Synodontis clarias (Linne, 1758)	2	2	2	1	1	1	1	2	1	3	1	2
Synodontis couterti (Pellegrin, 1906)	3	3	4	4	4	4	4	4	4	4	4	3
Synodontis filamentosus (Boulenger, 1901)	3	3	4	4	4	1	4	4	4	4	1	3
Mugiliformes												
Mugilidae												
Liza falcipinnis (Valenciennes, 1836)	2	2	3	2	2	2	2	3	4	4	1	2
*Mugil cephalus (*Linnaeus, 1758)	3	4	4	4	4	4	4	4	4	4	3	3
Synbranchiformes												
Mastacembelidae												
Caecomastacembelus decorsei (Pellegrin, 1919)	3	3	4	3	4	4	4	4	4	4	4	3
Pleuronectiformes												
Citharidae												
Citharus lingualata (Linnaeus, 1758)	3	4	4	4	3	3	2	4	3	4	4	3
Cynoglossidae												
Cynoglossus senegalensis (Kaup, 1858)	2	3	3	4	3	2	2	3	3	2	2	2
Gonorychiformes												
Phractolaemidae												
Phractolaemus ansorgii (Boulenger, 1901)	3	3	4	4	4	4	4	4	4	4	3	2
Decapoda												
Palaemonidae												
Macrobrachium vollenhoveni(Herklots,1857)	3	3	2	2	1	2	1	1	1	3	3	3
Macrobrachium macrobrachion (Herklots,1857)	1	1	2	2	2	2	2	2	2	4	4	4
Portunidae												
Callinectes amnicola (DeRocheburne, 1883)	3	3	2	1	1	2	1	1	1	3	3	3

1 = very abundant, 2 = abundant, 3 = few, and 4 = rare.

REFERENCES

Abohweyere PO, Falaye EA (2008). Population Parameters of Macrobrachium vollenhovenii in the Lagos-Lekki Lagoon System, Nigeria . J. Fish. Int. 3: 27-33.

Benech V, Duran JR, Quensiere J (1983). Fish communities of Lake Chad and associated rivers and flood pains. In Lake Chad, ecology and productivity of a shallow tropical system (Garmuze, J.P., Durand, J.R. and Leveque, C. eds) Dr. Junk, The Hague: pp. 293-356.

Blaber SJM, Blaber TG (1980). Factors affecting the distribution of juvenile estuarine and inshore fish. J. Fish Biol. 17:143-162.

Blaber SJM, Milton DA (1990). Species composition, community structure and zoogeography of fishes of mangroves in the Solomon Islands. Mar. Biol. 105:259-268.

Caldeira RMA, Russell P, Amorim A (2001). Evidence of an unproductive coastal front in Bala D' Abra, an embayment on the south east of Madeira island, Portugal. Bull. Mar. Sci. 69(3):1057-1072.

Emmanuel BE (2009). The artisanal fishing gears, crafts technology and their efficiency in the Lekki lagoon, Nigeria. Ph.D thesis, University of Lagos. p. 256.

Emmanuel BE, Chukwu LO, Azeez LO (2008a). Cast net design characteristics, catch composition and selectivity in tropical open lagoon. Afr. J. Biotechnol. 7 (12):2081-2089.

Emmanuel BE, Chukwu LO, Azeez LO (2008b). Gill net selectivity and catch rates of pelagic fish in tropical coastal lagoonal ecosystem. Afr. J. Biotechnol. 7(21):3962-3971

Emmanuel BE, Kusemiju K (2005). Variations in castnet catch in a tropical brackishwater pond. J. Sci. Technol. Environ. 5(1&2):6-14.

Emmanuel BE, Onyema IC (2007). The plankton and fishes of a tropical creek in south-western, Nigeria. Turk. J. Fish. Aquat. Sci. 7:105-114

Fagade SO (1969). Studies on the biology of some fishes and fisheries of the Lagos Lagoon. Ph.D. Thesis, University of Lagos, Nigeria.

Fagade SO, Olaniyan CIO (1974). Seasonal distribution of the fish fauna of the Lagos Lagoon. Bull. De I..I. F. A. N. T., Ser. A 36(1):244.252.

FAO (1969). Fisheries Survey in the Western and Mid-western Regions of Nigeria. FAO, Rome, SF74/NIR6.

Hedgepth JW (1957). Treatise on marine ecology and paleoecology. Geol. Soc. Am. Memoir 67(I):129-157

Hesse R, Allee WC, Schmidt KP (1951). Ecological animal geography. John Wiley and Sons, New York. p. 715.

Holden M, Reed W (1991). West African Water Fishes. West African Nature Handbooks. Longman Group Ltd., Harlow. p. 68.

Ita EO (1993). Inland Fishery Resources of Nigeria. F.A.O. CIFA Occasional Paper No. 20, P 120.

Koutrakis ET, Kokkinakis AK, Eleftheradis A, Argyropoulou MD (2000). Seasonal changes in distribution and abundance of the fish fauna in the two estuarine system of Strymonkos gulf (Mcedonea, Greece). Belg. J. Zool. 130 (Supplement):43-50

Kusemiju K (1973). A study of the catfishes of Lekki lagoon with particular reference to the species *Chrysichthys walkeri* Bagridae. Ph.D Thesis. University of Lagos. p. 188.

Lenanton RCJ (1982). Alternative non-estuarine nursery habitats for some commercially and recreationally important fish species of south – Western Australia. Austr. J. Mar. Fresh Water Resour. 33: 881-900.

Leveque C, Paugy D, Teugel GG (1991). Annotated checklist of the freshwater fishes of the Nilo-Sudan river basins in Africa. Rev. Hydrobiol. Trop. 24(2): 131.154.

Macan TT (1963). Fresh water ecology. The Gamelot Press Ltd, London. p. 338.

Motwani MP, Kanwai Y (1970). Fish and fisheries of the Coffer-dammed right channel of the River Niger at Kainji. Kainji Lake Studies. (1):27-48.

Oguzie FA (1997). A key to some of freshwater fishes of Nigeria (Adopted from Boulenger and Welman). Dept of Fisheries, University of Benin, Benin City. pp. 1-14.

Robertson AI, Duke NC (1987). Mangroves as nursery sites: Comparisons of the abundance and species composition of fish and crustaceans in mangroves and other nearshore habitats in tropical Australia. Marine Biol. 96:193-205.

Saliu JK, Joy O, Catherine O (2007). Condition factor, Fat and protein content of five fish species in Lekki Lagoon, Nigeria. Life Sci. J. 4: 54- 57.

Schneider W (1990). FAO Species Identification Sheets for Fishery Purposes. Field Guide to the Commercial Marine Resources of the Gulf of Guinea. FAO Regional Office for Africa,Rome.

Solarin BB (1998). The hydrobiology, fishes and fisheries of the Lagos lagoon, Nigeria. Ph.D Thesis. University of Lagos. p. 235.

Welman JB (1948). Preliminary Survey of the Freshwater Fisheries of Nigeria. Govt. Printer, Lagos, Nigeria. P 88.

Zander CD, Nieder J, Martin K (1999). Vertical distribution patterns. In: Intertidal fishes: Life in two worlds Horn, M.H., Martin, K.L.M. and Chotkowski, M.A. eds. Academic Press. San Diego pp. 26-53.

Diet composition, ontogenetic dietary shifts and morphometric relationships of the Nile barb *Barbus bynni* (Forsskål, 1775) (Pisces: Cyprinidae) in Lake Abaya, Ethiopia

Elias Dadebo[1], Seble Bancha[2], Abinet Wolde-Senbet[2] and Yosef Teklegiorgis[3]

[1]Department of Biology, Hawassa University, P. O. Box 5, Hawassa, Ethiopia.
[2]Department of Biology, Wolaita Sodo University, P. O. Box 138, Wolaita Sodo, Ethiopia.
[3]Department of Animal and Range Sciences, Hawassa University, P. O. Box 336, Hawassa, Ethiopia.

Food and feeding habits of *Barbus bynni* (Forsskål, 1775) were studied in Lake Abaya, Ethiopia, from January to February, 2010 (dry season) and May to June, 2010 (wet season). From a total of 292 fish samples, 251 (86.0%) contained food in their guts. Macrophytes, insects and detritus were the dominant food categories both during the dry and wet seasons. Macrophytes and detritus occurred in 78.1 and 73.3% of the guts, and constituted 30.4 and 29.3% of the total food volume, respectively while insects occurred in 72.5% of the guts and accounted for 29.2% of the total volume. Zooplankton occurred in 37.5% of the guts but they constituted only 4.6% of the total volume of food categories. No seasonality in diet was observed, however, there was an ontogenetic change in diet. In juveniles, the contribution of food of animal origin was relatively high, whereas in adults the contribution of plant materials was dominating. The relationships between total length (TL) and total weight (TW) of males, females and both sexes combined were curvilinear and statistically significant ($P < 0.001$). The relationships between TL and fork length (FL), TL and standard length (SL) and FL and SL were all linear and statistically significant ($P < 0.001$). The growth pattern of *B. bynni* in Lake Abaya was allometric because the slope of the relationship (b) was 3.25, which is a value considerably higher than 3.

Key words: *Barbus bynni*, diet composition, feeding habits, Lake Abaya, morphometric relationships.

INTRODUCTION

The Nile barb *Barbus bynni* (Forsskål, 1775) is a benthopelagic freshwater fish widely distributed in the Nile system and in lakes that were once connected to the Nile (Golubtsov and Krysanov, 1993; Bailey, 1994). In Ethiopia, the species is widely distributed in the Southern Rift Valley Lakes, Abaya and Chamo, the

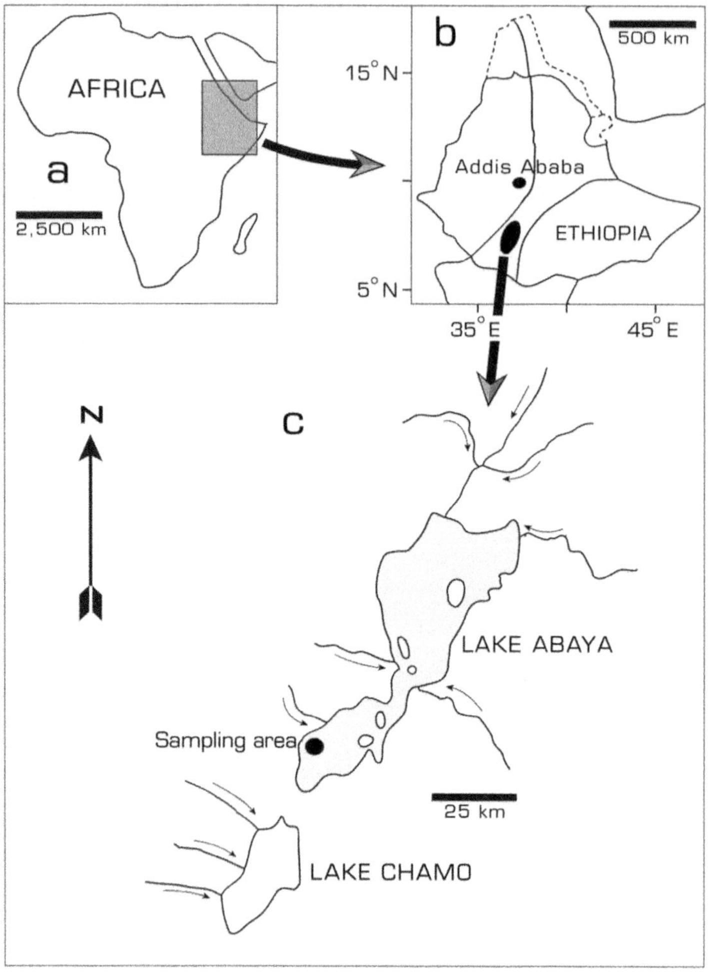

Figure 1. Map of Africa with horn of Africa region highlighted (a), map of Ethiopia with the southern rift valley lakes area indicated (b), maps of Lakes Abaya and Chamo with the sampling area in Lake Abaya shown (c).

Omo River and the rivers that drain into the Nile system (Golubtsov et al., 1995; Getahun, 2007; Golubtsov and Habteselassie, 2010).

African barbs belong to the genus *Barbus* and are divided into small and large barbs. The division is supported by cytogenetics, the smaller barbs having 50 chromosomes while the large barbs have 150 (Berrebi and Valiushok, 1998). *B. bynni* belongs to the large *Barbus* group sensu (Banister, 1973).

Although *B. bynni* is commercially important in Lake Abaya, no study has been conducted on the biology and ecology of the species. No published information is available on the food and feeding habits and morphometric relationships of *B. bynni* in African water bodies in general and in Lake Abaya in particular.

The aim of this work was, therefore, 1) to investigate the food and feeding habits of *B. bynni* in Lake Abaya, and 2) to provide basic information on morphometric relationships of this species in Lake Abaya.

MATERIALS AND METHODS

Study area

Lake Abaya (Latitude: 5°30' - 6°30' N and Longitude: 36°30' - 38°30'E) is the largest Ethiopian Rift Valley Lake located in the southern part of the Ethiopian Rift Valley, east of the Guge Mountains, at an elevation of 1,285 m (Figure 1). The Bilate River that rises from the southern slope of the Gurage Mountains is the major influent of the lake (Baxter, 2002). The town of Arbaminch lies on its southwestern shore. The lake has a maximum length of 60 km, width of 20 km, and a catchment area of 17,300 km^2 (Tudorancea et al., 1999). The surface area of the lake is 1,162 km^2, with a maximum depth of 13 m (Baxter, 2002).

There are a number of islands in this lake, the largest being Aruro;

others include Gidicho, Welege, Galmaka, and Alkal. The lake is reddish in color due to high load of suspended clay particles. Lake Abaya does not always have an outflow, but in some years, it overflows into Lake Chamo. Nech Sar National Park, known for its wild life and bird life surrounds the lake, which is also fished by local people. The ichthyo-fauna of Lake Abaya, and also that of the neighbouring Lake Chamo, is Soudanian species (Beadle, 1981). The fish species are more diverse than that of the other rift valley lakes of the country probably due to a previous inter-connection of Lakes Abaya, Chamo, Chew Bahir and Turkana with the Nile system (Beadle, 1981). There are 21 fish species found in Lakes Abaya and Chamo and the inflowing rivers (Golubtsov and Habteselassie, 2010). The commercially important species are *Oreochromis niloticus* (Linnaeus, 1758), *Labeo horie* (Heckel, 1847), *Bagrus docmak* (Forsskål, 1775), *Clarias gariepinus* (Burchell, 1802) and *Lates niloticus* (Linnaeus, 1758).

According to Tudorancea et al. (1999), the lake has a secchi disc depth of 19 cm and a pH of 8.7 with conductivity (K_{25}) of 800 µS/cm. The area in which the lake lies is mainly volcanic in origin while the waters of the lake are permanently turbid due to heavy colloidal suspension of ferric oxide and the high turbidity of the lake greatly reduces primary productivity and fish production (Schröder, 1984).

Fish sampling and measurements

Sampling was done four times, twice during the dry season (January and February) and twice during wet season (May and June). In the shallow littoral area, a beach seine (25 m long and 3 m wide with a mesh size of 0.6 cm) was used for sampling, while Nordic survey multi-mesh monofilament nylon gillnets (Appelberg et al., 1995) were used in the pelagic area of the lake (Figure 1). The multi-mesh gillnets consisted of twelve randomly distributed panels of the mesh sizes 5, 6.25, 8, 10, 12.5, 15.5, 19.5, 25, 29, 35, 43, and 53 mm (bar mesh). Each panel was 2.5 m long, and hence the total length (TL) of each net was 30 m. The nets were 1.5 m deep with a total area of 45 m^2. The depth of the water where the gillnets were set ranged between 3 to 5 m about 1.5 km inward from the littoral sampling station. They were set around 7.00 am and pulled around 3.00 pm. A total of 292 fish were collected. Fish were then removed from the gillnets shortly after landing. The TL, fork length (FL) and standard length (SL) were measured to the nearest mm, while total weight (TW) was measured to the nearest gram.

Food and feeding

Since *B. bynni* has no distinct stomach, contents of the entire gut were taken and preserved in 5% formalin immediately after morphometric measurements were taken. The gut contents were examined and prey items identified under a dissecting microscope (LEICA MS5, magnification 40x) and compound microscope (LEICA DME, magnification 100x), in the Biology Laboratory at Hawassa University. The relative importance of food items was determined using the following methods:

Frequency of occurrence

The number of gut samples containing one or more of a given food items was expressed as a percentage of all non-empty guts examined (Bagenal and Braum, 1978). The proportion of the population that feeds on certain food items was estimated by this method.

Volumetric analysis

Food items that were found in the guts were sorted into different food categories, and the water displaced by a group of items in each category was measured in a partially filled graduated cylinder (Bowen, 1983). The volume of water displaced by each category of food item was expressed as a percentage of the total volume of the gut contents (Bowen, 1983). The importance of different food categories was determined by categorizing the TL of the fish into four size classes (I: 10.0 to 19.9 cm, II: 20.0 to 29.9 cm, III: 30.0 to 39.9 cm and IV: 40.0 to 49.9 cm) and the total volume of food in each size class was determined. The volumetric contribution of each of food category was then expressed as a percentage of the total volume of food consumed in each size class.

Statistical analysis

Chi-square test was used to compare frequency of occurrences of the different food categories during the dry and wet seasons (Sokal and Rohlf, 1995). Similarly, the non-parametric Mann-Whitney's U test was used to compare the volume of the different food categories consumed during the dry and wet seasons since the data did not satisfy the assumption of equal variance to employ a parametric test.

Dietary overlap between different length-classes was calculated using Schoener Diet Overlap Index (SDOI) (Schoener, 1970; Wallace, 1981):

$$\alpha = 1\text{-}0.5\ (\textstyle\sum_{i=1}^{n} |pxi - pyi|) \qquad (1)$$

Where α is percentage overlap, SDOI, between length classes *x* and *y*, *pxi* and *pyi* are proportions of food category (type) *i* used by *x* and *y*, and *n* is the total number of food categories. Overlap in the index is generally considered to be biologically significant when α value exceeds 0.60 (Mathur, 1977).

Morphometric relationships

Length-weight relationship was calculated using the following formula:

$$TW = aTL^b \qquad (2)$$

Where TW is total weight (g), TL is total length (cm), and *a* and *b* are intercept and slope of the length-weight regression equation, respectively.

A linear relationship was used to determine the relationships between TL and FL, TL and SL and FL and SL.

RESULTS AND DISCUSSION

Diet composition

B. bynni ranging from 11.6 to 50.0 cm TL and 14.2 to 1,200 g TW were considered for food and feeding habit study. The longest female in the sample was 50 cm TL and 1,200 g TW, whereas the longest male fish was 35.8 cm TL and 359.2 g TW. The size frequency distribution of the fish during both dry and wet seasons is given in

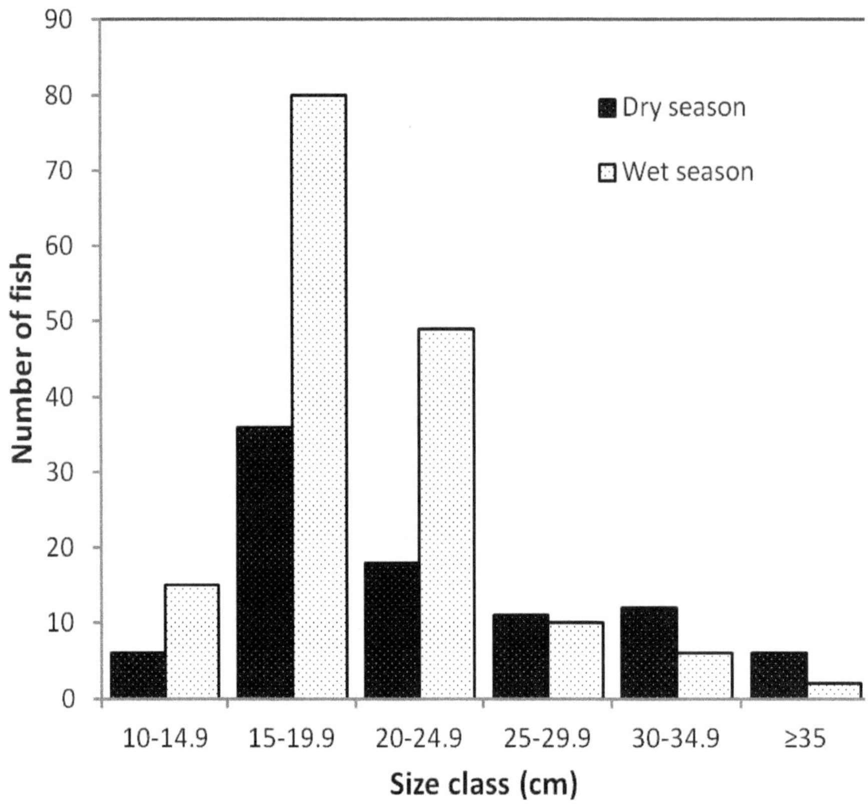

Figure 2. Size frequency distribution of *B. bynni* during the dry (n = 89) and wet (n = 162) seasons in Lake Abaya.

Figure 2. Out of the total number of 292 examined fish, 41(14%) had empty guts. The food categories found in the guts of 251 individuals of *B. bynni* included macrophytes, detritus, insects, zooplankton, phytoplankton, ostracods, gastropods and nematodes (Table 1). Macrophytes, detritus and insects were the most important food categories of *B. bynni* in Lake Abaya. Zooplankton were also encountered relatively frequently, while the remaining food categories (ostracods, nematodes, gastropods and phytoplankton) occurred in few guts (Table 1).

Macrophytes occurred in 78.1% of the examined guts and volumetrically they accounted for 30.4% of the total food volume (Table 1). Detritus occurred in 73.3% of the guts and accounted for 29.2% of the total volume, while insects occurred in 72.5% of the guts and accounted for 29.1% of the total volume. The digestibility of plant food categories could be an important factor in their relatively higher volume, because food of plant origin may resist digestion while soft bodied invertebrates and zooplankton may be lost quickly due to digestive processes. Zooplankton occurred in relatively high number of guts (37.5%), but their volumetric contribution was small (4.6%). Nematodes, ostracods, gastropods and phytoplankton occurred in 4.0 to 21.5% of the guts and collectively accounted for only 6.5% of the total food volume (Table 1). Among the three zooplankton groups identified in the guts of *B. bynni*, *Daphnia* sp. occurred in 19.9%, and accounted for 3.2% of the total volume, followed by *Cerriodaphnia* sp. and rotifers that occurred in 10 and 9.2%, and constituted 0.6 and 0.7% of the total food volume, respectively. Among the insects, Chironomidae larvae occurred in 43.0% of the guts and accounted for 11.3% of the total volume, followed by Ephemeroptera occurring in 22.3% and constituting 9.3% of the total volume (Table 1). Coleoptera and Plecoptera were also relatively important food items, occurring in 25.1 and 22.7% of the guts respectively and accounting for 3.5 and 3.3% of the total volume, respectively. The remaining insects were of minor importance, with occurrence in few guts, and with low contribution to the total gut content. Among the three phytoplankton categories identified, diatoms accounted for 1.1%, followed by blue green algae (0.7%), and green algae (0.4%) by volume.

Apparently, there are very few studies done concerning the food and feeding habits of *B. bynni*. Froese and Pauly(2010) have reported that three sub-species of *B. bynni*: *B. bynni bynni*, *Barbus bynni occidentalis* and *B.*

Table 1. Frequency of occurrence and volumetric contribution of food items consumed by *B. bynni* in Lake Abaya (n = 251).

Food item	Frequency of occurrence		Volumetric contribution	
	Frequency	Percent	Volume (ml)	Percent
Macrophytes	196	78.1	28.5	30.4
Detritus	184	73.3	27.4	29.2
Insects	182	72.5	27.3	29.1
Chironomid larvae	108	43.0	10.5	11.2
Ephemeroptera	56	22.3	8.7	9.3
Coleoptera	63	25.1	3.3	3.5
Plecoptera	57	22.7	3.1	3.3
Diptera	33	13.1	0.6	0.6
Hymnoptera	5	2.0	0.5	0.5
Chironomid pupa	5	2.0	0.4	0.4
Anisoptera	6	2.4	0.2	0.2
Zooplankton	94	37.5	4.3	4.6
Daphnia	50	19.9	3.0	3.2
Ceriodaphnia	25	10.0	0.6	0.6
Rotifers	23	9.2	0.7	0.7
Phytoplankton	31	12.4	2.1	2.2
Blue green algae	15	6.0	0.7	0.7
Green algae	4	1.6	0.4	0.4
Diatoms	12	4.8	1.0	1.1
Ostracods	54	21.5	2.0	2.1
Gastropods	10	4.0	1.4	1.5
Nematodes	51	20.3	0.7	0.7

Note that the percentage volume of the major food categories in bold adds up to 100%.

bynni waldroni feed on almost the same types of food items despite the difference in their distribution. *B. bynni bynni* is distributed in the Nile system, in Egypt, Ethiopia, Kenya and Sudan (Lévêque and Daget, 1984; Bailey, 1994; Seegers et al., 2003; Getahun, 2007). This sub-species feeds on detritus, zoobenthos, crustaceans, insects and gastropods. The above feeding habit of *B. bynni bynni* agrees well with the results of the present study, but unlike many other studies on other *Barbus* species in Ethiopia, no piscivorous individuals were encountered in *B. bynni* from Lake Abaya. Admassu and Dadebo (1997) studied the feeding habits of *Barbus* sp. in Lake Hawassa and reported that the species is polyphagous, but according to these authors, the species tended to become piscivorous above 30 cm TL. Nagelkerke et al. (1994) studied the barbs of Lake Tana and reported that 8 out of 15 species were piscivorous, while the remaining 7 species were benthivorous and polyphagous. Likewise, the polyphagous feeding habit of *Barbus* species was reported by several authors from different fresh waters of the world.

De Graaf (2003) conducted a detailed study on the feeding habits of piscivorous barbs and found a high degree of piscivory even though piscivory is rare in cyprinids due to lack of teeth, stomach and the associated digestive enzymes (de Graaf et al., 2010). In the absence of specialized piscivores many *Barbus* species may consume relatively large proportions of fish prey, especially at later stages during their ontogeny. Desta et al. (2006) suggested at least two feeding forms of *Lasiurus intermedius* in Lake Hawassa, Ethiopia: those that prey up on fish, and those that are polyphagous. In another extensive study on the feeding habits of a small barb, *Barbus paludinosus*, Desta et al. (2008) reported ostracods, zooplankton, mollusks, aquatic insects and fish to be the most important food categories.

Seasonal variation in the diet

The frequency of occurrence of zooplankton and phytoplankton in the diet significantly varied during the dry (n = 89) and wet (n = 162) seasons ($P < 0.05$). Similarly, the volumetric contribution of ostracods significantly differed during the two seasons ($P < 0.05$). On the other hand, both frequency and volumetric contributions of all other food items did not significantly differ during the two seasons (Table 2). Among the three

Table 2. Relative contribution of different food items in the diet of *B. bynni* during the dry (n = 89) and wet (n = 162) seasons.

Food item*	Frequency of occurrence (%)		Volumetric contribution (%)	
	Dry season	Wet season	Dry season	Wet season
Detritus	72.5[a]	73.5[a]	31.7[a]	24.6[a]
Macrophytes	84.3[a]	76.5[a]	30.1[a]	30.8[a]
Zooplankton	27.5[a]	39[b]	3.7[a]	6.3[a]
Rotifers	9.8	9	0.3	1.5
Cereodaphnia	9.8	10	0.5	1.0
Daphnia	9.8	22.5	2.9	3.7
Insects	78.4[a]	71[a]	29.8[a]	28.1[a]
Chironomidae larvae	68.6	36.5	13.9	6.2
Chironomidae pupa	2.0	2	0.1	1.1
Diptera	-	16.5	-	2.0
Plecoptera	9.8	26	1.5	6.7
Coleoptera	15.7	27.5	2.5	5.5
Ephemeroptera	31.4	20	11.0	6.0
Anisoptera	2.0	2.5	0.2	0.1
Hymnoptera	2.0	2	0.5	0.5
Phytoplankton	31.4[b]	7.5[a]	2.2[a]	2.3[a]
Blue green algae	21.6	2	1.2	0.1
Green algae	3.9	1	0.5	0.1
Diatom	5.9	4.5	0.6	2.1
Ostracods	17.6[a]	22.5[a]	0.7[a]	4.7[b]
Gastropods	2.0[a]	4.5[a]	1.8[a]	1.2[a]
Nematodes	2.0[a]	25[b]	0.01[a]	2.0[b]

Note that the volume of the major food categories in bold adds up to 100%. *Values of respective food categories under the same category given different superscript letters are significantly different ($\alpha < 0.05$).

major food categories of *B. bynni*, detritus composed slightly higher volume during the dry season (31.7%) than in the wet season (26.4%). The contribution of macrophytes in the diet during the dry and the wet seasons (30.1 and 30.9%, respectively) and insects (29.7 and 28.1%, respectively) was fairly comparable (Table 2). Also, the overall volume of phytoplankton in the guts was comparable during the dry (2.2%) and wet (2.3%) seasons. The importance of zooplankton in the diet increased during the wet season (6.3%) compared to the dry season (3.7%). Similarly, ostracods and nematodes were consumed considerably more during the wet season than in the dry season, while gastropod consumption remained fairly constant during the two seasons (Table 2).

Although the total volume of insects in the diet remained similar during the dry and wet seasons, the contribution of the different taxa considerably varied during the two periods (Table 2). During the dry season, Chironomidae larvae and Ephemeroptera composed most of the insect bulk consumed by *B. bynni* in Lake Abaya (13.9 and 11.0%, respectively). Whereas their contribution dropped by about 50% in the wet season (6.2 and 6% by volume, respectively).

On the other hand, the contribution of Plecoptera increased from 1.5 to 6.7% by volume, and that of Coleoptera increased from 2.5 to 5.5% by volume during the wet season compared to the dry season (Table 2). Comparing the two seasons, *B. bynni* consumed a fairly diverse group of insects during the wet season, whereas only two insect taxa composed most of the insect bulk consumed during the dry season.

Among zooplankton, *Daphnia* sp. was the most important during both seasons, but its contribution increased in the wet season than in the dry season. Also, the significance of the other two less important zooplankton (rotifers and *Cerriodaphnia* sp.) relatively increased in the wet season than in the dry season. Considering phytoplankton taxa, diatoms made up nearly all algal volume eaten during the wet season whereas, blue greens together with some greens and diatoms shared the dry season algae diet (Table 2).

The results of the present study have shown no significant difference on the feeding habits of *B. bynni* during the dry and wet seasons although there were some variations in the proportion of certain food categories during the two seasons. The proportions of macrophytes, zooplankton, ostracods and nematodes were higher during the wet season while the proportions of detritus, insects, phytoplankton and gastropods were higher during the dry season. This difference in the proportions of the different food categories could be due to the difference in the availability and emergence of various food items during dry and wet seasons. Kolding (1989) observed seasonal migration of *B. bynni* from inshore areas to offshore habitats in Lake Turkana (Kenya). According to this author, during the wet season with high turbidity the population tends to migrate to inshore areas with progressive offshore movement as the water clears during the dry season (Kolding, 1989). Admassu and Dadebo (1997) studying *Barbus* sp. in Lake Hawassa (Ethiopia) and Dadebo et al. (2013) working on *L. intermedius* in Lake Koka (Ethiopia) reported seasonal variation in the diet of the studied fish species. The present study is in agreement with the above works in that insects were more important during the dry season while macrophytes and other invertebrates were more important during the wet season (Admassu and Dadebo, 1997; Dadebo et al., 2013). In general, seasonality did not affect the feeding behavior and the type of prey organisms of *B. bynni* in Lake Abaya because the majority of the food items showed no significant seasonal differences.

Ontogenetic dietary shift

SDOI revealed no significant variation in the diet of the different size classes because percentage overlap was above 60% in all combinations of size classes (I and II: 97.2%; I and III: 75.5%, I and IV: 62.4%; II and III: 73.9%; II and IV: 64.4%; III and IV: 86.3%). In general, the contribution of animal origin food was high in the diet of smaller size classes whereas this component declined as the fish grew in size (Figure 3). In fish below 40 cm TL, insects were important in the diets and the contributions of detritus and macrophytes remained more or less unchanged with size of fish. However, in the diets of *B. bynni* above 40 cm TL, the fish depended only on detrital matter and macrophytes for its nourishment (Figure 3). Among the minor food items, zooplankton were relatively important in the diet of fish less than 40 cm TL, but they were absent in the diets of larger fish. Similarly, the significance of phytoplankton, ostracods, gastropods and nematodes was visible in the diets of fish below 40 cm TL, whereas they were absent in the diets of larger fish (Figure 3).

Ontogenetic diet shift has been shown to occur during the life of many fish species, and prey size is generally positively correlated with fish size (Desta, 2007). A fundamental characteristic of fish is that individuals increase considerably in size, which is usually associated with changes in food resource use (Desta et al., 2006). Based on the results of the present study, slight ontogenetic dietary change was noted. Small fish (<20 cm TL) and intermediate size fish (20 to 40 cm TL) predominantly fed on insects, followed by detritus and macrophytes. The composition of the diet of large fish (>40 cm TL) was different from the lower size classes because large fish exclusively fed on only food of plant origin, detritus and macrophytes comprising the total volume of the food categories.

Morphometric relationships

The relationships between TL and TW of males (11.6 to 35.8 cm TL; 14.2 to 359.2 g TW), females (13.4 to 50.0 cm TL; 15.9 to 1,200 g TW) and both sexes combined was expressed by the following regression equations:

Males: $TW = 0.0033TL^{3.277}$, $R^2 = 0.989$, $n = 144$, $p < 0.001$ (3)

Females: $TW = 0.0041\ TL^{3.207}$, $R^2 = 0.981$, $n = 148$, $p < 0.001$ (4)

Combined: $TW = 0.0036TL^{3.245}$, $R^2 = 0.986$, $n = 292$, $p < 0.001$ (5)

The relationships between TL and FL ($FL= 0.866TL - 0.316$, $R^2 = 0.997$, $n = 292$, $p < 0.001$), TL and SL ($SL = 0.806TL - 0.949$, $R^2 = 0.995$, $n = 292$, $p < 0.001$) and FL and SL ($SL = 0.929FL - 0.632$, $R^2 = 0.995$, $n = 292$, $p < 0.001$) were all linear and statistically significant. As far as our knowledge goes, no study was found on morphometric relationships of *B. bynni* in African water bodies. Admassu and Dadebo (1997) studying a related species (*Barbus* sp.) in Lake Hawassa found slightly allometric growth pattern with a *b* value of 3.14

Conclusions

The food categories consumed by *B. bynni* were, macrophytes, detritus, insects, zooplankton, phytoplankton, ostracods, gastropods, and nematodes suggesting that *B. bynni* are a polyphagous fish species in Lake Abaya. No seasonality in diet was observed, however, there was an ontogenetic change in diet. Smaller size classes (<19.9 cm TL) predominantly relied on insects, detritus and macrophytes while medium sized fish (20 to 39.9 cm TL), depended mainly on insects, macrophytes and detritus. On the other hand, larger fish (>40 cm TL) fed only on detritus and macrophytes. The growth pattern of *B. bynni* in Lake Abaya was found to be allometric because the slope of the relationship (b) was 3.25, which is a value considerably higher than 3.

Conflict of Interests

The authors have not declared any conflict of interests.

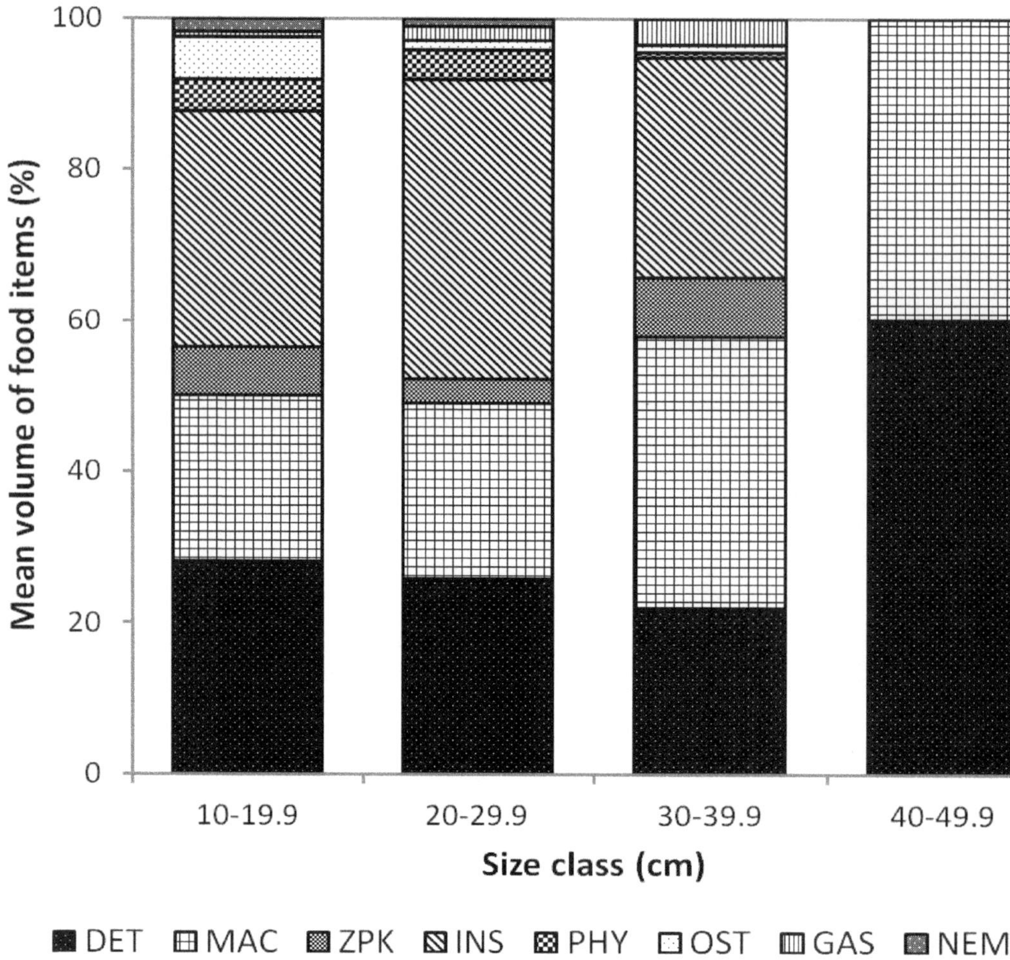

Figure 3. Volumetric contributions (%) of the different food categories in the diet of different size classes of *B. bynni* in Lake Abaya, as an average for the dry and wet season.

REFERENCES

Admassu D, Dadebo E (1997). Diet composition, length-weight relationship and condition factor of Barbus sp. (Rüppell, 1836) (Pisces: Cyprinidae) in Lake Hawassa, Ethiopia. SINET: Ethiop. J. Sci. 20(1):13-30.

Appelberg M, Berger HM, Hesthagen T, Kleiven E, Kurkilahti M, Raitaniemi J, Rask M (1995). Development and inter-calibration of methods in Nordic freshwater fish monitoring. Water Air Soil Pollut. 85: 401-406.http://dx.doi.org/10.1007/BF00476862

Bagenal TB, Braum E (1978). Eggs and early life History. In: Methods for assessment of fish production in fresh waters, (Bagenal, T.ed.). Blackwell Scientific Publications, London. pp. 165-201.

Bailey RG (1994). Guide to the fishes of the River Nile in the Republic of the Sudan. J. Nat. Hist. 28(4):937-970. http://dx.doi.org/10.1080/00222939400770501

Banister KE (1973). A revision of the large Barbus (Pisces: Cyprinidae) of East and Central Africa. Studies of African Cyprinidae. Pt. II, Bull. Brit. Mus. Nat. Hist. Zool. 26(1):1-148.

Baxter RM (2002). Lake morphology and chemistry. In: Ethiopian Rift Valley Lakes, (Taylor, W.D and Tudorancea, C., eds). Backhuys Publishers, Leiden.

Beadle LC (1981). The inland waters of tropical Africa. An introduction to tropical limnology, 2nd ed. New York: Longmans, 475 pages.

Berrebi P, Valiushok D (1998). Genetic divergence among morphotypes of Lake Tana (Ethiopia) barbs. Biol. J. Linn. Soc. 64:369-384.http://dx.doi.org/10.1111/j.1095-8312.1998.tb00338.x

Bowen SH (1983). Quantitative description of the diet. In: Fisheries Techniques, (Nielsen, L.A and Johnson, D. L. eds). Bethesda, Maryland. pp. 325-336.

De Graaf M (2003). Lake Tana's piscivorous Barbs (Cyprinidae, Ethiopia): Ecology, Evolution, Exploitation. Ph.D. Thesis, Experimental Zoology Group, Wageningen University, AH Wageningen, the Netherlands.

De Graaf M, van de Weerd GH, Osse JWM, Sibbing FA (2010). Diversification of prey capture techniques among the piscivores in Lake Tana's (Ethiopia) Labeobarbus species flock (Cyprinidae). Afr. Zool. 45(1):32-40. http://dx.doi.org/10.3377/004.045.0117

Dadebo E, Tesfahun A, Teklegiorgis Y (2013). Food and feeding habits of the African big barb *Labeobarbus intermedius* (Rüppell, 1836) (Pisces: Cyprinidae) in Lake Koka, Ethiopia. E3 J. Agric. Res. Dev. 3(4):49-58.

Desta Z, Borgstrøm R, Rosseland BO, Gebre-Mariam Z (2006). Major difference in mercury concentrations of the African big barb, Barbus intermedius (R.) due to shifts in trophic position. Ecol. Freshwat. Fish. 15:532-543. http://dx.doi.org/10.1111/j.1600-0633.2006.00193.x

Desta Z (2007). Food web structure and mercury transfer patterns in fish community of L. Hawassa, Ethiopia. Ph.D. Thesis. Norwegian

University of Life Science, Ås.

Desta Z, Borgstrøm R, Gebre-Mariam Z, Rosseland BO (2008). Habitat use and trophic position determine mercury concentration in the straight fin barb *Barbus paludinosus*, a small fish species in Lake Hawassa, Ethiopia. J. Fish Biol. 73:477-497. http://dx.doi.org/10.1111/j.1095-8649.2008.01920.x

Froese R, Pauly D (2010). List of freshwater fishes of Ethiopia. Fish Base, World Wide Electronic Publications.

Getahun A (2007). An overview of the diversity and conservation status of the Ethiopian freshwater fauna. J. Afrotrop. Zool. Special Issue: pp. 87-96.

Golubtsov AS, Krysanov EV (1993). Karyological study of some cyprinid species from Ethiopia. The ploidy difference between large and small Barbus of Africa. J. Fish. Biol. 42(3):445-455. http://dx.doi.org/10.1111/j.1095-8649.1993.tb00347.x

Golubtsov AS, Darkove AA, Dgebyadze YY, Mina MV (1995). An artificial key to fish species of Gambela region. Joint. Ethio-Rassian Biological Expedition. Artistic Printing Press, Addis Ababa.

Golubtsov AS, Habteselassie R (2010). Fish faunas of the Chamo-Abaya and Chew Bahir basins in southern portion of the Ethiopian Rift Valley: origin and prospects for survival. Aquat. Ecosys. Health Manag. 13(1):47–55. http://dx.doi.org/10.1080/14634980903578506

Kolding J (1989). The fish resources of Lake Turkana and their environment. PhD Thesis, Department of Fisheries Biology, University of Bergen, Norway.

Lévêque C, Daget J (1984). Cyprinidae. In: Check-list of the freshwater fishes of Africa (Daget, J., Gosse, J.P. and Thys Vnden Audenaerde, D.F.E, eds). (CLOFFA). Vol.1.ORSTOM, MRAC Tervuren and Paris, France. , pp. 217-342.

Mathur D (1977). Food habits and competitive relationships of Bandfin shiner in Halawakee Creek, Alabama. Am. Midland Nat. 97:89-100. http://dx.doi.org/10.2307/2424687

Nagelkerke LAJ, Sibbing F A, Vanden Boogaart JGM, Lammens EHRR, Osse JWM (1994). The barbs (Barbus Sp.) of Lake Tana: A forgotten species flock? Environ. Biol. Fish 39:1-22. http://dx.doi.org/10.1007/BF00004751

Seegers L, Devose L, Okeyo DO (2003). Annotated checklist of the freshwater fishes of Kenya (excluding the lacustrine haplochromine) from Lake Victoria. J. East Afr. Nat. Hist. 92:11-47. http://dx.doi.org/10.2982/0012-8317(2003)92[11:ACOTFF]2.0.CO;2

Schoener TW (1970). Non-synchronous spatial overlap of lizards in patchy habitats. Ecology 51:408-418. http://dx.doi.org/10.2307/1935376

Schröder R (1984). An attempt to estimate the fish stock and sustainable yield of Lake Ziway and Lake Abaya, Ethiopia Rift valley. Arch. Hydrobiol. (Supp.). 69(3):411-444.

Sokal RR, Rohlf FJ (1995). Biometry: The principles and practices of statistics in biological research 3rd ed. New York, Freeman and Company, P. 859.

Tudorancea C, Gebre-Mariam Z, Dadebo E (1999). Limnology in Ethiopia. In: Limnology in developing countries (Wetzel R.G and Gopel. B. eds). pp. 63-118.

Wallace RK (1981). An assessment of diet-overlap indexes. Trans. Am. Fish. Soc. 110: 72-76. http://dx.doi.org/10.1577/1548-8659(1981)110<72:AAODI>2.0.CO;2

Monthly spatial occurrence of phytoplankton and zooplankton in River Ogun, Abeokuta, Ogun State, Southwest Nigeria

Dimowo, Benjamin Onozeyi

Department of Aquaculture and Fisheries Management, College of Environmental Resources Management, Federal University of Agriculture, Abeokuta, Ogun state, Nigeria.

The aim of this study was to investigate the monthly spatial occurrence of phytoplankton and zooplankton in River Ogun, Abeokuta, Southwest Nigeria. This was carried out for seven months between December, 2011 and June, 2012 in 4 stations. A total of 41 species of phytoplankton and 16 zooplankton species from 5 classes respectively were recorded. Zooplankton was dominated by Cladocera throughout the study period while phytoplankton was dominated by blue green algae (Cyanophyta or Cyanobacteria). The dominance of Cyanophyta in this river is similar to findings by Sekadende and co-workers, Ogato, Deng and co-worker, and Shakila and co-worker. The dominance of Cladocera in this river is similar to findings by Ude and co-workers, and Ogbuagu and co-worker.

Key words: Monthly, spatial, occurrence, phytoplankton, zooplankton, River Ogun.

INTRODUCTION

The study of plankton (phytoplankton and zooplankton) is very important because they serve as a base upon which the aquatic ecosystem is supported. Phytoplankton (singular, phytoplankter) are primary producers while zooplankton (singular, zooplankter) are secondary producers. Phytoplankton serve as food to zooplankton which in turn serve as food to almost all larval forms (either meroplanktononic or holoplanktonic in nature) in natural surface water. Many Scientists including Abohweyere, 1990; Adakole et al., 2008; Adeogun et al., 2005; Adejuwon and Adelakun, 2012; Adesalu and Nwankwo, 2008; Agbaire and Obi, 2009; Ajuonu, et al., (2011); Atobatele and Ugwumba, 2008; Ayeni et al., 2011; Balarabe, 2001; Bwala et al., 2010; Chia et al., 2011; Chinedu et al., 2011; Davies and Ansa, 2010; Erondu and Chindah, 1991; Essien-Ibok et al., 2010; Ezekiel et al., 2011; Ezeribe et al., 2012; Fafioye et al., 2005; Ibrahim, 2009; Imoobe and Egborge, 1997; Ladipo et al., 2011; Mood, 2004; Muhibbu-din et al., 2011; Mustapha and Omotosho, 2005; Nkwoji et al., 2010; Offem et al., (2011); Ogbuagu and Ayoade, 2012; Onuoha and Vyverman, 2010; Onyema, 2007; Oso and Fagbuaro, 2008; Ovie, 1995; Ude et al., 2011; etc have worked on the various aspects of ecosystem studies of rivers, reservoirs, lakes, creeks and estuaries in Nigeria. However, not much work has been carried out on the plankton composition and distribution of River Ogun. This study is therefore aimed at investigating the monthly spatial occurrence of phytoplankton and zooplankton in River Ogun so as to contribute to the existing knowledge

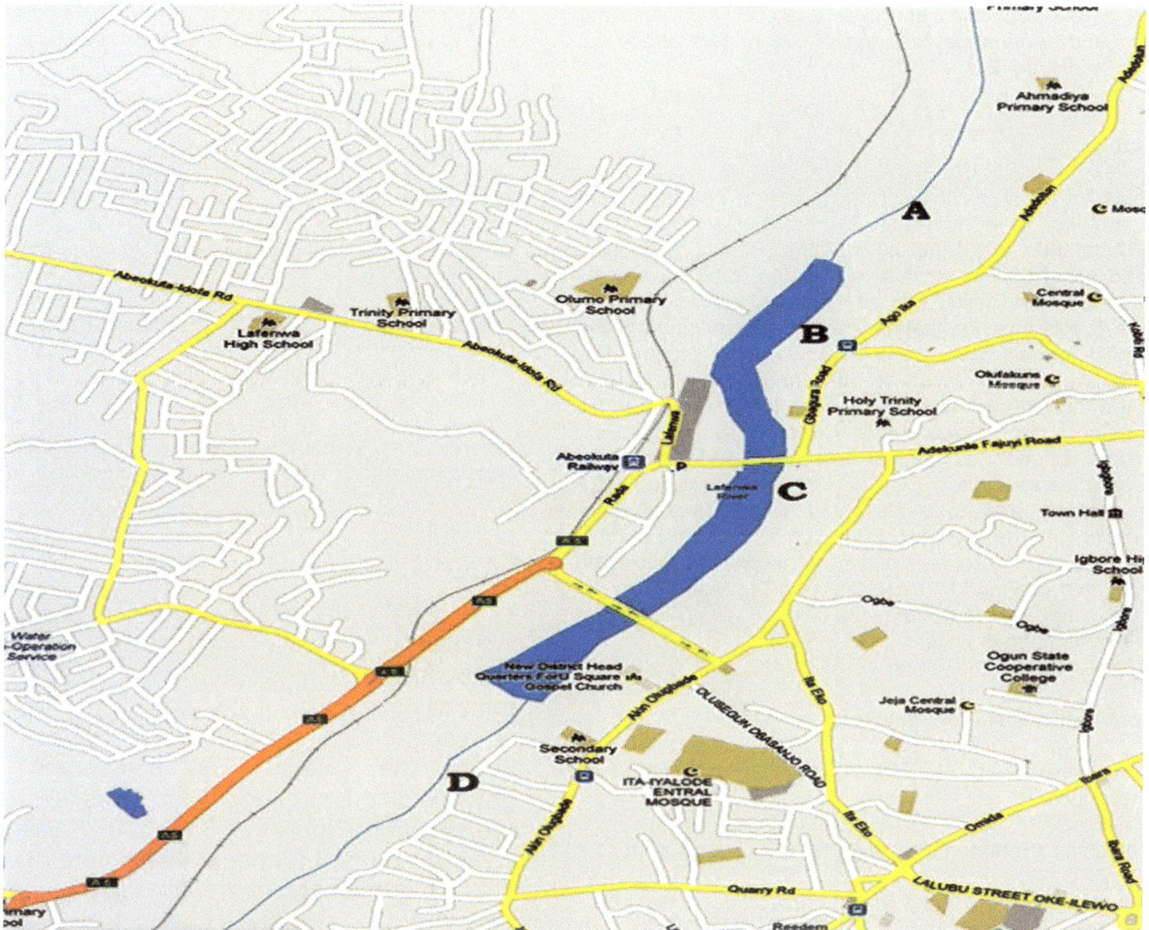

	STATIONS
A	IBEREKODO
B	AGO IKA
C	ENU GADA
D	OFF PEPSI BUSTOP

Figure 1. Map of River Ogun showing the sampling stations. Source: Google Maps, 2012.

on plankton ecology and distribution.

MATERIALS AND METHODS

Study area

Ogun State is a state in South-western Nigeria. It borders Lagos State to the South, Oyo and Osun states to the North, Ondo State to the east and the republic of Benin to the west. Abeokuta is the capital and largest city in the state (NBS, 2012).

Geography of the River

River Ogun (Figure 1) is one of the main rivers in the southwestern part of Nigeria with a total area of 22.4 km^2 and a fairly large flow of about 393 m^3sec G1 during the wet season (Oketola et al., 2006). It has coordinates of 3°28"E and 8°41"N from its source in Oyo state to 3°25"E and 6°35"N in Lagos where it enters the Lagos lagoon (Ayoade et al., 2004). There are two seasons in Ogun river basin. The dry season starts from November and ends in March while the wet season starts from April and ends in October. Mean annual rainfall ranges from 900 mm in the north to 2000 mm towards the south. The estimates of total annual potential evapotranspiration have been put between 1600 and 1900 mm (Bhattacharya and Bolaji, 2010). The water is used for agriculture, transportation, human consumption, various industrial activities and domestic purposes (Ayoade et al., 2004; Oketola et al., 2006). It also serves as a raw water supply to the Ogun state water corporation which treats it before dispensing it to the public. Along its course, it constantly receives effluents from breweries, slaughterhouses, dyeing industries, tanneries and domestic wastewater before finally discharging to Lagos lagoon (Ayoade et al., 2004; Oketola et al., 2006). A 100 km^2 area around River Ogun has an approximate population of 3637013 (0.03637 persons per square meter) and an average elevation of 336 meters above the sea (Travel Journals, 2012).

Experimentation

Water samples for physico-chemical and plankton analyses were collected monthly for seven consecutive months (December 2011 to June 2012) at the four sampling stations. Plankton samples were collected using 55 µm mesh size standard plankton net by trawling horizontally and filling into air tight 120 ml well labeled plastic bottles. It was ensured that fixation and preservation of the samples with 4% formalin was done within 5 min of collection. The fixed samples were then taken to the laboratory for further analysis. In the laboratory, the samples were identified under the microscope using suitable keys such as Edmondson (1959) etc. Abundance was reported as number of cells or individuals per drop. pH, air temperature (°C), water temperature (°C), conductivity (µs/cm) and total dissolved solids (mg/L) were conducted *in-situ* with the use of HANNA Combo pH and EC multi meter Hi 98129 and Mercury-in-glass thermometer. Dissolved oxygen (mg/L), nitrate (mg/L), phosphate (mg/L), alkalinity (mg/l) and hardness (mg/L) were determined *ex-situ* using standard methods for examination of water (APHA, 1998).

Plankton community structure analysis

Three indices were used for Plankton Community Structure Analysis. These include:

1. Species diversity (Shannon and Weaver, (1949) and Simpson, (1949)
2. Species equitability or evenness (Pielou 1966)
3. Species richness (Margalef (1951) and Menhimick (1964)

Shannon and Weaver diversity index (1948): The Shannon and Weaver diversity index (Hs) is given by:

$$Hs = \frac{N \text{ Log } N - \Sigma Pi \text{ Log } Pi}{N}$$

Where: Hs = Shannon and Weaver diversity Index; N = Total number of individuals in S species.
i = Counts denoting the ith species ranging from 1 – n, and Pi = Proportion that the ith species represents in terms of numbers of individuals with respect to the total number of individuals in the sampling space as whole.

Simpsons dominance index (Ogbeibu, 1949): It is given by:

$$C = \sum \left(\frac{n_i}{N}\right)^2$$

Where: C = Simpsons dominance index; n_i = Total number of organisms of a particular species;
N = Total number of organisms of all species; Pielou species equitability or evenness index (1966): It is given by:

$$j = \frac{Hs}{\text{Log}_2 S}$$

Where: j = Equitability index; Hs = Shannon Weaver's index, and S = Number of species in a population.

Margalef Species Richness Index (d) (1951): The Species richness index (d) was used to evaluate the community structure.

$$d = \frac{S - 1}{\ln N}$$

Where: d = Species richness index; S = Number of species in a population, and N = Total number of individuals in S species.
Menhinick's Index: it is given by:

$$D = \frac{S}{(N)^{1/2}}$$

Where: D = Menhinick's Index; S = Number of species in a population, and N = Total number of individuals in S species.

RESULTS

The result of the physico-chemical parameters is shown in Table 1 while results of phytoplankton and zooplankton composition are shown in Tables 2 and 3. A total of 41 genera of phytoplankton from 5 classes and 16 zooplankton genera from 5 classes were recorded during the studies. The percentage distribution of phytoplankton and zooplankton classes at all the stations are shown in Figures 2, 3, 4, 5, 6, 7, 8 and 9. Generally, Cyanophyta and Cladocera which are phytoplankton and zooplankton classes were dominant throughout the study period.

Physicochemical parameters

Water temperature ranged between 26°C (Station A in December, 2011) and 39.3°C (Station C in May, 2012). Dissolved oxygen ranged between 1.2 mg/l (Stations B and C in January, 2012) and 9 mg/l (Station A in May, 2012). Conductivity ranged between 76 µs/cm (Station A in February, 2012) and 276 µs/cm (Station B in June, 2012). TDS was between the values of 38 ppm (Station A in February, 2012) and 138 ppm (Station B in June, 2012). Transparency values ranged between 20 cm (Station A in June 2012) and 102.5 cm (Station A in March, 2012). Alkalinity concentrations ranged between 1 mg $CaCO_3$/L (Station A in December, 2011) and 18 mg $CaCO_3$/L (Stations A, B and C in April, 2012). Hardness concentrations also ranged between 32 mg $CaCO_3$/L (Station A in March, 2012) and 244 mg $CaCO_3$/L (Station A in April, 2012). Furthermore, Hydrogen ion concentration (pH) values ranged between 7.19 (Station B in January, 2012) and 9.46 (Station A in December, 2011). Nitrates concentration also ranged between 0.36 mg/L (Station C in December, 2011) and 135.61mg/L (Station A in June, 2012). Phosphate concentrations ranged from 0.01 mg/L (Station A in February, 2012) and

Table 1. Physicochemical parameters of the water samples.

Date	December, 2011				January, 2012				February, 2012				March, 2012			
Station	A	B	C	D	A	B	C	D	A	B	C	D	A	B	C	D
Water temperature (°C)	26	27	27	28	28.3	30.5	30.4	30.9	29.5	30.1	31.1	30.6	29.9	29.9	30.1	30
DO (mg/L)	3.25	2.6	3.15	2.45	3.5	1.2	1.2	5.2	5.5	3.2	1.9	2.7	5.5	2.5	4.1	4.8
Conductivity (µcm)	163	184	180	195	89	147	137	152	76	111	103	106	78	118	95	105
TDS (ppm)	82	93	90	98	43	74	67	75	38	54	50	53	42	58	47	52
Transparency (cm)	77	85	54	75	90	80	60	50	85	57.5	77.5	50.75	102.5	35	65	45
Alkalinity	1	6	7	6	3.5	5	4	5	3	6	4	6	6	4	6	4
Hardness	52	56	52	60	32	54	46	50	34	70	44	42	32	48	54	194
pH	9.46	9.03	8.87	9.23	8.34	7.19	7.39	8.32	8.92	7.48	7.62	7.8	7.91	7.33	7.61	7.96
Nitrate	1	0.45	0.36	0.54	4.9	1.72	0.91	1.27	1.63	5.08	2.18	1.45	43.9	60.49	66.34	61.46
Phosphate	0.14	0.06	0.09	0.09	0.19	0.09	0.05	0.13	0.01	0.067	0.015	0.011	0.43	0.708	0.637	0.843
Air temperature (°C)	29	24	24	28	34	35	36	36	30	31	32	34	30	31	31	32

	April, 2012				May, 2012				June, 2012			
	A	B	C	D	A	B	C	D	A	B	C	D
Water temperature (°C)	30.6	32.6	32.6	32.6	28.6	29.2	39.3	29.71	26.8	27	27.1	26.6
DO (mg/l)	7.5	5.8	6.3	6.6	9	6.7	7.2	8	7.8	3.1	7.1	8.5
Conductivity (µcm)	94	133	123	128	109	148	136	134.11	132	276	143	160
TDS (ppm)	47	66	61	64	54	78	67	66.96	66	138	71	80
Transparency (cm)	82.5	55	47.5	60	60	30	50	30	20	25	27.5	25
Alkalinity	18	18	18	17	3	7	4	7	5	6	5	6
Hardness	244	76	50	50	42	58	62	62	44	104	54	66
pH	8.96	7.65	7.9	8.16	8.39	7.9	8.02	8.23	8.58	8.6	8.91	8.8
Nitrate	16.59	41.95	28.29	17.56	44.88	32.2	31.22	65.37	135.61	92.68	109.27	116.1
Phosphate	0.179	0.332	0.108	0.143	0.081	0.09	0.161	0.108	0.43	0.35	0.323	0.368
Air Temperature (°C)	33	34	35	31	30	32	31	31	29	27	29	27

0.843 mg/L (Station D in March, 2012). Finally, air temperature ranged between 24°C (Stations B and C in December, 2011) and 36°C (Stations C and D in January, 2012)

Phytoplankton community structure

For the 4 stations studied in Table 2, diversity (S) ranged between 2 (Station B in December, 2011 and Station A in February, 2012) and 12 (Stations A and D in June, 2012). Abundance of cells per drop (N) of the samples was between 3 individuals per drop (Station A in February, 2012) and 37 individuals per drop (Stations A and D in June, 2012). Furthermore, the values for Shannon-Weaver Index (Hs) ranged between 0.24 (Station A in March, 2012) and 1.22 (Stations A and D in June, 2012). Menhinick Index (D) on the other hand ranged between 0.58 (Station B in December, 2011) and 2.46 (Station B in February, 2012). Margalef Index (d) values also ranged between 0.4 (Station B in December, 2011) and 3.34 (Station B in February). Pielou Equitability Index (j) values were between 0.15 (Station A in March, 2012) and 1.03 (Station B in December, 2011). Simpson's Dominance Index (C) also ranged between 0.09 (Station D in June, 2012) and 0.56 (Station A in February, 2012).

Zooplankton community structure

For the 4 stations studied in Table 3, diversity (S) ranged between 2 (Stations A and D in January, 2012; Station A in March, 2012 and Station A in April, 2012) and 8 (Station A in December, 2011). Abundance of individuals per drop (N) of the samples was between 5 individuals per drop (Station D in January, 2012; Station A in March, 2012 and Station A in May, 2012) and 29 individuals per drop (Stations C in December, 2011). Furthermore, the values for Shannon-Weaver Index (Hs) ranged between 0.5 (Station B in February, 2012) and 1.26 (Station C in December, 2011). Menhinick Index (D) on the other hand ranged between 0.82 (Station A in January, 2012) and 1.89 (Station A in December, 2011). Margalef Index (d) values also ranged between 0.56 (Station A in January, 2012) and 2.42 (Station A in December, 2011). Pielou Equitability Index (j) values were between 0.24 (Station D in March, 2012 and Station C in May, 2012) and 0.68 (Station A in January, 2012). Simpson's Dominance Index (C) also ranged between 0.16 (Station A in December, 2011) and 0.68 (Station D in January, 2012).

Table 2. Monthly spatial variation in the phytoplankton composition and abundance (number of cells per drop) in some parts of River Ogun.

Date	Dec, 2011				Jan, 2012				Feb, 2012				March, 2012				April, 2012				May, 2012				June, 2012			
Station	**A**	**B**	**C**	**D**	**A**	**B**	**C**	**D**	**A**	**B**	**C**	**D**	**A**	**B**	**C**	**D**	**A**	**B**	**C**	**D**	**A**	**B**	**C**	**D**	**A**	**B**	**C**	**D**
Euglenophyta																												
Phacus	0	0	0	0	0	3	2	0	0	0	0	0	0	0	0	0	0	0	0	0	0	0	0	0	0	0	0	0
Euglena	0	0	0	0	0	1	1	0	0	0	1	0	0	0	0	0	0	1	2	0	0	0	0	0	0	0	0	0
Menoidium	0	0	0	0	0	0	0	0	0	1	0	0	0	0	0	0	0	0	0	0	0	0	0	0	0	0	0	0
Chlorophyta																												
Gonatoxygon	0	0	2	1	0	0	0	0	0	0	0	0	0	0	0	0	0	0	0	0	0	0	0	1	0	0	0	0
Ankistrodesmus	0	7	0	1	1	0	0	0	0	0	2	0	1	0	0	0	0	0	0	0	2	0	1	2	2	3	4	3
Closterium	0	0	7	1	0	3	0	0	0	0	0	2	0	1	0	0	0	0	0	2	0	0	2	0	1	0	0	0
Ulothrix	0	0	0	0	0	0	0	0	0	0	0	0	0	0	0	0	1	0	0	0	0	0	0	0	0	0	0	0
Spirotaenia	0	0	0	2	0	3	0	0	0	1	0	0	0	0	0	0	0	0	0	0	3	0	0	0	0	0	0	0
Penium	0	0	0	1	0	0	2	0	0	3	0	0	0	0	0	0	0	0	0	0	0	0	0	1	2	0	0	0
Pleurotaenium	0	0	0	0	1	0	0	0	0	0	0	0	0	0	0	0	0	1	0	0	0	0	0	0	0	0	0	0
Protococcus	0	0	0	0	0	5	0	0	0	0	0	0	0	0	0	1	0	0	0	0	0	0	0	0	3	0	0	0
Spirogyra	0	0	0	0	0	1	1	0	0	1	0	0	0	0	0	0	0	0	0	1	0	0	0	0	0	0	0	0
Zygnema	0	0	0	0	0	0	0	0	0	0	0	0	0	0	0	0	0	0	0	0	4	0	0	0	0	1	2	4
Mougeotia	0	0	0	0	0	0	0	0	0	0	0	0	0	0	0	0	0	0	0	0	0	1	0	0	0	0	0	0
Netrium	0	0	0	0	0	0	0	0	0	0	0	0	0	0	0	0	0	0	0	0	0	0	0	1	0	0	0	0
Tetraspora	0	0	0	0	0	0	0	0	0	4	0	0	2	0	0	0	0	0	0	0	0	0	0	0	0	0	0	0
Closteriopsis	1	0	0	0	0	0	0	0	0	1	0	0	0	0	0	0	0	0	1	0	2	0	0	0	0	1	0	0
Chrysophyta																												
Pinnularia	3	0	0	0	0	0	1	0	0	0	0	0	0	0	0	0	0	0	0	0	0	0	0	0	0	0	0	0
Nitzchia	1	0	4	0	0	0	0	0	0	0	0	0	0	2	0	0	0	4	0	2	3	0	5	4	10	3	2	3
Navicula	0	0	0	0	0	0	0	0	1	2	1	3	0	3	3	2	0	0	0	0	0	0	2	0	2	0	1	2
Ophiocytium	0	0	5	0	0	0	0	0	0	0	0	0	0	0	0	0	0	0	0	0	1	0	2	0	0	0	1	0
Eunotia	0	0	2	0	0	0	0	0	0	0	0	0	0	0	0	0	0	0	1	1	0	0	0	0	0	0	0	0
Chrysophyta																												
Cyclotella	0	0	0	0	0	0	0	0	0	0	0	0	0	0	0	5	5	0	0	0	0	0	0	0	0	0	0	0
Oscillatoria	0	0	0	0	0	1	2	0	0	0	3	0	0	0	0	0	0	2	2	1	0	2	0	0	0	0	0	0
Synedra	0	0	0	0	0	0	0	0	0	1	0	0	0	0	0	0	0	0	0	2	1	0	0	2	0	0	0	0
Tabellaria	0	0	0	0	0	0	0	0	0	0	0	0	0	0	0	0	0	0	0	0	0	2	1	0	3	4	0	0
Chlorogibba	0	0	0	0	0	0	2	1	0	0	0	0	0	0	0	0	0	0	0	0	0	0	0	0	0	0	0	0
Monallantus	0	0	0	0	0	0	0	2	0	0	0	0	0	0	0	0	0	0	0	0	0	0	0	0	0	0	0	0
Gloeochloris	0	0	0	0	0	0	2	1	0	0	0	0	0	0	0	0	0	1	0	0	0	0	0	0	0	0	0	0
Gomphonema	0	0	0	0	0	0	0	0	0	2	0	0	0	0	0	0	0	0	0	0	0	0	0	0	0	0	0	0
Fragillaria	0	0	0	0	0	0	0	0	0	0	0	0	0	0	0	0	0	0	0	0	0	1	1	0	4	2	0	0
Diatoma	0	0	0	0	0	0	0	0	0	0	0	1	0	0	0	0	0	0	0	0	0	0	0	0	0	0	0	0
Pyrrophyta																												
Peridinium	0	0	0	0	0	1	1	0	0	0	1	0	0	0	0	0	0	1	0	1	0	0	0	0	0	0	0	0
Chilomonas	0	0	0	0	0	0	0	0	0	0	0	0	1	0	0	0	0	0	0	0	1	0	0	0	0	0	0	0
Cyanophyta																												
Phormidium	2	5	0	0	0	0	0	0	0	0	0	0	0	0	0	0	0	0	3	0	0	0	0	0	0	0	0	2
Spirulina	0	0	5	0	0	0	0	0	0	0	0	0	0	1	0	0	0	0	0	1	0	0	0	0	0	0	0	0

Table 2. Contd

Anabaena	0	0	0	0	1	0	0	3	2	0	0	0	0	0	0	0	0	0	0	0	0	0	0	0	4	2	1	0
Aphanocapsa	0	0	0	14	10	0	0	0	0	0	0	0	0	0	1	2	2	0	0	0	3	0	2	3	1	2	3	0
Merismopedia	0	0	0	0	0	0	0	0	0	0	0	0	0	0	0	0	0	0	0	0	0	6	2	0	2	1	2	0
Coelosphaerium	0	0	0	0	2	0	0	10	0	3	5	3	0	4	1	0	0	0	0	0	7	0	4	0	3	4	4	0
Polycystis	0	0	0	0	0	0	0	0	0	1	1	5	0	0	4	2	2	1	3	4	2	2	2	1	0	5	4	0
N	7	12	25	20	15	18	14	17	3	20	14	14	4	11	9	12	10	11	12	15	29	14	24	15	37	28	24	**37**
S	4	2	6	6	5	8	9	5	2	11	7	5	3	5	4	5	4	7	6	9	11	6	11	8	12	11	10	12
Hs	0.50	1.03	1.21	1.07	0.94	0.85	0.53	1.02	0.28	0.73	0.72	0.90	0.24	0.72	0.69	0.79	0.76	0.50	0.69	0.60	1.07	0.81	0.90	0.69	1.22	1.04	0.96	1.22
D	1.51	0.58	1.20	1.34	1.29	1.89	2.41	1.21	1.15	2.46	1.87	1.34	1.50	1.51	1.33	1.44	1.26	2.11	1.73	2.32	2.04	1.60	2.25	2.07	1.97	2.08	2.04	1.97
d	1.54	0.40	1.55	1.67	1.48	2.42	3.03	1.41	0.91	3.34	2.27	1.52	1.44	1.67	1.37	1.61	1.30	2.50	2.01	2.95	2.97	1.89	3.15	2.58	3.05	3.00	2.83	3.05
J	0.25	1.03	0.47	0.41	0.41	0.28	0.17	0.44	0.28	0.21	0.26	0.39	0.15	0.31	0.34	0.34	0.38	0.18	0.27	0.19	0.31	0.31	0.26	0.23	0.34	0.30	0.29	0.34
C	0.31	0.51	0.20	0.51	0.48	0.17	0.12	0.40	0.56	0.12	0.21	0.24	0.38	0.26	0.33	0.26	0.34	0.21	0.19	0.15	0.13	0.26	0.12	0.16	0.13	0.11	0.13	0.09

Table 3. Monthly spatial variation in the zooplankton composition and abundance (number of individuals per drop) in some parts of River Ogun.

Date	Dec, 2011				Jan, 2012				Feb, 2012				March, 2012				April, 2012				May, 2012				June, 2012			
Site	**A**	**B**	**C**	**D**	**A**	**B**	**C**	**D**	**A**	**B**	**C**	**D**	**A**	**B**	**C**	**D**	**A**	**B**	**C**	**D**	**A**	**B**	**C**	**D**	**A**	**B**	**C**	**D**
Rotifera																												
Platyias	2	0	4	0	4	0	1	0	0	0	0	0	0	0	0	0	0	0	0	0	0	0	0	0	0	0	0	0
Cupelopagis	4	3	0	1		3	4	1	1	1	0	2	0	0	0	1	0	0	0	1	0	2	0	0	3	0	4	2
Keratella	4	5	10	0	0	0	4	0	0	0	0	0	0	0	0	0	0	0	0	0	0	0	0	0	0	0	0	0
Testudinella	1	0	2	4	0	0	0	0	0	0	0	0	0	1	0	2	0	1	0	1	0	0	0	0	0	0	0	0
Protozoa																												
Amoeba	1	2	0	2	0	0	0	0	4	0	0	0	0	2	0	1	0	0	0	0	0	0	0	0	0	0	0	0
Oikomonas	0	0	0	0	0	2	0	0	0	0	0	0	0	0	0	0	0	1	3	1	0	0	0	0	0	0	0	0
Paramecium	0	0	0	1	2	4	0	4	0	0	0	0	0	0	0	0	1	0	2	1	0	0	0	0	0	0	0	0
Blepharisma	3	0	3	1	0	0	0	0	0	0	1	0	0	0	0	0	0	0	0	0	0	0	0	0	0	0	0	0
Copepoda																												
Diaptomus	0	2	3	0	0	2	0	0	1	0	1	3	0	0	0	0	0	4	0	5	0	0	0	0	0	0	0	0
Canthocamptus	0	0	0	0	0	0	0	0	2	1	0	0	0	0	0	0	0	0	0	0	0	0	0	0	0	0	0	0
Cladocera																												
Polyphemus	0	0	3	7	0	0	0	0	0	0	0	0	0	0	1	0	0	2	0	4	0	0	1	5	4	3	2	1
Ceriodaphnia	2	0	4	0	0	4	0	0	0	0	0	0	0	0	0	0	0	0	0	0	0	0	0	0	0	0	0	0
Scapholeberis	1	0	0	0	0	0	0	0	0	1	0	1	0	4	0	4	0	0	0	0	2		5	3	5	4	1	4
Bosmina	0	0	0	0	0	0	0	0	0	0	0	0	2	0	1	0	5	2	0	0	3	2	1	5	0	4	6	7
Eurycerus	0	0	0	0	0	0	0	0	0	4	4	0	3	0	4	1	1	0	4	0	0	0	1	0	3	2	1	2
Ostracoda																												
Cypridiopsis	0	0	0	0	0	0	0	0	0	0	0	0	0	0	0	0	0	0	0	0	0	4	1	0	4	0	2	1
N	18	12	29	16	6	15	9	5	8	7	6	6	5	7	6	9	7	10	9	13	5	8	9	13	19	13	16	17
S	8	4	7	6	2	5	3	2	4	4	3	3	2	3	3	5	3	5	3	6	2	3	5	3	5	4	6	6

Table 3. Contd

Hs	0.85	0.88	1.26	0.91	0.68	0.94	0.80	0.58	0.60	0.50	0.54	0.54	0.58	0.64	0.54	0.57	0.64	0.65	0.80	0.75	0.58	0.72	0.57	1.00	1.09	0.93	0.91	0.96
D	1.89	1.15	1.30	1.50	0.82	1.29	1.00	0.89	1.41	1.51	1.22	1.22	0.89	1.13	1.22	1.67	1.13	1.58	1.00	1.66	0.89	1.06	1.67	0.83	1.15	1.11	1.50	1.46
d	2.42	1.21	1.78	1.80	0.56	1.48	0.91	0.62	1.44	1.54	1.12	1.12	0.62	1.03	1.12	1.82	1.03	1.74	0.91	1.95	0.62	0.96	1.82	0.78	1.36	1.17	1.80	1.76
J	0.28	0.44	0.45	0.35	0.68	0.41	0.50	0.58	0.30	0.25	0.34	0.34	0.58	0.40	0.34	0.24	0.40	0.28	0.50	0.29	0.58	0.46	0.24	0.63	0.47	0.46	0.35	0.37
C	0.16	0.29	0.19	0.28	0.56	0.22	0.41	0.68	0.34	0.39	0.50	0.39	0.52	0.43	0.50	0.28	0.55	0.26	0.36	0.27	0.52	0.38	0.36	0.35	0.21	0.27	0.24	0.26

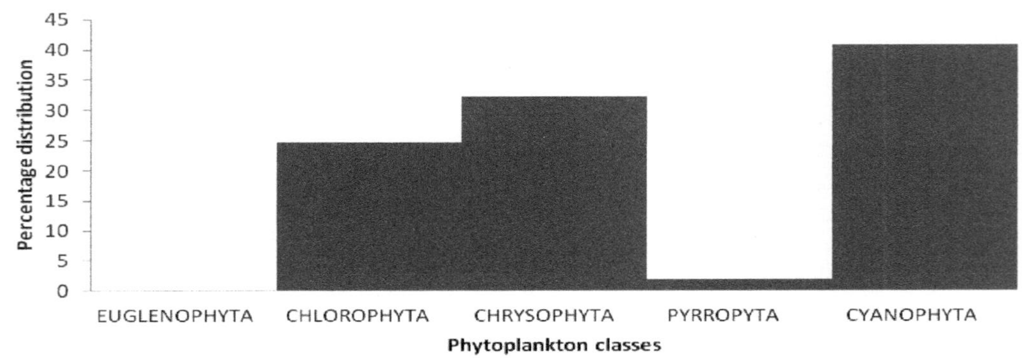

Figure 2. Percentage distribution of phytoplankton classes in station A.

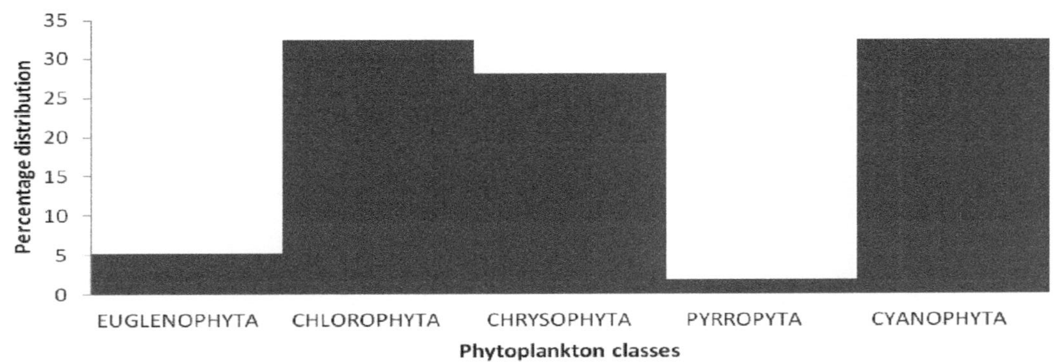

Figure 3. Percentage distribution of phytoplankton classes in station B.

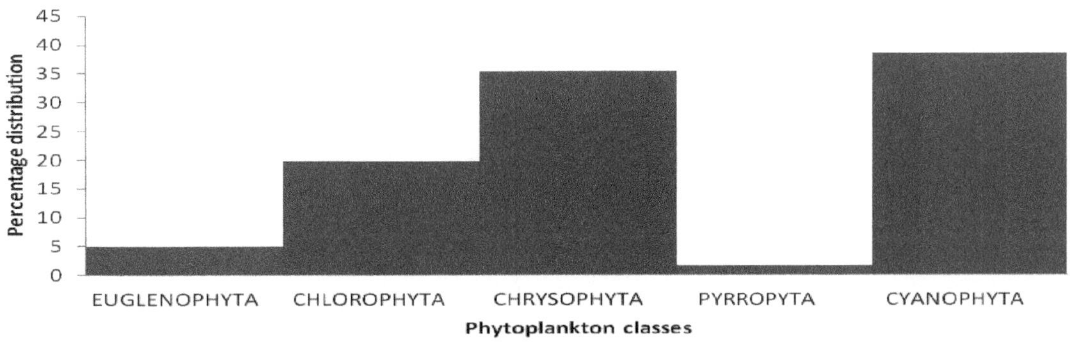

Figure 4. Percentage distribution of phytoplankton classes in station C.

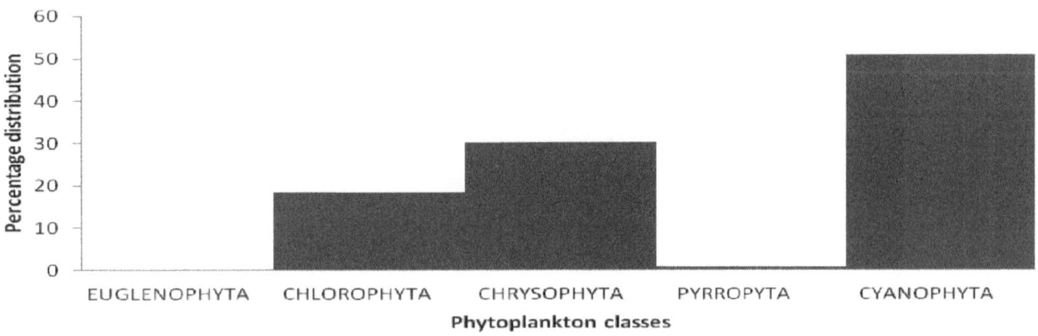

Figure 5. Percentage distribution of phytoplankton classes in station D.

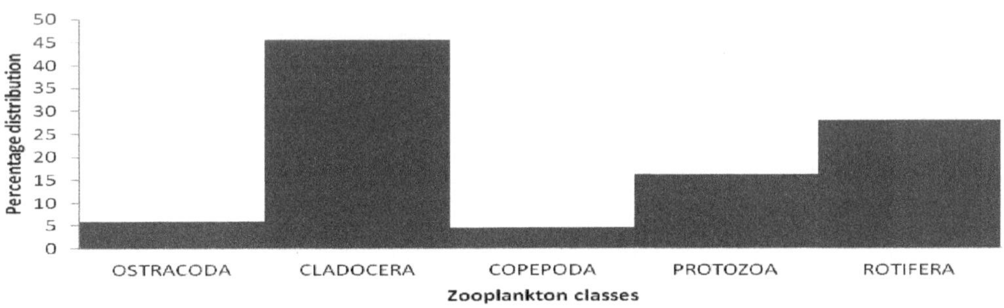

Figure 6. Percentage distribution of zooplankton classes in station A.

DISCUSSION

For the 4 Stations, water temperature was lowest in Station A in December, 2011 and was highest in Station C in May, 2012. It was observed that for the first four months of this study, the dissolved oxygen values were very critically low, that is, less than 5 mg/l. This could have been caused by pollution due to anthropogenic activities carried out around the water resource. Jaji et al. (2007) reported that pollution of Ogun river water along its course is evidenced by the high concentrations of pollution indicators, nutrients and trace metals above the acceptable limit.

In the frequency distribution of phytoplankton classes, Cyanophyta was predominant in all the stations excluding station B where there was a tie between Cyanophyta and Chlorophyta. The dominance of Cyanophyta in this river is similar to findings by Sekadende et al. (2004), Ogato (2007), Deng et al. (2007), and Shakila and Natarajan (2012).

In the frequency distribution of zooplankton classes, Cladocera was predominant in all the stations. This is shown in Figures 6, 7, 8 and 9. The dominance of Cladocera in this river is similar to findings by Ude et al.

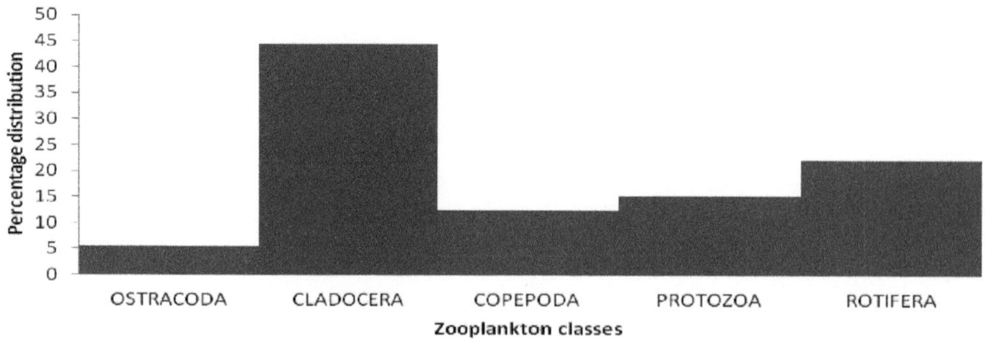

Figure 7. Percentage distribution of zooplankton classes in station B.

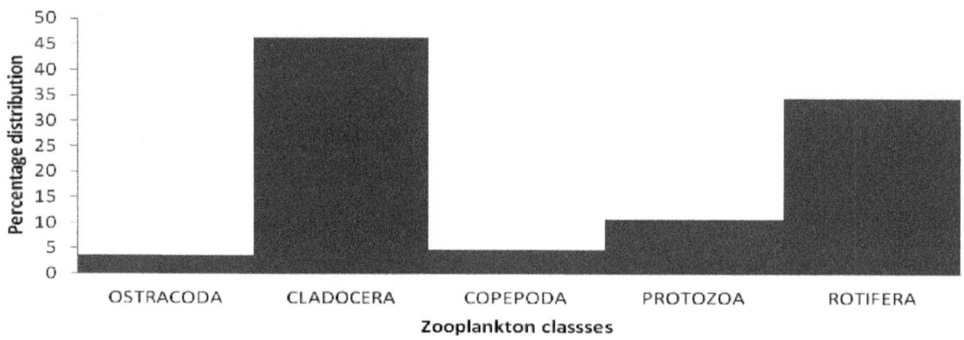

Figure 8. Percentage distribution of zooplankton classes in station C.

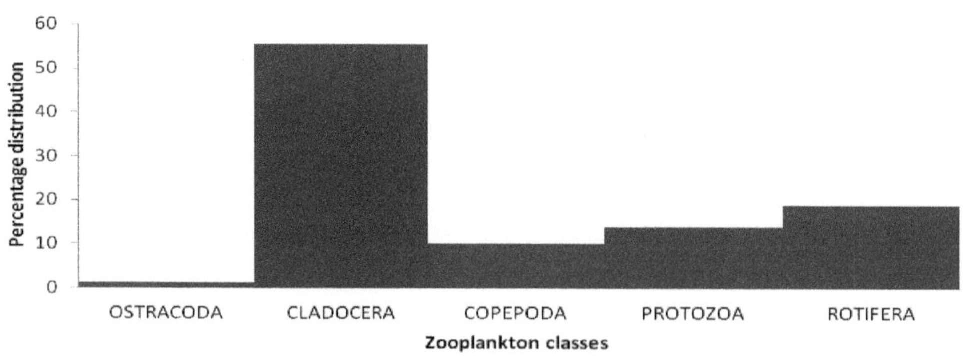

Figure 9. Percentage distribution of zooplankton classes in station D.

(2011) and Ogbuagu and Ayoade (2012).

ACKNOWLEDGEMENTS

We thank all the members of staff especially Dr. W. O. Alegbeleye of the Department of Aquaculture and Fisheries Management, Federal University of Agriculture, Abeokuta, Ogun state, Nigeria for their collaborative spirit. We thank Mrs. Binyotubor of the National Institute of Freshwater Fisheries Research for her support. We also appreciate Mr. Thomas Okaku Dimowo for the statistical analyses.

REFERENCES

Abohweyere PO (1990). Study of limnological parameter and potential fish yield in Kigera reservoir (extensive system) in Kainji, New Bussa, Nigeria; J. Aqua. Sci. 5: 53-58.

Adakole JA, Abulode DS, Balarabe ML (2008). Assessment of Water Quality of a Man-Made Lake in Zaria, Nigeria; In Proceedings of Taal, 2007: The 12th World Lake Conference: 1373-1382.

Adejuwon JO, Adelakun MO (2012). Physiochemical and bacteriological analysis of surface water in Ewekoro Local Government Area of Ogun State, Nigeria: Case study of Lala, Yobo and Agodo Rivers; Int. J. Water Res.Environ. Eng. 4(3): 66-72. ISSN 1991-637X.

Adeogun 0A, Fafioye OO, Olaleye BA, Ngobili GO (2005). The relationship between some physicochemical parameters and plankton composition on fish production in ponds. In the 19th annual conference of the Fisheries Society of Nigeria (FISON), Ilorin, Nigeria.

Adesalu TA, Nwankwo DI (2008). Effect of Water quality Indices on Phytoplankton of a Sluggish Tidal Creek in Lagos, Nigeria; Pak. J. Biol. Sci. 11(6): 836-844. ISSN 1028-8880.

Agbaire PO, Obi CG (2009). Seasonal Variations of Some Physico-Chemical Properties of River Ethiope Water in Abraka, Nigeria; J. Appl. Sci. Environ. Manage. 13(1): 55–57. ISSN 1119-8362.

Ajuonu N, Ukaonu SU, Oluwajoba EO, Mbawuike BE, Williams AB, Myade EF (2011). The abundance and distribution of plankton species in the bonny estuary; Nigeria; Agriculture and Biology Journal of North America. ISSN Print: 2151-7517, ISSN Online: 2151-7525

American Public Health Association (APHA) (1998). Standard methods for the examination of water and wastewater. 20th ed., Washington D.C

Atobatele OE, Ugwumba OA (2008). Seasonal variation in the physicochemistry of a small tropical reservoir (Aiba Reservoir, Iwo, Osun, Nigeria). Afr. J. Biotechnol. 7(12): 1962-1971. ISSN 1684–5315.

Ayeni AO, Balogun II, Soneye ASO (2011). Seasonal Assessment of Physico-chemical Concentration of Polluted Urban River: A Case of Ala River in Southwestern-Nigeria; Res. J. Environ. Sci., 5: 22-35. ISSN 1819-3412.

Ayoade AA, Sowunmi AA, Nwachukwu HI (2004). Gill asymmetry in Labeo ogunensis from Ogun River, Southwest Nig. Rev. Biol. Trop. 52(1):171–175.

Balarabe ML (2001). Effect of limnological characteristics on zooplankton composition and distribution in Dumbi and Kwangila ponds Zaria, Nigeria. Phd Thezis, Department of Biological Sciences, Ahmadu Bello University, Zaria.

Bhattacharya AK, Bolaji GA (2010). Fluid flow interactions in Ogun River, Nigeria. IJRRAS 2(2):173-180.

Bwala RL, Ovie SI, Ajayi O, Haruna A (2010). Preliminary Study On The Limnology And Plankton Abundance In Relation To Fish Production In Some NIFFR Reservoirs; Report and Opinion 2(6). http://www.sciencepub.net/report.

Chia MA, Bakol S, Alonge SO, Adamu AK (2011). Green algal interactions with physicochemical parameters of some manmade ponds in Zaria, northern Nigeria; Rev. bras. Bot. 34(3). São Paulo; ISSN 0100-8404.

Chinedu SN, Nwinyi OC, Adetayo YO, Eze VN (2011). Assessment of water quality in Canaanland, Ota, Southwest Nigeria. Agric. Biol. J. North Ame. ISSN Print: 2151-7517, ISSN Online: 2151-7525.

Davies OA, Ansa E (2010). Comparative Assessment of Water Quality Parameters of Freshwater Tidal Earthen Ponds and Stagnant Concrete tanks for Fish Production in Port Harcourt, Nigeria; I.J.S.N., 1(1): 34-37. ISSN 0973 -3140.

Deng D, Xie P, Zhou Q, Yang H, Guo L (2007). Studies on temporal and spatial variations of phytoplankton in Lake Chaohu, J. Integrative Plant Biol. 49(4):409-418.

Edmondson WT (ed.) (1959). Freshwater Biology; John Wiley Sons Inc; New York: ISBN 0-471-23298-X.

Erondu ES, Chindah AC (1991). Variations in the physico-chemical features and phytoplankton of the New Calabar River at Aluu, Rivers state, Nigeria. NIOMR Technical Paper No. 75; ISBN 978 – 2345 – 078.

Essien-Ibok MA, Akpan AW, Udo MT, Chude LA, Umoh IA, Asuquo IE (2010). Seasonality in the Physical and Chemical Characteristics of Mbo River, Akwa Ibom State, Nigeria; Nig. J. Agric., Food Environ. 6(1 and 2):60-72.

Ezekiel EN, Ogamba EN, Abowei JFN (2011). The Zooplankton Species Composition and Abundance in Sombreiro River, Niger Delta, Nigeria. Asian J. Agric. Sci. 3(3):200-204; ISSN: 2041-3890

Ezeribe AI, Oshieke KC, Jauro A (2012). Physico-Chemical Properties of Well Water Samples from some Villages in Nigeria with cases of Stained and Mottle Teeth; Sci. World J. 7 (1); ISSN 1597-634.

Fafioye OO, Olurin KB, Sowunmi AA (2005). Studies on the physico-chemical parameters of Omi water body of Ago-Iwoye, Nigeria. Afr. J Biotechnol. 4(9): 1022-1024. ISSN 1684–5315.

Google Maps (2012). "Map of River Ogun"; (www.maps.google.com.ng/maps?hl=en&ll=7.153952,3.329973&spn=0.033213,0.055189&t=m&z=14) Retrieved 28th July, 2012.

Ibrahim S (2009). A Survey of Zooplankton diversity of Challawa River, Kano and Evaluation of some of its Physico-Chemical conditions; Bayero J. Pure Appl. Sci. 2(1):19–26.

Imoobe TOT, Egborge ABM (1997). The Composition, Distribution and Seasonal variation of Crustacea in Jamieson River, Southwest Nigeria. Tropical Freshwater Biology 6:49–63.

Jaji MO, Bamgbose O, Odukoya OO, Arowolo TA (2007). Water quality assessment of Ogun river, Southwest Nigeria. Environmental Monitoring Assessment 133(1-3):473-482.

Ladipo MK, Ajibola VO, Oniye SJ (2011). Seasonal Variations in Physicochemical Properties of Water in Some Selected Locations of the Lagos Lagoon. Sci. World J. 6(4):2011. ISSN 1597-6343.

Margalef R (1951). Diversidad de especies en las cominidales naturales. Publ. Inst. Biol. Appl. 9:5-27.

Mood YFO (2004). An assessment of water quality in integrated poultry-cum-fish reservoir, NIFFRI New Bussa. In: P.A. Araoye(ed), Conference Proceeding of Fisheries Society of Nigeria (FISON), Ilorin, 29th November – 3rd December, 2004.

Muhibbu-din OI, Aduwo AO, Adedeji AA (2011). Study of physiochemical parameter of effluent impacted stream in Obafemi Awolowo University, Ile-Ife, Nigeria; A Dissertation submitted to the Department of Zoology, Obafemi Awolowo University, Ile Ife, Osun State, Nigeria.

Mustapha MK, Omotosho JS (2005). An Assessment of the Physico-Chemical Properties of Moro Lake; Afr. J. Appl. Zool. Environ. Biol. 7: 73–77. ISSN 1119-023X.

National Bureau of Statistics (NBS) (2012). State Information; (www.nigerianstat.gov.ng/information/details/Ogun).

Nkwoji JA, Onyema IC, Igbo JK (2010). Wet Season Spatial Occurrence of Phytoplankton and Zooplankton in Lagos Lagoon,Nigeria. Sci. World J. 5(2). ISSN 1597-6343.

Offem BO, Ayotunde EO, Ikpi GU, Ochang SN, Ada FB (2011). Influence of Seasons on Water Quality, Abundance of Fish and Plankton Species of Ikwori Lake, South-Eastern Nigeria. Fisheries and Aquaculture Journal, 2011: FAJ-13.

Ogato T (2007). Dynamics of phytoplankton in relation to physico-chemical factors in Lake Bishoftu, Ethiopia, M.Sc. Thesis, School of Graduate Studies; Addis Ababa University, Ethiopia.

Ogbeibu AE (2005). Biostatistics: A practical approach to reseach and data handling. Mindex Publishing Company limited, Benin City, Nigeria.

Ogbuagu DH, Ayoade AA (2012). Seasonal Dynamics in Plankton Abundance and Diversity of a Freshwater Body in Etche, Nigeria. Environment Natural Resour. Res. 2(2):48 - 59. ISSN 1927-0488 E-ISSN 1927-0496.

Oketola AA, Osibanjo O, Ejelonu BC, Oladimeji, YB, Damazio OA (2006). Water Quality Assessment of River Ogun around the Cattle Market of Isheri, Nig.. J. Appl. Sci. 6:511-517.

Onuoha PC, Vyverman W (2010). Hydrochemistry and Plankton Dynamics of Eutrophic Tropical Lagoon; Nature and Science 2010: 8(9).

Onyema IC (2007). The Phytoplankton Composition, Abundance and Temporal Variation of a Polluted Estuarine Creek in Lagos, Nigeria; Turkish Journal of Fisheries and Aquatic Sciences 7:89-96.

Oso JA, Fagbuaro O (2008). An Assessment of the Physico-Chemical Properties of a Tropical Reservoir, Southwestern, Nigeria. J. Fish. Int. 3(2):42-45. ISSN: 1817-3381

Ovie SI (1995). Zooplankton species richness and Sorenson's index of similarity for Asa Lake and some other freshwater ecosystem in Nigeria. National Institute for Freshwater Fisheries annual report. pp. 57-62.

Sekadende BC, Mbonde ASE, Shayo S, Lyimo TJ (2004) Phytoplankton species diversity and abundance in satellite lakes of Lake Victoria basin (Tanzanian side), Tanz. J. Sci. 31:1.

Shakila H, Natarajan S (2012). Phytoplankton Diversity and its Relationship to the Physico–Chemical Parameters in the Temple Pond of Thiruporur, Chennai. Int. J. Environ. Biol. 2(2):81-83.

Shannon CE, Weaver W (1949). The mathematical theory of communication. Urban University, Illinois Press, Illinois, P. 125.

Travel Journals (2012). River Ogun, Nigeria on World map Coordinates and short facts. (www.Traveljournals.net) Retrieved 21st May, 2012.

Ude EF, Ugwu LLC, Mgbenka BO (2011). Evaluation of Zooplankton Diversity in Echara River, Nigeria. Continental. J. Biol. Sci. 4(1):1-5.

Chemical composition of five marine microalgae that occur on the Brazilian coast

Giselle S. Costard[1], Rodrigo R. Machado[1], Elisabete Barbarino[1], Ricardo C. Martino[2] and Sergio O. Lourenço[1]*

[1]Universidade Federal Fluminense, Departamento de Biologia Marinha, Niterói, RJ, Brazil.
[2]Fundação Instituto de Pesca do Estado do Rio de Janeiro, Rio de Janeiro, RJ, Brazil.

Five native species of marine microalgae were batch cultured and their chemical composition was measured in different growth phases. The highest cell yield was recorded in *Chlorella* sp. *Chlorella* sp. and *Bellerochea* sp. showed a reduction of the mean cell volume throughout growth. All microalgae showed protein as the main chemical component, with peaks in the exponential growth phase. The same was found for chlorophyll. Carbohydrates (*Chlorella* sp.), ash (*Bellerochea* sp. and *Chlorella* sp.), and lipids (*Rhodomonas* sp. and *Thalassiosira* sp.) were relatively high in some of the species, at least in one of the growth phases. The carbohydrate content increased throughout growth, but no clear relationship among growth, total lipid and ash content was identified. Total saturated fatty acids were higher in the exponential growth phase of all species, decreasing throughout growth. *Rhodomonas* sp. showed high levels of polyunsaturated fatty acid (PUFA), but low percentages of some essential fatty acids. *Bellerochea* sp., *Chaetoceros* sp. and *Chlorella* sp. showed low percentages of total PUFA, but all essential fatty acids were present, except for the green alga. Current results may contribute to selecting strains that possess suitable chemical composition and fast growth, useful characteristics for the sustainable use of native species in aquaculture.

Key words: Microalgae, chemical composition, native species, growth, mariculture.

INTRODUCTION

Capture fisheries and aquaculture supplied the world with about 142 million tons of food fish in 2008. Of this total, aquaculture accounted for 46% (FAO, 2010). Production of bivalves in hatcheries is undoubtedly related to the quality and the quantity of the supplied microalgae (Rico-Villa et al., 2006). With the development of the mariculture in the last two decades, microalgal cultivation has increasing its importance due the fundamental use of microalgae for feeding many animal species in aquaculture (Spolaore et al., 2006). There is particular reference to mollusk and crustacean rearing, and production of live food for fish larvae (Ponis et al., 2008), larvae of red shrimp *Farfantepenaeus brasiliensis* (Gaxiola et al., 2010), or specifically juvenile *Artemia* as prey for *Octopus vulgaris* paralarvae (Seixas et al., 2008). The growth of bivalves is dependent on the algae species and their nutritional value (Reitan, 2011).

Microalgae contain essential nutrients which determine the quality, survival, growth and resistance to disease of cultured animal species (Gouveia et al., 2008). Microalgae are some of the most important feed sources in aquaculture due to their nutritional value and their ability to synthesize and accumulate great amounts of ω3-polyunsaturated fatty acid (PUFA) (Patil et al., 2007), such as 22:6n-3 (DHA) and 20:5n-3 (EPA), two of the most important essential fatty acids required for gametogenesis (Ehteshami et al., 2011). Currently, 30% of the world microalgal production is sold for animal feed applications (Spolaore et al., 2006). The knowledge of the chemical composition of food-species has a key role for mariculture (Thompson et al., 1996; Laing and

*Corresponding author. E-mail: solourenco@id.uff.br.

Psimopoulos, 1998; Southgate et al., 1998; Leonardos and Lucas, 2000; Rivero-Rodríguez et al., 2007), especially in larviculture (Pettersen et al., 2010). Many studies on the use of both monoalgal and combination of different species have been carried out in the last decades (Ben-Amotz et al., 1987; Laing et al., 1990; Carić et al., 1993; Thompson et al., 1996; Southgate et al., 1998; Tang et al., 2006). In this context, a variety of microalgal species has been studied, since many species are not suitable for use in commercial aquaculture systems (Enright et al., 1986a).

The microalgal culture system in hatcheries could be inadequate in qualitative or quantitative terms, and the quality of biomass produced could be subject to a nutritional drift (Ponis et al., 2006). In particular, the number of good quality microalgae currently available in hatcheries is limited, and several species which were previously used in commercial hatcheries have now been discarded due to their poor nutritional value (Ponis et al., 2006). Moreover, the availability of good quality seeds could be limited, partly due to the scarce knowledge of the requirements of juvenile stages, as occur with scallop *Nodipecten subnodosus* in Mexico (Cerón-Ortiz et al., 2009). In addition, some species may be edible and digestible but poorly assimilated by prey, like *Nitzschia* for the pearl oyster *Pinctada fucata* (Hashimoto et al., 2008). The nutritional value of the dietary algae is not only dependent on the chemical composition but also on factors such as the capability of bivalves to ingest and digest the algae and to assimilate their nutrients (Lora-Vilchis and Maeda-Martinez, 1997).

In Brazil, strains from foreign countries are largely used because they are world-wide recognized and because they are also chemically known. However, this procedure has been stimulating the introduction of many non-indigenous species (Assad and Bursztyn, 2000), and this option is tremendously contested in many cases (Renaud et al., 1999). There is a concern about the need of screening studies with native species, which could reduce the use of non-indigenous microalgae in commercial aquaculture (Brown et al., 1998; McCausland et al., 1999; Gouda et al., 2006; Ponis et al., 2006). Ideally, the chemical composition of potentially useful native species should be evaluated to assess the nutritional properties of them (Gouda et al., 2006). Despite this issue is obviously important with many positive consequences, experimental data in this field are needed in Brazil to make it feasible. Native species could generate a less environmental impact in the location where the system is operating, due more ecological compatibility and high level of fitness to local and/or regional characteristics.

The chemical content of microalgae can vary with culture age and with changes in environmental conditions (Fernández-Reiriz et al., 1989), and culture conditions (Araújo and Garcia, 2005). The effect of variation of these parameters on many algae species has been studied in order to better understand their physiology, as well as to answer specific and relevant questions for mass culture and nutrition of bivalves and others herbivores (Uriarte et al., 1993). Data on the chemical composition of microalgae may also vary widely due to differences of the methods of measurement used (Barbarino and Lourenço, 2005), the physiological state of the microalgae (Fernández-Reiriz et al., 1989), as well as to the experimental conditions applied (Lourenço et al., 2002), like temperature (Durmaz et al., 2009), light intensity (Lourenço et al., 2008), medium cultivation (Huerlimann et al., 2010) or in outdoors conditions (Banerjee et al., 2011). In addition, due to the interaction of the organisms with the culture medium, a batch culture is under a continuous chemical change. These variations reflect on the cell metabolism and consequently on their chemical composition (Lourenço et al., 2002). Thus, the chemical composition of a given species may vary widely under different growth conditions, and such changes may be related to the growth phase of the culture (Fernández-Reiriz et al., 1989; Lourenço et al., 1997). However, studies focusing sampling in different growth phases are relatively scarce; most of the papers report the chemical profile of given species in a fixed momentum of the growth phase, ignoring the continuing process of interaction between microalgae and the medium.

The present study was designed to evaluate the chemical composition of five native microalgae in different growth phases, characterizing their chemical profile under standardized conditions. In addition, quantification of ash content and cell biovolume were also performed to increase information for a better interpretation of the dynamics of the microalgae in cultures.

MATERIALS AND METHODS

Microalgae

Five species were studied; all of them were obtained from the Elizabeth Aidar Microalgae Culture Collection, Department of Marine Biology, Fluminense Federal University, Brazil. The following strains were used: *Bellerochea* sp. Van Heurck 1885 (Heterokontphyta–Biddulphiales; strain SP1), *Chaetoceros* sp. Ehrenberg 1844 (Heterokontphyta–Bacillariales; strain SP1), *Chlorella* sp. M. Beijerinck 1890 (Chlorophyta–Chlorococcales; strain SP1), *Rhodomonas* sp. Karsten 1898 (Cryptophyta–Cryptomonadales; strain PB1) e *Thalassiosira* sp. Cleve 1873 (Heterokontphyta–Bacillariales; strain UB7). All strains were isolated from Brazilian coastal waters and were selected because of their fast growth responses and taxonomical diversity. The strains tested here do not form chains in cultures, always showing isolated individual cells.

Culture conditions

The species were grown in batch cultures, with three replicates (n = 3), in 3.0 L borosilicate round flasks, filled with 2.0 L of Conway culture medium (Walne, 1966). The seawater was diluted with distilled water to a salinity of 31 psu. Seawater (2.0 L) was autoclaved at 121ºC for 30 min at 1 atmospheric pressure, and enriched with nutrients, which had been a utoclaved separately. Nutrients were added to the seawater before inoculation. Each cultures were exposed to a saturating irradiance (180 µmol photons m^{-2} s^{-1}, measured with a Biospherical Instruments quantum meter, model QLS100) provided from below by fluorescent lamps (Sylvania daylight tubes), on a 12:12 h light:dark cycle. Mean temperatures were 21±2°C. Cultures were continuously bubbled with filtered air at a rate of 2.0 L min^{-1} $flask^{-1}$, provided by aquarium air pumps. Daily sampling for cell counts was carried out four hours after the start of the light period. Cells were counted in Fuchs-Rosenthal or Neubauer chambers, depending on the cell size. Cell counts were also carried at the preparation of the experiments ("day 0") and the growth rates were calculated daily. Starter cultures of 50 to 100 ml in mid-exponential growth phase were inoculated into the medium. The initial cell densities of cultures were 3.0 x 10^4 cell ml^{-1} for *Bellerochea* sp., *Rhodomonas* sp. and *Chaetoceros* sp.; 1.0 x 10^4 cell ml^{-1} for *Thalassiosira* sp. and 8.0 x 10^4 cell ml^{-1} for *Chlorella* sp. In the experiments, differences in initial cell densities are due the remarkable differences in microalgae cell volumes (Lourenço et al., 2002).

Preparation of samples for chemical analyses

Sampling for chemical analysis was done three times: in the exponential growth phase, in the transitional phase (period between the end of exponential and early stationary growth phases) and in the stationary growth phase for all species, as indicated in Figure 1. Measured volumes (800 ml) were sampled from the cultures and centrifuged at 7000*g* (centrifuge Sigma, model σ-15), at room temperature, for 10 min, once or several times, to obtain highly concentrated material. All supernatants obtained from each centrifuged sample were collected, and any cell losses were quantified by means of cell counts of the supernatant. The centrifuged cell samples were frozen at -20°C and then freeze-dried, weighed, and stored in vacuum desiccators, sheltered from light for later analyses of protein, carbohydrate, and lipid/fatty acids. For chlorophyll *a*, total carotenoid analyses and ash content samples were obtained by filtering 10 to 30 ml culture onto Whatman GF/F glass fibre filters. The filters were kept at -20°C in flasks containing silica gel until analysis, one to ten days after sampling.

Measurements of biovolume

The cell volumes were measured using the equations provided by Hillebrand et al. (1999), assuming a suitable geometrical shape for each species. Mean volumes were based on measurements of 30-70 cells for each species in each culture flask, under a Hund (Wilovert S model) inverted microscope, giving three mean values used for statistics (n=3).

Chemical analyses

Protein extraction followed the method proposed by Barbarino and Lourenço (2005). The Lowry et al. (1951) method was used to evaluate protein in the samples, with bovine serum albumin as a protein standard. Spectrophotometric determinations were done at 750 nm, 35 min after the start of the chemical reaction.

Carbohydrates were extracted with 80% H_2SO_4 according to Myklestad and Haug (1972). Total carbohydrate concentration was determined spectrophotometrically by the phenol-sulfuric acid method (DuBois et al., 1956), using glucose as a standard. Total lipids were extracted according to Folch et al. (1957), and determined gravimetrically after total solvent evaporation.

Aliquots of the final lipid extracts in chloroform were used to prepare the fatty acid methyl esters (FAME) according to Metcalfe and Schmitz (1961). Samples were saponified with 50% KOH and the methyl-esters of fatty acids were prepared by esterification with 7% B_3F_3 in methanol. The separation of the FAME was carried out in a gas chromatograph Shimadzu, model GC-15 A, equipped with a flame and ionization detector, and a capillary fused silica column (Omegawax 320 x 30 m x 0.32 mm ID). The flow rate of hydrogen (carrier gas) was 40 ml min^{-1}. A Shimadzu, model C-R4, recorder and integrator was used to obtain the chromatograms and quantify the FAME. Standard mixtures of fatty acid methyl esters (Supelco, Bellefont, USA) were used to obtain relative retention times and to identify the fatty acids in the samples. Samples for fatty acid analysis were kept for no more than two weeks in N_2 atmosphere at -20°C until analysis in order to avoid possible oxidation.

Pigment extraction was performed in 90% acetone at 4°C for 20 h. Spectrophotometric determinations of chlorophyll *a* and total carotenoids were carried out as described by Lorenzen (1967) and by Parsons et al. (1984), respectively.

The ash content of the samples was measured gravimetrically, using previously weighed glass fibre filters, according to AOAC (1990). Samples in the filters were burned in a muffle furnace (Quimis, model Q-318M) at a 450°C for 24 h.

Statistical analysis

The results for chemical composition (except FA) and cell volumes were analyzed by one-way analysis of variance (ANOVA), comparing the three measurements (in different growth phases) along time for each species, using a significance level α = 0.05 (Zar, 1996). Tukey's multiple comparison test followed the ANOVA, where applicable.

RESULTS

Microalgal growth and final yield

The growth curves of all microalgae are demonstrated in Figure 1. The exponential growth phase lasted at least for four days for all species. The stationary growth phase typically lasted four to five days, and the transitional phase occurred for two to three days, depending on the species. *Chlorella* sp. showed the higher final yield, with ca. 10.5x10^6 cell ml^{-1}. *Thalassiosira* sp. and *Rhodomonas* sp. showed the lowest values, ca. 6.4 x 10^5 and 1.3x10^6 cell ml^{-1}, respectively. For the diatoms *Bellerochea* sp. and *Chaetoceros* sp. the final yields were similar, varying around 6.2 x 10^6 cell ml^{-1}.

Cell volume

Significant differences in cell volume were detected throughout growth for all species (Table 1). *Thalassiosira* sp. and *Rhodomonas* sp. showed the largest cell volumes and *Bellerochea* sp. and *Chaetoceros* sp. the lowest. *Chlorella* sp., *Chaetoceros* sp. and *Bellerochea* sp. showed a decrease cell volume throughout growth ($0.0005 \leq P \leq 0.049$). Conversely, *Thalassiosira* sp. and *Rhodomonas* sp. increased their cell volume from exponential to stationary growth phase ($0.0053 \leq P \leq 0.044$).

Gross chemical composition

Significant differences in the protein content were

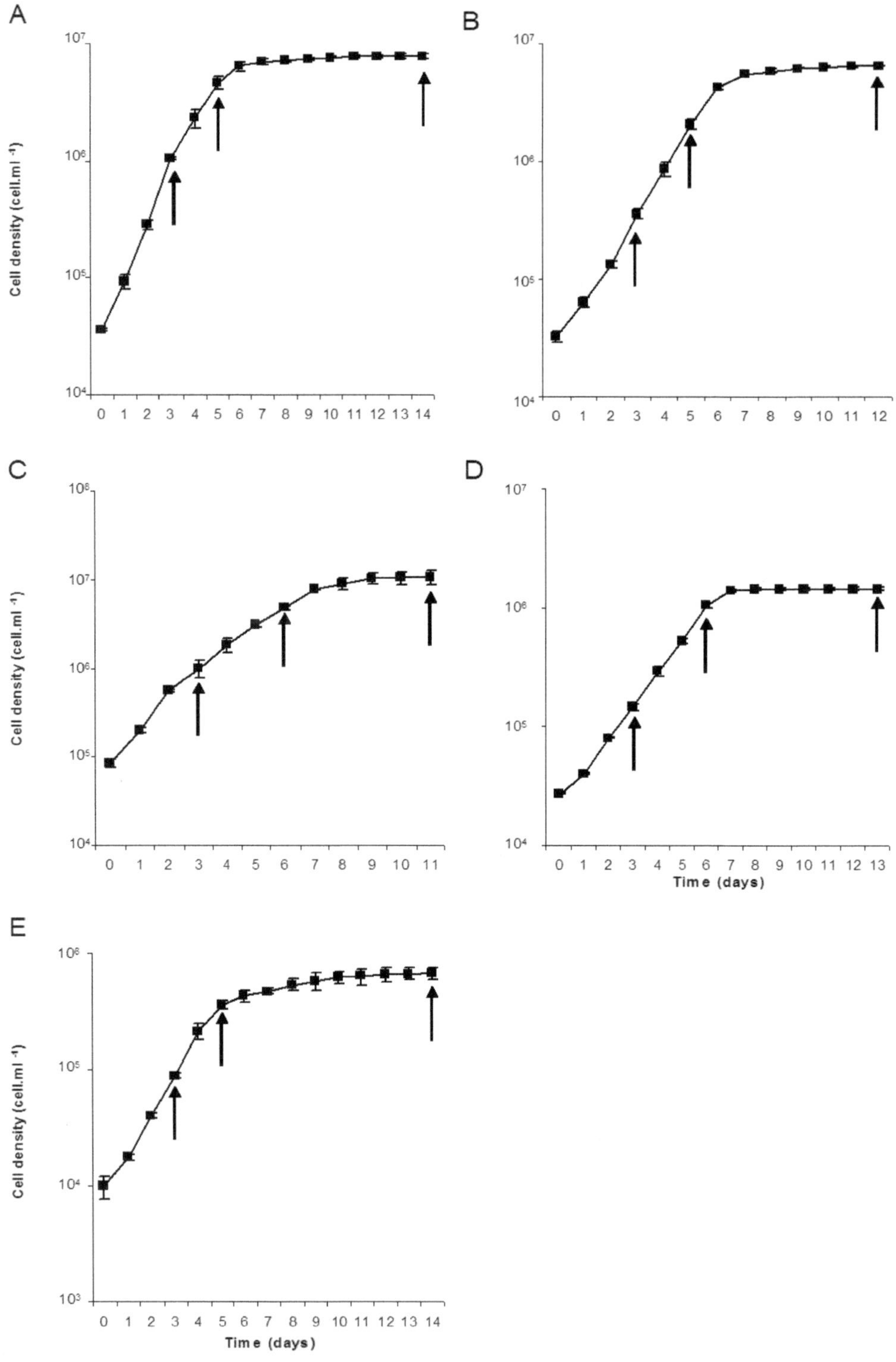

Figure 1. Growth curves *Bellerochea* sp. (A), *Chaetoceros* sp. (B), *Chlorella* sp. (C), *Rhodomonas* sp. (D), and *Thalassiosira* sp. (E) cultured with Conway culture medium. Each curve represents the mean of three replicates ± SD (n = 3). Arrows indicate the times of sampling for chemical analyses and cell measurements.

detected for all species along time (Table 2). All species achieved maximum levels of protein in the exponential growth phase, decreasing to the stationary growth phase ($0.001 \leq P \leq 0.04$), except for *Thalassiosira* sp., which showed maximum values in the transitional growth phase.

Changes in carbohydrate concentrations were detected in all experiments, with increasing values with time, achieving the highest concentrations in stationary growth

Table 1. Cell volumes of five marine microalgae in different growth phases. Values are expressed in μm^3 and represent the mean of three replicates ± SD (n = 3) (each culture flask is considered as a replicate)*. The determination of mean values for each culture flask included measurements in at least 30 cells, randomly chosen.

Growth phases	*Bellerochea* sp.	*Chaetoceros* sp.	*Chlorella* sp.	*Rhodomonas* sp.	*Thalassiosira* sp.
	*	*	*	*	*
Exponential	46.2 ± 4.41^a	68.1 ± 16.2^a	83.7 ± 7.12^a	226 ± 34.9^a	1475 ± 76.1^a
Transitional	43.9 ± 2.36^a	45.1 ± 2.21^a	51.4 ± 2.85^b	179 ± 6.89^b	1579 ± 164^a
Stationary	32.0 ± 3.35^b	30.3 ± 3.89^b	53.7 ± 4.90^b	235 ± 14.9^a	1707 ± 56.2^b

*Mean values significantly different: *P<0.05, a>b; ** P<0.05, a>b>c. Identical superscript letters (a,a; b,b) or absence of letters indicate that mean values are not significantly different.

Table 2. Concentrations of hydrosoluble protein, total carbohydrate, total lipid and total ash of five microalgae in different growth phases. The values are expressed in pg . $cell^{-1}$ and represent the mean of three replicates ± SD (n = 3)*.

Growth phase/species	*Bellerochea* sp.	*Chaetoceros* sp.	*Chlorella* sp.	*Rhodomonas* sp.	*Thalassiosira* sp.
			Protein		
	*	*	*	*	**
Exponential	21.9 ± 2.68^a	18.0 ± 0.28^a	25.1 ± 2.00^a	69.6 ± 4.43^a	115 ± 1.56^b
Transitional	21.9 ± 2.38^a	12.9 ± 0.48^b	9.93 ± 1.84^b	52.6 ± 6.39^b	163 ± 23.6^a
Stationary	14.2 ± 1.84^b	10.5 ± 1.52^b	6.07 ± 1.14^b	44.9 ± 6.52^b	73.1 ± 2.60^c
			Carbohydrate		
	*	*	**	**	**
Exponential	1.49 ± 0.02^b	1.61 ± 0.15^a	2.46 ± 0.19^c	3.95 ± 0.23^c	15.0 ± 0.87^c
Transitional	1.82 ± 0.17^b	0.93 ± 0.09^b	5.85 ± 1.13^b	5.06 ± 0.29^b	26.4 ± 2.63^b
Stationary	3.01 ± 0.39^a	1.50 ± 0.09^a	7.09 ± 0.84^a	8.60 ± 0.48^a	51.4 ± 5.46^a
			Lipid		
	**	**	*	*	
Exponential	6.96 ± 0.14^c	17.5 ± 4.01^a	12.6 ± 1.66^a	17.6 ± 1.81^b	26.8 ± 3.52
Transitional	13.9 ± 1.65^a	6.49 ± 0.99^b	1.82 ± 0.45^b	23.2 ± 7.14^b	29.0 ± 5.95
Stationary	9.87 ± 0.79^b	3.73 ± 0.54^c	1.82 ± 0.21^b	39.2 ± 3.85^a	39.6 ± 7.18
			Ash		
		**	*	**	*
Exponential	2.32 ± 0.25	3.05 ± 0.75^a	4.53 ± 0.91^b	6.75 ± 0.96^a	8.10 ± 1.53^b
Transitional	2.32 ± 0.52	1.14 ± 0.08^c	4.64 ± 0.46^b	3.45 ± 0.63^c	11.9 ± 1.53^a
Stationary	2.12 ± 0.31	2.32 ± 0.18^b	7.54 ± 1.21^a	5.38 ± 0.35^b	10.4 ± 0.80^a

*Mean values significantly different: *p < 0.05, a > b; **p < 0.05, a > b > c. Identical superscript letters (a,a; b,b) or absence of letters indicate that mean values are not significantly different.

phase (Table 2) (0.001≤P< 0.05), except for *Chaetoceros* sp. (maximum values in transitional growth phase).

Variations of total lipid showed different trends for the species throughout growth (Table 2). *Chaetoceros* sp. and *Chlorella* sp. showed a decrease in lipid concentrations from exponential to stationary growth phase (0.0004≤P≤0.03), but a significant increase of lipid was detected in *Rhodomonas* sp. during growth (with maximum values in stationary growth phase). For *Bellerochea* sp. maximum lipid concentrations were measured in transitional growth phase (P=0.018), while *Thalassiosira* sp. did not show differences in mean values of lipids during the experiment (P=0.63). Ash content of *Bellerochea* sp. did not vary during growth (Table 2), but in *Chlorella* sp. and *Thalassiosira* sp.a significant increase was detected from exponential to stationary growth phase. On the other hand, *Rhodomonas Rhodomonas* sp. and *Chaetoceros* sp. showed maximum

Table 3. Concentrations of chlorophyll *a*, chlorophyll *b* / chlorophyll *c* and total carotenoid of five microalgae in different growth phases. The values are expressed in pg . cell^{-1} and represent the mean of three replicates ± SD (n = 3)*.

Species	Growth phases	Chlorophyll *a*	Chlorophyll *b*	Chlorophyll *c*	Total carotenoid
		**		*	**
	Exponential	0.56 ± 0.05^{b}	–	0.26 ± 0.02^{a}	0.23 ± 0.01^{b}
Bellerochea sp.	Transitional	0.93 ± 0.38^{a}	–	0.39 ± 0.16^{a}	0.42 ± 0.18^{a}
	Stationary	0.25 ± 0.03^{c}	–	0.11 ± 0.02^{b}	0.12 ± 0.02^{c}
		*		*	*
	Exponential	0.35 ± 0.04^{a}	–	0.09 ± 0.01^{a}	0.13 ± 0.02^{b}
Chaetoceros sp.	Transitional	0.40 ± 0.04^{a}	–	0.09 ± 0.01^{a}	0.19 ± 0.02^{a}
	Stationary	0.20 ± 0.01^{b}	–	0.04 ± 0.01^{b}	0.11 ± 0.01^{b}
		*	**		**
	Exponential	0.64 ± 0.11^{a}	0.41 ± 0.09^{a}	–	0.37 ± 0.05^{a}
Chlorella sp.	Transitional	0.12 ± 0.03^{b}	0.08 ± 0.02^{b}	–	0.10 ± 0.01^{b}
	Stationary	0.04 ± 0.01^{b}	0.02 ± 0.00^{c}	–	0.04 ± 0.02^{c}
		*		*	
	Exponential	1.38 ± 0.12^{a}	–	0.59 ± 0.08^{a}	0.99 ± 0.13
Rhodomonas sp.	Transitional	1.12 ± 0.19^{a}	–	0.47 ± 0.08^{a}	0.92 ± 0.11
	Stationary	0.49 ± 0.10^{b}	–	0.31 ± 0.06^{b}	0.94 ± 0.09
		**		**	*
	Exponential	4.18 ± 0.18^{b}	–	1.47 ± 0.06^{b}	1.70 ± 0.07^{b}
Thalassiosira sp.	Transitional	7.16 ± 0.74^{a}	–	2.57 ± 0.26^{a}	2.52 ± 0.26^{a}
	Stationary	2.06 ± 0.14^{c}	–	0.85 ± 0.07^{c}	1.54 ± 0.12^{b}

*Mean values significantly different: *P<0.05, a>b; ** P<0.05, a>b>c. Identical superscript letters (a,a; b,b) or absence of letters indicate that mean values are not significantly different.

ash content in exponential growth phase, with lower values in the other phases (P≤0.0004). *Chaetoceros* sp., *Rhodomonas* sp., *Thalassiosira* sp. and *Chlorella* sp. showed a decrease in chlorophyll *a* throughout growth (0.0001≤P≤0.027) (Table 3). For *Bellerochea* sp. maximum values for chlorophyll *a* were detected in transitional growth phase. Similar trends (as described for chlorophyll *a*) were found for chlorophyll *b* and total carotenoid (Table 3), except for total carotenoid in *Rhodomonas* sp., with no difference in values along time.

Fatty acids

All species showed high percentages of saturated fatty acids (SFA) (Table 4). Most species tended to show maximum SFA in exponential growth phase, except *Rhodomonas* sp., with similar values in the three growth phases. The FA 12:0 was the most abundant SFA (0.001≤P≤0.013), except for *Bellerochea* sp., with higher concentrations of 14:0 (P<0.01). The FA 16:0 has been observed in high concentrations in *Chlorella* sp. and *Chaetoceros* sp.

The microalgae tended to show concentrations of monounsaturated fatty acids (MFA) progressively higher throughout growth (Table 4), with maximum values in stationary growth phase, except for *Chlorella* sp. (maximum values in transitional growth phase). The three diatoms and *Rhodomonas* sp. showed remarkable increase of MFA from transitional to stationary growth phase. Sixteen-carbon MFA were the most abundant FAs in all species (p < 0.01), especially the isomer 16:1n-7 (palmitoleic acid). High concentrations of 18:1n-9 was also detected in *Chlorella* sp. and in some samples of *Chaetoceros* sp.

Variations in polyunsaturated fatty acids (PUFA) were detected in the microalgae (Table 4). Lower percentages of PUFA were found in the diatoms *Bellerochea* sp. and *Chaetoceros* sp. (P<0.05), with no variations during growth for *Bellerochea* sp. (P>0.21). Arachidonic acid – AA (20:4n-6) has not been detected in *Thalassiosira* sp., and it occurred in low concentrations in *Bellerochea* sp., *Chaetoceros* sp. and *Chlorella* sp., but an increase of AA Concentrations occurred during growth of *Rhodomonas* sp. All species showed the eicosapentaenoic acid – EPA (20:5n-3), but in *Chlorella* sp. this FA was detected

Table 4. Fatty acid composition of five marine microalgae in different growth phases. Data are given as percentage of total fatty acids and refer to the actual results obtained for three replicates ± SD (n = 3), without corrections for 100% recovery. SFA = saturated fatty acids. MFA = monounsaturated fatty acids. PUFA = polyunsaturated fatty acids. n.d. = not detected. * Samples of *Bellerochea* sp. in the exponential growth phase were lost.

Fatty acid	*Bellerochea* sp.		*Chaetoceros* sp.			*Chlorella* sp.			*Rhodomonas* sp.			*Thalassiosira* sp.		
	Transitional	Stationary	Exponential	Transitional	Stationary	Exponential	Transitional	Stationary	Exponential	Transitional	Stationary	Exponential	Transitional	Stationary
12:0	2.50 ± 0.14	1.22 ± 0.31	51.0 ± 0.01	29.2 ± 1.05	21.5 ± 6.49	42.7 ± 5.94	28.9 ± 13.0	22.5 ± 0.12	45.4 ± 5.33	28.8 ± 5.09	26.1 ± 0.07	52.6 ± 7.39	39.9 ± 2.21	21.0 ± 0.06
14:0	14.7 ± 1.82	15.9 ± 2.36	10.6 ± 0.04	6.71 ± 0.06	9.79 ± 0.01	2.14 ± 1.46	1.72 ± 0.26	1.77 ± 0.12	3.05 ± 0.79	6.80 ± 0.86	11.7 ± 0.13	1.24 ± 0.44	2.91 ± 0.55	15.5 ± 0.46
15:0	2.31 ± 0.04	1.07 ± 0.21	2.72 ± 0.02	1.83 ± 0.05	2.12 ± 0.26	n.d.	1.72 ± 0.84	1.59 ± 0.08	1.07 ± 0.05	1.39 ± 0.81	1.24 ± 0.02	1.13 ± 0.10	1.45 ± 0.06	1.68 ± 0.02
16:0	2.43 ± 0.25	6.77 ± 5.39	11.9 ± 0.01	9.85 ± 0.16	5.41 ± 1.12	3.16 ± 0.84	5.59 ± 2.04	10.7 ± 0.01	2.95 ± 1.23	5.33 ± 2.41	5.46 ± 0.13	1.42 ± 0.31	3.04 ± 0.04	1.88 ± 0.01
17:0	5.83 ± 1.15	3.50 ± 1.50	0.46 ± 0.01	n.d.	0.54 ± 0.11	n.d.	0.77 ± 0.00	4.54 ± 1.16	0.84 ± 0.35	0.74 ± 0.18	6.73 ± 0.01	0.65 ± 0.16	0.73 ± 0.64	n.d.
18:0	2.31 ± 0.78	1.43 ± 0.36	0.71 ± 0.01	0.41 ± 0.01	1.54 ± 0.50	1.19 ± 0.34	4.90 ± 1.58	6.23 ± 0.32	0.62 ± 0.14	1.07 ± 0.19	0.74 ± 0.01	2.61 ± 2.03	0.66 ± 0.01	0.43 ± 0.00
20:0	0.32 ± 0.55	0.22 ± 0.21	n.d.	n.d.	n.d.	n.d.	n.d.	0.14 ± 0.01	n.d.	n.d.	n.d.	n.d.	n.d.	n.d.
22:0	n.d.	0.25 ± 0.12	n.d.	0.34 ± 0.01	n.d.	n.d.	n.d.	0.44 ± 0.02	n.d.	n.d.	0.13 ± 0.00	n.d.	n.d.	n.d.
∑ SFA	30.4 ± 4.73	30.4 ± 10.5	77.4 ± 0.10	48.4 ± 1.34	40.9 ± 8.49	49.2 ± 8.58	43.6 ± 17.7	47.8 ± 1.81	53.9 ± 7.89	44.1 ± 9.54	52.0 ± 0.37	59.6 ± 10.4	48.7 ± 3.51	40.5 ± 0.52
16:1n-7	8.26 ± 0.07	23.2 ± 2.87	0.47 ± 0.03	3.12 ± 0.01	17.2 ± 3.29	n.d.	1.54 ± 0.38	1.53 ± 0.02	1.33 ± 0.19	1.90 ± 0.25	15.1 ± 0.16	1.33 ± 0.39	1.62 ± 0.02	21.6 ± 0.00
16:1n-9	0.90 ± 0.06	0.90 ± 0.25	0.43 ± 0.00	1.24 ± 0.03	0.77 ± 0.40	n.d.	1.57 ± 0.47	0.34 ± 0.01	0.73 ± 0.42	0.63 ± 0.02	1.73 ± 0.01	0.78 ± 0.17	n.d.	2.53 ± 0.01
18:1n-9	0.39 ± 0.02	0.29 ± 0.05	1.06 ± 0.03	4.50 ± 0.08	1.07 ± 0.65	5.19 ± 0.86	15.4 ± 4.74	1.10 ± 0.01	1.08 ± 0.04	1.66 ± 0.81	0.53 ± 0.06	0.72 ± 0.18	0.83 ± 0.01	0.42 ± 0.00
18:1n-7	0.17 ± 0.18	0.27 ± 0.16	0.45 ± 0.01	0.35 ± 0.00	0.39 ± 0.08	n.d.	0.93 ± 0.00	0.85 ± 0.01	n.d.	n.d.	0.32 ± 0.01	n.d.	n.d.	n.d.
20:1n7+9	0.09 ± 0.15	0.30 ± 0.06	0.40 ± 0.00	n.d.	n.d.	n.d.	n.d.	0.13 ± 0.00	0.95 ± 0.28	0.62 ± 0.00	0.24 ± 0.01	n.d.	0.93 ± 0.05	n.d.
22:1	n.d.	0.17 ± 0.08	n.d.	n.d.	n.d.	n.d.	n.d.	0.13 ± 0.02	n.d.	n.d.	0.12 ± 0.00	n.d.	n.d.	n.d.
∑ MFA	9.81 ± 0.48	24.8 ± 3.47	2.81 ± 0.07	9.21 ± 0.12	19.4 ± 4.42	5.19 ± 0.86	19.5 ± 5.59	4.09 ± 0.07	4.09 ± 0.93	4.81 ± 1.08	18.0 ± 0.25	2.83 ± 0.74	3.38 ± 0.08	24.5 ± 0.01
18:2n-6	0.65 ± 0.36	0.28 ± 0.04	0.66 ± 0.01	0.42 ± 0.01	0.41 ± 0.07	n.d.	2.25 ± 0.75	7.48 ± 0.35	0.90 ± 0.38	3.48 ± 1.28	0.46 ± 0.08	0.56 ± 0.06	0.66 ± 0.00	0.41 ± 0.01
18:3n-6	n.d.	n.d.	n.d.	0.26 ± 0.01	0.71 ± 0.07	n.d.	0.70 ± 0.00	0.84 ± 0.00	n.d.	n.d.	n.d.	n.d.	n.d.	n.d.
18:3n-3	5.40 ± 0.18	5.23 ± 1.57	1.15 ± 0.01	1.06 ± 0.01	4.81 ± 0.16	3.05 ± 1.25	7.45 ± 2.90	8.92 ± 0.06	6.98 ± 0.84	7.07 ± 3.84	0.86 ± 0.01	7.35 ± 1.19	4.93 ± 0.13	0.52 ± 0.03
18:4n-3	1.03 ± 0.60	0.23 ± 0.13	0.50 ± 0.01	1.14 ± 0.02	0.63 ± 0.22	n.d.	7.60 ± 2.95	5.56 ± 0.01	4.20 ± 0.58	13.6 ± 3.43	4.72 ± 0.01	1.00 ± 0.27	1.81 ± 0.04	3.90 ± 0.60
20:2n-6	6.18 ± 0.82	3.33 ± 0.42	n.d.	3.55 ± 0.03	4.19 ± 0.77	8.70 ± 1.95	5.93 ± 1.53	3.47 ± 0.01	4.92 ± 0.56	2.75 ± 0.73	1.24 ± 0.00	6.14 ± 0.68	3.83 ± 0.11	0.98 ± 0.01
20:4n-6	0.48 ± 0.09	0.74 ± 0.08	4.87 ± 0.04	0.68 ± 0.03	n.d.	n.d.	n.d.	0.14 ± 0.01	n.d.	n.d.	0.33 ± 0.00	n.d.	n.d.	n.d.
20:3n-3	0.31 ± 0.02	0.11 ± 0.05	n.d.	n.d.	n.d.	n.d.	n.d.	n.d.	0.29 ± 0.10	0.54 ± 0.15	0.37 ± 0.03	0.47 ± 0.00	0.64 ± 0.01	n.d.
20:4n-3	0.45 ± 0.09	0.25 ± 0.14	0.41 ± 0.01	0.36 ± 0.01	0.58 ± 0.18	n.d.	n.d.	0.19 ± 0.00	0.69 ± 0.16	1.83 ± 0.67	2.36 ± 0.01	0.64 ± 0.00	1.35 ± 0.03	3.98 ± 0.01
20:5n-3	1.55 ± 0.29	1.90 ± 0.51	1.53 ± 0.02	4.01 ± 0.02	1.95 ± 0.43	n.d.	n.d.	0.66 ± 0.01	0.93 ± 0.06	3.78 ± 1.5	1.54 ± 0.00	1.05 ± 0.12	2.81 ± 0.06	0.55 ± 0.01
22:4n-9	1.37 ± 0.06	1.10 ± 0.12	1.49 ± 0.02	1.22 ± 0.01	1.08 ± 0.21	3.36 ± 2.20	2.30 ± 1.00	1.36 ± 0.03	1.28 ± 0.06	1.25 ± 0.52	1.33 ± 0.01	1.42 ± 0.02	2.36 ± 0.12	n.d.
22:4n-6	n.d.	n.d.	n.d.	n.d.	n.d.	n.d.	n.d.	0.16 ± 0.03	n.d.	n.d.	0.91 ± 0.00	n.d.	n.d.	n.d.
22:5n-6	n.d.	n.d.	n.d.	n.d.	n.d.	n.d.	0.71 ± 0.00	0.35 ± 0.04	n.d.	n.d.	0.34 ± 0.00	n.d.	n.d.	n.d.
22:5n-3	0.10 ± 0.18	0.53 ± 0.38	n.d.	0.49 ± 0.08	n.d.	n.d.	n.d.	0.17 ± 0.00	n.d.	n.d.	0.18 ± 0.00	n.d.	n.d.	n.d.
22:6n-3	1.15 ± 0.15	1.75 ± 0.60	1.09 ± 0.01	0.49 ± 0.03	2.46 ± 0.42	n.d.	n.d.	0.44 ± 0.11	2.06 ± 0.97	2.61 ± 1.46	1.78 ± 0.00	2.91 ± 1.76	2.22 ± 0.11	0.42 ± 0.00
∑ PUFA	18.7 ± 2.84	15.5 ± 4.04	10.6 ± 0.12	13.2 ± 0.23	16.8 ± 2.53	15.1 ± 5.40	26.9 ± 9.13	29.7 ± 0.66	22.3 ± 3.71	36.9 ± 13.6	16.4 ± 0.15	21.5 ± 4.10	20.6 ± 0.61	10.8 ± 0.67

in low percentage in stationary growth phase only. EPA tended to occur in low percentages in all species, typically lower than 2%, except for some samples of *Chaetoceros* sp., *Rhodomonas* sp. and *Thalassiosira* sp. The FA docoeicosapentaenoic acid – DHA (22:6n-3) was found in all species in low percentages. Only *Rhodomonas* sp., *Chaetoceros* sp. and *Thalassiosira* sp. showed more than 2% of DHA in some observations. Other PUFA showed percentages relatively high in some species, such as 18:3n-3 (*Chlorella* sp.), 18:4n-3 (*Rhodomonas* sp.) and 22:4n-9 (*Chlorella* sp.). The FA 20:2n-6 presented high measurements for at least one observation for all species.

DISCUSSION

Growth, final yield and cell volume

Small-sized species (*Chlorella* sp., *Chaetoceros* sp. and *Bellerochea* sp.) showed higher growth rates in exponential growth phase (data not shown), whereas greater species (*Rhodomonas* sp. and *Thalassiosira* sp.) exhibited lower rates. These trends are in accordance with many studies (Lourenço et al., 2002; Lavín and Lourenço, 2005; Borges-Campos et al., 2010), and show the critical importance of surface : volume relationship for phytoplankton. In this study, *Rhodomonas* sp. and *Thalassiosira* sp., species of larger biovolume, showed an increasing biovolume throughout growth. These characteristics are supported by findings of Hellebust (1976), who points that the increase in mean cell volume during growth reflects the accumulation of reserve products (especially long-chain polysaccharides) in the stationary growth phase, when growth conditions are unfavorable. Silva et al. (2009) observed the same trend for the cryptomonad *Rhodomonas* sp. in a study of nitrogen reduction in the culture media. However, for *Bellerochea* sp., *Chaetoceros* sp. and *Chlorella* sp. a decrease of cell volume occurred throughout the experiments. The same trend was reported for *Tetraselmis gracilis* by Lourenço et al. (1997), and it was interpreted as a result of a progressive reduction of resources (nutrients) available to the cells, since each individual could acquire a smaller amount of nutrients, allowing the biosynthesis of a small cellular mass. Such trend could also be related to the changes in the chemical profile of the species.

Considering final cell yield, species of smaller cell volumes (*Chlorella* sp., *Chaetoceros* sp. and *Bellerochea* sp.) showed higher values, while the larger cells (*Rhodomonas* sp. and *Thalassiosira* sp.) exhibited lower final yields, corroborating the studies of Borges-Campos et al. (2010) with *Synechococcus subsalsus* and *Chlorella minutissima*. These results were expected, especially taking into account the previous discussion for growth rates. However, among the three smaller species the behavior of *Chlorella* sp. compared to *Bellerochea* sp. and *Chaetoceros* sp. was remarkable. Even showing the larger cell volume among the three smaller species, *Chlorella* sp. achieved a final yield significantly higher than the other two species, suggesting a high efficiency of nutrient conversion into organic matter, according to Kaplan et al. (1986). Moreover, a possible cause for difference in final cell yield might be the demand for silicon. N and Si concentrations in Conway medium are 1,178 e 377 µM, respectively, with an atomic ratio N:Si of ca. 3.12. Some need an atomic ratio N:Si = 2.0, or even lower to get an optimum growth and cell yield (Brzezinski, 1985). As this hypothesis could not have been tested in this study, its consideration is speculative, but plausible. If *Bellerochea* sp. and *Chaetoceros* sp. had a high demand for Si, it could affect negatively their growth in the experiments, with smaller production of cells and biomass.

Chemical composition of microalgae

The evaluation of protein per cell revealed maximum protein content in the exponential growth phase, decreasing throughout growth. Fernández-Reiriz et al. (1989) found the same trend for the prasinophycean *Tetraselmis suecica*, but the diatoms *Phaeodactylum tricornutum* and *Chaetoceros calcitrans* did not change their protein content over growth, such as other diatoms tested by Knuckey et al. (2002). Fernández-Reiriz et al. (1989) also reported a peak in protein content of the cryptomonad *Rhodomonas* in transitional growth phase as well as Silva et al. (2009). Aidar et al. (1994) also reported higher protein concentrations per cell in exponential growth phase of the prasinophycean *T. gracilis* and the diatom *Cyclotella caspia*, decreasing protein per cell in the subsequent measurements over growth. Seixas et al. (2008, 2009) found a considerable high protein percentage (62 and 55%, d.w.), respectively, in *Rhodomonas lens* obtained in semi-continuous culture. All species showed a decreasing concentration of chlorophyll. Lourenço et al. (2004) reported this same trend for the most of the species cultivated with Conway media. Chlorophyll concentration is strongly affected by nitrogen availability, with variations similar to those described for protein (Fábregas et al., 1987).

For carbohydrates different trends were found, compared to protein. For all species, except *Chaetoceros* sp., increasing concentrations of carbohydrates were found throughout growth. This finding is in accordance with many other studies (Fernández-Reiriz et al., 1989; Lourenço et al., 1997; Knuckey et al., 2002). According to Enright et al. (1986a), when the rate of cell division in a phytoplankton culture is limited by nutrients, cells alter their metabolism and convert energy to produce reserve substances. Typically, at the end of a batch culture, with a frequent nutrient limitation, the synthesis of both protein and chlorophyll decreases, and the concentration of lipid and/or carbohydrate increases. There is a coupling between protein and carbohydrates in algal cells, which also reflects the budget of carbon and nitrogen available to the cells (Geider et al., 1993; Turpin, 1991).

The microalgae exhibited changes in ash content throughout growth, except *Bellerochea* sp. (with no change over growth). Knuckey et al. (2002) found the same behavior with the diatom *Nitszchia paleacea*, which did not change its chemical profile during different growth phases. On the other hand, *Chlorella* sp. and *Thalassiosira* sp. exhibited an increase in ash content during growth. Fernández-Reiriz et al. (1989) found the same trend for the prasinophycean *T. suecica*, but for the diatoms *P. tricornutum* and *C. calcitrans* and the cryptomonad *Rhodomonas* sp. lower ash content was reported in the stationary growth phase. In the current study, a lower ash content for the cryptomonad *Rhodomonas* sp. was reported in transitional growth phase.

Changes in lipid concentration of *Chlorella* sp. and *Rhodomonas* sp. are similar to those reported by Fernández-Reiriz et al. (1989) with the prasinophycen *T. suecica* and with the cryptomonad *Rhodomonas* sp. Conversely, for those species a decreasing concentration of lipid per cell has been reported during growth. For the diatoms *C. calcitrans* and *P. tricornutum*, Fernández-Reiriz et al. (1989) found increasing concentrations of lipid throughout growth. The same was reported by Knuckey et al. (2002), who found an increase in lipid concentration in eight out of ten diatoms over growth. Knuckey et al. (2002) also reported that only the diatoms *Entomoneis punctulata* and *Papiliocellulus simplex* showed decreasing concentrations of lipid per cell over growth.

PUFA concentrations of *Bellerochea* sp., *Chaetoceros* sp., *Rhodomonas* sp. and *Thalassiosira* sp. showed lower values when compared to the diatoms and cryptomonads studied by Renaud et al. (1999) and to the cryptomonads studied by Seixas et al. (2009). *Chaetoceros* sp. showed higher values of PUFA compared to the results of Chen (2012) obtained with *Chaetoceros muelleri.* Knuckey et al. (2002) reported the same trends for MFA and PUFA described here for another *Thalassiosira* species, but changes in SFA content were opposite to those described in the present study. Knuckey et al. (2002) also found in *E. punctulata* exactly the same variations for SFA, MFA and PUFA reported here for *Bellerochea* sp. Chen (2012) cultivated 12 species of marine diatoms in three different environments and concluded that these conditions altered the fatty acid composition of the diatom species, emphasizing that production and storage of lipids is species-specific.

Fatty acids play a key role in aquaculture, since they are very important for the growth of marine organisms. Because of this, some microalgal species may be potentially indicated as useful for feeding marine animals, while others show to be unsuitable as food-species, if the essential PUFA lack in them. The microalgae may have superior lipid stability compared with traditional PUFAs, because they are naturally rich in antioxidant carotenoids and vitamins and because lipids are bioencapsulated by the algal cell wall (Patil et al., 2007). Regarding FA content, *Bellerochea* sp. and *Chaetoceros* sp., for instance, could be considered promising species, since they possess the three essential FA (AA, EPA e DHA) in all growth phases. *Chlorella* sp. should not be indicated for use in marine aquaculture as a monoalgal diet, since it is a poor source of PUFA, showing the three essential FA in low concentrations in stationary growth phase only. Concerns for the use of *Rhodomonas* sp. are essentially the same described for *Chlorella* sp. The cryptomonad exhibits the three essential FA in low concentrations.

Differences among the current study and others in the literature are evident. A large part of the data on chemical composition of marine microalgae is obtained using f/2 culture medium (Guillard, 1975), such as Enright et al. (1986a, b), Brown et al. (1998), McCausland et al. (1999), Renaud et al. (1999), Knuckey et al. (2002), Lafarga-De la Cruz et al. (2006), among others. However, in the current study, the microalgae were cultured with Conway culture medium (Walne, 1966), a richer culture medium. The f/2 medium contains 880 μM $NaNO_3$ and 36 μM NaH_2PO_4, while Conway medium contains 1,176 μM $NaNO_3$ and 128 μM NaH_2PO_4 (Walne, 1966). Despite changes in the chemical composition of microalgae reflect genetic differences, such variations are also influenced by culture conditions. Chemical composition of microalgae may change largely when the species are cultured under contrasting conditions, as well as in different growth phases (Brown et al., 1997). Thus, the detection of high concentration of protein and low concentration of carbohydrate in the experiments (even with the chlorophycean *Chlorella* sp.) may be an effect of high nitrogen concentration and other components of the Conway culture medium.

Conclusion

Data on the chemical composition of microalgal strains useful in tropical and subtropical mariculture are scarce in the literature (McCausland et al., 1999; Martínez-Fernández et al., 2006; Silva et al., 2009). This study provides basic information on the chemical composition of five native Brazilian microalgae. Possible uses of this knowledge are wide and could stimulate other studies. Besides the information on the chemical profile of the species (a tremendous gap in Brazil), the results may be useful for the selection of suitable native species in mariculture, to diminish the current dependence of Brazilian producers for exotic microalgal strains.

All strains tested in this study seem to be potentially useful as food-species in mariculture, and further studies on them in larger systems are required, as well as studies focusing the feeding of animals, such as oyster larvae and post-larvae. Since microalgae may be limited by one or more essential nutrients, a mixed algal cultures supplies a better equilibrium of nutrient properties, contributing for a better success in mariculture (Brown et al., 1997; Rico-Villa et al., 2006) probably because their combined nutrient is more likely to meet to nutritional requirements of the target species (Cerón-Ortiz et al., 2009). This is what could be interpreted for *Chlorella* sp., which presented a low FA content, but high percentages of other substances, such as carbohydrates. Thus, data on biovolume, cell density and chemical composition of microalgae (used all together) may provide important subsidies for choosing suitable diets for larvae of marine organisms, contributing to decrease their mortality.

We have been performing further studies with the same five native microalgae to evaluate their use in mariculture. In one of them, the conservation of them as live concentrated pastes was tested. In a second study, the micro algae were tested in trials as food-species for mangrove oyster larvae (*Crassotrea rhizophora*). These promising new results will be published soon.

ACKNOWLEDGEMENTS

This work was supported by the National Council for the Development of Science and Technology (CNPq). The authors G.S.C., R.R.M. and S.O.L. thank CNPq for their research fellowships. Thanks are due to Dr. Aguinaldo N. Marques Júnior and Dr. Emmanoel Vieira da Silva Filho (Universidade Federal Fluminense) for offering us laboratory facilities.

REFERENCES

Aidar E, Gianesella-Galvão SMF, Sigaud TCS, Asano CS, Liang TH, Rezende KRV, Oishi MK, Aranha FJ, Milani GM, Sandes MAL (1994). Effects of light quality on growth, biochemical composition and photosynthetic production in *Cyclotella caspia* Grunow and *Tetraselmis gracilis* (Kylin) Butcher. J. Exp. Mar. Biol. Ecol. 180:175-187.

AOAC - Association of Official Analytical Chemists (1990). Official Methods of Analysis. 15th ed. Washington, DC.

Araújo SC, Garcia VMT (2005). Growth and biochemical composition of the diatom *Chaetoceros* cf. *wighamii* brightwell under different temperature, salinity and carbon dioxide levels. I. Protein, carbohydrates and lipids. Aquaculture 246:405-412.

Assad LT, Bursztyn M (2000). Aqüicultura Sustentável. In: Valenti WC, Poli CR, Pereira JA, Borghetti JR (eds). Aqüicultura no Brasil: Bases para um Desenvolvimento Sustentável. CNPq / Ministério da Ciência e Tecnologia, Brasília, pp.33-72.

Banerjee S, Hew WE, Khatoon H, Shariff M, Yusoff FM (2011). Growth and proximate composition of tropical marine *Chaetoceros calcitrans* and *Nannochloropsis oculata* cultured outdoors and under laboratory conditions. Afr. J. Biotechnol. 10:1375-1383.

Barbarino E, Lourenço SO (2005). An evaluation of methodologies for extraction and quantification of protein of marine macro- and microalgae. J. Appl. Phycol. 17:447-460.

Ben-Amotz A, Fishler R, Schneller A (1987). Chemical composition of dietary species of marine unicellular algae and rotifers with emphasis on fatty acids. Mar. Biol. 95:31-36.

Borges-Campos V, Barbarino E, Lourenço SO (2010). Crescimento e composição química de dez espécies de microalgas marinhas em cultivos estanques. Cienc. Rural. 40:339-347.

Brown MR, Jeffrey SW, Volkman JK, Dunstan GA (1997). Nutritional properts of microalgae for mariculture. Aquaculture 151:315-331.

Brown MR, McCausland MA, Kowalski K (1998). The nutritional value of four Australian microalgal strains fed to Pacific oyster *Crassostrea gigas* spat. Aquaculture 165:281-293.

Brzezinski MA (1985). The Si-C-N ratio of marine diatoms - interspecific variability and the effect of some environmental variables. J. Phycol. 21:347-357.

Carić M, Sanko-Njire J, Skaramuca B (1993). Dietary effects of different feeds on the biochemical composition of the rotifer (*Brachionus plicatilis* Muller). Aquaculture 110:141-150.

Cerón-Ortiz AN, Cordero B, Arredondo-Vega BO, Voltolina D (2009). Effect of algal diet and temperature on survival, growth and biochemical composition of spat of the lion's paw scallop *Nodipecten subnodosus*. Aquaculture 298:64-69.

Chen YC (2012). The biomass and total lipid content and composition of twelve species of marine diatoms cultured under various environments. Food Chem. 131:211-219.

DuBois M, Gilles KA, Hamilton JK, Rebers PA, Smith F (1956). Colorimetric method for the determination of sugars and related substances. Anal. Chem. 18:350-356.

Durmaz Y, Donato M, Monteiro M, Gouveia L, Nunes ML, Gama Pereira T, Gökpmar S, Bandarra NM (2009). Effect of temperature on α-tocopherol, fatty acid profile, and pigments of *Diacronema vlkianum* (Haptophyceae). Aquacult. Int. 17:391-399.

Ehteshami F, Christianus A, Rameshi H, Harmin SA, Saad CR (2011). Proximate and fatty acid composition of the gonads of wild versus hatchery-conditioned *Pinctada margaritifera* broodstock. Aquacult. Nutr. 17:e675-e682.

Enright CT, Newkirk GF, Craigie JS, Castell JD (1986a). Evaluation of phytoplankton as diets for juvenile *Ostrea edulis* L. J. Exp. Mar. Biol. Ecol. 96:1-13.

Enright CT, Newkirk GF, Craigie JS, CAstell JD (1986b). Growth of juvenile *Ostrea edulis* L. fed *Chaetoceros gracilis* Schutt. J. Exp. Mar. Biol. Ecol. 96:15-26.

Fábregas J, Toribio L, Abalde J, Cabezas B, Herrero C (1987). Approach to biomass production of the marine microalga *Tetraselmis suecica* (Kylin) Butch using common garden fertilizer and soil extract as cheap nutrient supply in batch cultures. Aquacult. Eng. 6:141-150.

FAO-Food and Agriculture Organization of the United Nations (2010). The state of world fisheries and aquaculture 2010. FAO Fisheries and Aquaculture Department, Rome, p.197.

Fernández-Reiriz MJ, Perez-Camacho A, Ferreiro MJ, Blanco J, Planas M, Campos MJ, Labarta U (1989). Biomass production and variation in the biochemical profile (total protein, carbohydrates, RNA, lipids and fatty acids) of seven marine microalgae. Aquaculture 83:17-37.

Folch J, Lees M, Sloane-Stanley GM (1957). A simple method for the isolation and purification of total lipids from animal tissues. J. Biol. Chem. 226:497-509.

Gaxiola G, Gallardo P, Simôes N (2010). A red shrimp, *Farfantepenaeus brasiliensis* (Latreille, 1817), larvae feeding regime based on live food. J. World Aquacult. Soc. 41:402-410.

Geider R, La Roche J, Greene R, Olaizola M (1993). Response of the photosynthetic apparatus of *Phaeodactylum tricornutum* (Bacillariophyceae) to nitrate, phosphate, or iron starvation. J. Phycol. 29:755-766.

Gouda R, Kenchington E, Hatcher B, Vercaemer B (2006). Effects of locally-isolated micro-phytoplankton diets on growth and survival of sea scallop (*Placopecten magellanicus*) larvae. Aquaculture 259:169-180.

Gouveia L, Batista AP, Sousa I, Raymundo A, Bandarra NM (2008). Microalgae in novel foods products. In: Food Chemistry Research Developments (ed. KN Papadopoulos). Nova Science Publishers, pp. 75-111.

Guillard RRL (1975). Culture of phytoplankton for feeding marine invertebrates. In: Culture of Marine Invertebrate Animals (eds. W.L. Smith, M.H. Chanley). Plenum Publishing Corporation, pp. 26-60.

Hashimoto T, Hyodoh K, Hirose T, Nishikawa S, Katano T, Nakano S-i (2008). Evaluation of three phytoplankton species as food for the pearl oyster *Pinctada fucata*. Aquacult. Int. 16:309-318.

Hellebust JA (1976). Effect of salinity on photosynthesis and mannitol synthesis in the green flagellate *Platymonas suecica*. Can. J. Bot. 54:1735-1741.

Hillebrand H, Dürselen C-D, Kirschtel D (1999). Biovolume calculation for pelagic and benthic microalgae. J. Phycol. 35:403-424.

Huerlimann R, de Nyz R, Heimann K (2010). Growth, lipid content, productivity, and fatty acid composition of tropical microalgae for scale-up production. Biotechnol. Bioeng. 107:245-257.

Kaplan D, Richmond AE, Dubinsky Z (1986). Algal nutrition. In: Handbook of Microlgal Mass Culture. (ed. A. Richmond) CRC Press, pp.147-197.

Knuckey RM, Brown MR, Barrett SM, Hallegraeff GM (2002). Isolation of new nanoplanktonic diatom strains and their evaluation as diets for juvenile Pacific oysters (*Crassostrea gigas*). Aquaculture. 211:253-274.

Lafarga-De la Cruz F, Valenzuela-Espinoza E, Millán-Núñez R, Trees CC, Santamaría-del-Ángel E, Núñez-Cebrero F (2006). Nutrient uptake, chlorophyll *a* and carbon fixation by *Rhodomonas* sp. (Cryptophyceae) cultured at different irradiance and nutrient concentrations. Aquacult. Eng. 35:51-60.

Laing I, Child AR, Janke A (1990). Nutritional value of dried algae diets for larval of Manila clam (*Tapes philippinarum*). J. Mar. Biol. Assoc. U.K. 70:1-12.

Laing I, Psimopoulous A (1998). Hatchery cultivation of king scallop (*Pecten maximus*) spat with cultured and bloomed algal diets. Aquaculture 169:55-68.

Lavín PL, Lourenço SO (2005). An evaluation of the accumulation of the intracellular inorganic nitrogen pools by marine microalgae in batch cultures. Braz. J. Oceanogr. 53:55-67.

Leonardos N, Lucas IAN (2000). The nutritional value of algae grown under different culture conditions for *Mytilus edulis* L. larvae. Aquaculture. 182:301-315.

Lora-Vilchis MC, Maeda-Martinez NA (1997). Ingestion and digestion index of catarina scallop *Argopecten ventricosus-circularis*, Sowerby II, 1842, veliger larvae with ten microalgae species. Aquacult. Res. 28:905-910.

Lorenzen CJ (1967). Determination of chlorophyll and phaeophytin: spectrophotometric equations. Limnol. Oceanogr. 12:343-346.

Lourenço SO, Barbarino E, Bispo MGS, Borges DA, Coelho- Gomes C, Lavín PL, Santos F (2008). Effects of light intensity on growth, inorganic nitrogen storage, and gross chemical composition of four marine microalgae in batch cultures. XI Congresso Brasileiro de Ficologia e Simpósio Latino-Americano sobre Algas Nocivas. Museu Nacional. Série Livros 30:203-214.

Lourenço SO, Barbarino E, Lavín PL, Lanfer Marquez UM, Aidar E (2004). Distribution of intracellular nitrogen in marine microalgae: Calculation of new nitrogen-to-protein conversion factors. Eur. J. Phycol. 39:17-32.

Lourenço SO, Barbarino E, Mancini-Filho J, Schinke KP (2002). Effects of different nitrogen sources on growth and biochemical profile of ten marine microalgae under batch cultures. Phycologia 41:158-168.

Lourenço SO, Lanfer Marquez UM, Mancini-Filho J, Barbarino E, Aidar E (1997). Changes in biochemical profile of *Tetraselmis gracilis*. I. Comparison of two culture media. Aquaculture. 148:153-168.

Lowry OH, Rosebrough NJ, Farr AL, Randall RJ (1951). Protein measurement with the folin phenol reagent. J. Biol. Chem. 193:265-275.

Martínez-Fernández E, Acosta-Salmón H, Southgate PC (2006). The nutritional value of seven species of tropical microalgae for black-lip pearl oyster (*Pinctada margaritifera*, L.) larvae. Aquaculture 257:491-503.

McCausland MA, Brown MR, Barrett SM (1999). Evaluation of live microalgae and microalgal pastes as supplementary food for juvenile Pacific oysters (*Crassostrea gigas*). Aquaculture 174:323-342.

Metcalfe AP, Schmitz AA (1961). The rapid preparation of fatty acids for gas chromatographic analysis. Anal. Chem. 33:363-364.

Myklestad S, Haug A (1972). Production of carbohydrates by the marine *Chaetoceros affinis* var. *willei* (Gran.) Hustedt. I. Effect of the concentration of nutrients in the culture medium. J. Exp. Mar. Biol. Ecol. 9:125-136.

Parsons TR, Maita Y, Lalli CM (1984). A Manual of Chemical and Biological Methods for Seawater Analysis. Pergamon Press, Oxford.

Patil V, Källqvist T, Olsen E, Vogt G, Gislerød HR (2007). Fatty acid composition of 12 microalgae for possible use in aquaculture feed. Aquacult. Int. 15:1-9.

Pettersen AK, Turchini GM, Jahangard S, Ingram BA, Sherman CDH (2010). Effects of different dietary microalgae on survival, growth, settlement and fatty acid composition of blue mussel (*Mytilus galloprovincialis*) larvae. Aquaculture 309:115-124.

Ponis E, Parise G, Chini Zittelli G, Lavista F, Robert R, Tredici MR (2008). *Pavlova lutheri*: Production, preservation and use as food for *Crassostrea gigas* larvae. Aquaculture 282:97-103.

Ponis E, Probert I, Véron B, Mathieu M, Robert R (2006). New microalgae for the Pacific oyster *Crassostrea gigas* larvae. Aquaculture 253:618-627.

Reitan KI (2011). Digestion of lipids and carbohydrates from microalgae (*Chaetoceros muelleri* Lemmermann and *Isochrysis* aff. *galbana* clone T-ISO) in juvenile scallops (*Pecten maximus* L.). Aquac. Res. doi: 10.1111/j.1365-2109.2010.02745.x

Renaud SM, Thinh L-V, Parry DL (1999). The gross chemical composition and fatty acid composition of 18 species of tropical Australian microalgae for possible use in mariculture. Aquaculture 170:147-159.

Rico-Villa B, Le Coz JR, Mingant C, Robert R (2006). Influence of phytoplankton diet mixtures on microalgae consumption, larval development and settlement of the Pacific oyster *Crassostrea gigas* (Thunberg). Aquaculture 256:377-388.

Rivero-Rodríguez S, Beaumont AR, Lora-Vilchis MC (2007). The effect of microalgal diets on growth, biochemical composition, and fatty acid profile of *Crassostrea corteziensis* (Hertlein) juveniles. Aquaculture 263:199-210.

Seixas P, Rey-Méndez M, Valente LMP, Otero A (2008). Producing juvenile *Artemia* as prey for *Octopus vulgaris* paralarvae with different microalgal species of controlled biochemical composition. Aquaculture 283:83-91.

Seixas P, Coutinho P, Ferreira M, Otero A (2009). Nutritional value of the cryptophyte *Rhodomonas lens* for *Artemia* sp. J. Exp. Mar. Biol. Ecol. 381:1-9.

Silva AF, Lourenço SO, Chaloub RM (2009). Effects of nitrogen starvation on the photosynthetic physiology of a tropical marine microalgae *Rhodomonas* sp. (Cryptophyceae). Aquacult. Bot. 91:291-297.

Southgate PC, Beer AC, Duncan PF, Tamburri R (1998). Assessment of the nutritional value of three species of tropical microalgae, dried *Tetraselmis* and a yeast-based diet for larvae of the blacklip pearl oyster, *Pinctada margaritifera* (L.). Aquaculture 162:247-257.

Spolaore P, Joannis-Cassan C, Duran E, Isambert A (2006). Commercial applications of microalgae. J. Biosci. Bioeng. 101:87-96.

Tang B, Liu B, Wang G, Zhang T, Xiang J (2006). Effects of various algal diets and starvation on larval growth and survival of *Meretrix meretrix*. Aquaculture 254:526-533.

Thompson PA, Guo M-x, Harrison PJ (1996). Nutritional value of diets that vary in fatty acid composition for larval Pacific oysters (*Crassostrea gigas*). Aquaculture 143:379-391.

Turpin DH (1991). Effects of inorganic N availability on algal photosynthesis and carbon metabolism. J. Phycol. 27:14-20.

Uriarte I, Farías A, Hawkins AJS, Bayne BL (1993). Cell characteristics and biochemical composition of *Dunaliella primolecta* Butcher conditioned at different concentrations of dissolved nitrogen. J. Appl. Phycol. 5:447-453.

Walne PR (1966). Experiments in the large scale culture of the larvae of *Ostrea edulis*. Fishery Invest., London 25:1-53.

Zar JH (1996). Biostatistical analysis, 3rd ed. Prentice Hall, Inc., Upper Saddle River.

Bioactivity of venom extracted from the sea anemone *Anthopleura asiatica* (Cnidaria: Anthozoa): Toxicity and Histopathological studies

Ramkumar S.[1]*, Arun Sudhagar S.[2] and Venkateshvaran K.[1]

[1]Fisheries Resources, Harvest and Post Harvest Division, Central Institute of Fisheries Education, Mumbai, India.
[2]Aquatic Environment and Health Management Division, Central Institute of Fisheries Education, Mumbai, India.

The bioactivity of the venom from a locally available sea anemone, *Anthopleura asiatica* collected from the Mumbai coast was studied. The crude venom from sixty sea anemones (Sixty numbers) was extracted in aqueous medium. The protein content of the crude venom was 4.3459±0.027 mg/ml. The crude venom was found to be lethal at 1ml when injected intra-peritoneal to kasuali strain male albino mice (20±2 g). The crude venom was partially purified by anion exchange chromatography using a step-wise gradient of 0.1-1.0M NaCl and 10 fractions each of 15 ml, F1 – F10 were collected. Fractions F8, F9 and F10 exhibited lethality to mice, upon envenomation. The symptoms of toxicity observed in the mice indicated that the venom affected the central nervous, cardiovascular and urinary systems. Histopathological study revealed accumulation of polymorphic nuclear cells with necrosis in brain, hemolysis in heart, occlusion with hemolysed blood in kidney and necrosis, vacuolation with pleomorphic nuclear material and hemolysis in liver.

Key words: Sea anemone, venom, toxicity, histopathology.

INTRODUCTION

The sea anemones which are exclusively marine anthozoan invertebrates possess specialized stinging cells, the nematocysts in their body surface (Alsen et al., 1982; Bosmans and Tytgat., 2007). These cells contain venom which is used to paralyze the prey, to defense against predators and to fight territorial disputes (Lotan et al., 1996; Yanagihara et al., 2002). This venom contains variety of active compounds, including potential toxins affecting voltage-gated Na^+ and K^+ channels, acid-sensing ion channels, pore-forming toxins (actinoporins) and protease inhibitors (Belmonte et al., 1994; Beress et al., 1975; Diochot et al., 2003). Of 9000 species of cnidarins, some 70 species have been reported to be toxic to man (Picken and Skaer., 1996). A few species of sea anemones possess highly toxic venoms hazardous for humans, among which *Phyllodiscus semoni* venom caused unexplained acute renal failure in the victim (Mizuno et al., 2000) and in most of the cases venom causes local inflammations, pain and in few cases edema. Through ages studies have proved that sea anemone toxins have hemolytic activity, cardiac toxicity, neurotoxicity and nephrotoxicity (Bunc et al., 1999; Wang et al., 2004; Kawai et al., 2004; Mizuno et al., 2007).The sea anemone from Indian waters have been poorly studied especially with regard to their toxicity. The present study was therefore undertaken to gain a better understanding of the toxicity of the venom extracted from the intertidal sea anemone *Anthopleura asiatica* collected from the coast of Mumbai.

Corresponding author. E-mail: ramfrm@gmail.com.

Abbreviations: PBS, Phosphate buffer saline; **BSA,** bovine serum albumin.

MATERIALS AND METHODS

Sample collection and storage

The sea anemone *A. asiatica* (Uchida, 1958) was collected from the

Table 1. Protein content of crude venom and partially purified fractions of *Anthopleura asiatica* (values are mean ± SE of triplicate sets).

S/N	Sample	Protein (mg/ml)
1.	Crude venom	4.3459±0.027
2.	F1	0.6106±0.004
3.	F2	0.5531±0.024
4.	F3	0.8291±0.004
5.	F4	0.6220±0.003
6.	F5	0.3222±0.005
7.	F6	0.5886±0.002
8.	F7	0.3626±0.004
9.	F8	1.2388±0.058
10.	F9	0.9303±0.036
11.	F10	0.8573±0.006

Khardhanda Beach (Mumbai) during low tide (0.52 m). They were transported live to the laboratory and stored immediately at -20°C for further analysis.

Preparation of crude venom

Sixty anemones weighing 25 g each were homogenized in three volumes of distilled water. The homogenate was filtered and centrifuged at 10,000 rpm at 4°C for 15 min. The supernatant collected was dialyzed against D-glucose. The dialyzed solution was further centrifuged at 16,000 rpm for 20 min at 4°C. The supernatant was collected and stored at -20°C for further use.

Partial purification of crude extract

Partial purification of the crude extract was carried out using DEAE cellulose Anion Exchange Chromatography (Shiomi et al., 1987). Ten fractions were collected in a stepwise gradient with 0.1 to 1.0 M NaCl in phosphate buffer saline (PBS). The collected fractions were stored at -20°C for further use.

Protein estimation

The protein content of Crude venom extract and partially purified fractions were obtained by Lowery's method (Lowery et al., 1951) using bovine serum albumin (BSA) as a standard.

Mice bioassay for lethality

Clinically healthy Kausauli strain male albino mice of mean weight 20±2 g procured from M/s. Haffkine Biopharma, Mumbai, were maintained in healthy condition in the laboratory, following the codal formalities of the Institute's Ethical Committee. Mice in triplicate sets were challenged intraperitoneally with 0.25, 0.50, 0.75 and 1.0 ml of the crude venom. In the case of partially purified fractions, 1ml of the each fraction was injected. A control was maintained in each case by injecting an equal volume of PBS (pH 7.4). The time of injection and lethality, and behavioral changes were also recorded. Autopsy of the dead mice was performed to study the anatomical and histopathological changes.

Histopathology

Brain, heart, liver and kidney were dissected out from the mice that died upon envenomation. The dissected organs were fixed in 10% formalin for a minimum period of 24 h and further processed according to Lefkowitch (1987). Prepared sections were examined and photographed under a microscope (LAMBOMED, CX II).

RESULTS AND DISSCUSSION

In the present study, toxic proteins have been isolated from the body of *A. asiatica* using distilled water as the extraction medium (Sokotun et al., 2007). The protein content of the crude venom of *A. asiatica* was found to be 4.3459±0.027 mg/ml which is higher than the previous work in anemones from Indian waters (Ravindran et al., 2010). The results are presented in the Table 1.

Lethality was observed in mice injected intraperitoneally with crude venome, F8, F9 and F10, similar to proteins isolated from the sea anemone *Actineria villosa* (Oshiro et al., 2004). This reveals the instance of toxicity in sea anemone. The dose and time at which the crude and fractionated proteins were toxic has been tabulated (Table 2). The potency of the presently extracted toxins is comparable with other toxins extracted from various other tropical and temperate sea anemones (Bruhn et al., 2001; Honma, et al., 2003).

Immediately upon being injected with the venom, the mice exhibited the following symptoms: palpitation, restlessness, exploring around the cage, standing on hind legs, flexing of the muscles, stretching of body, wiping the mouth with forelimbs, excess defecation and micturition, and hairs were found to stand nape. As time advanced and the degree of illness worsened, the symptoms observed were tonic convulsions, paralysis of forelimbs, opaqueness of eyes, dragging of hind limbs, violent jumping, lying on back, gasping for the breath and foaming from the mouth. Finally there was bleeding from the mouth, violent convulsions and then death. This may be due to the action of anemone toxin on the organ systems. Similar observations have also been reported by Gondran et al. (2002). However, the mice envenomated with other fractions did not show any of the above symptoms and found to be normal.

Histopathological observations revealed that, the toxic extract had similar effects on the organs of mice viz, brain, heart, kidney and liver irrespective of the nature of its purity.

Brain: Hemolysis in the cerebral vessels was not common. Accumulation of polymorphic nucleus cells with necrotic changes was common to observe. Area of sponges was discernable at certain regions (Figure 1). This can be attributed to the fact that the sea anemones have been reported to possess neurotoxin of polypeptide nature (Bruhn et al., 2001; Zaharenko et al., 2008).

Heart: Ventricle was totally blocked with hemolysed blood. Blood vessels were also enlarged with hemolysed

Table 2. Toxicity of crude venom and partially purified extracts of *Anthopleura asiatica* injected intraperitoneal to male albino mice of 20 ± 2 g weight (values are mean ± SE of triplicate sets).

S/N	Sample	Dose (ml)	Death time (Min)	Remark
1	Crude Venom	0.25	-	Non lethal
		0.50	-	Non lethal
		0.75	-	Non lethal
		1.0	645±17	Lethal
2	F1	1.0	-	Non lethal
3	F2	1.0	-	Non lethal
4	F3	1.0	-	Non lethal
5	F4	1.0	-	Non lethal
6	F5	1.0	-	Non lethal
7	F6	1.0	-	Non lethal
8	F7	1.0	-	Non lethal
9	F8	1.0	725±22	Lethal
10	F9	1.0	685±12	Lethal
11	F10	1.0	684±19	Lethal
12	FU	1.0	-	Non lethal

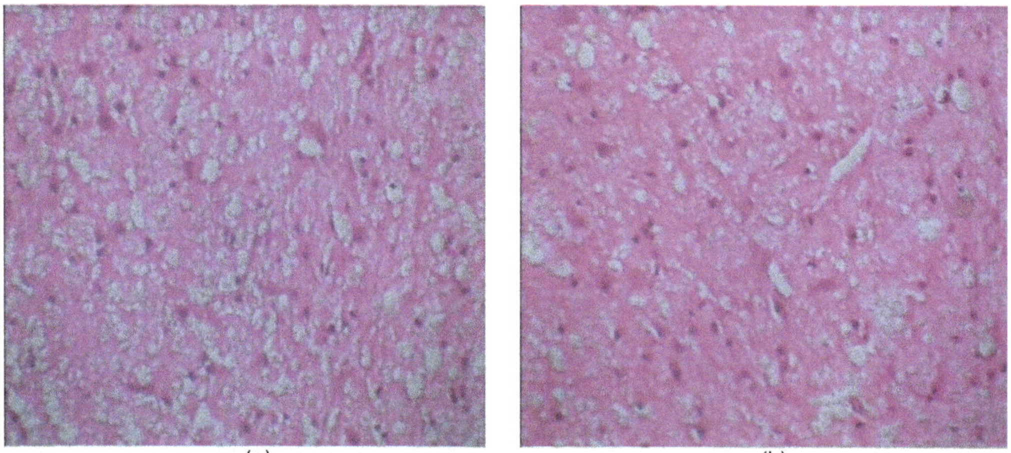

(a) (b)

Figure 1. Effect of toxins from Sea Anemone, *A. Asiaticaon;* the brain of male albino mice, arrows indicating a) spongiosis in brain b) necrotic cells with polymorphic nuclei (H&E, ×400).

blood (Figure 2). It implies acute cardiotoxicity of the sea anemone toxins. Instances of the cardiotoxicity by anemone toxins have been reported earlier from *Anthopleura sp* (Ouyang et al., 2005) and *Anthopleura elegantissima* (Bruhn et al., 2001).

Kidney: The changes observed in kidney included, tubules filled with homogenous mucus like substance; larger blood vessels showed occlusion with hemolysed blood; no changes were observed in glomeruli (Figure 3). According to the report of Luciano et al. (2004), venom toxins were localized in the damaged kidney tissues, this might be due to glomerular filtration which leads to high degree of localization of the venom in the kidneys. The kidneys may also be a secondary site of damage due to hemolysis in heart. The hemoglobin thus released reaches the tubular filtrate of the kidney. If the amount of hemoglobin in blood is excess to the reabsorbing capacity of the proximal tubules, it would lead to precipitation of unabsorbed hemoglobin. This would further block the tubules, eventually causing renal damage (Ravindran et al., 2010; Chen et al., 1976). Recent study in *P. semoni* revealed that, envenomation caused acute renal damage in rats as a result of

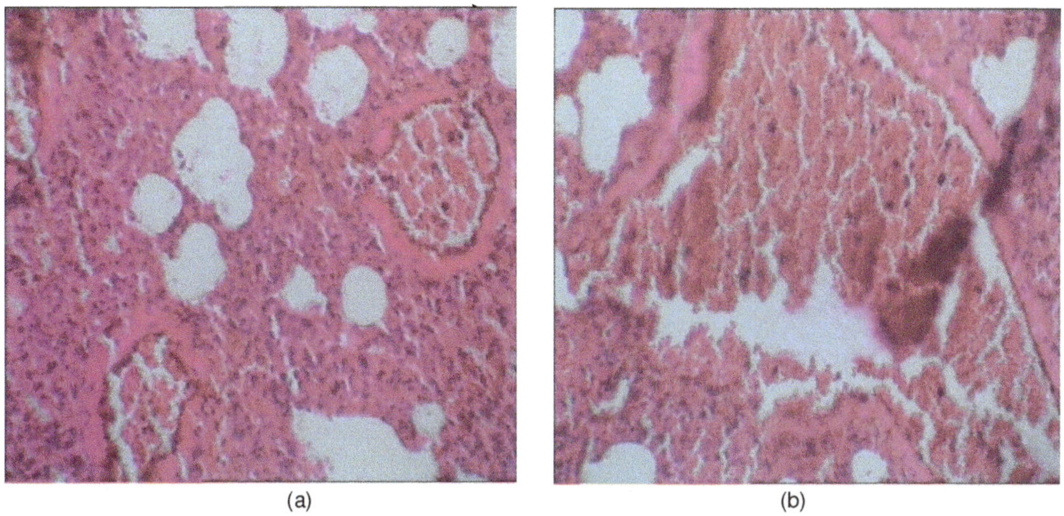

Figure 2. Effect of toxins from sea anemone, *A. asiatica* on the heart of male albino mice, arrows indicating a) enlarged blood vessels (H&E, X 100) b) hemolysed blood in ventricle (H&E, ×400).

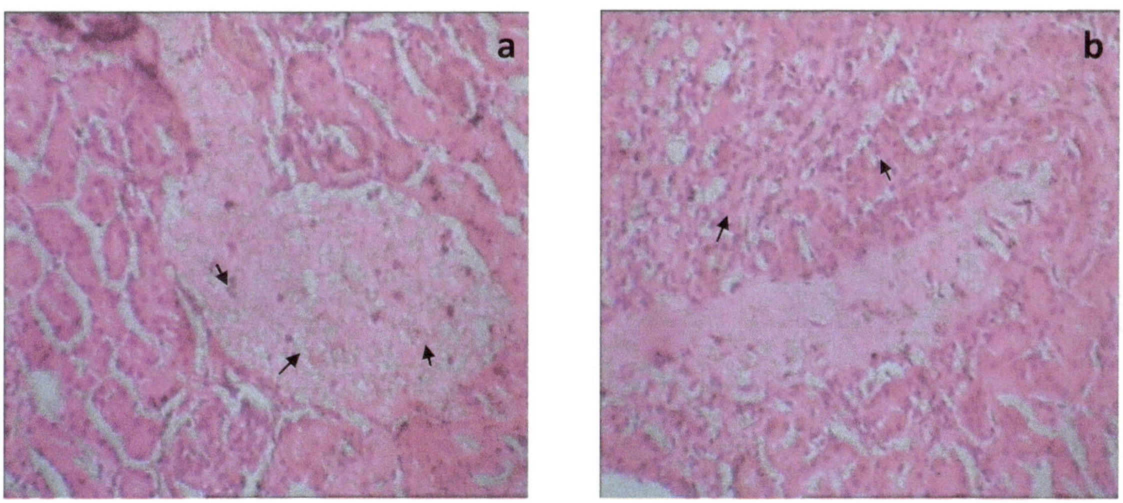

Figure 3. Effect of toxins from sea anemone, *A. asiatica*on the kidney of male albino mice, arrows indicating a) homogenous mucus like substance (H&E, ×400) b) occlusions with hemolysed blood (H&E, ×100).

induction of complement activation (Mizuno et al., 2007).

Liver: In the present study distinctive changes were noticed in the liver. They were occlusion with hemolysed blood in the central vein; engorgement of blood vessels with hemolysed blood; necrosis with fragmented nuclei in the hepatocytes around the central veins; vacuolation with pleomorphic nuclear material in liver tissue. In few cases nuclear chromatin was markedly dissolved to show heterochromatisation (Figure 4). These changes might have been due to detoxification of the toxin in liver. As no recent studies are available in sea anemone in this context, some representative groups from marine invertebrates are used to discuss the findings of the present study. Plancitoxin obtained from *Acanthaster planci* (Crown-of-thorns, an echinoderm) induced apoptosis in hepatocytes of envenomated mice (Ota et al., 2006).

Conclusion

This study demonstrates that the sea anemone, *A. asiatica* produces one or more toxins that may have direct effects on cardiac system, nervous system, hepatic system and excretory system in mammalian model. Further studies could explore and characterize the active agents responsible for toxicity, in order to reveal the bio-pharmacological properties of the toxin.

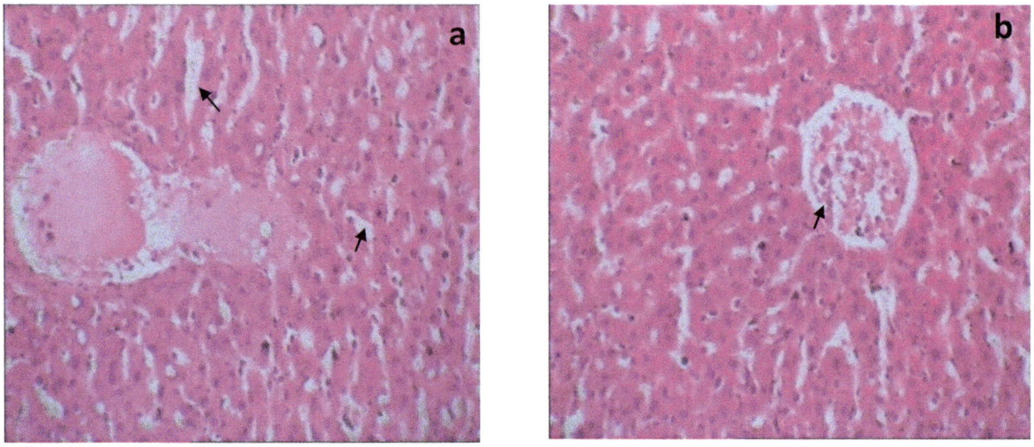

Figure 4. Effect of toxins from sea anemone, *A. asiatica* on the liver of male albino mice, arrows indicating a) vacuolation with pleomorphic nuclear material in hepatocytes b) hemolysed blood in the central vein (H&E, ×400).

ACKNOWLEDGEMENTS

This paper is a tribute to the co-author (Late) Dr K. Venkateshvaran, Principal Scientist, CIFE, Mumbai. The authors thank Central Institute of Fisheries Education, Mumbai for facilities provided. The authors also thank Kanagarajan and Baskara Doss for their technical assistance.

REFERENCES

Alsen C, Peters T, Scheufler E (1982). Studies on the mechanism of the positive inotropic effect of ATX II (*Anemonia sulcata*) on isolated guinea pig atria. J. Cardiovasc. Pharmacol., 4: 63–69.

Belmonte G, Menestrina G, Pederzolli C, Krizaj I, Gubensek F, Turk T, Macek P (1994). Primary and secondary structure of a pore-forming toxin from the sea anemone, *Actina equina*, and its association with lipid vesicles. Biochimica et Biophysica Acta., 1192: 197-204.

Beress L, Beress R, Wunderer G (1975). Purification of three polypeptides with neuro and cardiotoxic activity from the sea anemone *Anemonia sulcata*. Toxiconomy, 13: 359.

Bosmans F, Tytgatb J (2007). Sea anemone venom as a source of insecticidal peptides acting on voltage-gated Na^+ channels. Toxiconomy, 49: 550–560.

Bruhn T, Schaller C, Schulze C, Sanchez-Rodriguez J, Dannmeier C, Ravens U, Heubach JF, Eckhardt K, Schmidtmayer J, Schmidt HI (2001). Isolation and characterisation of five neurotoxic and cardiotoxic polypeptides from the sea anemone *Anthopleura elegantissima*. Toxiconomy, 39: 693-702.

Bunc M, Drevensek G, Budihna M, Suput D (1999). Effects of equinatoxin II from *Actinia equina* (L.) on isolated rat heart: The role of direct cardiotoxic effects in equinatoxin II lethality. Toxiconomy, 37: 109–123.

Chen WY, Yen TS, Chen JT, Hsieh BS, Hsu HC (1976). Acute renal failure due to ingestion of raw bile of grass carp (*Ctenopharyngodon idellus*). J. Formosan Med. Assoc., 75: 149–157.

Diochot S, Loret E, Bruhn T, Beress L, Lazdunski, M (2003). APETx1, a new toxin from the sea anemone *Anthopleura elegantissima*, blocks voltage-gated human ether-a-go-go-related gene potassium channels. Mol. Pharmacol., 64: 59–69.

Gondran M, Eckeli AL, Miges PV, Gabilan NH, Rodrigues ALS (2002). The crude extract from the sea anemone, *Bundosoma caissarum* elicits convulsions in mice: Possible involvement of the glutamatergic systems. Toxiconomy, 40: 1667.

Honma T, Iso T, Ishida M, Nagashima Y, Shiomi K (2003). Occurrence of type III sodium channel peptide toxins in two species of sea anemones. Toxiconomy, 41: 637.

Kawai N, Konno K (2004). Molecular determinants of two neurotoxins that regulate sodium current inactivation in rat hippocampal neurons. Neurosci. Lett., 361: 44–46.

Lefkowitch HJ (1987). Histopathology of disease. Churchill Living Stone publisher, New York, p. 233.

Lotan A, Fishman L, Zlotkin E (1996). Toxin compartmentation and delivery in the cnidaria: The nematocyst's tubule as a multiheaded poisonous arrow. J. Exp. Zool., 275: 444–451.

AL, Randall RJ (1951). Protein measurement with folin phenol reagent. J. Biol. Chem., 193: 265.

Luciano MN, da Silva PH, Chaim OM, dos Santos VL, Franco CR, Soares MF (2004). Experimental evidence for a direct cytotoxicity of *Loxosceles intermedia* (brown spider) venom in renal tissue. J. Histochem. Cytochem., 52: 455–467.

Mizuno M, Nishikawa K, Yuzawa Y, Kanie T, Mori H, Araki Y, Hotta N, Matsuo S (2000). A case report of acute renal failure following a sting presumably by a sea anemone. Am. J. Kidney Dis., 36: 10.

Mizuno M, Nozaki M, Morine N, Suzuki N, Nishikawa BK, Morgan P, Matsuo S (2007). A Protein Toxin from the Sea Anemone *Phyllodiscus semoni* Targets the Kidney and Causes a Severe Renal Injury with Predominant Glomerular Endothelial Damage. Am. J. Pathol., 171(2): 402-414.

Oshiro N, Kobayashi C, Iwanaga S, Nozaki M, Namikoshi M, Spring J, Nagai J(2004). A new membrane-attack complex/perforin (MACPF) domain lethal next term toxin from the nematocyst venom of the Okinawan previous term sea anemone next term *Actineria villosa*. Toxiconomy, 43(2): 225-228.

Ota E, Nagashima Y, Shiomi K (2006). Caspase-independent apotosis induced in rat liver cells by plancitoxin I, the major lethal factor from the crown-of-thorns starfish *Acathaster planci* venom. Toxiconomy, 48: 1002–1010.

Ouyang P, Xie JG, Liu J, Xiao WX, Liu YB, Wang L, Liang D, Wang YH, Xu DL, Liu YL, Xu AL (2005). Effects of a novel recombinant polypeptide from the sea anemone *Anthopleura* sp. on left ventricular function in dogs with acute cardiac insufficiency. Di Yi Jun Yi Da Xue Xue Bao., 25(1): 37–39.

Picken LER, Skaer RJ (1996). A review of researches on nematocysts. In: W.J. Rees (ed.), The cnidaria and Their Evolution. Symp. Zool. Soc. London, Acad., New York, 16: 19-50.

Ravindran VS, Kannan L, Venkateshvaran K (2010). Biological activity of sea anemone proteins: I. Toxicity and Histopathology. Indian J. Exp. Biol., 47: 1225-1232.

Shiomi K, Takamiya M, Yamanaka H, kikuchi T (1987). Purification of a lethal factor in the skin secretion from the oriental catfish *Plotosus*

lineatus. Nippon suisan gakkaishi, 53: 1275.
Sokotun IN, Il'ina AP, Monastyrnaya MM, Leychenko EV, Es'kov AA, Anastuk SD, Kozlovskaya EP (2007). Proteinase Inhibitors from the Tropical Sea Anemone *Radianthus macrodactylus*: Isolation and Characteristic. Biokhimiya, 72(3): 368-374.
Wang L, Ou J, Peng L, Zhong X, Du J, Liu Y (2004). Functional expression and characterization of four novel neurotoxins from sea anemone *Anthopleura sp.* Biochem. Biophys. Res. Commun., 313: 163–170.
Yanagihara AA, Kuroiwa JMY, Oliver LM, Chung JJ, Kunkel DD (2002). Ultrastructure of a novel eurytele nematocyst of *Carybdea alata* Reynolds (Cubozoa Cnidaria). Cell Tiss. Res., 308: 307–318.
Zaharenko AJ, Ferreira WA, Oliveira JS, Richardson M, Pimenta DC, Konno K, Portaro FCV, de Freitas JC (2008). Proteomics of the neurotoxic fraction from the sea anemone *Bunodosoma cangicum* venom: Novel peptides belonging to new classes of toxins. Comp. Biochem. Physiol. D., 3(3): 219-225.

Polyculture of seabass with tilapia for the utilization of brown fields in the coastal areas of Cox's Bazar, Bangladesh

Md. Mostafa Monwar , A. K. M. Ruhul Amin Sarker and Nani Gopal Das

Institute of Marine Sciences and Fisheries, University of Chittagong, Chittagong-4331, Bangladesh.

Disease outbreak in shrimp culture sector has abandoned the vast coastal areas of Bangladesh. These abandoned areas treated as 'brown fields' could be utilized for finfish culture using low cost inputs. This type of culture practice has the potential to be successful in the coastal waters due to having seasonally fluctuating salinity of 0 to 15 ppt. The present study was conducted to determine the growth and survival of seabass (*Lates calcarifer*) and tilapia (*Oreochromis niloticus*) using different stocking ratios without giving any supplementary feed to seabass. Tilapia was stocked at 1 individual/m^2. Tilapia fry was expected to provide food for seabass in the selected culture ponds. The stocking ratio of tilapia and seabass were 4:1, 5:1 and 6:1 in ponds with three replicates for each treatment. During 3 months culture period, the highest harvest weight of seabass and tilapia was recorded as 74.3 and 49.1 g/m^2, respectively and the highest survival rate of seabass and tilapia was recorded as 78.3 and 61.9%, respectively. The highest growth rate of seabass was recorded as 14.67 g/day and the highest biomass of seabass and tilapia (collectively) was found as 116.9 g/m^2. The present study reveals that seabass-tilapia polyculture may be a good solution to utilize the coastal brown fields of Cox's Bazar, Bangladesh.

Key words: Polyculture, seabass (*Lates calcarifer),* tilapia (*Oreochromis niloticus),* brown fields.

INTRODUCTION

Due to unplanned and intensive shrimp farming, the coastal ecosystem has degraded and prone to disease outbreak during the last three decades which has left many small and large coastal ponds unutilized. These barren, unproductive abandoned shrimp farming areas are generally treated as brown fields. Though there is an opportunity to utilize these ponds, alternative culture practice has not yet been developed. To compensate the loss especially for the existing resources and employment generation of the poor farmers, it is necessary to develop an alternative technology such as finfish culture in this sector.

Seabass (*Lates calcarifer*) can be cultured in a variety of culture systems using marine water, brackish water and freshwater (Harpaza et al., 2005; FAO, 2006). In Asia, seabass may be cultured in brackish water ponds with tilapia (FAO, 2006). In Thailand, seabass has been cultured for decades in association with shrimp, mullet and milkfish (Rauangpanit et al., 1984). Moreover, attempts have been made to compare the culture practice of seabass with tilapia in Thailand and achieved mixed results (personal communication: Bart, AIT and Turner, NSF. Thailand).

Farming based on supplementary feed was conducted

by using different types of feeds like trash fish, crustacean pellets, formulated feed, and live food (Corre and Hassan, 1995; Biswas et al., 2010; Boonyaratpalin et al., 1998; Bermudes et al., 2010). There are only a few studies published on culture of tilapia-seabass combination. Fortes (1985) indicated that seabass-tilapia combination of 1:15 ratio appears very promising result, whereas Tesorero (1995) reported 1:20 as the optimum ratio of seabass and tilapia in polyculture which had some advantages of among different stocking ratios. However, the additional crop of seabass in the seabass-tilapia combination made tilapia culture more attractive. Fortes and Genodepa (1997) showed a ratio of 1:5 was effective in an aqua farm condition. In Bangladesh, a preliminary result is also available from the study of a polyculture system of tilapia and seabass (Hossain et al., 1997). It was found that seabass shows a fast growth in the coastal environment of Cox's Bazar region (Das, 2000).

So, there is an opportunity to develop a new culture practice of tilapia and seabass in the existing brown fields and derelict areas. Tilapia is considered as a low-input species, as it can be cultured without supplementary feeding. It is a prolific breeder often resulting in over population. It tolerates a wide range of salinities (0 to 30 ppt) and can breed at salinity as high as 15 ppt (Watanabe et al., 1985). Similarly, seabass is an excellent cultivable species in variable water salinities ranging from 0 to 30 ppt (James and Marichamy, 1986; Kungvankij et al., 1986; Mukhopadhyay and Karmakar, 1981). However, polyculture of tilapia and seabass has the potential to be successful in Bangladesh because coastal areas of Cox's Bazar have seasonal fluctuation of salinity (5 to 15 ppt). So this polyculture system would be effective as tilapia utilize plankton in the culture ponds, while seabass consume tilapia fry as a live feed.

MATERIALS AND METHODS

Nine ponds of different sizes (169 to 805 m^2) owned by the marginal farmers were selected in the coastal area of Chakaria, Cox's Bazar, Bangladesh. Ponds were dried and renovated with respect to dykes, depth, slope, bottom elevation, supply and drainage facilities. Soil pH was assessed prior to application of treatments. These ponds were limed at a rate of 1000 to 2000 kg/ha. Then inorganic fertilizer was applied (100 to 200 kg/ha; Urea: TSP, 2:1) followed by organic fertilizer at a rate of 4 to 5 ton/ha (cow dung: chicken manure, 3:1). Ponds were filled with water up to 1.2 m. After 3 to 5 days when the colour of the water turned into green, ponds were stocked with adult tilapia (*Oreochromis niloticus;* female: male = 3:1) of size 50 to 70 g. Tilapia was stocked at 1 indivisual/m^2. They were acclimatized by gradually increasing salinity to the pond conditions before stocking to their respective culture ponds.

Recruitment of tilapia fry was observed 3 weeks after stocking of adult tilapia. In the meantime, the fingerlings of seabass (12 to 18 cm with 30 to 70 g) were stocked into the culture ponds such that tilapia: seabass was 4:1, 5:1 and 6:1 in three replicates. The fingerlings were collected from the local fry collectors of the coastal area during the months of June to July and acclimatized by gradually decreasing salinity before stocking. Water parameters were taken twice a month from the experimental ponds. Both soil and water pH was recorded by soil pH meter and pen pH meter, respectively; salinity and secchi depth by refractometer and secchi disc, respectively; dissolved oxygen (DO), free CO_2, alkalinity and hardness measured by titration methods followed by APHA (1976). Fish samples were taken monthly by seine net for determination of weight and survival rate during the culture period.

A randomized block design (RBD) for three treatments (Seabass: Tilapia ratio) with three replicates was applied. Three stocking ratios were analyzed with one-way analysis of variance (ANOVA) and the differences were compared at 5% level of significance.

RESULTS

The initial soil pH of nine ponds as recorded was 6.4, 6.0, 5.5, 5.7, 6.1, 6.2, 4.2, 5.7 and 5.2, respectively. As the initial soil pH of ponds, 1, 2, 5 and 6 was found satisfactory (above 6), liming was not applied to those ponds. It was found that after treatment with lime, the pH value of both soil and water has been improved to a certain extent in most of the ponds except in Pond 7 where water pH retained in between 4.8 to 5.5. Though the project site was in coastal area but variation of water salinity was not too much (0 to 4 ppt) during observation period and DO was found in acceptable range of 4 to 8.9 mg/L in most of the ponds. Analysis of water in respect to free CO_2, alkalinity and hardness showed no detrimental impact to the water conditions of culture ponds (Table 1).

In this study, seabass attained very fast growth more than 1300 g within 90 days culture period in Pond 1 that shows the growth rate of 10 to 15 g/day (Table 2). This is the first record of such growth in our coastal environment that is yet not available in the scientific world. The satisfactory results are available from most of the culture ponds (Ponds 1 to 6, 8 and 9) that may be due to their favourable ecological conditions. The highest harvest weight of seabass and tilapia was recorded as 74.27 and 49.09 g/m^2, respectively and the highest survival rate of seabass and tilapia was recorded as 78.33 and 61.90%, respectively. The highest growth rate of seabass was 14.67 g/day and the highest biomass of seabass and tilapia (collectively) was found as 116.85 g/m^2. The harvested data showed that after 90 days of culture period, the biomass of seabass and tilapia (collectively) varies within the range of 771.4 to 1168.5 kg/ha among the nine culture ponds (Tables 2 to 4).

The growth of seabass was very smooth and sharp especially when the supply of tilapia fry was quite enough for the requirement of the juveniles' seabass during first 2 months culture period. Then the growth of seabass was sluggish due to increase of food consumption rate of growing seabass which was also proved from the records of reduced number of tilapia fry in the culture ponds and in that period the mouth of young seabass was sufficiently wide even to engulf certain big size tilapia due to their cannibalistic habit. That is why the decision of harvest was made after 3 months culture period when it was found that the difference in size between the fish

Table 1. Physico-chemical parameters of the different fish culture ponds.

Item	P*-1	P-2	P-3	P-4	P-5	P-6	P-7	P-8	P-9
Water temperature (°C)	29 - 32	29 - 32	29 - 32	29 - 32	30 - 32	30 - 32	30 - 32	30 - 32	30 - 32
Soil pH	6.2 - 6.8	6.1 - 6.8	5.5 - 5.7	5.7 - 5.9	6.1 - 6.4	6.2 - 6.4	4.2 - 5.2	5.7 - 6.4	5.2 - 6.3
Water pH	7.5 - 8.9	7.0 - 8.7	6.8 - 9.1	6.3 - 7.4	6.9 - 8.7	7.4 - 8.7	4.8 - 5.2	5.5 - 6.7	5.6 - 7.6
Salinity (ppt)	0 - 5	0 - 6	0 - 3	0 - 2	0 - 2	0 - 2	0 - 5	0 - 4	0 - 4
DO (mg/L)	4.2 - 6.1	3.9 - 6.7	5.0 - 6.3	4.0 - 5.4	5.2 - 6.7	5.7 - 6.9	4.4 - 7.5	3.9 - 6.4	4.1 - 8.9
Free CO_2 (mg/L)	4 - 14	7 - 15	4 - 18	10 - 13	3 - 10	3 - 10	11 - 39	8 - 18	6 - 10
Alkalinity (mg/L)	48 - 76	42 - 78	26 - 54	30 - 46	30 - 46	34 - 54	24 - 26	34 - 62	28 - 50
Hardness (mg/L)	75 - 136	88 - 152	87 - 122	72 - 145	36 - 52	26 - 51	58 - 181	44 - 145	99 - 165
Secchi depth (cm)	25 - 29	20 - 29	23 - 24	28 - 59	15 - 18	15 - 18	92 - 96	32 - 96	30 - 73

*P = Pond No. Soil and water pH by pH meter, salinity and secchi depth by refractometer and secchi disc, respectively; DO, free CO_2, alkalinity and hardness determined by titration methods followed by APHA (1976).

Table 2. Production in experimental ponds with stocking ratio of seabass-tilapia at 1:4.

Item	Pond 1		Pond 4		Pond 8	
Area (m^2)	805		286		240	
Fish type	Seabass	Tilapia	Seabass	Tilapia	Seabass	Tilapia
No. of fish stocked	201	805	72	286	60	240
No. of fish recovered	120	294	54	119	47	104
Survival rate (%)	59.7	36.5	75.0	41.6	78.3	43.3
Initial weight (g)	35 - 70	50 - 70	35 - 55	55 - 70	30 - 55	50 - 70
Final weight (g)	450 - 1390	85 - 150	200 - 700	85 - 140	225 - 750	80 - 135
Biomass (kg)	43.20	29.19	21.24	12.18	12.20	9.88
Total biomass (kg)	72.39		33.42		22.08	
Production (kg/ha)	899.26		1168.5		920	

within the population of the same ponds varied significantly due to food scarcity of seabass. This was due to avoid a further deteriorating condition which could have resulted in greater loss of fish biomass due to cannibalistic habit.

The size range of seabass was quite wide during harvest especially in the ponds with the highest stocking of 4:1. There was also a higher percentage of fish below 200 g in the culture ponds 1, 4 and 8 in comparison to other ponds where stocking density was 5:1 or 6:1. In this situation, it would be expected that the occurrence of cannibalism would be much greater in the higher stocking (4:1) compared to lower stocking (6:1) of seabass. Considering the factors such as the fingerlings were collected from same area, pond preparation and management were the same, water quality was almost similar, stocking density among three categories was not so wide and feed type and source were also similar, it is therefore, likely that the differences in biomass of production in all three categories of ponds were more or less same (Tables 2 to 4).

It is evident from the one-way ANOVA (Table 5) that there is no significant difference ($p > 0.05$) among the different treatments on the production of total biomass. That means there is no significant difference ($p > 0.05$) among the different levels of stocking density. On the other hand, it can comment that the effects of all proportions of stocking densities are the same that is, no benefit can be gained from applying different stocking densities in order to get higher product of biomass, while it is evident from the same task that there is a significant ($p < 0.05$) effect of pond conditions on the production. It might happen due to the inherent physicochemical and biological parameters of the pond soil and water.

DISCUSSION

Though the growth record of seabass is excellent and highly satisfactory, there was an abrupt size variation of seabass in the culture ponds. Those fishes which were to be stunted to less than 200 g during harvest were transferred to another pond for an additional culture period of 1 or 2 months. The growth of seabass was

Table 3. Production in experimental ponds showing stocking ratio of seabass-tilapia at 1:5.

Item	**Pond 3**		**Pond 5**		**Pond 7**	
Area (m^2)	385		504		169	
Fish type	Seabass	Tilapia	Seabass	Tilapia	Seabass	Tilapia
No. of fish stocked	77	385	101	504	34	169
No. of fish recovered	46	160	58	177	18	104
Survival rate (%)	59.7	41.6	57.4	35.1	52.9	61.90
Initial weight (g)	30 - 60	55 - 70	30 - 55	55-65	30 - 40	55 - 70
Final weight (g)	350 - 1000	85 - 135	150 - 750	85 - 140	275 - 600	80 - 130
Biomass (kg)	18.12	18.90	24.27	18.59	6.47	7.02
Total biomass (kg)	37.02		42.86		13.49	
Production (kg/ha)	961.6		850.3		798.2	

Table 4. Production in experimental ponds showing stocking ratio of seabass-tilapia at 1:6

Item	**Pond 2**		**Pond 6**		**Pond 9**	
Area (m^2)	583		530		175	
Fish type	Seabass	Tilapia	Seabass	Tilapia	Seabass	Tilapia
No of fish stocked	97	583	88	530	9	175
No of fish recovered	46	203	53	172	16	82
Survival rate (%)	67.0	34.8	60.2	32.5	55.7	46.9
Initial weight (g)	30 - 65	55 - 70	30 - 60	50 - 70	30 - 40	55 - 65
Final weight (g)	500 - 1200	85 - 130	475 - 1150	85 - 135	250 - 570	85 - 130
Biomass (kg)	26.72	22.54	22.10	20.41	5.19	8.31
Total biomass (kg)	49.26		42.51		13.50	
Production (kg/ha)	844.9		802.1		771.4	

significantly higher during the first 2 months in comparison to the later month, which could be explained by the presence of more tilapia to be prayed upon the first phase of the culture period (Advance, 1984). The decision to harvest was made after 3 months culture period when it was found that the difference in size between fish within the population of same pond varied significantly ($p < 0.05$). This was done to avoid a further deteriorating condition that could have resulted in greater loss of fish very frequently due to shortage of food and cannibalism. The variation in size of seabass was found wide during harvest due to high stocking densities that caused tremendous effect on growth and survival rate. Seabass performed a tremendous growth record in the present study showing daily average growth rate of more than 10 g by using tilapia fry only in the culture ponds. Danakusumah and Ismail (1986) reported that average individual daily growth was 1.3 to 1.5 gm for those cultured with formulated feed and 2.65 to 3.58 gm for those with trash fish. Sugama and Eda (1986) showed that the average daily growth of 4.62 to 6.05 g in net cage culture with trash fish. Growth and production are dependent on the amount of supplied feed (Bardach et al., 1972). ICAR (2010) reported that seabass attain 450 to 950 g after 270 days of culture from seabass tilapia polyfarming at Kakdwip, India. Similar types of work were also performed by the KAU (2012). Awang (1986) stated that seabass were reared about 5 to 9 months in cages to raise market size of about 500 g (1.55 to 3.0 g/day) using trash fish. Genodepa (1986) also harvested 351.5 g of seabass with a stocking size of 221.5 g having a culture period of 94 days (daily growth rate of 1.3 g) using trash fish in a monoculture pond.

Conclusion

The newly developed concept of an alternative eco-friendly technology on 'integrated tilapia-seabass polyculture' is treated as a milestone in the brackish water aquaculture in Bangladesh. The implications of this research are to bring back the fertility of the coastal land,

Table 5. ANOVA table for biomass of seabass.

Source of variance	Sum of square	Degree of freedom	Mean sum of square	Observed F	Calculated F at 5%
Treatment	145.59	2	72.80	2.08	6.94
Block	788.72	2	394.36	11.27	6.94
Error	140.01	4	35.00		

utilization of vast unutilized or underutilized derelict coastal areas, low cost but profitable and quick return on the investment, easy tech and risk-covered in comparison to shrimp culture, eco-friendly affordable environmentally sustainable technology and employment opportunity for the marginal farmers. To meet the increasing demand of animal protein as well as to raise the economic progress of the country, the policy makers among government, non-government and other related organizations should take immediate steps to extend the eco-friendly technology of tilapia-seabass polyculture for the increase of fish production especially in the coastal areas which were abandoned by shrimp farming.

ACKNOWLEDGEMENTS

Gratitude and deep appreciation to the Support for Universities Fisheries Education and Research (SUFER) project, DFID Dhaka for allocation of a research grand fund and the SUFER Staff for their constant encouragement, valuable suggestion and sincerity for successful completion of the research project. The support and assistance of Mr. S. M. Nazer Hossain, Executive Director and other staff members of Integrated Social Development Effort (ISDE) - Bangladesh, partner NGO of the project are greatly acknowledged.

REFERENCES

Advance, I (1984). Effect of various food types on growth, survival and production of seabass in brackishwater earthen ponds. Iloilo, Philippines. M.S. Thesis, University of the Philippines in the Visayas. p. 40.

APHA (1976). Standard methods for the examination of water and waste water, American Public Health association, Washington D.C. USA. p. 119.

Awang AB (1986). Status of seabass (*Lates calcarifer*) culture in Malaysia. Eds: Copland and Grey. Proceeding of an International Workshop held at Darwin, Australia, 24-26 Sep., 1986 on Management of Wild and Culture Sea Bass/Barramundi. pp. 165-167.

Bardach JE, Ryther JH, McLarney WO (1972). Aquaculture, the farming and husbandry of freshwater and marine organism, Wiley-Interscience, New York. p. 868.

Bermudes M, Glencross B, Austen K, Hawkins W (2010). The effects of temperature and size on the growth, energy budget and waste outputs of barramundi (*Lates calcarifer*). Aquaculture 306:160-166.

Biswas G, Thirunavukkarasu AR, Sundaray JK, Kailasam M (2010). Optimization of feeding frequency of Asian seabass (*Lates calcarifer*) fry reared in net cages under brackishwater environment. Aquaculture 305:26-31.

Boonyaratpalin M, Suraneiranat, P, Tunpibal T (1998). Replacement of fish meal with various types of soybean products in diets for the Asian seabass, *Lates calcarifer*. Aquaculture 161:67-78.

Corre V, Hassan H (1995). Feeds and the feeding strategies for low-cost seabass production. International Information System for the Agricultural Sciences and Technology (AGRIS). UP Res. Digest. 2(1):354.

Danakusumah E, Ismail A (1986). Culture of seabass (*Lates calcarifer*) in Earthen Brackishwater Ponds. Eds: Copland and Grey. Proceeding of an International Workshop held at Darwin, Australia, 24-26 Sep., 1986 on Management of Wild and Culture Seabass. pp. 156-157.

Das NG (2000). Development of breeding technology of Bhetki, *Lates calcarifer*. Final Report, Contract Research Project, Bangladesh Agricultural Research Council (BARC), Dhaka. p. 18.

FAO (2006). Cultured Aquatic Species Information Programme. Text by Rimmer MA In: FAO Fisheries and Aquaculture Department, Rome. Available at: http://www.fao.org/fishery/culturedspecies/Lates_calcarifer/en

Fortes RD (1985). Development of culture technique of seabass, *Lates calcarifer*, Iloilo, Philippines. BAC Technical 85-01:80.

Fortes RD, Genodepa J (1997). Pond Culture: Seabass with tilapia, Aquaculture News, SEAFDEC Asian Aquaculture.19:4-17.

Genodepa JG (1986). Seabass (Lates calcarifer) research at the Brackish Water Aquaculture Centre, Philippines. Eds: Copland and Grey. Proceeding of an International Workshop held at Darwin, Australia, 24-26 Sep., 1986 on Management of Wild and Culture Seabass. pp. 161-164.

Harpaza ST, Hakima Y, Slosmana T, Eroldogana OT (2005). Effects of adding salt to the diet of Asian seabass (*Lates calcarifer*) reared in fresh or salt water recirculating tanks, on growth and brush border enzyme activity. Aquaculture 248:315-324.

Hossain MA, Sultana N, Islam, SQ, Haque KA, Alamgir M (1997). Determination of optimum ratio of seabass, *Lates calcarifer* and *Tilapia sp.* for their mix culture, Bangladesh J.Zool. 25(1):9-14.

ICAR (2010). Demonstration of polyfarming of Asian seabass and tilapia. Indian Council of Agricultural Research (ICAR). Available at: http://www.ciba.res.in/index.php?option=com_content&view=article&id=371&Itemid=72&lang=en.

James PSBR, Marichamy R (1986). Status of Seabass (*Lates calcarifer*) Culture in India. Eds: Copland and Grey. Proceeding of an International Workshop held at Darwin, Australia, 24-26 Sep., 1986 on Management of Wild and Culture Seabass. pp. 74-79.

KAU (2012). Culture Fisheries: Sea Fishes, Pond culture of seabass. Kerala Agricultural University (KAU), Centre for E learning. Available at: http://www.celkau.in/Fisheries/CultureFisheries/Seafishes/Seabass/growout_sys.aspx

Kungvankij P, Pudadera BJ, Tiro, IBJ, Potestas IO (1986). Biology and culture of seabass (*Lates calcarifer*), NACA Training Manual Serial. 3:70.

Mukhopadhyay MK, Karmakar HK (1981). Effect of salinity on food intake, growth and conversion efficiency in juveniles of *Lates calcarifer*. J. Inland Fisheries Soc. India 13(1):8-14.

Rauangpanit N, Maneewong S, Pechmanee T (1984). Fry Production of Sea bass, *Lates calcarifer* at National Institute of Coastal Aquaculture in 1983. In. Report of Thailand and Japan Coastal Aquaculture Research Project (April 1981, March 1984) 1:7-12.

Sugama K, Eda H (1986). Preliminary study on rearing of giant sea perch, *Lates calcarifer* in the floating net cages, Research Station for Coastal aquaculture, Bojonegara, Indonesia. p. 9.

Tesorero LB (1995). Effects of different seabass-tilapia ratios on fish yield and the efficiency of seasbass as biological control of tilapia young in polyculture. International Information System for the Agricultural Sciences and Technology (AGRIS). UP Research Digest. 2(1):376.

Watanabe WO, Kuo CM, Huang MC (1985). Salinity tolerance of Nile tilapia fry (*Oreochromis niloticus*) spawned and hatched at various salinities. Aquaculture 48:159-176.

Enzyme activity in the Nile perch gut: Implications to Nile perch culture

Namulawa V. T.[1], Kato C. D.[2], Rutaisire J.[1], Britz P. J.[3], Beukes N.[4], Pletschke BI[4] and Whiteley C.[4]

[1]Aquaculture Research and Development Center, P. O. Box, 530, Kampala, Uganda.
[2]College of Veterinary Medicine, Animal Resources and BioSecurity, P. O. Box 7026, Kampala, Uganda.
[3]Department of Ichthyology and Fisheries, Rhodes University, P. O. Box 94 Grahamstown, South Africa.
[4]Department of Biochemistry, Microbiology and Biotechnology, Rhodes University, Grahamstown, South Africa.

Nile perch (*Lates niloticus)* is a high-value freshwater fish of great social economic importance in Africa's great lakes region. High demand for this fish has caused its extensive harvest, resulting into a decline in the species stocks and consequently financial loss to the Nile perch industry. Several strategies have been suggested to reinstate the fish population, including the domestication of Nile perch, however, this will greatly depend on the ability of this fish to digest artificial diets, since these diets greatly determine the survival of fish in aquaculture systems. In this study gut enzymes were assayed using standard procedures. Results indicated presence of amylase and trypsin, which is an indicator that carbohydrates and proteins have to be included in the artificial diets that will be formulated for this species. The study further revealed a variation in enzyme activity in the juvenile stages of this fish, indicating that the most critical stage in the nutrition of the Nile perch is the juvenile stage. This variation reveals insights in the requirement for different diets formulations for the different juvenile developmental stages of Nile perch. These findings make Nile perch a potential aquaculture candidate, since carbohydrates and proteins contribute substantively to artificial diets in cultured fish.

Key words: Aquaculture, aquafeeds, enzyme activity, Nile perch.

INTRODUCTION

Artificial feeds are an important aquaculture input, and their efficacy depends on the ability of the cultured fish to digest and hence utilize the feeds for growth and development (Jauncey et al., 2007). It is for this reason that the ability of any aquaculture candidate to digest artificial diets is investigated before it is considered for culture (Kolkovski, 2001). Therefore, information about the characteristics of the digestive enzymes is important because it provides insights into the ability of a fish species to digest dietary components (proteins, carbohydrates, vitamins, minerals and lipids). This research is therefore vital to the artificial diet formulation and manufacturing process for any new aquaculture species (Britz, 1995), since it guides the selection of suitable ingredients to constitute the diets. Several suitable artificial diets have been developed following detailed studies of the functioning of the digestive systems in fish. For example, the formulation of appropriate diets in *Anaba testudineus*, followed an *in vitro* assessment of protein digestibility of different feed ingredients by the crude enzyme extracts obtained from the fish gut (Ali et al., 2009), and in *Labeo rohita*, where diets were formulated following an assessment of digestive enzymes of fingerlings fed on different diets

(Debnath et al., 2007).

Due to the previously limited interest in Nile perch aquaculture, no published information is available in relation to the physiology of the digestive system of this fish in this respect. The development of artificial diets for carnivorous fish such as Nile perch therefore continues to be a challenge because of the lack of knowledge of the specificity and activity of digestive enzymes (D'Abramo, 2002; Oliveira and Cyrino, 2004). The ability to utilize artificial diets in other carnivorous fish has been widely studied and research reveals that digestion is greatly attributed to the presence of digestive enzymes that can efficiently hydrolyze the different components especially the proteins and carbohydrates which occur in the artificial diets (Caruso et al., 2009). Proteins contribute a large percentage of artificial diets in fish; where it serves as a major source of amino acids, constituting between 40 to 56% of the diets (Cerda et al., 1994; Garcia-Ortega et al., 2001; Razzaque et al., 2008; Mukherjee et al., 2011). Carbohydrates on the other hand, occur in diets as binders, additives and sources of energy and are therefore frequently utilized in artificial feed development (Ovie et al., 2005). Several studies have therefore focused on the proteolytic and amyolytic ability of the fish gut to utilize artificial diets (Walford and Lam, 1993; Garcia-Carreno et al., 2002; Matsumiya et al., 2006). This study used similar approaches toestablish the specific activity of amylases and trypsin in the gastro-intestinal tract of Nile perch, given that the activities of these enzymes in the Nile perch gut will determine its ability to efficiently utilize the formulated artificial diets. Several studies engaged in weaning candidate aquaculture species onto artificial diets begin with the larval stages (Herbert and Graham, 2003; Vega-Orellana et al., 2006; Cassiano et al., 2009). This approach was not possible with Nile perch in this study, because of the inability to obtain *L. niloticus* hatchlings during the period of this study. In the absence of fish weaned on to artificial diets, we commenced the study by testing the ability of wild Nile perch juveniles to digest ingredients that constitute artificial diets. Strategies towards the culture of Nile perch require information regarding its ability to digest artificial diets; this can be obtained by investigating the nature and abundance the gut enzymes.

The purpose of this study was to characterize amylase and trypsin-like activity present in Nile perch gut and to identify their specific location. The results from this study provide information to guide the formulation of artificial diets for the culture of Nile perch.

PROCEDURES

Sample collection

Wild Nile perch were collected live using beach seines from the northern side of Lake Victoria in Uganda near Kiggungu Fish landing site located at 32° 26′ 15″E, 00° 2′ 49″N . The fish were starved for twenty-four hours in a tank containing fresh water and continuously aerated with oxygen. Fish samples were divided into eight size classes namely; 1 - 5, 6 -10, 11- 15, 16 – 20, 21 - 25, 26 - 30, 31 – 35 and 36 - 40 cm of total length. This classification was adapted from Hopson (1972) and Hughes (1990) Nile perch length-growth models.

Preparation of enzyme extracts

Fish samples were individually measured and weighed and dissected on ice (German et al., 2010) to access the gut. The guts were sectioned into the oesophagus, stomach, liver, pyloric ceaca, anterior intestine, middle intestine and hind intestine (Féral, 1989; García-Carreño et al., 2002). Each of the gut sections from the different samples were separately mixed with distilled water and homogenized. Extracts for enzyme assays were then placed in 2 ml microcentrifuge tubes and centrifuged at 10 rpm for 10 min and stored at - 40°C until needed.

Amylase activity

Amylase activity was measured using modified methods adapted from Tongsiri et al. (2010) and Xiong et al. (2011) by estimating the reducing sugar produced using dinitrosalicylic acid (DNS). The reaction mixture consisted of 400 µl of 4% (v/v) warmed soluble starch as substrate and citric acid-Na_2HPO_4 buffer (at pH 6 for the stomach extracts and pH 7 for the intestine, liver and oesophagus extracts) plus 100 µl of homogenate. All gut extracts were assayed at 20°C for 60 min. The enzyme assays were terminated by adding 150 µl of each reaction mixture to 300 µl of DNS and heating at 100°C for 5 min. The mixture was then cooled on ice for 5 min, 250 µl from each aliquot was transferred to a 96 well plate and absorbance was measured at 540 nm. A standard curve wave was developed with maltose (0.1 to 1.0 mg/ml), where one unit of enzyme was expressed as the mg of maltose produced per minute. Amylase specific activity was presented as amylase unit per mg of total protein.

Trypsin-like activity

Trypsin-like activity was measured using the Protease Fluorescent Detection Kit (Sigma PF0100) (Boitano et al., 2011; Fu et al., 2011). In this assay, 20 µl of Fluorescein Isothiocyanate (FITC) - labeled casein was mixed with 20 µl of incubation buffer (20 mM sodium phosphates with 150 mM sodium chloride, pH 7.6) and 10 µl of homogenate which were incubated at 37°C for 60 min. The blank sample constituted 20 µl of incubation buffer, 20 µl of FITC-Casein substrate and 10 µl of distilled water. To each aliquot of the reaction mixture, 150 µl of 0.6 M trichloroactic acid (TCA) solution was then added and incubated at 37°C in the dark for 30 min. The mixture was centrifuged for 10 min at 10,000 x g; 10 µl of the supernatant was then mixed with 1 ml of assay buffer, the fluorescence intensity with excitation wavelength at 485 nm and emission wavelength of 535 nm was recorded. Trypsin-like activity was equivalent to the amount of the reference protease (trypsin) that produced a significant amount of fluorescence above the value obtained with the blank sample, after 60 min at 37°C. One unit of protease was defined as the amount of trypsin (µg) detected per minute at 37°C, while trypsin specific activity was presented as the enzyme activity per total mg of protein.

Total protein

Protein was assayed using a modified Bradford protein assay (Bradford, 1976), using bovine serum albumin (BSA) as a standard.

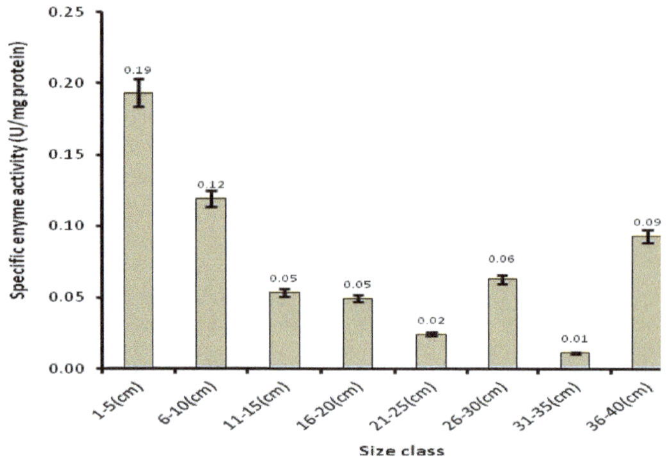

Figure 1. Amylase specific activity of gut extracts from Nile perch samples of different size classes.

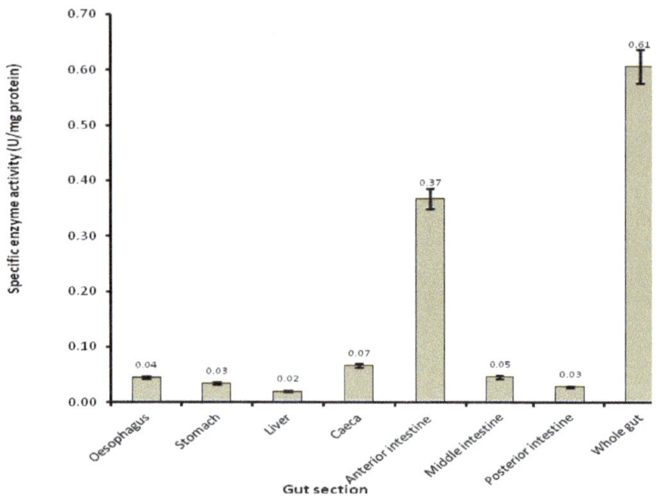

Figure 2. Amylase specific activity in the different gut sections of the Nile perch.

Bradford reagent (240 µl) was mixed to 10 µl of homogenate and incubated at ambient room temperature for 10 min; the absorbance was measured at 595 nm.

Statistical analysis

The comparisons of enzyme specific activity in the different size classes and gut regions were performed using the Kruskal-Wallis non-parametric test (SPSS version 16); and a probability level of $p<0.05$ was considered statistically significant.

RESULTS

Amylase specific activity

The quantitative results for the amylase assay of the different Nile perch gut extracts are presented in Figures 1 and 2. Results indicate a significant difference ($p<0.05$) of enzyme activity amongst the different size classes and gut regions. The specific activity decreased with increasing fish length. Gut extracts from the youngest fish (1 to 5 cm) showed higher amylase specific activity (0.19 U/mg protein) compared to the older fish (6 to 40 cm). Most amylase specific activity was observed in gut extracts from the anterior intestine (0.37 U/mg protein) compared to the other sections of the gut (Figure 2). The lowest activity was observed in the liver (0.02 U/mg protein), stomach (0.03 U/mg protein) and posterior intestine (0.03 U/mg protein) extracts.

Trypsin-like activity

Trypsin-like activity was observed in all gut extracts (Figures 3 and 4); but with a significant difference ($p<0.05$) between the different size classes and amongst the different gut regions. Activity decreased with increasing fish length. Activity of the gut extracts from the youngest fish (1 to 5 cm) was high (0.75 U/mg protein) compared to that in the older fish (6 to 40 cm) (Figure 3). Trypsin-like specific activity was highest in the anterior intestine (0.36 U/mg protein) and lowest in the oesophagus, liver and ceaca compared to the other sections of the gut (Figure 4).

DISCUSSION

Enzyme activity has been reported in the gut of several fish species (Al-Tameemi et al., 2010; Lazzari et al., 2010; Chaudhuri et al., 2012). Studies have revealed that herbivorous and omnivorous fishes tend to have high amylase activity in comparison to the carnivores, since the former need it to breakdown the polysaccharides that dominate their natural diets (Hidalgo et al., 1999). No wonder, the amylase activity levels observed in Nile perch in this study are much lower than those observed in some herbivorous and omnivorous fish (Table 1). However, different amylase specific activity levels have been reported in several carnivorous fish including *Pseudoplatystoma corruscans* and *Glyptosternum maculatum* (Table 1); and these vary from those observed in Nile perch in this study. This variance in results could have been caused by the difference in the analytical procedures used in the different studies or the difference in amylase quantities in the gastrointestinal tracts of different fishes.

The occurrence of amylase in predatory fish cannot be well explained, since very little or no carbohydrates are found in their diet. Some studies (Natalia et al., 2004) have, however, proposed that its role is in the digestion of glycogen (a source of energy in animals). In the case of Nile perch; amylase activity is observed in the oesophagus, liver, stomach, and in all intestinal extracts

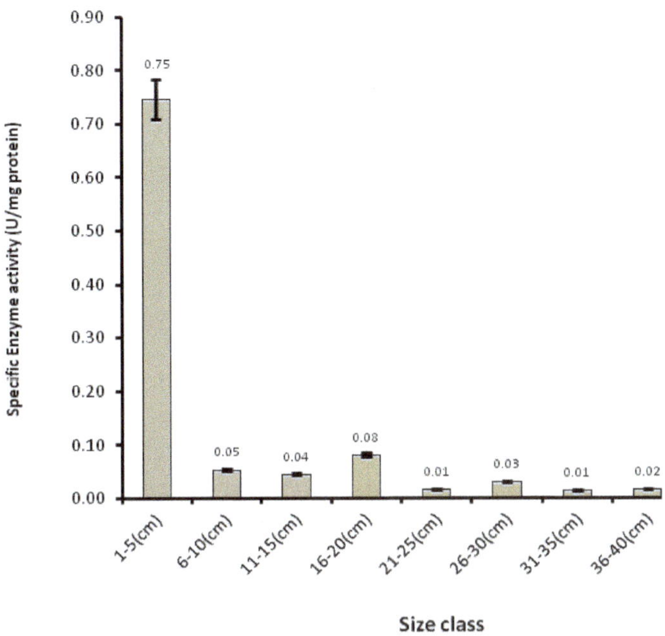

Figure 3. Trypsin-like specific activity of gut extracts of Nile perch samples of different size classes.

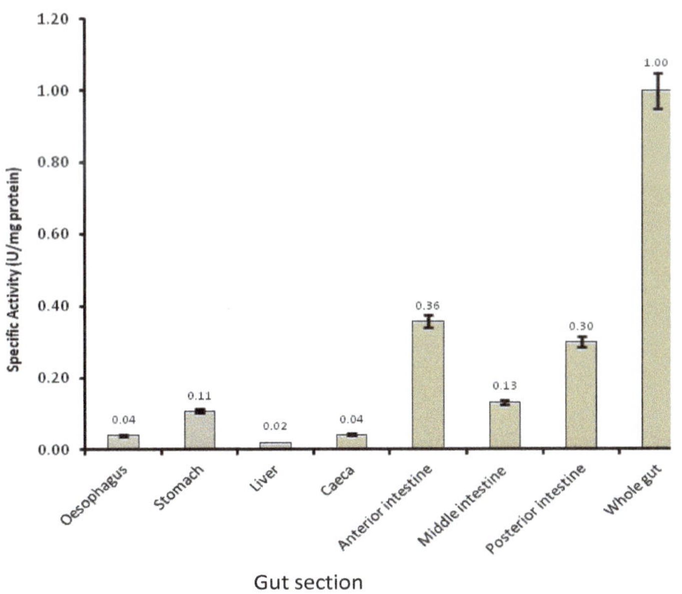

Figure 4. Trypsin-like specific activity in the different gut sections of the Nile perch.

(Figure 2). The amylase activity observed in the oesophagus (Figure 2) of Nile perch suggests that digestion of carbohydrates occurs in this region. The amylase activity observed in this region could be a mainfestation of saliva amylase in the bucal cavity, which is observed in some mammals that masticate their food (Meisler and Ting, 1993). However, the need for amylase in the bucal cavity and oeophagus of this fish might not be necessary given that mastication does not occur in Nile perch due to its non-flexible toungue (Namulawa et al., 2011). Amylase activity could instead be important to ease digestion in the subsequent regions of the gut. Activity of the same enzyme was also observed with the liver extracts from this fish (0.02 ± 0.002 U/mg protein); this amylase activity is lower than that recorded in extracts from other gut sections. The stomach extracts also showed activities levels lower (0.03 ± 0.002 U/mg protein) than those observed in the oesophagus (0.04 ± 0.005 U/mg protein); this observation is in agreement with reports from other carnivorous fish such as *Scleropages formosus* (Natalia et al., 2004), *Pseudoplatystoma corruscans* (Lundstedt et al., 2004), and in *Glyptosternum maculatum* (Xiong et al., 2011). The presence of amylase in the stomach of Nile perch and other carnivores can however not be explain, given that natural diet in Nile perch is rich in protein (Hopson, 1972).

The anterior intestine section extracts showed the highest amylase activity level (0.37 ± 0.005 U/mg protein) compared to that observed in the other gut section extracts. This level of activity in this fish was not expected, given that high amylase activity levels are synonymous with omnivorous and herbivours fishes and not carnivores. However, similar observations have been reported in other carnvorous fish species such as *Uranoscopus scaber* (Papoutsoglou and Lyndon, 2003), and *Glyptosternum maculatum* (Xiong et al., 2011).The presence of amylase in the anterior intestine suggests that starch digestion and glucose absorption mainly occur in this section of the gut and this gives prospects of artificial feed digestion in this fish. Further analysis also showed that amylase activity decreased in the hind intestine section extracts (0.03 U/mg protein). Similar observations have been reported in *Clarias gariepinus* by Uys and Hecht (1987), who described this phenomenon as the descending proximo-distal gradient (Kuz'mina, 1985). This could suggest that the hind intestine has limited participation in the chemical digestion of carbohydrates in Nile perch. Analyses from this study showed that the gut extracts of the youngest fish (1 to 5 cm) displayed the highest amylase activity (0.19 U/mg protein); compared to the gut extracts from the other size classes (Figure 2); as observed in *Trachinotus ovatus* (Youjun et al., 2011). Activity of this enzyme decreased with increasing fish length, suggesting a reduction in carbohydrase hydrolysis with increasing age. This observation is similar to that made in *Pagrus pagrus* (Suzer et al., 2007) and in *Anoplarchus purpurescens* (German et al., 2004) where their diets shift from small invertebrates to a carnivorous diet. The high carbohydrate activity levels observed in these fishes is probably due to the need for such enzymes to digest the tough exoskeleton in the small invertebrates that is constituted of complex carbohydrates (Kleinow, 1993;

Table 1. Comparative amylase activity of gut extracts of fish species of differing feeding behaviour.

Fish species	Gut section	Activity (U/mg protein)	Feeding behaviour	References
Pseudoplatystoma corruscans	Stomach	0.012±0.010 -0.18±0.016	Carnivorous	Lundstedt et al. (2004)
P. corruscans	Anterior Intestine	0.023±0.008 - 0.053±0.010	Carnivorous	Lundstedt et al. (2004)
P. corruscans	Middle intestine	0.013±0.005 - 0.022±0.016	Carnivorous	Lundstedt et al. (2004)
P. corruscans	Posterior intestine	0.003±0.002 - 0.029±0.019	Carnivorous	Lundstedt et al. (2004)
Glyptosternum maculatum	Stomach	0.0032±0.0007	Carnivorous	Xiong et al. (2011)
G. maculatum	Anterior Intestine	0.0062±0.0009	Carnivorous	Xiong et al. (2011)
G. maculatum	Middle intestine	0.0023±0.0005	Carnivorous	Xiong et al. (2011)
G. maculatum	Posterior intestine	0.0062±0.007	Carnivorous	Xiong et al. (2011)
Cyprinous carpio	Whole gut	72.73±8.46	Omnivorous	Hidalgo et al. (1999)
Carassius auratua	Whole gut	75.47±15.76	Omnivorous	Hidalgo et al. (1999)
Tinca tinca	Whole gut	19.37±2.67	Omnivorous	Hidalgo et al. (1999)
Sparus aurata	Whole gut	1.75±0.28	Carnivorous	Hidalgo et al. (1999)
Oncorhynchus mykiss	Whole gut	1.30±0.07	Carnivorous	Hidalgo et al. (1999)
Anguilla Anguilla	Whole gut	1.40±0.07	Carnivorous	Hidalgo et al. (1999)
Sparisoma cretense	Whole gut	35.9±2.21	Herbivorous	Papoustsoglou and Lyndon (2003)
Uranoscopus scaber	Whole gut	0.202±0.04	Carnivorous	Papoustsoglou and Lyndon (2003)
P. bogaraveo	Stomach	0.13±0.01	Omnivorous	Caruso et al. (2009)
P. bogaraveo	Pyloric ceca	0.273±0.03	Omnivorous	Caruso et al. (2009)
P. bogaraveo	Intestine	0.195±0.020	Omnivorous	Caruso et al. (2009)

Table 2. Comparative trypsin-like activity of gut extracts of fish species of differing feeding behaviour.

Fish species	Gut section	Activity (U/mg protein)	Feeding behaviour	References
P. corruscans	Stomach	2.70±0.45 - 4.20±0.60	Carnivorous	Lundstedt et al. (2004)
P. corruscans	Anterior Intestine	0.24±0.05 - 0.41±0.06	Carnivorous	Lundstedt et al. (2004)
P. corruscans	Middle intestine	0.22±0.04 - 0.75±0.14	Carnivorous	Lundstedt et al. (2004)
P. corruscans	Posterior intestine	0.19±0.05 - 0.53±0.22	Carnivorous	Lundstedt et al. (2004)
G. maculatum	Anterior Intestine	3.18±0.25	Carnivorous	Xiong et al. (2011)
G. maculatum	Middle intestine	1.52±0.23	Carnivorous	Xiong et al. (2011)
G. maculatum	Posterior intestine	1.76±0.21	Carnivorous	Xiong et al. (2011)
Cyprinous carpio	Whole gut	2.50±0.62	Omnivorous	Hidalgo et al. (1999)
Carassius auratua	Whole gut	2.01±0.11	Omnivorous	Hidalgo et al. (1999)
Tinca tinca	Whole gut	1.70±0.29	Omnivorous	Hidalgo et al. (1999)
Sparus aurata	Whole gut	0.81±0.24	Carnivorous	Hidalgo et al. (1999)
Oncorhynchus mykiss	Whole gut	3.44±0.45	Carnivorous	Hidalgo et al. (1999)
Anguilla anguilla	Whole gut	0.46±0.05	Carnivorous	Hidalgo et al. (1999)
Polyodon spathula	Whole gut	1.90±0.01	Omnivore	Ji et al. (2012)
Symphysodon aequifasciata	Intestine	0.79±0.05	Carnivorous	Chong et al. (2002)
S. aequifasciata	Stomach	0.09±0.03	Carnivorous	Chong et al. (2002)

Segers, 2008). The decrease in amylase activity observed in Nile perch in this study could similarly be caused by the dietary shift that occurs in this fish; from rotifers and small arthropods in the juvenile to a fish diet as Nile perch matures (Hopson, 1972).

This study shows trypsin-like activity in all the Nile perch gut extracts (Figure 3). Trypsin-like activity has been reported in the guts of several carnivorous fish (Table 2), since it contributes 40 to 50% in protein digestion that makes up much of their diet (Natalia et al., 2004). Studies have also reported protease activity in the guts of herbivorous fish (Table 2); which is attributed to the fact that herbivorous fish require this enzyme for the lysis of algal cell before amylase activities proceed (Benitez and Tiro, 1982; Zemk-White et al., 1999). Despite the presence of proteases in some herbivorous fish species, the ratio of proteases to amylases in the herbivores remains lower than that observed in their carnivore counterparts (Natalia et al., 2004). Studies have also shown high protease activity in some omnivores such when compared to that observed in some carnivores (Table 2); suggesting that difference in enzyme activity cannot efficiently be used in fish classification (Kuz'mina and Kuz'mina, 1990; Hidalgo et al., 1999).

The trypsin-like activity was observed in the oesophagus (Figure 3) of Nile perch in this study and this may be important in the facilitation of digestion in the subsequent regions of the gut; since the food residence time in the oesophagus is too low to enable chemical digestion, given that Nile perch quickly swallows its prey (Hopson, 1972). On the other hand, trypsin-like activity in the liver in this study is relatively low (Figure 3), possibly due to the fact that this enzyme is largely inactive (trypsinogen) in the pancreas which is located in the liver of this fish. Yet the stomach extract trypsin-like activity (Figure 3) was a little higher, probably due to other digestive enzymes and the ambient acidic environment in this region suggesting that the initial trypsin-like activity occurs in this region before digestion progresses to the intestinal region.

Observations in this study showed highest trypsin-like activity levels in the anterior intestine extracts (0.368 ± 0.042 U/mg protein) compared to those observed in extracts from other sections of the gut. This corresponds with reports from *Symphysodon aequifasciata* (Chong et al., 2002), *Siganus canaliculatus* and *Lates calcarifer* (Sabapathy and Teo, 1993), *Engraulis encrasicholus* (Martinez and Seera, 1989), *Scophthalmus maximus* (Munilla-Moran and Stark, 1990), and *Leptosynapta galliennei* (Féral, 1989). This suggests that the anterior intestine is the primary site for the hydrolysis of proteins to amino acids prior to absorption (Uscanga et al., 2010) in other fish and in the Nile perch.

Further observations made from this study reveled higher trypsin-like activity in the gut extract from the youngest fish (1 to 5 cm; 0.045 U/mg protein) compared to that observed in gut extracts from fish of older size classes (Figure 3). Similarly, decrease in Trypsin-like activity as fish increases in length and age has been reported in the *Coregonus sp* (Lauff and Hofer, 1984). It is probable that the action of trypsin is very crucial during digestion in the younger fish, especially because their stomachs are not yet fully developed for pepsin activity as observed in *Pagrus pagrus* larvae (Suzer et al., 2007); and when the fish matures pepsin becomes more important in the breakdown of long-chain polypeptide chains in the acidic stomach (Xiong et al., 2011), than trypsin. The high trypsin-like activity observed in the juveniles at this stage in could therefore be important to maximize protein digestion efficiency since young fish have higher specific metabolic demands in comparison to older fish (Mehner, 1996). This observation is emphasized by the fact that protein is a critical requirement for the growth and development of the fish at this stage (Izquierdo, 2004). This is probably the reason why gut extracts from the mature fishes in this study exhibited lower trypsin-like activity compared to the younger ones.

Conclusion

This study demonstrates the presence of amylases and proteases in the gut of Nile perch fish. The existence of these enzymes implies that Nile perch has the ability to digest carbohydrates and proteins; this makes it able to digest the most important ingredients that make up artificial diets. In this way Nile perch qualifies as a possible aquaculture candidate with criteria that it is able to utilize aqua feeds. The high protease activity observed in the youngest juveniles suggests that protein is critical at this development stage; therefore the larval feed should contain substantive amounts of proteins, probably between 45 and 55% as observed in other carnivorous larvae (Kpogue et al., 2013). In the same light, the high amylase activity observed in the juveniles suggests that it is much easier to wean Nile perch juveniles onto artificial diets, because at this stage the gut can easily utilize the carbohydrates that form much of the formulated aquafeeds on the market.

ACKNOWLEDGEMENT

Authors appreciate the World Bank funding through the Uganda National Council for Science and Technology, which made this work possible.

REFERENCES

Ali H, Haque MM, Chowdhury MR, Shariful MI (2009). Invitro protein digestibity of different feed ingredients in Thai koi (*Anabas testudineus*). J. Bangladesh Agric. Univ. 7(1):205-210.

Al-Tameemi R, Aldubaikul A, Salman NA (2010). Comparative study of α-amylase activity in three Cyprinid species of different feeding habits from Southern Irq. Turk. J. Fish Aquat. Sci. 10:411-414.

Benitez LV, Tiro LB (1982). Studies on the digestive proteases of the milkfish *Chanos chanos*. Mar. Biol. 71(3):309-315.

Boitano S, Flynn AN, Scherwood CL, Schulz SM, Hoffman J, Gruzinova I, Daines MO (2011). *Alternaria alternata* serine proteases induce lung inflammation and airway epithelial cell activation via PAR_2. Am. J. Physiol. Lung cell Mol. Physiol. 300(4):L605-L614.

Bradford MM (1976). A rapid senstive method for the quantification of microgram quantities of protein utilizing the principle of protein-dye binding. Anal. Biochem. 72:248-254.

Britz PJ (1995). The nutritional requirements of haliotis midae and development of a practical diet for abalone aquaculture. Grahamstown: PhD thesis, Rhodes university, South Africa.

Caruso G, Denaro MG, Genovese L (2009). Digestive enzymes in some Teleost species of interest for Mediterranean aquaculture. Open Fish Sci. J. 2:74-86.

Cassiano EJ, Ohs CL, Hill JE (2009). Candidate species for Florida Aquaculture: Pig fish *Orthropristis chrysoptera*. University of Florida, Florida.

Cerda J, Carrillo M, Zanuy S, Ramos J, Higuera M (1994). Influence of nutritional composition of diet on sea bass, *Dicentrarchus labrax* L, reproductive performance, egg and larval quality. Aquaculture 128:345-361.

Chaudhuri A, Mukherjee S, Homechaudhuri S (2012). Diet composition and digestive enzyme activity in carnivorous fishes inhabiting Mudflats of Indian Sundarban estuaries. Turk. J. Fish Aquat. Sci. 12:265-275.

Chong SC, Hashim R, Chow-Yang L, Ali AB (2002). Partial characterization and activities of proteases from the digestive tract of discus fish (*Symphysodo aequifasciata*). Aquaculture 203(3-4):321-333.

D'Abramo LR (2002). Challenges in developing successful formulated feeds for culture of larval fish and crustaceans. In L. E. Cruz-Suarez, D. Ricque-Marie, M. Tapia-Salazar MG. Gaxiola-Cortes, & N. Simoes, *Avances en Nutricion Acuicola VI. Memorias del VI Simposium Internacional de Nutricion Acuicola.* (al de Septiembre del 2002). Cancum Quintana Roo, Mexico. P. 3.

Debnath D, Pal AK, Sahu NP, Yengkokpam S, Baruah K, Choudhury D, Venkateshwarlu G (2007). Digestive enzymes and metabolic profile of *Labeo rohita* fingerlings fed diets with different crude protein levels. Comp. Biochem. Physiol. B: Biochem. Mol. Biol. 146:107-114.

Féral JP (1989). Activity of the principal digestive enzymes in the detritivorous apadus holothuroid *Leptosynapta galliennei*. Mar. Biol. 101:367-379.

Fu Q, Berbée M, Wang W, Boerma M, Wang J, Schmid HA, Jensen M (2011). Preclinical evaluation of SOM 230 as a radiation mitigator in a mouse model: Post-exposure time window and mechanism of action. Radiat. Res. 175(6):728–735.

García-Carreño FL, Albuqergue-Cavalcanti C, Navarretedel MA, Zaniboni-Filho E (2002). Digestive proteinases of *Brycon orbignyanus* (Characidae, Teleostei): Characteristics and effects of protein quality. Comp. Biochem. Physiol. B.132:343-352.

Garcia-Ortega A, Huisman EA, Sorgeloos P, Verreth J (2001). Evaluation of protein quality in microbound starter diets made with decapsulated cycts of Artemia and fish meal for fish larvae. J. World Aquac. Soc. 32(3):317-329.

German DP, Horn MH, Gawlicka A (2004). Digestive enzyme activities in herbivorous and carnivorous Pricklebach fishes (Teleostei: Stichaeidae): Ontogenetic, dietary, and phylogenetic effects. Physiol. Biochem. Zool. 77(5):789-804.

German DP, Nagle BC, Villeda JM, Ruiz AM, Thompson AW, Bald SC, Evans DH (2010). Evolution of herbivory in a carnivorous clade of minnows (Teleostei: Cyprinidae): effects of gut size and digestive physiology. Physiol. Biochem. Zool. 83:1-18.

Herbert R, Graham P (2003). Using artemia, frozen zooplankton and artifical food for weaning fingerlings of the freshwater fish Golden perch *Macquaria ambigua ambigua* (Percichthidae). Asian Fish Sci. 16:85-90.

Hidalgo MC, Urea E, Sanz A. (1999). Comapartive study of digestive enzymes in fish with different nutritional habits. Proteolytic and amylase activities. Aquaculture 170:267-283.

Hopson AJ (1972). The study of Nile perch (*Lates niloticus*) (Pisces: centropomidae) in Lake Chad. Overseas Research Publication, .19:93. Overseas Development Administration Her Majesty's Stationary Office, Commonwealth Office, London.

Hughes FN (1990). Growth and reproduction of the Nile perch, *Lates niloticus*, an introduced predator in the Nyanza gulf, Lake Victoria, East Africa. Environ. Biol. Fish, 33:299-305.

Izquierdo MS (2004). Nutritional requirements for finfish larvae. The second hatchery feeds and technology workshorp, September 30 – October, 1, 2004, Sydney.

Jauncey K, Sorensen PL, Areola F (2007). A short handbook – Catfish feed for Nigeria. Materials storage, handling, stock and quality control, ingredients, feed formulation, feed use (feed husbandry) and technical aspects of feed manufacture. Sponsored by the CDE, Brussels.

Ji H, Sun HT, Xiong DM (2012). Studies on activity, distribution, and zymogram of protease, -amylase, and lipase in the paddle fish *Polyodon spathula*. Fish Physiol. Biochem. 38(3):603-613.

Kleinow W (1993). Biochemical studies on *Brachionus plicatilis*: hydrolytic enzymes, integument protein and composition of trophi. Hydrobiologia 255/256:1-12.

Kolkovski S (2001). Digestive enzymes in fish larvae and juveniles-implications and applications to formulated diets. Aquculture 200:181-201.

Kpogue D, Gangbazo H, Fiogbe E (2013). A preliminary study on the dietary protein requirement of *Parachanna obscura* (Günther, 1861) larvae. Turk. J. Fish Aquat. Sci. 13:111-117.

Kuz'mina VV (1985). Distribution of carbohydrase activity along the intestine of several species of freshwater fish. J. Ichthyol. 25(4):137-143.

Kuz'mina VV, Kuz'mina YG (1990). Level of proteolytic activity in some species of the fish from the Volga barin. J. Ichthyol. 30:25-35.

Lauff M, Hofer R (1984). Proteolytic enzymes in fish development and the importance of dietary enzymes. Aquaculture 37:335-346.

Lazzari R, Neto R, Pedron F, Loro V, Pretto A, Gioda CR (2010). Protein sources and digestive enzyme activities in jundiá (*Rhamdia quelen*). Sci. Agric. 67(3):259-266.

Lundstedt LM, Melo JF, Moraes G (2004). Digestive enzymes and metabolic profile of *Pseudoplatystoma corruscans* (Teleostei: Siluriformes) in response to diet composition. Comp. Biochem. Physiol. B. 137(3):331-339.

Martinez A, Seera JL (1989). Proteolytic activities of the digestive tract of anchovy (*Engraulis encrasicholus*). Comp. Biochem. Physiol. B. 93(1):61-66.

Matsumiya M, Yasuyuki A, Atsunobu H, Subaratnam M, Kramer KJ (2006). Substrate specificity of chitinase from species of fish, Greenling, *Hexagrammos otakii*, and common Mackerel, *Scomber japanicas*, and the insect Tobacco hornworm *Manduca sexta*. Biosci. Biotechnol. Biochem. 70(4):971-979.

Mehner T (1996). Predation of age-0 fish on copepod population in a Baltic Sea inlet as estimated by two bioenergetics models. J. Plankton Res. 18(8):1323-1340.

Meisler MH, Ting C (1993). The remarkable evolutionary history of the human amylase genes. Crit. Rev. Oral Biol. Med. 4(3/4):503-509.

Mukherjee S, Parial D, Khatoo N, Chaudhuri A, Senroy S, Homechaudhuri S, Pal R (2011). Effect of formulated algal diet on growth performance of *Labeo rohita* Hamilton. J. Algal Biomass Utln. 2(4):1 - 9.

Munilla-Moran R, Stark JR (1990). Metabolism in marine flatfish-VI. Effect of nutritional state on digestion in turbot, *Scophthalumus maximus* (L.). Comp. Biochem. 95(3):625-634.

Namulawa VT, Kato CD, Nyatia E, Britz P, Rutaisire J. (2011). Histomorphological description of the digestive system of Nile perch (*L. niloticus*). Int. J. Morphol. 29(3):723-732.

Natalia Y, Hashim R, Ali A, Chong A (2004). Characterization of digestive enzymes in a carnivorous ornamental fish, the Asian bony tongue *Scleropagages formosus* (Osteoglossidae). Aquaculture 233:305-320.

Oliveira AM, Cyrino AM (2004). Attractants in plant protein-based diets for the carnivorous Largemouth bass *Micropterus salmoides*. Sci. Agric. 61(3):326-331.

Ovie SO, Sadiku SO, Ovie S (2005). Protein-sparing activity of lipid and carbohydrate in the Giant African Mudfish, *H. longifilis* diets. J. Appl. Sci. Environ. Manag. 9(3):109-113.

Papoutsoglou ES, Lyndon AR (2003). Distribution of & -amylase along the alimentary tract of two Mediterranean fish species, the parrot fish *Sparisoma cretense* L. and the stargazer, *Uranoscopus scaber* L. Med. Mar. Sci. 4(2):115-124.

Razzaque MA, Mazid MA, Islam MN, Mansur MA (2008). Culture possibility of scheilbeid catfish using formulated feed in a natural pond. J. Bangladesh Agric. Univ. 6(2):375-380.

Sabapathy U, Teo LH (1993). A quantitative study of some digestive enzymes in the rabbitfish *Siganus canaliculatus* and sea bass, *Lates calcarifer*. J. Fish Biol. 42(4):595-602.

Seger H (2008). Global diversity of rotifers (Rotifera) in fresh water. Hydrobiologia. 595:49-59.

Suzer C, Kamaci HO, Coban D, Saka S, Firat K, Ozkara B, Ozkara A (2007). Digestive activity of the red porgy (*Pagrus pagrus* L) during larval development under culture condition. Aquacult. Res. 38:1778–1785.

Tongsiri S, Mang-Amphan K, Peerapornpisal Y (2010). Characterisation of amylase, cellulase, and proteinase enzyme in stomach and intestine of the Mekong Giant Catfish fed with various diets consisting of Spirulina. Curr. Res. J. Biol. Sci. 2(4):268-274.

Uscanga A, Moyano FJ, Alvarez CA (2010). Assessment of enzymatic efficency on protein digestion in tilapia *Oreochromis niloticus*. Fish Physiol. Biochem. 36:1079-1085.
Uys W, Hecht T (1987). Assays on the digestive enzymes of Sharptooth Catfish, *Clarias gariepinus* (Pisces: Clariidae). Aquaculture 63:310-313.
Vega-Orellana OM, Fracalossi DM, Sugai JK (2006). Dourado (*Salminus brasiliensis*) larviculture: Weaning and ontogenetic development of digestive proteinases. Aquaculture 252:484-493.
Walford J, Lam TJ (1993). Development of digestive tract and proteolytic enzyme activity in Seabass (*Lates calcarifer*) larvae and juveniles. Aquaculture 109:187-205.
Xiong DM, Xie CX, Zhang HJ, Liu HP (2011). Digestie enzymes along the digestive tract of a carnivorous fish *Glyptosternum maculatum* (Sisoridae, Siluriformes). J. Anim. Physiol. Anim. Nutr. 95:56-64.
Youjun U, LUO Qi L, LI Jia'er L, Chunyan F (2011). Study on digestive enzymes activities of *Trachinotus ovatus* I . Comparison of their distribution in different digestive organs of adult and young fish. S. China Fish Sci. 7(1):50-55.
Zemk-White WL, Clements KD, Harris PJ (1999). Acid lysis of macroalgae by marine herbivoros fishes: myth or digestive mechanism. J. Exp. Mar. Biol. Ecol. 233:95-113.

Permissions

All chapters in this book were first published in IJFA, by Academic Journals; hereby published with permission under the Creative Commons Attribution License or equivalent. Every chapter published in this book has been scrutinized by our experts. Their significance has been extensively debated. The topics covered herein carry significant findings which will fuel the growth of the discipline. They may even be implemented as practical applications or may be referred to as a beginning point for another development.

The contributors of this book come from diverse backgrounds, making this book a truly international effort. This book will bring forth new frontiers with its revolutionizing research information and detailed analysis of the nascent developments around the world.

We would like to thank all the contributing authors for lending their expertise to make the book truly unique. They have played a crucial role in the development of this book. Without their invaluable contributions this book wouldn't have been possible. They have made vital efforts to compile up to date information on the varied aspects of this subject to make this book a valuable addition to the collection of many professionals and students.

This book was conceptualized with the vision of imparting up-to-date information and advanced data in this field. To ensure the same, a matchless editorial board was set up. Every individual on the board went through rigorous rounds of assessment to prove their worth. After which they invested a large part of their time researching and compiling the most relevant data for our readers.

The editorial board has been involved in producing this book since its inception. They have spent rigorous hours researching and exploring the diverse topics which have resulted in the successful publishing of this book. They have passed on their knowledge of decades through this book. To expedite this challenging task, the publisher supported the team at every step. A small team of assistant editors was also appointed to further simplify the editing procedure and attain best results for the readers.

Apart from the editorial board, the designing team has also invested a significant amount of their time in understanding the subject and creating the most relevant covers. They scrutinized every image to scout for the most suitable representation of the subject and create an appropriate cover for the book.

The publishing team has been an ardent support to the editorial, designing and production team. Their endless efforts to recruit the best for this project, has resulted in the accomplishment of this book. They are a veteran in the field of academics and their pool of knowledge is as vast as their experience in printing. Their expertise and guidance has proved useful at every step. Their uncompromising quality standards have made this book an exceptional effort. Their encouragement from time to time has been an inspiration for everyone.

The publisher and the editorial board hope that this book will prove to be a valuable piece of knowledge for researchers, students, practitioners and scholars across the globe.

List of Contributors

A. A. Adedeji
Department of Zoology, Obafemi Awolowo University, Ile-Ife, Nigeria

I. F. Adeniyi
Department of Zoology, Obafemi Awolowo University, Ile-Ife, Nigeria

O. R. Adetokunbo
Department of Zoology, Obafemi Awolowo University, Ile-Ife, Nigeria

O. Olumuyiwa Ajala
Department of Pure and Applied Biology, Ladoke Akintola University of Technology, Ogbomoso, Oyo State, Nigeria

O. Olatunde Fawole
Department of Pure and Applied Biology, Ladoke Akintola University of Technology, Ogbomoso, Oyo State, Nigeria

S. D. Yomi Alfred
Department of Agricultural Economics and Extension, Federal University of Technology, Akure, Ondo State, Nigeria

O. O. Odefadehan
Department of Agricultural Economics and Extension, Federal University of Technology, Akure, Ondo State, Nigeria

M. E. Ukut
Department of Agricultural Economics and Extension, Federal University of Technology, Akure, Ondo State, Nigeria

P. Laxmilatha
Central Marine Fisheries Research Institute, P. B. NO. 1603., Ernakulam, Cochin 682 014, India

S. Dinesh Kumar
Marine Planktonology and Aquaculture Laboratory, Department of Marine Science, School of Marine Sciences, Bharathidasan University, Tiruchirappalli-620 024, Tamil Nadu, India

P. Santhanam
Marine Planktonology and Aquaculture Laboratory, Department of Marine Science, School of Marine Sciences, Bharathidasan University, Tiruchirappalli-620 024, Tamil Nadu, India

S. Ananth
Marine Planktonology and Aquaculture Laboratory, Department of Marine Science, School of Marine Sciences, Bharathidasan University, Tiruchirappalli-620 024, Tamil Nadu, India

A. Shenbaga Devi
Marine Planktonology and Aquaculture Laboratory, Department of Marine Science, School of Marine Sciences, Bharathidasan University, Tiruchirappalli-620 024, Tamil Nadu, India

R. Nandakumar
Marine Planktonology and Aquaculture Laboratory, Department of Marine Science, School of Marine Sciences, Bharathidasan University, Tiruchirappalli-620 024, Tamil Nadu, India

B. Balaji Prasath
Marine Planktonology and Aquaculture Laboratory, Department of Marine Science, School of Marine Sciences, Bharathidasan University, Tiruchirappalli-620 024, Tamil Nadu, India

S. Jeyanthi
Marine Planktonology and Aquaculture Laboratory, Department of Marine Science, School of Marine Sciences, Bharathidasan University, Tiruchirappalli-620 024, Tamil Nadu, India

T. Jayalakshmi
Marine Planktonology and Aquaculture Laboratory, Department of Marine Science, School of Marine Sciences, Bharathidasan University, Tiruchirappalli-620 024, Tamil Nadu, India

P. Ananthi
Marine Planktonology and Aquaculture Laboratory, Department of Marine Science, School of Marine Sciences, Bharathidasan University, Tiruchirappalli-620 024, Tamil Nadu, India

M. T. Salawu
Department of Biological Sciences, Osun State University, Osun State, Nigeria

O. A. Morenikeji
Department of Zoology, University of Ibadan, Oyo State, Nigeria

A. A. Sowunmi
Department of Zoology, University of Ibadan, Oyo State, Nigeria

A. B. Odaibo
Department of Zoology, University of Ibadan, Oyo State, Nigeria

C. D. Kato
School of Biosecurity, Biotechnical and Laboratory Sciences, Makerere University P. O. Box 7062, Kampala Uganda

E. Nyatia
School of Biosecurity, Biotechnical and Laboratory Sciences, Makerere University P. O. Box 7062, Kampala Uganda

E. Matovu
School of Biosecurity, Biotechnical and Laboratory Sciences, Makerere University P. O. Box 7062, Kampala Uganda

Z. Uni
Department of Animal Sciences, Faculty of Agriculture, Food and Environment, the Hebrew University of Jerusalem, P. O. Box 12, Rehovot 76100, Israel

O. Kedar
Department of Animal Sciences, Faculty of Agriculture, Food and Environment, the Hebrew University of Jerusalem, P. O. Box 12, Rehovot 76100, Israel

Y. Hakim
Department of Animal Sciences, Faculty of Agriculture, Food and Environment, the Hebrew University of Jerusalem, P. O. Box 12, Rehovot 76100, Israel

B. Levavi-Sivan
Department of Animal Sciences, Faculty of Agriculture, Food and Environment, the Hebrew University of Jerusalem, P. O. Box 12, Rehovot 76100, Israel

J. Rutaisire
Aquaculture Research and Development Centre, P. O. Box 530, Kampala, Uganda

Confred G. Musuka
School of Natural Resources, Copperbelt University, P. O. Box 21692, Kitwe, Zambia

Felix F. Musonda
School of Natural Resources, Copperbelt University, P. O. Box 21692, Kitwe, Zambia

Carlos Henrique dos Anjos dos Santos
Universidade Nilton Lins (UNINILTONLINS), Laboratório de Genética Aplicada à Aquicultura and Biologia Molecular, Avenida Professor Nilton Lins, Flores, 69058-030, Manaus, Amazonas, Brasil Avenida André Araújo, 2936, Aleixo, 69060-001, Manaus, Amazonas, Brasil
Instituto Nacional de Pesquisas da Amazônia (INPA), Laboratório de Ecofisiologia e Evolução Molecular (LEEM),

Carolina Sousa de Sá Leitão
Instituto Nacional de Pesquisas da Amazônia (INPA), Laboratório de Ecofisiologia e Evolução Molecular (LEEM),
Avenida André Araújo, 2936, Aleixo, 69060-001, Manaus, Amazonas, Brasil

Maria de Nazaré Paula e Silva
Instituto Nacional de Pesquisas da Amazônia (INPA), Laboratório de Ecofisiologia e Evolução Molecular (LEEM),
Avenida André Araújo, 2936, Aleixo, 69060-001, Manaus, Amazonas, Brasil

Vera Maria Fonseca de Almeida e Val
Universidade Nilton Lins (UNINILTONLINS), Laboratório de Genética Aplicada à Aquicultura and Biologia Molecular, Avenida Professor Nilton Lins, Flores, 69058-030, Manaus, Amazonas, Brasil
Instituto Nacional de Pesquisas da Amazônia (INPA), Laboratório de Ecofisiologia e Evolução Molecular (LEEM),
Avenida André Araújo, 2936, Aleixo, 69060-001, Manaus, Amazonas, Brasil

Gregory E. Odo
Department of Zoology, University of Nigeria, Nsukka, Nigeria

S. U. Onoja
Department of Home Science and Nutrition, University of Nigeria, Nsukka, Nigeria

C. Onyishi Grace
Department of Zoology, University of Nigeria, Nsukka, Nigeria

Jakeline G. França
Aquaculture Center, UNESP, Via de Acesso Prof. Paulo Donato Castellane, 14884-900 Jaboticabal, SP, Brazil

Maria J. T. Ranzani-Paiva
Fisheries Institute/APTA/SAA SP, Avenida Francisco Matarazzo, 05001-900 São Paulo, SP, Brazil

Julio V. Lombardi
Fisheries Institute/APTA/SAA SP, Avenida Francisco Matarazzo, 05001-900 São Paulo, SP, Brazil

Solange de Carvalho
Aquaculture Center, UNESP, Via de Acesso Prof. Paulo Donato Castellane, 14884-900 Jaboticabal, SP, Brazil

Francisco Filipak-Neto
Departament of Celular Biology, University Federal do Paraná, 81.531-990 Curitiba, PR, Brazil

Ciro A. Oliveira-Ribeiro
Departament of Celular Biology, University Federal do Paraná, 81.531-990 Curitiba, PR, Brazil

Huong N. T. Thanh
Department of Science Technology and International Cooperation, Vietnam Directorate of Fisheries, Vietnam

Hien L. Thuy
Department of Petroleum Microbiology, Institute of Biotechnology (IBT), Vietnamese Academy of Science and Technology, Vietnam

Wenresti G. Gallardo
Aquaculture and Aquatic Resources Management, Asian Institute of Technology, Thailand

Hai N. Thanh
Shanghai Ocean University, Shanghai, China
Vietnam Institute of Fisheries Economics and Planning, Ha Noi, Vietnam

Firooz Fadaeifard
Department of Aquatic Animal Disease and Health, Islamic Azad University, Shahrekord Branch, Shahrekord - Iran

Mehdi Raissy
Department of Aquatic Animal Disease and Health, Islamic Azad University, Shahrekord Branch, Shahrekord - Iran

Mostafa Faghani
Department of Animal Science, Islamic Azad University, Shahrekord Branch, Shahrekord - Iran

Alireza Majlesi
Faculty of Veterinary Medicine, Islamic Azad University, Shahrekord Branch, Shahrekord - Iran

Gholamhossein Nodeh Farahani
Department of Chemistry, Islamic Azad University, Shahrekord Branch, Shahrekord - Iran

Daniel N. Akongyuure
Department of Oceanography and Fisheries, University of Ghana, P. O. Box LG 99, Legon, Accra - Ghana

Patrick K. Ofori-Danson
Department of Oceanography and Fisheries, University of Ghana, P. O. Box LG 99, Legon, Accra - Ghana

Francis K. E. Nunoo
Department of Oceanography and Fisheries, University of Ghana, P. O. Box LG 99, Legon, Accra - Ghana

Oguntuase Bukola Gbemisola
Department of Ecotourism and Wildlife Management, Federal University of Technology, Akure, Ondo State, Nigeria

O. T. Adebayo
Department of Fisheries and Aquaculture Technology, Federal University of Technology, Akure, Ondo State, Nigeria

H. Miliou
Department of Applied Hydrobiology, Faculty of Animal Science and Aquaculture, Agricultural University of Athens, Iera Odos 75, 11855, Athens, Hellas

K. Kapiris
Hellenic Centre for Marine Research, Institute of Marine Biological Recourses, Agios Kosmas Hellinikon, 16604, Athens, Hellas

I. Castritsi-Catharios
Department of Zoology - Marine Biology, Faculty of Biology, National and Kapodistrian University of Athens, Panepistimiopolis, 15784, Athens, Hellas

F. E. Ajagbe
Department of Zoology, Faculty of Science, University of Lagos, Lagos, Nigeria

A. O. Osibona
Department of Marine Sciences, Faculty of Science, University of Lagos, Lagos, Nigeria

A. A. Otitoloju
Department of Zoology, Faculty of Science, University of Lagos, Lagos, Nigeria

Lianfeng Liu
College of Food Science and Engineering, Ocean University of China, Qingdao 266003, P. R. China

Zhaohui Zhang
College of Food Science and Engineering, Ocean University of China, Qingdao 266003, P. R. China

Qian Liu
College of Food Science and Engineering, Ocean University of China, Qingdao 266003, P. R. China

Bofeng Yang
College of Food Science and Engineering, Ocean University of China, Qingdao 266003, P. R. China

Jinfa Huang
College of Food Science and Engineering, Ocean University of China, Qingdao 266003, P. R. China

Xin Gao
College of Food Science and Engineering, Ocean University of China, Qingdao 266003, P. R. China

TOMBI Jeannette
Laboratory of Parasitology and Ecology, Faculty of Science, University of Yaounde I, P. O. BOX 812, Yaounde, Cameroon

BILONG BILONG Charles Félix
Laboratory of Parasitology and Ecology, Faculty of Science, University of Yaounde I, P. O. BOX 812, Yaounde, Cameroon

Mayavan Veeramuthu Rajeswari
Centre of Advanced Study in Marine Biology, Faculty of Marine Sciences, Annamalai University, Parangipettai 608502, India

Thangavel Balasubramanian
Centre of Advanced Study in Marine Biology, Faculty of Marine Sciences, Annamalai University, Parangipettai 608502, India

Seth Mensah Abobi
Department of Fisheries and Aquatic Resources Management, University for Development Studies, P. O. Box TL 1882 Tamale, Ghana

Werner Ekau
Leibniz Center for Tropical Marine Ecology (ZMT), Fahrenheitstr. 6 D-28359 Bremen, Germany

Ajani Funmilola
Department of Animal Science and Fisheries Management, Bowen University, Iwo, Osun state, Nigeria

Adeyemo Olanike Kudirat
Department of Veterinary Public Health and Preventive Medicine, University of Ibadan, Ibadan, Nigeria

B. E. Emmanuel
Marine Sciences Department, University of Lagos, Akoka, Lagos, Nigeria

A. O. Osibona
Marine Sciences Department, University of Lagos, Akoka, Lagos, Nigeria

Elias Dadebo
Department of Biology, Hawassa University, P. O. Box 5, Hawassa, Ethiopia

Seble Bancha
Department of Biology, Wolaita Sodo University, P. O. Box 138, Wolaita Sodo, Ethiopia

Abinet Wolde-Senbet
Department of Biology, Wolaita Sodo University, P. O. Box 138, Wolaita Sodo, Ethiopia

Yosef Teklegiorgis
Department of Animal and Range Sciences, Hawassa University, P. O. Box 336, Hawassa, Ethiopia

Benjamin Onozeyi Dimowo
Department of Aquaculture and Fisheries Management, College of Environmental Resources Management, Federal University of Agriculture, Abeokuta, Ogun state, Nigeria

Giselle S. Costard
Universidade Federal Fluminense, Departamento de Biologia Marinha, Niterói, RJ, Brazil

Rodrigo R. Machado
Universidade Federal Fluminense, Departamento de Biologia Marinha, Niterói, RJ, Brazil

Elisabete Barbarino
Universidade Federal Fluminense, Departamento de Biologia Marinha, Niterói, RJ, Brazil

Ricardo C. Martino
Fundação Instituto de Pesca do Estado do Rio de Janeiro, Rio de Janeiro, RJ, Brazil

Sergio O. Lourenço
Universidade Federal Fluminense, Departamento de Biologia Marinha, Niterói, RJ, Brazil

S. Ramkumar
Fisheries Resources, Harvest and Post Harvest Division, Central Institute of Fisheries Education, Mumbai, India

S. Arun Sudhagar
Aquatic Environment and Health Management Division, Central Institute of Fisheries Education, Mumbai, India

K. Venkateshvaran
Fisheries Resources, Harvest and Post Harvest Division, Central Institute of Fisheries Education, Mumbai, India

Md. Mostafa Monwar
Institute of Marine Sciences and Fisheries, University of Chittagong, Chittagong-4331, Bangladesh

A. K. M. Ruhul Amin Sarker
Institute of Marine Sciences and Fisheries, University of Chittagong, Chittagong-4331, Bangladesh

Nani Gopal Das
Institute of Marine Sciences and Fisheries, University of Chittagong, Chittagong-4331, Bangladesh

V. T. Namulawa
Aquaculture Research and Development Center, P. O. Box, 530, Kampala, Uganda

C. D. Kato
College of Veterinary Medicine, Animal Resources and BioSecurity, P. O. Box 7026, Kampala, Uganda

J. Rutaisire
Aquaculture Research and Development Center, P. O. Box, 530, Kampala, Uganda

P. J. Britz
Department of Ichthyology and Fisheries, Rhodes University, P. O. Box 94 Grahamstown, South Africa

N. Beukes
Department of Biochemistry, Microbiology and Biotechnology, Rhodes University, Grahamstown, South Africa

BI Pletschke
Department of Biochemistry, Microbiology and Biotechnology, Rhodes University, Grahamstown, South Africa

C. Whiteley
Department of Biochemistry, Microbiology and Biotechnology, Rhodes University, Grahamstown, South Africa